Number Theory for Computing

Springer
Berlin
Heidelberg
New York
Barcelona
Hong Kong
London
Milan
Paris
Tokyo

Song Y. Yan

Number Theory for Computing

Second Edition

Foreword by Martin E. Hellman

With 26 Figures, 78 Images, and 33 Tables

Song Y. Yan
Computer Science
Aston University
Birmingham B4 7ET
UK
s.yan@aston.ac.uk

ACM Computing Classification (1998): F.2.1, E.3-4, D.4.6, B.2.4, I.1.2

AMS Mathematics Subject Classification (1991): 11Axx, 11T71, 11Yxx, 11Dxx, 11Z05, 68Q25, 94A60

Library of Congress Cataloging-in-Publication Data applied for

Die Deutsche Bibliothek – CIP-Einheitsaufnahme

Yan, Song Y.:
Number theory for computing: with 32 tables/Song Y. Yan. – 2. ed., rev. and extended. – Berlin; Heidelberg; New York; Barcelona; Hong Kong; London; Milan; Paris; Tokyo: Springer, 2002

ISBN 978-3-642-07710-4

This work is subject to copyright. All rights are reserved, whether the whole or part of the material is concerned, specifically the rights of translation, reprinting, reuse of illustrations, recitation, broadcasting, reproduction on microfilm or in any other way, and storage in data banks. Duplication of this publication or parts thereof is permitted only under the provisions of the German Copyright Law of September 9, 1965, in its current version, and permission for use must always be obtained from Springer-Verlag. Violations are liable for prosecution under the German Copyright Law.

Springer-Verlag Berlin Heidelberg New York,
a member of Springer Science+Business Media
http://www.springer.de

© Springer-Verlag Berlin Heidelberg 2010
Printed in Germany

The use of general descriptive names, trademarks, etc. in this publication does not imply, even in the absence of a specific statement, that such names are exempt from the relevant protective laws and regulations and therefore free for general use.

Cover Design: KünkelLopka, Heidelberg

Foreword

Modern cryptography depends heavily on number theory, with primality testing, factoring, discrete logarithms (indices), and elliptic curves being perhaps the most prominent subject areas. Since my own graduate study had emphasized probability theory, statistics, and real analysis, when I started working in cryptography around 1970, I found myself swimming in an unknown, murky sea. I thus know from personal experience how inaccessible number theory can be to the uninitiated. Thank you for your efforts to ease the transition for a new generation of cryptographers.

Thank you also for helping Ralph Merkle receive the credit he deserves. Diffie, Rivest, Shamir, Adleman and I had the good luck to get expedited review of our papers, so that they appeared before Merkle's seminal contribution. Your noting his early submission date and referring to what has come to be called "Diffie-Hellman key exchange" as it should, "Diffie-Hellman-Merkle key exchange", is greatly appreciated.

It has been gratifying to see how cryptography and number theory have helped each other over the last twenty-five years. Number theory has been the source of numerous clever ideas for implementing cryptographic systems and protocols while cryptography has been helpful in getting funding for this area which has sometimes been called "the queen of mathematics" because of its seeming lack of real world applications. Little did they know!

Stanford, 30 July 2001

Martin E. Hellman

Preface to the Second Edition

Number theory is an experimental science.

J. W. S. CASSELS (1922–)
Professor Emeritus of Mathematics, The University of Cambridge

If you teach a course on number theory nowadays, chances are it will generate more interest among computer science majors than among mathematics majors. Many will care little about integers that can be expressed as the sum of two squares. They will prefer to learn how Alice can send a message to Bob without fear of eavesdropper Eve deciphering it.

BRAIN E. BLANK, Professor of Mathematics
Washington University, St. Louis, Missouri

The success of the first edition of the book encouraged me to produce this second edition. I have taken this opportunity to provide proofs of many theorems, that had not been given in the first edition. Some additions and corrections have also been included.

Since the publication of the first edition, I have received many communications from readers all over the world. It is my great pleasure to thank the following people for their comments, corrections and encouragements: Prof. Jim Austin, Prof. Friedrich L. Bauer, Dr. Hassan Daghigh Dr. Deniz Deveci, Mr. Rich Fearn, Prof. Martin Hellman, Prof. Zixin Hou, Mr. Waseem Hussain, Dr. Gerard R. Maze, Dr. Paul Maguire, Dr. Helmut Meyn, Mr. Robert Pargeter, Mr. Mok-Kong Shen, Dr. Peter Shiu, Prof. Jonathan P. Sorenson, and Dr. David L. Stern. Special thanks must be given to Prof. Martin Hellman of Stanford University for writing the kind Foreword to this edition and also for his helpful advice and kind guidance, to Dr. Hans Wössner, Mr. Alfred Hofmann, Mrs. Ingeborg Mayer, Mrs. Ulrike Stricker, and Mr. Frank Holzwarth of Springer-Verlag for their kind help and encouragements during the preparation of this edition, and to Dr. Rodney Coleman, Prof. Glyn James, Mr. Alexandros Papanikolaou, and Mr. Robert Pargeter for proofreading the final draft. Finally, I would like to thank Prof. Shiing-Shen Chern,

Director Emeritus of the Mathematical Sciences Research Institute in Berkeley for his kind encouragements; this edition is dedicated to his 90th birthday!

Readers of the book are, of course, very welcome to communicate with the author either by ordinary mail or by e-mail to `s.yan@aston.ac.uk`, so that your corrections, comments and suggestions can be incorporated into a future edition.

Birmingham, February 2002 S. Y. Y.

Preface to the First Edition

Mathematicians do not study objects, but relations among objects; they are indifferent to the replacement of objects by others as long as relations do not change. Matter is not important, only form interests them.

HENRI POINCARÉ (1854–1912)

Computer scientists working on algorithms for factorization would be well advised to brush up on their number theory.

IAN STEWART
Geometry Finds Factor Fast
Nature, Vol. 325, 15 January 1987, page 199

The theory of numbers, in mathematics, is primarily the theory of the properties of integers (i.e., the whole numbers), particularly the positive integers. For example, Euclid proved 2000 years ago in his *Elements* that there exist infinitely many prime numbers. The subject had long been considered as the *purest* branch of mathematics, with very few applications to other areas. However, recent years have seen considerable increase in interest in several central topics of number theory, precisely because of their importance and applications in other areas, particularly in computing and information technology. Today, number theory has been applied to such diverse areas as physics, chemistry, acoustics, biology, computing, coding and cryptography, digital communications, graphics design, and even music and business[1]. In particular, congruence theory has been used in constructing perpetual calendars, scheduling round-robin tournaments, splicing telephone cables, devising systematic methods for storing computer files, constructing magic squares, generating random numbers, producing highly secure and reliable encryption schemes and even designing high-speed (residue) computers. It is specifically worthwhile pointing out that computers are basically finite machines; they

[1] In his paper [96] in the *International Business Week*, 20 June 1994, pp. 62–64, Fred Guterl wrote: "Number Theory, once the esoteric study of what happens when whole numbers are manipulated in various ways, is becoming a vital practical science that is helping solve tough business problems".

have finite storage, can only deal with numbers of some finite length and can only perform essentially finite steps of computation. Because of such limitations, congruence arithmetic is particularly useful in computer hardware and software design.

This book takes the reader on a journey, starting at elementary number theory, going through algorithmic and computational number theory, and finally finishing at applied number theory in computing science. It is divided into three distinct parts:

(1) Elementary Number Theory,

(2) Computational/Algorithmic Number Theory,

(3) Applied Number Theory in Computing and Cryptography.

The first part is mainly concerned with the basic concepts and results of divisibility theory, congruence theory, continued fractions, Diophantine equations and elliptic curves. A novel feature of this part is that it contains an account of elliptic curves, which is not normally provided by an elementary number theory book. The second part provides a brief introduction to the basic concepts of algorithms and complexity, and introduces some important and widely used algorithms in computational number theory, particularly those for primality testing, integer factorization, discrete logarithms, and elliptic curve discrete logarithms. An important feature of this part is that it contains a section on quantum algorithms for integer factorization and discrete logarithms, which cannot be easily found, so far, in other texts on computational/algorithmic number theory. This part finishes with sections on algorithms for computing $\pi(x)$, for finding amicable pairs, for verifying Goldbach's conjecture, and for finding perfect and amicable numbers. The third part of the book discusses some novel applications of elementary and computational number theory in computing and information technology, particularly in cryptography and information security; it covers a wide range of topics such as secure communications, information systems security, computer organisations and design, error detections and corrections, hash function design, and random number generation. Throughout the book we follow the style "Definition-Theorem-Algorithm-Example" to present our material, rather than the traditional Hardy–Wright "Definition-Theorem-Proof" style [100], although we do give proofs to most of the theorems. We believe this is the most suitable way to present mathematical material to computing professionals. As Donald Knuth [121] pointed out in 1974; "It has often been said that a person does not really understand something until he teaches it to someone else. Actually a person does not really understand something until he can teach it to a computer." The author strongly recommends readers to implement all the algorithms and methods introduced in this book on a computer using a mathematics (computer algebra) system such as Maple in order to get a better understanding of the ideas behind the algorithms and methods. A small number of exercises is also provided in some sections, and it is worthwhile trying all of them.

The book is intended to be self-contained with no previous knowledge of number theory and abstract algebra assumed, although some familiarity with first-year undergraduate mathematics will be helpful. The book is suitable either as a text for an undergraduate/postgraduate course in *Number Theory/Mathematics for Computing/Cryptography*, or as a basic reference researchers in the field.

Acknowledgements

I started to write this book in 1990 when I was a lecturer in the School of Mathematical and Information Sciences at La Trobe University, Australia. I completed the book when I was at the University of York and finalized it at Coventry and Aston Universities, all in England. I am very grateful to Prof. Bertram Mond and Dr. John Zeleznikow of the School of Mathematical and Information Sciences at La Trobe University, Dr. Terence Jackson of the Department of Mathematics and Prof. Jim Austin of the Department of Computer Science at the University of York, Prof. Glyn James, Mr. Brian Aspinall and Mr. Eric Tatham of the School of Mathematical and Information Sciences at Coventry University, and Prof. David Lowe and Dr. Ted Elsworth of Computer Science and Applied Mathematics at Aston University in Birmingham for their many fruitful discussions, kind encouragement and generous support. Special thanks must be given to Dr. Hans Wössner and Mr. Andrew Ross at Springer-Verlag Berlin/Heidelberg and the referees of Springer-Verlag, for their comments, corrections and suggestions. During the long period of the preparation of the book, I also got much help in one way or another from, whether they are aware of it or not, Prof. Eric Bach of the University of Wisconsin at Madison, Prof. Jim Davenport of the University of Bath, Prof. Richard Guy of the University of Calgary, Prof. Martin Hellman of Stanford University, Dr. David Johnson of AT&T Bell Laboratories, Prof. S. Lakshmivarahan of the University of Oklahoma, Dr. Ajie Lenstra of Bell Communication Research, Prof. Hendrik Lenstra Jr. of the University of California at Berkeley, Prof. Roger Needham and Dr. Richard Pinch of the University of Cambridge, Dr. Peter Pleasants of the University of the South Pacific (Fiji), Prof. Carl Pomerance of the University of Georgia, Dr. Herman te Riele of the Centre for Mathematics and Computer Science (CWI), Amsterdam, and Prof. Hugh William of the University of Manitoba. Finally, I would like to thank Mr. William Bloodworth (Dallas, Texas), Dr. John Cosgrave (St. Patrick's College, Dublin), Dr. Gavin Doherty (Rutherford Appleton Laboratory, Oxfordshire), Mr. Robert Pargeter (Tiverton, Devon), Mr. Alexandros Papanikolaou (Aston University, Birmingham), and particularly Prof. Richard Brent (Oxford University Computing Laboratory), Dr. Rodney Coleman (Université Joseph Fourier, Grenoble) and Prof. Glyn James (Coventry University) for reading the various versions of the book. As communicated by Dr. Hans Wössner: nothing is perfect and nobody is perfect. This book and the author are no exception. Any comments, corrections and suggestions from readers of the book are especially very welcome and can be sent to the author either by ordinary mail or by e-mail to `s.yan@aston.ac.uk`.

Birmingham, February 2000 S. Y. Y.

Table of Contents

Notation

All notation should be as simple as the nature of the operations to which it is applied.

CHARLES BABBAGE (1791–1871)

Notation	Explanation
$\mathbb{N}$	set of natural numbers: $\mathbb{N} = \{1, 2, 3, \cdots\}$
$\mathbb{Z}$	set of integers (whole numbers): $\mathbb{Z} = \{0, \pm n : n \in \mathbb{N}\}$
$\mathbb{Z}^+$	set of positive integers: $\mathbb{Z}^+ = \mathbb{N}$
$\mathbb{Z}_{>1}$	set of positive integers greater than 1: $\mathbb{Z}_{>1} = \{n : n \in \mathbb{Z} \text{ and } n > 1\}$
$\mathbb{Q}$	set of rational numbers: $\mathbb{Q} = \left\{\frac{a}{b} : a, b \in \mathbb{Z} \text{ and } b \neq 0\right\}$
$\mathbb{R}$	set of real numbers: $\mathbb{R} = \{n + 0.d_1d_2d_3 \cdots : n \in \mathbb{Z},\ d_i \in \{0, 1, \cdots, 9\}$ and no infinite sequence of 9's appears$\}$
$\mathbb{C}$	set of complex numbers: $\mathbb{C} = \{a + bi : a, b \in \mathbb{R} \text{ and } i = \sqrt{-1}\}$
$\mathbb{Z}/n\mathbb{Z}$	also denoted by $\mathbb{Z}_n$, residue classes modulo n; a ring of integers; a field if n is prime
$(\mathbb{Z}/n\mathbb{Z})^*$	multiplicative group; the elements of this group are the elements in $\mathbb{Z}/n\mathbb{Z}$ that are relatively prime to n: $(\mathbb{Z}/n\mathbb{Z})^* = \{[a]_n \in \mathbb{Z}/n\mathbb{Z} : \ \gcd(a, n) = 1\}$.
$\mathbb{F}_p$	finite field with p elements, where p is a prime number
$\mathbb{F}_q$	finite field with $q = p^k$ a prime power
$\mathcal{K}$	(arbitrary) field
$\mathcal{R}$	ring

$\mathcal{G}$	group
$\lvert\mathcal{G}\rvert$	order of group $\mathcal{G}$
B_n	Bernoulli numbers: $\binom{n+1}{1} B_n + \cdots + \binom{n+1}{n} B_1 + B_0 = 0$
F_n	Fermat numbers: $F_n = 2^{2^n} + 1,\ n \geq 0$
M_p	Mersenne primes: $M_p = 2^p - 1$ is prime whenever p is prime
$\sqrt{x}$	square root of x
$\sqrt[k]{x}$	kth root of x
$\sim$	asymptotic equality
$\approx$	approximate equality
∞	infinity
$\Longrightarrow$	implication
$\Longleftrightarrow$	equivalence
$\square$	blank symbol; end of proof
$\sqcup$	space
Prob	probability measure
$\lvert S\rvert$	cardinality of set S
$\in$	member of
$\subset$	proper subset
$\subseteq$	subset
$\star, *$	binary operations
$\oplus$	binary operation (addition); exclusive or (XOR)
$\odot$	binary operation (multiplication)
$f(x) \sim g(x)$	$f(x)$ and $g(x)$ are asymptotically equal
$(\mathcal{G}, *) \cong (\mathcal{H}, \star)$	$(\mathcal{G}, *)$ and $(\mathcal{H}, \star)$ are isomorphic
$\perp$	undefined
e_k	encryption key
d_k	decryption key
$E_{e_k}(M)$	encryption process $C = E_{e_k}(M)$, where M is the plaintext
$D_{d_k}(C)$	decryption process $M = D_{d_k}(C)$, where C is the ciphertext

$f(x)$	function of x
f^{-1}	inverse of f
$\binom{n}{i}$	binomial coefficient
$\int$	integration
$\mathrm{Li}(x)$	logarithmic integral: $\mathrm{Li}(x) = \int_2^x \frac{\mathrm{d}t}{\ln t}$
$\sum_{i=1}^{n} x_i$	sum: $x_1 + x_2 + \cdots + x_n$
$\prod_{i=1}^{n} x_i$	product: $x_1 x_2 \cdots x_n$
$n!$	factorial: $n(n-1)(n-2)\cdots 3 \cdot 2 \cdot 1$
x^k	x to the power k
kP	$kP = \underbrace{P \oplus P \oplus \cdots \oplus P}_{k \text{ summands}}$, where P is a point (x, y) on an elliptic curve $E: \ y^2 = x^3 + ax + b$
$\mathcal{O}_E$	the point at infinity on an elliptic curve E over a field
e	the transcendental number $e = \sum_{n \geq 0} \frac{1}{n!} \approx 2.7182818$
$\log_b x$	logarithm of x to the base b ($b \neq 1$): $x = b^{\log_b x}$
$\log x$	binary logarithm: $\log_2 x$
$\ln x$	natural logarithm: $\log_e x$
$\exp(x)$	exponential of x: $e^x = \sum_{n \geq 0} \frac{x^n}{n!}$
$a \mid b$	a divides b
$a \nmid b$	a does not divide b
$p^\alpha \mid\mid n$	$p^\alpha \mid n$ but $p^{\alpha+1} \nmid n$
$\gcd(a, b)$	greatest common divisor of (a, b)
$\mathrm{lcm}(a, b)$	least common multiple of (a, b)
$\lfloor x \rfloor$	the greatest integer less than or equal to x
$\lceil x \rceil$	the least integer greater than or equal to x
$x \bmod \mathrm{n}$	remainder: $x - n \left\lfloor \frac{x}{n} \right\rfloor$
$x = y \bmod n$	x is equal to y reduced to modulo n
$x \equiv y \pmod{\mathrm{n}}$	x is congruent to y modulo n
$x \not\equiv y \pmod{\mathrm{n}}$	x is not congruent to y modulo n

$[a]_n$	residue class of a modulo n
$+_n$	addition modulo n
$-_n$	subtraction modulo n
$\cdot_n$	multiplication modulo n
$x^k \bmod n$	x to the power k modulo n
$kP \bmod n$	kP modulo n
$\mathrm{ord}_n(a)$	order of an integer a modulo n; also denoted by $\mathrm{ord}(a, n)$
$\mathrm{ind}_{g,n}a$	index of a to the base g modulo n; also denoted by $\mathrm{ind}_g a$ whenever n is fixed
$\pi(x)$	number of primes less than or equal to x: $\pi(x) = \sum_{\substack{p \le x \\ p \text{ prime}}} 1$
$\tau(n)$	number of positive divisors of n: $\tau(n) = \sum_{d \mid n} 1$
$\sigma(n)$	sum of positive divisors of n: $\sigma(n) = \sum_{d \mid n} d$
$s(n)$	sum of proper divisors of n: $s(n) = \sigma(n) - n$
$\phi(n)$	Euler's totient function: $\phi(n) = \sum_{\substack{0 \le k < n \\ \gcd(k,n)=1}} 1$
$\lambda(n)$	Carmichael's function: $\lambda(n) = \mathrm{lcm}\,(\lambda(p_1^{\alpha_1})\lambda(p_2^{\alpha_2}) \cdots \lambda(p_k^{\alpha_k}))$ if $n = \prod_{i=1}^{k} p_i^{\alpha_i}$
$\mu(n)$	Möbius function
$\zeta(s)$	Riemann zeta-function: $\zeta(s) = \prod_{n=1}^{\infty} \frac{1}{n^s}$, where s is a complex variable
$\left(\frac{a}{p}\right)$	Legendre symbol, where p is prime
$\left(\frac{a}{n}\right)$	Jacobi symbol, where n is composite
Q_n	set of all quadratic residues of n
$\overline{Q}_n$	set of all quadratic nonresidues of n
J_n	$J_n = \left\{a \in (\mathbb{Z}/n\mathbb{Z})^* : \left(\frac{a}{n}\right) = 1\right\}$
$\tilde{Q}_n$	set of all pseudosquares of n: $\tilde{Q}_n = J_n - Q_n$
$K(k)_n$	set of all kth power residues of n, where $k \ge 2$
$\overline{K(k)}_n$	set of all kth power nonresidues of n, where $k \ge 2$

$[q_0, q_1, q_2, \cdots, q_n]$	finite simple continued fraction
$C_k = \dfrac{P_k}{Q_k}$	k-th convergent of a continued fraction
$[q_0, q_1, q_2, \cdots]$	infinite simple continued fraction
$[q_0, q_1, \cdots, q_k, \overline{q_{k+1}, q_{k+2}, \cdots, q_{k+m}}]$	periodic simple continued fraction
$\mathcal{P}$	class of problems solvable in deterministic polynomial time
$\mathcal{NP}$	class of problems solvable in nondeterministic polynomial time
$\mathcal{RP}$	class of problems solvable in random polynomial time with one-sided errors
$\mathcal{BPP}$	class of problems solvable in random polynomial time with two-sided errors
$\mathcal{ZPP}$	class of problems solvable in random polynomial time with zero errors
$\mathcal{O}(\cdot)$	upper bound: $f(n) = \mathcal{O}(g(n))$ if there exists *some* constant $c > 0$ such that $f(n) \le c \cdot g(n)$
$o(\cdot)$	upper bound that is not asymptotically tight: $f(n) = \mathcal{O}(g(n))$, $\forall c > 0$ such that $f(n) < c \cdot g(n)$
$\Omega(\cdot)$	low bound: $f(n) = \Omega(g(n))$ if there exists a constant c such that $f(n) \ge \frac{1}{c} \cdot g(n)$
$\Theta(\cdot)$	tight bound: $f(n) = \Theta(n)$ if $f(n) = \mathcal{O}(g(n))$ and $f(n) = \Omega(g(n))$
$\mathcal{O}(N^k)$	polynomial-time complexity measured in terms of arithmetic operations, where $k > 0$ is a constant
$\mathcal{O}\left((\log N)^k\right)$	polynomial-time complexity measured in terms of bit operations, where $k > 0$ is a constant
$\mathcal{O}\left((\log N)^{c \log N}\right)$	superpolynomial complexity, where $c > 0$ is a constant
$\mathcal{O}\left(\exp\left(c\sqrt{\log N \log\log N}\ \right)\right)$	subexponential complexity, $\mathcal{O}\left(\exp\left(c\sqrt{\log N \log\log N}\ \right)\right) = \mathcal{O}\left(N^{c\sqrt{\log\log N/\log N}}\right)$
$\mathcal{O}(\exp(x))$	exponential complexity, sometimes denoted by $\mathcal{O}(e^x)$
$\mathcal{O}(N^\epsilon)$	exponential complexity measured in terms of bit operations; $\mathcal{O}(N^\epsilon) = \mathcal{O}\left(2^{\epsilon \log N}\right)$, where $\epsilon > 0$ is a constant
CFRAC	Continued FRACtion method (for factoring)
ECM	Elliptic Curve Method (for factoring)

NFS	Number Field Sieve (for factoring)
QS/MPQS	Quadratic Sieve/Multiple Polynomial Quadratic Sieve (for factoring)
ECPP	Elliptic Curve Primality Proving
DES	Data Encryption Standard
AES	Advanced Encryption Standard
DSA	Digital Signature Algorithm
DSS	Digital Signature Standard
RSA	Rivest-Shamir-Adleman
WWW	World Wide Web

1. Elementary Number Theory

The elementary theory of numbers should be one of the very best subjects for early mathematical instruction. It demands very little previous knowledge, its subject matter is tangible and familiar; the processes of reasoning which it employs are simple, general and few; and it is unique among the mathematical sciences in its appeal to natural human curiosity.

G. H. HARDY (1877–1947)

This chapter introduces the basic concepts and results of the elementary theory of numbers. Its purpose is twofold:

- Provide a solid foundation of elementary number theory for *Computational, Algorithmic*, and *Applied Number Theory* of the next two chapters of the book.
- Provide independently a self-contained text of *Elementary Number Theory for Computing*, or in part a text of *Mathematics for Computing*.

1.1 Introduction

In this section, we shall first give a brief review of the fundamental ideas of number theory and then present some mathematical preliminaries of elementary number theory.

1.1.1 What is Number Theory?

Mathematics is the Queen of the sciences, and number theory is the Queen of mathematics.

C. F. GAUSS (1777–1855)

Number theory, in mathematics, is primarily the theory of the *properties* of integers (whole numbers), such as parity, divisibility, primality, additivity and multiplicativity, etc. To appreciate the intrinsic mathematical beauty of the theory of numbers, let us first investigate some of the properties of the integers (the investigation is by no means complete; more detailed discussions will be given later in the book).

(I) Parity. Perhaps the simplest property of an integer is its parity, that is, whether it is odd or even. By definition, an integer is odd if dividing it by 2 leaves a remainder of 1; otherwise, it is even. Of course, if the binary representation of an integer is readily available for inspection, division by 2 can be avoided, since we need only look to see if the integer's rightmost bit is a 1 (indicating oddness), or a 0 (indicating evenness). Two integers m and n have the same parity if both m and n are even or odd, otherwise, they have opposite parity. Some well-known results, actually already known to Euclid[1], about the parity property of integers are as follows:

(1) The sum of two numbers is even if both are even, or both are odd. More generally, the sum of n even numbers is even, the sum of n odd numbers is even if n is even and the sum of n odd numbers is odd if n is odd.

(2) The difference of two numbers is even if both have the same parity. More generally, the difference of n even numbers is even, the difference of n odd numbers is even if n is even and the difference of n odd numbers is odd if n is odd.

(3) The product of two numbers is even if at least one of them is even. More generally, the product of n numbers is even if at least one of them is even.

That is,

$$\underbrace{\text{even} \pm \text{even} \pm \text{even} \pm \cdots \pm \text{even}}_{n \text{ even numbers, } n \text{ is even}} = \text{even},$$

[1]

Euclid (about 350 B.C.) was the author of the most successful mathematical textbook ever written, namely his thirteen books of *Elements*, which has appeared in over a thousand different editions from ancient to modern times. It provides an introduction to plane and solid geometry as well as number theory. For example, some properties of the parity of integers are given in Propositions 21–29 of Book IX, Euclid's algorithm for computing the greatest common divisor of two and three positive integers is found in Book VII Proposition 2 and Proposition 3, respectively, and his proofs for the infinitude of primes and a sufficient condition for even numbers to be perfect are found in Book IX Proposition 20 and Proposition 36, respectively. The "Axiom-Definition-Theorem-Proof" style of Euclid's work has become the standard for formal mathematical writing up to the present day. (All portrait images in this book, unless stated otherwise, are by courtesy of O'Connor and Robertson [177].)

$$\underbrace{\text{odd} \pm \text{odd} \pm \text{odd} \pm \cdots \pm \text{odd}}_{n \text{ odd numbers, } n \text{ is even}} = \text{even},$$

$$\underbrace{\text{odd} \pm \text{odd} \pm \text{odd} \pm \cdots \pm \text{odd}}_{n \text{ odd numbers, } n \text{ is odd}} = \text{odd},$$

$$\underbrace{\text{odd} \times \text{odd} \times \text{odd} \times \cdots \times \text{odd}}_{\text{all odd}} = \text{odd},$$

$$\underbrace{\text{even} \times \text{odd} \times \text{odd} \times \cdots \times \text{odd}}_{\text{at least one even}} = \text{even}.$$

Example 1.1.1. Following are some examples:

$$\begin{aligned}
&100 + 4 + 54 + 26 + 12 = 196, \\
&100 - 4 - 54 - 20 - 18 = 4, \\
&101 + 1 + 13 + 15 + 17 + 47 = 194, \\
&101 - 1 - 13 - 15 - 17 - 47 = 8, \\
&101 + 1 + 13 + 15 + 17 + 47 + 3 = 197, \\
&101 - 1 - 13 - 15 - 17 - 47 - 3 = 5, \\
&23 \times 67 \times 71 \times 43 = 4704673, \\
&23 \times 67 \times 72 \times 43 = 4770936.
\end{aligned}$$

It is worthwhile pointing out that the parity property of integers has important applications in error detection and correction codes, that are useful in computer design and communications. For example, a simple error detection and correction method, called *parity check*, works as follows. Let $x_1 x_2 \cdots x_n$ be a binary string (codeword), to be sent (from the main memory to the central processing unit (CPU) of a computer, or from a computer to other computers connected to a network). This code is of course in no way an error detection and correction code. However, if an additional bit 1 (respect to 0) is added to the end of the codeword when the number of 1's in the codeword is odd (respect to even), then this new code is error detecting. For instance, let the two codewords be as follows:

$$\begin{aligned}
C_1 &= 1101001001, \\
C_2 &= 1001011011,
\end{aligned}$$

then the new codewords will become

$$\begin{aligned}
C_1' &= 11010010011, \\
C_2' &= 10010110110.
\end{aligned}$$

These codes apparently have some error detecting function. For example, if after transmission C_2' becomes $C_2'' = 11010110110$, then we know there is an error in the transmitted code, since

$$1 + 1 + 0 + 1 + 0 + 1 + 1 + 0 + 1 + 1 + 0 = 7 \bmod 2 \neq 0.$$

(The notation $a \bmod n$ is defined to be the remainder when a is divided by n; for example, $10 \bmod 3 = 1$.) Of course, the new codes are still not error correction codes. However, if we arrange data in a rectangle and use parity bits for each row and column, then a single bit error can be corrected.

(II) Primality. A positive integer $n > 1$ that has only two distinct factors, 1 and n itself (when these are different), is called *prime*; otherwise, it is called *composite*. It is evident that a positive integer $n > 1$ is either a prime or a composite. The first few prime numbers are: $2, 3, 5, 7, 11, 13, 17, 19, 23$. It is interesting to note that primes thin out: there are eight up through 20 but only three between 80 and 100, namely 83, 89 and 97. This might lead one to suppose that there are only finitely many primes. However as Euclid proved 2000 years ago there are infinitely many primes. It is also interesting to note that 2 is the only even prime; all the rest are odd. The prime pairs $(3, 5)$, $(5, 7)$ and $(11, 13)$ are twin primes of the form $(p,\ p+2)$ where p and $p+2$ are prime; two of the largest known twin primes (both found in 1995) are: $570918348 \cdot 10^{5120} \pm 1$ with 5129 digits and $242206083 \cdot 2^{38880} \pm 1$ with 11713 digits. It is not known if there are infinitely many twin primes; however, it has been proved by J. R. Chen that there are infinitely many pairs of integers $(p,\ p+2)$, with p prime and $p+2$ a product of at most two primes. The triple primes are those prime triples of the form either $(p,\ p+2,\ p+4)$ or $(p,\ p+2,\ p+6)$. For example, $(3, 5, 7)$ is a prime triple of the form $(p,\ p+2,\ p+4)$, whereas the prime triples $(5, 7, 11)$, $(11, 13, 17)$, $(17, 19, 23)$, $(41, 43, 47)$, $(101, 103, 107)$, $(107, 109, 113)$, $(191, 193, 197)$, $(227, 229, 233)$, $(311, 313, 317)$, $(347, 349, 353)$, $(347, 349, 353)$ are all of the form $(p,\ p+2,\ p+6)$. It is amusing to note that there is only one prime triple of the form $(p,\ p+2,\ p+4)$, namely $(3, 5, 7)$; however, we do not know whether or not there are infinitely many prime triples of the form $(p,\ p+2,\ p+6)$. There are other forms of prime triples such as $(p,\ p+4,\ p+6)$; the first ten triples of this form are as follows: $(7, 11, 13)$, $(13, 17, 19)$, $(37, 41, 43)$, $(67, 71, 73)$, $(97, 101, 103)$, $(103, 107, 109)$, $(193, 197, 199)$, $(223, 227, 229)$, $(277, 281, 283)$, and $(307, 311, 313)$. Again, we also do not know whether or not there are infinitely many prime triples of this form. According to Dickson [65], the ancient Chinese mathematicians, even before Fermat (1601–1665), seem to have known that

$$p \in \text{Primes} \implies p \mid (2^p - 2). \tag{1.1}$$

However, there are some composites n that are not prime but satisfy the condition that $n \mid (2^n - 2)$; for example, $n = 341 = 11 \cdot 31$ is not prime, but $341 \mid (2^{341} - 2)$. It is not an easy task to decide whether or not a large number is prime. One might think that to test whether or not the number n is prime, one only needs to test all the numbers (or just the primes) up to $\sqrt{n}$. Note that the number n has about $\beta = \log n$ bits. Thus for a number n with β bits, this would require about $\exp(\beta/2)$ bit operations since $\sqrt{n} = \exp\left(\frac{1}{2}\log n\right) = \exp(\beta/2)$, and hence, it is inefficient and essentially useless for large values of n. The current best algorithm for primality testing needs at most $\beta^{c \log \log \beta}$ bit operations, where c is a real positive constant.

(III) Multiplicativity. Any positive integer $n > 1$ can be written uniquely in the following prime factorization form:

$$n = p_1^{\alpha_1} p_2^{\alpha_2} \cdots p_k^{\alpha_k}, \tag{1.2}$$

where $p_1 < p_2 < \cdots < p_k$ are primes, and $\alpha_1, \alpha_2, \cdots, \alpha_k$ are positive integers. This is the famous Fundamental Theorem of Arithmetic; it was possibly known to Euclid (around 350 B.C.), but it was first clearly stated and proved by Gauss (1777–1855). It can be very easy to factor a positive integer n if n is not very big; the following are the prime factorizations of n for $n = 1999, 2000, \cdots, 2010$:

$$\begin{array}{ll}
1999 = 1999 & 2000 = 2^4 \cdot 5^3 \\
2001 = 3 \cdot 23 \cdot 29 & 2002 = 2 \cdot 7 \cdot 11 \cdot 13 \\
2003 = 2003 & 2004 = 2^2 \cdot 3 \cdot 167 \\
2005 = 5 \cdot 401 & 2006 = 2 \cdot 17 \cdot 59 \\
2007 = 3^2 \cdot 223 & 2008 = 2^3 \cdot 251 \\
2009 = 7^2 \cdot 41 & 2010 = 2 \cdot 3 \cdot 5 \cdot 67.
\end{array}$$

However, it can be very difficult to factor a large positive integer (e.g., with more than 100 digits at present) into its prime factorization form – a task even more difficult than that of primality testing. The most recent and potentially the fastest factoring method yet devised is the Number Field Sieve (NFS), which can factor an integer N in approximately

$$\exp\left(c(\log N)^{1/3}(\log \log N)^{2/3}\right) \tag{1.3}$$

bit operations, where c is a positive real constant (an admissible value is $c = (64/9)^{1/3} \approx 1.9$, but this can be slightly lowered to $c = (32/9)^{1/3} \approx 1.5$ for some special integers of the form $N = c_1 r^t + c_2 s^u$: see Huizing [107]) and exp stands for the exponential function. By using the NFS, the 9th Fermat number $F_9 = 2^{2^9} + 1$, a number with 155 digits, was completely factored in 1990. (However, the 12th Fermat number $F_{12} = 2^{2^{12}} + 1$ has still not completely been factored, even though its five smallest prime factors are known.) The most recent record of NFS is perhaps the factorization, by a group led by Herman te Riele [206] in August 1999 of the random 155 digit (512 bit) number RSA-155, which can be written as the product of two 78-digit primes:

102639592829741105772054196573991675900716567808_
38066803341933521790711307779,

106603488380168454820927220360012878679207958575_
989291522270608237193062808643.

It is interesting to note that a number of recent proposals for cryptographic systems and protocols, such as the Rivest–Shamir–Adleman (RSA) public-key cryptography, rely for their security on the infeasibility of the integer factorization problem. For example, let M be a message. To encrypt the message M, one computes

$$C \equiv M^e \pmod{n}, \tag{1.4}$$

where e is the encryption key, and both e and n are *public*. (The notation $a \equiv b \pmod{n}$ reads "a is congruent to b modulo n". Congruences will be studied in detail in Section 1.6.) To decrypt the encrypted message C, one computes

$$M \equiv C^d \pmod{n}, \tag{1.5}$$

where d is the *private* decryption key satisfying

$$ed \equiv 1 \pmod{\phi(n)}, \tag{1.6}$$

where $\phi(n)$ is Euler's ϕ-function ($\phi(n)$, for $n \geq 1$, is defined to be the number of positive integers not exceeding n which are relatively prime to n; see Definition 1.4.6). By (1.6), we have $ed = 1 + k\phi(n)$ for some integer k. By Euler's theorem (see Theorem 1.244), $M^{\phi(n)} \equiv 1 \pmod{n}$, we have $M^{k\phi(n)} \equiv 1 \pmod{n}$. Thus,

$$C^d \equiv M^{ed} \equiv M^{1+k\phi(n)} \equiv M \pmod{n}. \tag{1.7}$$

For those who do not have the private key but can factor n, say, e.g., $n = pq$, they can find d by computing

$$d \equiv e^{-1} \pmod{\phi(n)} \equiv e^{-1} \pmod{(p-1)(q-1)}, \tag{1.8}$$

and hence, decrypt the message.

(IV) Additivity. Many of the most difficult mathematical problems are in additive number theory; Goldbach's conjecture is just one of them. On 7th June 1742 the German-born mathematician Christian Goldbach (1690–1764) wrote a letter (see Figure 1.1) to the Swiss mathematician Euler (then both in Russia), in which he proposed two conjectures on the representations of integers as the sum of prime numbers. These conjectures may be rephrased as follows:

(1) Every odd integer greater than 7 is the sum of three odd prime numbers.

(2) Every even integer greater than 4 is the sum of two odd prime numbers.

They may also be stated more strongly (requiring the odd prime numbers to be distinct) as follows:

(1) Every odd integer greater than 17 is the sum of three *distinct* odd prime numbers.

Figure 1.1. Goldbach's letter to Euler

(2) Every even integer greater than 6 is the sum of two *distinct* odd prime numbers.

The following are some numerical examples of these conjectures:

$9 = 3 + 3 + 3$
$11 = 3 + 3 + 5$
$13 = 3 + 3 + 7 = 3 + 5 + 5$
$15 = 3 + 5 + 7 = 5 + 5 + 5$
$17 = 3 + 3 + 11 = 3 + 7 + 7 = 5 + 5 + 7$
$19 = 3 + 3 + 13 = 3 + 5 + 11 = 5 + 7 + 7$
$21 = 3 + 5 + 13 = 3 + 7 + 11 = 5 + 5 + 11$
$= 7 + 7 + 7.$

$6 = 3 + 3$
$8 = 3 + 5$
$10 = 3 + 7 = 5 + 5$
$12 = 5 + 7$
$14 = 3 + 11$
$16 = 3 + 13 = 5 + 11$
$18 = 5 + 13 = 7 + 11$

It is clear that the second conjecture implies the first. As a result, the first became known as the little Goldbach conjecture (or the ternary Goldbach conjecture), whereas the second became known as the Goldbach conjecture (or the binary Goldbach conjecture). Euler believed the conjectures to be true but was unable to produce a proof. The first great achievement on the study of the Goldbach conjecture was obtained by the two great British mathematicians, Hardy[2] and Littlewood[3]; using their powerful analytic method [99] (known as the "Hardy-Littlewood-Ramanujan method", or the "Hardy-Littlewood method", the "circle method" for short) they proved in 1923 that

> If a certain hypothesis (a natural generalization of Riemann's hypothesis concerning the complex zeros of the ζ-function) is true, then every sufficiently large odd integer is the sum of three odd primes, and almost all even integers are sums of two primes.

---

[2]

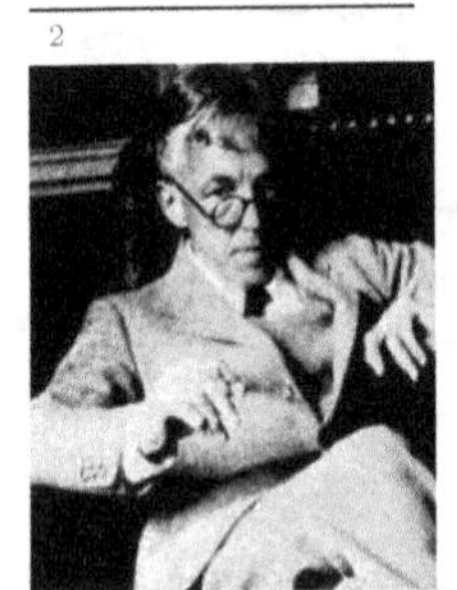

Godfrey Harold Hardy (1877–1947), was born in Cranleigh, England, and was admitted to Trinity College, Cambridge in 1896. He studied and taught there until 1919, at which date he was appointed as Savilian professor of geometry at Oxford. He spent about 10 years at Oxford and one year at Princeton, then he returned to Cambridge in 1931 and remained there until his death. Hardy collaborated with his friend John E. Littlewood, an eminent British mathematician also at Cambridge University, for more than 35 years – surely the most successful collaboration ever in mathematics! They wrote a hundred joint papers, with their last publication a year after Hardy's death. In the 1920s the eminent German mathematician Edmund Landau (1877–1938) expressed the view that "the mathematician Hardy-Littlewood was the best in the world, with Littlewood the more original genius and Hardy the better journalist". Someone once even jokingly said that "nowadays, there are only three really great English mathematicians: Hardy, Littlewood and Hardy-Littlewood". Hardy made significant contributions to number theory and mathematical analysis, and received many honours for his work, among them the prestigious Copley Medal of the Royal Society in 1947; he learnt of this award only a few weeks before his death. Hardy's book *An Introduction to the Theory of Numbers* [100] is classic and possibly the best in the field, and influenced several generations of number theorists in the world. Another book by Hardy *A Mathematician's Apology* [98] is one of the most vivid descriptions of how a mathematician thinks and the pleasure of mathematics.

[3]

John Edensor Littlewood (1885–1977), is best known for his 35 years collaboration with G. H. Hardy on summability, function theory and number theory. Littlewood studied at Trinity College, Cambridge. From 1907 to 1910 he lectured at the University of Manchester. He became a Fellow of Trinity College (1908) returning there in 1910. He was to become Rouse Ball professor of mathematics there in 1928. In World War I Littlewood also served in the Royal Garrison Artillery. Hardy once wrote of Littlewood that he knew of *no one else who could command such a combination of insight, technique and power*. Note that Littlewood also wrote a very readable book *A Mathematician's Miscellany* [144] (a collection of Littlewood's 15 articles in mathematics), published in line with Hardy's *A Mathematician's Apology*.

In 1937, without appealing to any form of Riemann's hypothesis, the great Russian mathematician I. M. Vinogradov[4] proved unconditionally that

> Every sufficiently large odd integer can be written as the sum of three odd prime numbers.

This is the famous Vinogradov's *Three-Prime Theorem* for the little Goldbach conjecture. As for the Goldbach conjecture, the best result is still Chen's theorem (see Chen [46], or Halberstam and Richert [97]), in honour of the Chinese mathematician J. R. Chen[5]:

> Every sufficiently large even integer can be written as the sum of a prime and a product of at most two primes.

Exercise 1.1.1. Let a representation of an even number as the sum of two distinct primes (i.e., $n = p_1 + p_2, n$ even, $p_1 < p_2$) or a representation of an odd number as the sum of three distinct primes (i.e., $n = p_1 + p_2 +$

---

[4]

Ivan Matveevich Vinogradov (1891–1983), a great Russian mathematician, studied at St Petersburg and obtained his first degree in 1914 and master's degree in 1915, respectively. Vinogradov taught at the State University of Perm from 1918 to 1920, and returned to St Petersburg and was promoted to professor at the State University of St Petersburg in 1925, becoming head of the probability and number theory section there. He moved to Moscow to become the first director of the Steklov Institute of Mathematics in 1934, a post he held until his death. Vinogradov used trigonometric sums to attack deep problems in analytic number theory, particularly the Goldbach conjecture.

[5]

Jing Run Chen (1933–1996), one of the finest mathematicians in China and a distinguished student of the eminent Chinese mathematician Loo Keng Hua (1909–1985), died on the 19th of March 1996 after fighting disease for many years. In about 1955 Chen sent Hua (then the Head of the Institute of Mathematics of the Chinese Academy of Sciences, Beijing), a paper on Tarry's problem, which improves Hua's own result on the problem. It was this paper that Hua decided to bring him from Xia Men University in a Southern China Province to the Institute in Beijing. Chen devoted himself entirely to mathematical research, particularly to some hard problems in number theory, such as Waring's problem, Goldbach's conjecture and the twin prime problem, and even during the *cultural revolution* (1966–1976), a very chaotic period over the long Chinese history, he did not stop his research in mathematics. During that difficult period, he worked on number theory, particularly on Goldbach's conjecture almost all day and all night, in a small dark room (about 6 square meters); there were no electric lights (he had to use the kerosene to light the room in the night), no table and no chairs in that room (he read and wrote by setting at the bed using a plate on his legs), just a single bed and his many books and manuscripts; It was in this room that he completed the final proof of the famous Chen's theorem. (Photo by courtesy of the Chinese Mathematical Society.)

p_3, n odd, $p_1 < p_2 < p_3$) be a *Goldbach partition* of n, denoted by $G(n)$. Let also $|G(n)|$ be the number of partitions of n. Then

$$G(100) = 3+97 = 11+89 = 17+83 = 29+71 = 41+59 = 47+53,$$

$$\begin{aligned} G(101) &= 3+19+79 = 3+31+67 = 3+37+61 = 5+7+89 \\ &= 5+13+83 = 5+17+79 = 5+23+73 = 5+29+67 \\ &= 5+37+59 = 5+43+53 = 7+11+83 = 7+23+71 \\ &= 7+41+53 = 11+17+73 = 11+19+71 = 11+23+67 \\ &= 11+29+61 = 11+31+59 = 11+37+53 = 11+43+47 \\ &= 13+17+71 = 13+29+59 = 13+41+47 = 17+23+61 \\ &= 17+31+53 = 17+37+47 = 17+41+43 = 19+23+59 \\ &= 19+29+53 = 23+31+47 = 23+37+41 = 29+31+41. \end{aligned}$$

Hence $|G(100)| = 6$, and $|G(101)| = 32$.

(1) Find the values for $|G(1000)|$ and $|G(1001)|$. (Hint: $|G(1001)| > 1001$.)

(2) List all the partitions of $G(1000)$ and $G(1001)$.

(3) Can you find any patterns from your above computation?

There are, of course, many other fascinating properties of positive integers that interest mathematicians. The following well-known story of the "Hardy–Ramanujan[6] taxi number" might also give us an idea of what number theory is. One day Hardy went to visit Ramanujan in a hospital in England. When he arrived, he idly remarked that the taxi in which he had ridden had the license number 1729, which, he said, seemed to him a rather uninteresting number. Ramanujan replied immediately that it is an interesting number, since it is the

---

[6] 

Srinivasa Ramanujan (1887–1920) was one of India's greatest mathematical geniuses. He made substantial contributions to the analytical theory of numbers and worked on elliptic functions, continued fractions, and infinite series. Despite his lack of a formal education, he was well-known as a mathematical genius in Madras (the place where he lived) and his friends suggested that he should send his results to professors in England. Ramanujan first wrote to two Cambridge mathematicians E. W. Hobson and H. F. Baker trying to interest them in his results but neither replied. In January 1913 Ramanujan then wrote to Hardy a long list of unproved theorems, saying that "I have had no university education but I have undergone the ordinary school course. After leaving school I have been employing the spare time at my disposal to work at mathematics." It did not take long for Hardy and Littlewood to conclude that Ramanujan was a man of exceptional ability in mathematics and decided to bring him to Cambridge. Ramanujan arrived in Cambridge in April 1914. Hardy was soon convinced that, in terms of natural talent, Ramanujan was in the class of Euler and Gauss. He worked with Hardy and made a series of outstanding breakthroughs in mathematics, and was elected a Fellow of the Royal Society at the age of just 31. It was Littlewood who said that every positive integer was one of Ramanujan's personal friends. But sadly, in May 1917, Ramanujan fell ill; he returned to India in 1919 and died in 1920, at the early age of 33.

smallest positive integer *expressible* as a sum of two positive cubes in *exactly* two different ways, namely, $1729 = 1^3 + 12^3 = 9^3 + 10^3$. (Ramanujan could have pointed out that 1729 was also the third smallest Carmichael number!) Hardy then naturally asked Ramanujan whether he could tell him the solution of the corresponding problem for fourth powers. Ramanujan replied, after a moment's thought, that he knew no obvious example, and supposed that the first such number must be very large. It is interesting to note that the solution to the fourth power was known to Euler [7]: $635318657 = 59^4 + 158^4 = 133^4 + 134^4$.

Exercise 1.1.2. Let $r(m, n, s)$ denote the smallest integer that can be expressed as a sum of m positive (not necessarily distinct) n-th powers in s different ways. Then we have

$$\begin{aligned}
r(2,2,2) &= 50 = 5^2 + 5^2 = 1^2 + 7^2\\
r(2,3,2) &= 1729 = 1^3 + 12^3 = 9^3 + 10^3\\
r(2,4,2) &= 635318657 = 59^4 + 158^4 = 133^4 + 134^4\\
r(6,4,4) &= 6625 = 1^4 + 2^4 + 2^4 + 2^4 + 2^4 + 9^4 = 2^4 + 2^4 + 2^4 + 3^4 + 7^4 + 8^4\\
&= 2^4 + 4^4 + 4^4 + 6^4 + 7^4 + 7^4 = 3^4 + 4^4 + 6^4 + 6^4 + 6^4 + 7^4.
\end{aligned}$$

Find an example for each of the following numbers:

$$\begin{array}{llll}
r(3,2,2), & r(4,2,2), & r(5,2,2), & r(3,3,2),\\
r(2,2,3), & r(3,4,2), & r(3,5,2), & r(3,6,2),\\
r(2,2,4), & r(3,3,3), & r(3,4,3), & r(5,5,3).
\end{array}$$

Finally, we wish to remark that number theory is not only the oldest subject of mathematics, but also a most active and lively branch in mathematics. It uses sophisticated techniques and deep results from almost all areas of modern mathematics; a good example would be the solution by Andrew Wiles[7] to the famous Fermat's Last Theorem (FLT), proposed by the great

[7] 
Andrew J. Wiles, a well-kown number theorist and algebraic geometer, was born in 1953 in Cambridge, England. He attended Merton College at the University of Oxford, starting from 1971, and received his BA there in 1974. He then went to Clare College at the University of Cambridge, earning his PhD there in 1980, under the supervision of John Coates. He emigrated to the U.S.A. in the 1980s and became a professor at Princeton University in 1982. Wiles was elected a Fellow of the Royal Society, London in 1989. He has recently received several prestigious awards in mathematics, including the Wolf Prize and the U.S. National Academy of Sciences award in 1996, for his proof of Fermat's Last Theorem. It is interesting to note that Wiles became interested in Fermat's Last Theorem at the age of ten, when he read the book *The Last Problem* (by Eric Temple Bell, 1962), a book with only one problem and no solution, in a Cambridge local library.

French mathematician Fermat[8] 350 years ago. Wiles proof of Fermat's Last Theorem employed almost all the sophisticated modern pure mathematical techniques.

It should also be noted that number theory has many different faces, and hence different branches. This means that number theory can be studied from e.g., an algebraic point of view, a geometrical point of view, or an analytical point of view. Generally speaking, number theory, as a branch of mathematics, can be broadly classified into the following sub-branches:

(1) Elementary number theory,
(2) Algebraic number theory,
(3) Analytic number theory,
  (i) Multiplicative number theory,
  (ii) Additive number theory,
(4) Geometric number theory,
(5) Probabilistic number theory,
(6) Combinatorial number theory,
(7) Logic number theory,
(8) Algorithmic/Computational number theory,
(9) Arithmetic algebraic geometry, and
(10) Applied number theory.

These sub-branches reflect, either the study of the properties of the integers from different points of view, or techniques used to solve the problems in number theory. For example, probabilistic number theory makes extensive use of probabilistic methods, whilst analytic number theory employs deep results in mathematical analysis in solving number-theoretic problems. Note that arithmetic algebraic geometry is a brand new subject of modern number theory, which is the study of arithmetic properties of elliptic (cubic) curves.

8

The great amateur French scientist Pierre de Fermat (1601–1665) led a quiet life practising law in Toulouse, and producing high quality work in number theory and other areas of mathematics as a hobby. He published almost nothing, revealing most of his results in his extensive correspondence with friends, and generally kept his proofs to himself. Probably the most remarkable reference to his work is his *Last Theorem* (called Fermat's Last Theorem (FLT)), which asserts that if $n > 2$, the equation $x^n + y^n = z^n$ cannot be solved in integers x, y, z, with $xyz \neq 0$. He claimed in a margin of his copy of Diophantus's book that he had found a beautiful proof of this theorem, but the margin was too small to contain his proof. Later on mathematicians everywhere in the world struggled to find a proof for this theorem but without success. The theorem remained open for more than 300 years and was finally settled in June 1995 by two English number theorists, Andrew Wiles, currently Professor at Princeton University, and Richard Taylor, a former student of Wiles and currently Professor at Harvard University; the original result of Wiles (with a hole in it) was first announced on 23 June 1993 at the Isaac Newton Institute in Cambridge.

This book, however, shall be mainly concerned with elementary and algorithmic number theory and their applications in computer science.

1.1.2 Applications of Number Theory

Number theory is usually viewed as the purest branch of pure mathematics, to be admired for its beauty and depth rather than its applicability. It is not well known that number theory has, especially in recent years, found diverse "real-world" applications, in areas such as

(1) Physics,
(2) Chemistry,
(3) Biology,
(4) Computing,
(5) Digital information,
(6) Communications,
(7) Electrical and electronic engineering,
(8) Cryptography,
(9) Coding theory,
(10) Acoustic, and
(11) Music.

It is impossible to discuss all the above applications of number theory. We only concentrate ourselves on the applications of number theory in computing. In the pas few decades, number theory has been successfully applied to the following computing-related areas:

(1) Computer architecture and hardware design,
(2) Computer software systems design,
(3) Computer and network security,
(4) Random number generation,
(5) Digital signal processing,
(6) Computer graphics and image processing,
(7) Error detection and correction,
(8) Faulty-tolerant computing,
(9) Algorithm analysis and design,
(10) Theory of Computation, and
(11) Secure computation and communications.

In this book, we, of course, cannot deal with all the applications of numbers theory in computing; instead, we shall only deal with the applications of number theory in the following three computing-related areas:

(1) Computer systems design,
(2) Information systems security, and
(3) Random number generation.

1.1.3 Algebraic Preliminaries

If you are faced by a difficulty or a controversy in science, an ounce of algebra is worth a ton of verbal argument.

J. B. S. HALDANE (1892–1964)

The concepts and results in number theory are best described in certain types of modern abstract algebraic structures, such as groups, rings and fields. In this subsection, we shall provide a brief survey of these three widely used algebraic structures. Let us first introduce some set-theoretic notation for numbers.

(1) The set of *natural numbers* (positive integers, or counting numbers) $\mathbb{N}$:

$$\mathbb{N} = \{1, 2, 3, \cdots\}. \tag{1.9}$$

Some authors consider 0 as a natural number. But like Kronecker[9], we do not consider 0 as a natural number in this book.

(2) The set of *integers* $\mathbb{Z}$ (the letter $\mathbb{Z}$ comes from the German word Zählen):

$$\mathbb{Z} = \{0, \pm 1, \pm 2, \pm 3, \cdots\}. \tag{1.10}$$

We shall occasionally use

(i) $\mathbb{Z}_{\geq 0}$ to represent the set of *nonnegative integers*:

$$\mathbb{Z}_{\geq 0} = \{0, 1, 2, 3, \cdots\}, \tag{1.11}$$

(ii) $\mathbb{Z}^+$ to represent the set of *positive integers*:

$$\mathbb{Z}^+ = \{1, 2, 3, \cdots\} = \mathbb{N}, \tag{1.12}$$

(iii) $\mathbb{Z}_{>1}$ to represent the set of positive integers greater than 1:

$$\mathbb{Z}_{>1} = \{2, 3, 4, \cdots\}. \tag{1.13}$$

9

Leopold Kronecker (1823–1891) studied mathematics at Berlin University, and did his doctoral thesis on algebraic number theory under Dirichlet's supervision. Kronecker was one of the few of his generation to understand and master Évariste Galois's theory, and is well known for his famous remark "Natural numbers are made by God, all the rest are man made." Kronecker believed that mathematics should deal only with finite numbers and with a finite number of operations.

(3) The set of all *residue classes* modulo a positive integer n, denoted by $\mathbb{Z}/n\mathbb{Z}$ (which is read "$\mathbb{Z}$ modulo n"):

$$\mathbb{Z}/n\mathbb{Z} = \{0, 1, 2, \cdots, n-1\} = \mathbb{Z}_n. \tag{1.14}$$

One of the main tasks in this chapter is to study the arithmetic in the set $\mathbb{Z}/n\mathbb{Z}$. Note that some authors use $\mathbb{Z}_n$ to denote the set of all residue classes modulo n.

(4) The set of *rational numbers* $\mathbb{Q}$:

$$\mathbb{Q} = \left\{ \frac{a}{b} \; : \; a, b \in \mathbb{Z} \text{ and } b \neq 0 \right\}. \tag{1.15}$$

(5) The set of *real numbers* $\mathbb{R}$:
$\mathbb{R}$ is defined to be the set of converging sequences of rational numbers or decimals; they may or may not repeat. There are two subsets within the set of real numbers: *algebraic numbers* and *transcendental numbers.* An algebraic number is a real number that is the root of a polynomial equation with integer coefficients; all rational numbers are algebraic, since a/b is the root of the equation $bx - a = 0$. An *irrational number* is a real number that is not rational. For example, $\sqrt{2} = 1.4142135\cdots$, $\pi = 3.1415926\cdots$ and $e = 2.7182818\cdots$ are all real numbers but not rational, and hence they are irrational. Some irrational numbers are algebraic; for example, $\sqrt{2}$ is the root of equation $x^2 - 2 = 0$, and hence $\sqrt{2}$ is an algebraic number. An irrational number that is not a root of a polynomial equation with integer coefficients (i.e., not algebraic, such as π and e) is a transcendental number. Thus, we have

$$\text{real number} \begin{cases} \text{rational} - \text{algebraic, e.g., } 5/4, 2/3, 20/7 \\ \text{irrational} \begin{cases} \text{algebraic, e.g., } \sqrt{2},\ 1+\sqrt{2} \\ \text{transcendental, e.g., } \pi, e \end{cases} \end{cases}$$

(6) The set of complex numbers $\mathbb{C}$:

$$\mathbb{C} = \{a + bi \; : \; a, b \in \mathbb{R} \text{ and } i = \sqrt{-1}\}. \tag{1.16}$$

Definition 1.1.1. A *binary operation* $\star$ on a set S is a *rule* that assigns to each ordered pair (a, b) of elements of S a unique element of S.

Example 1.1.2. Ordinary addition $+$ is a binary operation on the sets $\mathbb{N}$, $\mathbb{Z}, \mathbb{R}$, or $\mathbb{C}$. Ordinary multiplication $\cdot$ is another binary operation on the same sets.

Definition 1.1.2. A *group*, denoted by $\langle \mathcal{G}, \star \rangle$, or $(\mathcal{G}, \star)$, or simply $\mathcal{G}$, is a nonempty set $\mathcal{G}$ of elements together with a binary operation $\star$, such that the following axioms are satisfied:

(1) Closure: $a \star b \in \mathcal{G}, \ \ \forall a, \ b \in \mathcal{G}$.

(2) Associativity: $(a \star b) \star c = a \star (b \star c), \quad \forall a, b, c \in \mathcal{G}$.
(3) Existence of identity: There is a unique element $e \in \mathcal{G}$, called the identity, such that $e \star a = a \star e = a, \quad \forall a \in \mathcal{G}$.
(4) Existence of inverse: For every $a \in \mathcal{G}$ there is a unique element b such that $a \star b = b \star a = e$. This b is denoted by a^{-1} and called the inverse of a.
The group $\langle \mathcal{G}, \star \rangle$ is called a *commutative group* if it satisfies a further axiom:
(5) Commutativity: $a \star b = b \star a, \quad \forall a, b \in \mathcal{G}$.

A commutative group is also called an *Abelian group*, in honour of the Norwegian mathematician N. H. Abel[10].

Example 1.1.3. The set $\mathbb{Z}^+$ with operation $+$ is *not* a group, since there is no identity element for $+$ in $\mathbb{Z}^+$. The set $\mathbb{Z}^+$ with operation $\cdot$ is *not* a group; there is an identity element 1, but no inverse of 3.

Example 1.1.4. The set of all nonnegative integers, $\mathbb{Z}_{\geq 0}$, with operation $+$ is *not* a group; there is an identity element 0, but no inverse for 2.

Example 1.1.5. The sets $\mathbb{Q}^+$ and $\mathbb{R}^+$ of positive numbers and the sets $\mathbb{Q}^*$, $\mathbb{R}^*$ and $\mathbb{C}^*$ of nonzero numbers with operation $\cdot$ are Abelian groups.

Definition 1.1.3. $\mathcal{G}$ is said to be a *semigroup* with respect to the binary operation $\star$ if it only satisfies the group axioms (1) and (2) of Definition 1.1.2. $\mathcal{G}$ is said to be a *monoid* with respect to the binary operation $\star$ if it only satisfies the group axioms (1), (2) and (3).

Definition 1.1.4. If the binary operation of a group is denoted by $+$, then the identity of a group is denoted by 0 and the inverse a by $-a$; this group is said to be an *additive group*.

Definition 1.1.5. If the binary operation of a group is denoted by $*$, then the identity of a group is denoted by 1 or e; this group is said to be a *multiplicative group*.

Definition 1.1.6. A group is called a *finite group* if it has a finite number of elements; otherwise it is called an *infinite group*.

10

Many mathematicians have had brilliant but short careers; Niels Henrik Abel (1802–1829), is one of such mathematicians. Abel made his greatest contribution to mathematics at the age of nineteen and died in poverty, just eight years later, of tuberculosis. Charles Hermite (1822–1901), a French mathematician who worked in algebra and analysis, once said that Abel "has left mathematicians something to keep them busy for five hundred years"; it is certainly true that Abel's discoveries still have a profound influence on today's number theorists.

Definition 1.1.7. The order of a group $\mathcal{G}$, denoted by $|\mathcal{G}|$ (or by $\#(\mathcal{G})$), is the number of elements in $\mathcal{G}$.

Example 1.1.6. The order of $\mathbb{Z}$ is $|\mathbb{Z}| = \infty$.

Definition 1.1.8. A nonempty set $\mathcal{G}'$ of a group $\mathcal{G}$ which is itself a group, under the same operation, is called a *subgroup* of $\mathcal{G}$.

Definition 1.1.9. Let a be an element of a multiplicative group $\mathcal{G}$. The elements a^r, where r is an integer, form a subgroup of $\mathcal{G}$, called the subgroup generated by a. A group $\mathcal{G}$ is *cyclic* if there is an element $a \in \mathcal{G}$ such that the subgroup generated by a is the whole of $\mathcal{G}$. If $\mathcal{G}$ is a finite cyclic group with identity element e, the set of elements of $\mathcal{G}$ may be written $\{e, a, a^2, \cdots, a^{n-1}\}$, where $a^n = e$ and n is the smallest such positive integer. If $\mathcal{G}$ is an infinite cyclic group, the set of elements may be written $\{\cdots, a^{-2}, a^{-1}, e, a, a^2, \cdots\}$.

By making appropriate changes, a cyclic *additive group* can be defined. For example, the set $\{0, 1, 2, \cdots, n-1\}$ with addition modulo n is a cyclic group, and the set of all integers with addition is an infinite cyclic group.

Definition 1.1.10. A *ring*, denoted by $\langle \mathcal{R}, \oplus, \odot \rangle$, or $(\mathcal{R}, \oplus, \odot)$, or simply $\mathcal{R}$, is a set of at least two elements with *two* binary operations $\oplus$ and $\odot$, which we call addition and multiplication, defined on $\mathcal{R}$ such that the following axioms are satisfied:

(1) The set is *closed* under the operation $\oplus$:

$$a \oplus b \in \mathcal{R}, \quad \forall a,\ b \in \mathcal{R}, \tag{1.17}$$

(2) The associative law holds for $\oplus$:

$$a \oplus (b \oplus c) = (a \oplus b) \oplus c, \quad \forall a,\ b,\ c \in \mathcal{R}, \tag{1.18}$$

(3) The commutative law holds for $\oplus$:

$$a \oplus b = b \oplus a, \quad \forall a,\ b \in \mathcal{R}, \tag{1.19}$$

(4) There is a special (zero) element $0 \in \mathcal{R}$, called the additive identity of $\mathcal{R}$, such that

$$a \oplus 0 = 0 \oplus a = a, \quad \forall a \in \mathcal{R}, \tag{1.20}$$

(5) For each $a \in \mathcal{R}$, there is a corresponding element $-a \in \mathcal{R}$, called the additive inverse of a, such that:

$$a \oplus (-a) = 0, \quad \forall a \in \mathcal{R}, \tag{1.21}$$

(6) The set is closed under the operation $\odot$:

$$a \odot b \in \mathcal{R}, \quad \forall a,\ b \in \mathcal{R}, \tag{1.22}$$

(7) The associative law holds for $\odot$:

$$a \odot (b \odot c) = (a \odot b) \odot c, \quad \forall a, b, c \in \mathcal{R}, \tag{1.23}$$

(8) The operation $\odot$ is distributive with respect to $\oplus$:

$$a \odot (b \oplus c) = a \odot b \oplus a \odot c, \quad \forall a, b, c \in \mathcal{R}, \tag{1.24}$$

$$(a \oplus b) \odot c = a \odot c \oplus b \odot c, \quad \forall a, b, c \in \mathcal{R}. \tag{1.25}$$

From a group theoretic point of view, a ring is an Abelian group, with the additional properties that the closure, associative and distributive laws hold for $\odot$.

Example 1.1.7. $\langle \mathbb{Z}, \oplus, \odot \rangle$, $\langle \mathbb{Q}, \oplus, \odot \rangle$, $\langle \mathbb{R}, \oplus, \odot \rangle$, and $\langle \mathbb{C}, \oplus, \odot \rangle$ are all rings.

Definition 1.1.11. A *commutative ring* is a ring that further satisfies:

$$a \odot b = b \odot a, \quad \forall a, b \in \mathcal{R}. \tag{1.26}$$

Definition 1.1.12. A *ring with identity* is a ring that contains an element 1 satisfying:

$$a \odot 1 = a = 1 \odot a, \quad \forall a \in \mathcal{R}. \tag{1.27}$$

Definition 1.1.13. An *integral domain* is a commutative ring with identity $1 \neq 0$ that satisfies:

$$a, b \in \mathcal{R} \ \ \& \ \ ab = 0 \quad \Longrightarrow a = 0 \ \text{ or } \ b = 0. \tag{1.28}$$

Definition 1.1.14. A *division ring* is a ring $\mathcal{R}$ with identity $1 \neq 0$ that satisfies:

for each $a \neq 0 \in \mathcal{R}$, the equation $ax = 1$ and $xa = 1$ have solutions in $\mathcal{R}$.

Definition 1.1.15. A *field*, denoted by $\mathcal{K}$, is a division ring with commutative multiplication.

Example 1.1.8. The integer set $\mathbb{Z}$, with the usual addition and multiplication, forms a commutative ring with identity, but is not a field.

It is clear that a field is a type of ring, which can be defined more generally as follows:

Definition 1.1.16. A *field*, denoted by $\langle \mathcal{K}, \oplus, \odot \rangle$, or $(\mathcal{K}, \oplus, \odot)$, or simply $\mathcal{K}$, is a set of at least two elements with *two* binary operations $\oplus$ and $\odot$, which we call addition and multiplication, defined on $\mathcal{K}$ such that the following axioms are satisfied:

(1) The set is *closed* under the operation $\oplus$:

$$a \oplus b \in \mathcal{K}, \quad \forall a, \ b \in \mathcal{K}, \tag{1.29}$$

(2) The associative law holds for $\oplus$:

$$a \oplus (b \oplus c) = (a \oplus b) \oplus c, \quad \forall a,\ b,\ c \in \mathcal{K}, \tag{1.30}$$

(3) The commutative law holds for $\oplus$:

$$a \oplus b = b \oplus a, \quad \forall a,\ b \in \mathcal{K}, \tag{1.31}$$

(4) There is a special (zero) element $0 \in \mathcal{K}$, called the additive identity of $\mathcal{K}$, such that

$$a \oplus 0 = 0 \oplus a = a, \quad \forall a \in \mathcal{K}, \tag{1.32}$$

(5) For each $a \in \mathcal{K}$, there is a corresponding element $-a \in \mathcal{K}$, called the additive inverse of a, such that:

$$a \oplus (-a) = 0, \quad \forall a \in \mathcal{K}, \tag{1.33}$$

(6) The set is closed under the operation $\odot$:

$$a \odot b \in \mathcal{K}, \quad \forall a,\ b \in \mathcal{K}, \tag{1.34}$$

(7) The associative law holds for $\odot$:

$$a \odot (b \odot c) = (a \odot b) \odot c, \quad \forall a, b, c \in \mathcal{K} \tag{1.35}$$

(8) The operation $\odot$ is distributive with respect to $\oplus$:

$$a \odot (b \oplus c) = a \odot b \oplus a \odot c, \quad \forall a, b, c \in \mathcal{K}, \tag{1.36}$$

$$(a \oplus b) \odot c = a \odot c \oplus b \odot c, \quad \forall a,\ b,\ c \in \mathcal{K}. \tag{1.37}$$

(9) There is an element $1 \in \mathcal{K}$, called the multiplicative identity of $\mathcal{K}$, such that $1 \neq 0$ and

$$a \odot 1 = a, \quad \forall a \in \mathcal{K}, \tag{1.38}$$

(10) For each nonzero element $a \in \mathcal{K}$ there is a corresponding element $a^{-1} \in \mathcal{K}$, called the multiplicative inverse of a, such that

$$a \odot a^{-1} = 1, \tag{1.39}$$

(11) The commutative law holds for $\odot$:

$$a \odot b = b \odot a, \quad \forall a,\ b \in \mathcal{K}, \tag{1.40}$$

Again, from a group theoretic point of view, a field is an Abelian group with respect to addition and also the non-zero field elements form an Abelian group with respect to multiplication.

Figure 1.2 gives a Venn diagram view of containment for algebraic structures having two binary operations.

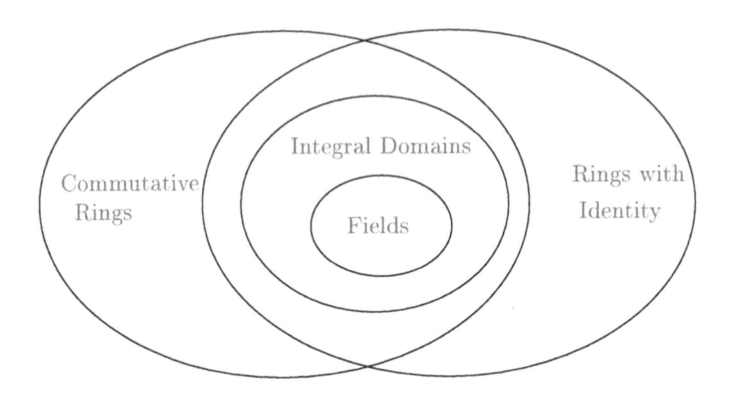

Figure 1.2. Containment of various rings

Example 1.1.9. Familiar examples of fields are the set of rational numbers, $\mathbb{Q}$, the set of real numbers, $\mathbb{R}$ and the set of complex numbers, $\mathbb{C}$; since $\mathbb{Q}$, $\mathbb{R}$ and $\mathbb{C}$ are all infinite sets, they are all infinite fields. The set of integers $\mathbb{Z}$ is a ring but *not* a field, since 2, for example, has no multiplicative inverse; 2 is not a unit in $\mathbb{Z}$. The only units in $\mathbb{Z}$ are 1 and -1. Another example of a ring which is not a field is the set $\mathcal{K}[x]$ of polynomials in x with coefficients belonging to a field $\mathcal{K}$.

Definition 1.1.17. A *finite field* is a field that has a finite number of elements in it; we call the number the *order* of the field.

The following fundamental result on finite fields was first proved by Évariste Galois[11]:

Theorem 1.1.1. There exists a field of order q if and only if q is a prime power (i.e., $q = p^r$) with p prime and $r \in \mathbb{N}$. Moreover, if q is a prime power, then there is, up to relabelling, only one field of that order.

A field of order q with q a prime power is often called a *Galois field*, and is denoted by $\mathrm{GF}(q)$, or just $\mathbb{F}_q$. Clearly, a Galois field is a finite field.

11

Évariste Galois (1811–1832), a French mathematician who made major contributions to the theory of equations (for example, he proved that the general quintic equation is not solvable by radicals) and groups before he died at the age of 21, shot in an illegal duel; he spent the whole night before the duel writing a letter containing notes of his discoveries. Galois's unpublished mathematical papers were copied and sent to Gauss, Jacobi and others by his brother and a friend. No record exists of any comment from Gauss and Jacobi. However when the papers reached Liouville (1809–1882), he announced in 1843 to the French Academy that he had found deep results in Galois's papers, and subsequently published Galois's work in 1846 in his Journal.

Example 1.1.10. The finite field $\mathbb{F}_5$ has elements $\{0, 1, 2, 3, 4\}$ and is described by the following addition and multiplication table (see Table 1.1):

Table 1.1. The addition and multiplication for $\mathbb{F}_5$

$\oplus$	0	1	2	3	4
0	0	1	2	3	4
1	1	2	3	4	0
2	2	3	4	0	1
3	3	4	0	1	2
4	4	0	1	2	3

$\odot$	1	2	3	4
1	1	2	3	4
2	2	4	1	3
3	3	1	4	2
4	4	3	2	1

The theory of groups, rings, and particularly finite fields plays a very important role in elementary, algorithmic and applied number theory, including cryptography and information security.

1.2 Theory of Divisibility

The primary source of all mathematics is the integers.

H. MINKOWSKI (1864–1909)

Divisibility has been studied for at least three thousand years. From before the time of Pythagoras, the Greeks considered questions about even and odd numbers, perfect and amicable numbers, and the primes, among many others; even today a few of these questions are still unanswered.

1.2.1 Basic Concepts and Properties of Divisibility

Definition 1.2.1. Let a and b be integers with $a \neq 0$. We say a divides b, denoted by $a \mid b$, if there exists an integer c such that $b = ac$. When a divides b, we say that a is a *divisor* (or *factor*) of b, and b is a *multiple* of a. If a does not divide b, we write $a \nmid b$. If $a \mid b$ and $0 < a < b$, then a is called a *proper divisor* of b.

Remark 1.2.1. We never use 0 as the left member of the pair of integers in $a \mid b$, however, 0 may occur as the right member of the pair, thus $a \mid 0$ for every integer a not zero. Under this restriction, for $a \mid b$, we may say that b is divisible by a, which is equivalent to say that a is a divisor of b. The notation $a^\alpha \parallel b$ is sometimes used to indicate that $a^\alpha \mid b$ but $a^{\alpha+1} \nmid b$.

Example 1.2.1. The integer 200 has the following positive divisors (note that, as usual, we shall be only concerned with positive divisors, not negative divisors, of an integer):

$$1, 2, 4, 5, 8, 10, 20, 25, 40, 50, 100, 200.$$

Thus, for example, we can write

$$8 \mid 200, \quad 50 \mid 200, \quad 7 \nmid 200, \quad 35 \nmid 200.$$

Definition 1.2.2. A divisor of n is called a *trivial divisor* of n if it is either 1 or n itself. A divisor of n is called a *nontrivial divisor* if it is a divisor of n, but is neither 1, nor n.

Example 1.2.2. For the integer 18, 1 and 18 are the trivial divisors, whereas 2, 3, 6 and 9 are the nontrivial divisors. The integer 191 has only two trivial divisors and does not have any nontrivial divisors.

Some basic properties of divisibility are given in the following theorem:

Theorem 1.2.1. Let a, b and c be integers. Then

(1) if $a \mid b$ and $a \mid c$, then $a \mid (b + c)$.
(2) if $a \mid b$, then $a \mid bc$, for any integer c.
(3) if $a \mid b$ and $b \mid c$, then $a \mid c$.

Proof.

(1) Since $a \mid b$ and $a \mid c$, we have

$$b = ma, \quad c = na, \qquad m, n \in \mathbb{Z}.$$

Thus $b + c = (m + n)a$. Hence, $a \mid (m + n)a$ since $m + n$ is an integer. The result follows.

(2) Since $a \mid b$ we have

$$b = ma, \qquad m \in \mathbb{Z}.$$

Multiplying both sides of this equality by c gives

$$bc = (mc)a$$

which gives $a \mid bc$, for all integers c (whether or not $c = 0$).

(3) Since $a \mid b$ and $b \mid c$, there exists integers m and n such that

$$b = ma, \qquad c = nb.$$

Thus, $c = (mn)a$. Since mn is an integer the result follows.

Exercise 1.2.1. Let a, b and c be integers. Show that

(1) $1 \mid a$, $a \mid a$, $a \mid 0$.
(2) if $a \mid b$ and $b \mid a$, then $a = \pm b$.
(3) if $a \mid b$ and $a \mid c$, then for all integers m and n we have $a \mid (mb + nc)$.
(4) if $a \mid b$ and a and b are positive integers, then $a < b$.

The next result is a general statement of the outcome when any integer a is divided by any positive integer b.

Theorem 1.2.2 (Division algorithm). For any integer a and any positive integer b, there exist unique integers q and r such that

$$a = bq + r, \quad 0 \leq r < b, \tag{1.41}$$

where a is called the *dividend*, q the *quotient*, and r the *remainder*. If $b \nmid a$, then r satisfies the stronger inequalities $0 < r < a$.

Proof. Consider the arithmetic progression

$$\cdots, -3b, -2b, -b, 0, b, 2b, 3b, \cdots$$

then there must be an integer q such that

$$gb \leq a \leq (q+1)b.$$

Let $a - qb = r$, then $a = bq + r$ with $0 \leq r < b$. To prove the uniqueness of q and r, suppose there is another pair q_1 and r_1 satisfying the same condition in (1.41), then

$$a = bq_1 + r_1, \quad 0 \leq r_1 < b.$$

We first show that $r_1 = r$. For if not, we may presume that $r < r_1$, so that $0 < r_1 - r < b$, and then we see that $b(q - q_1) = r_1 - r$, and so $b \mid (r_1 - r)$, which is impossible. Hence, $r = r_1$, and also $q = q_1$. □

Remark 1.2.2. Theorem 1.2.2 is called the division algorithm. An *algorithm* is a mathematical procedure or method to obtain a result (we will discuss algorithms and their complexity in detail in Chapter 2). We have stated in Theorem 1.2.2 that "there exist unique integer q and r" and this wording suggests that we have an *existence theorem* rather than an *algorithm*. However, it may be observed that the proof does provide a method for obtaining the integer q and r, since q and r can be obtained by the arithmetic division a/b.

Example 1.2.3. Let $b = 15$. Then

(1) when $a = 255$, $a = b \cdot 17 + 0$, so $q = 17$ and $r = 0 < 15$.
(2) when $a = 177$, $a = b \cdot 11 + 12$, so $q = 11$ and $r = 12 < 15$.

(3) when $a = -783$, $a = b \cdot (-52) + 3$, so $q = -52$ and $r = 3 < 15$.

Definition 1.2.3. Consider the following equation

$$a = 2q + r, \quad a, q, r \in \mathbb{Z}, \ 0 \leq r < q. \tag{1.42}$$

Then if $r = 0$, then a is *even*, whereas if $r = 1$, then a is *odd*.

Definition 1.2.4. A positive integer n greater than 1 is called *prime* if its only divisors are n and 1. A positive integer n that is greater than 1 and is not prime is called *composite*.

Example 1.2.4. The integer 23 is prime since its only divisors are 1 and 23, whereas 22 is composite since it is divisible by 2 and 11.

Prime numbers have many special and nice properties, and play a central role in the development of number theory. Mathematicians throughout history have been fascinated by primes. The first result on prime numbers is due to Euclid:

Theorem 1.2.3 (Euclid). There are infinitely many primes.

Proof. Suppose that $p_1, p_2, \cdots, p_k$ are all the primes. Consider the number $N = p_1 p_2 \cdots p_k + 1$. If it is a prime, then it is a new prime. Otherwise, it has a prime factor q. If q were one of the primes p_i, $i = 1, 2, \cdots, k$, then $q \mid (p_1 p_2 \cdots p_k)$, and since $q \mid (p_1 p_2 \cdots p_k + 1)$, q would divide the difference of these numbers, namely 1, which is impossible. So q cannot be one of the p_i for $i = 1, 2, \cdots, k$, and must therefore be a new prime. This completes the proof. □

Remark 1.2.3. The above proof of Euclid's theorem is based on the modern algebraic language. For Euclid's original proof, translated in English, see Figure 1.3.

Two other related elementary results about the infinitude of primes are as follows.

Theorem 1.2.4. If n is an integer ≥ 1, then there is a prime p such that $n < p \leq n! + 1$.

Proof. Consider the integer $N = n! + 1$. If N is prime, we may take $p = N$. If N is not prime, it has some prime factor p. Suppose $p \leq n$, then $p \mid n!$; hence, $p \mid (N - n!)$, which is ridiculous since $N - n! = 1$. Therefore, $p > n$. □

Theorem 1.2.5. Given any real number $x \geq 1$, there exists a prime between x and $2x$.

PROPOSITION 20.

Prime numbers are more than any assigned multitude of prime numbers.

Let *A*, *B*, *C* be the assigned prime numbers;
I say that there are more prime numbers than *A*, *B*, *C*.

For let the least number measured by *A*, *B*, *C* be taken,
and let it be *DE*;
let the unit *DF* be added to *DE*.

Then *EF* is either prime or not.

First, let it be prime;
then the prime numbers *A*, *B*, *C*, *EF* have been found which are more than *A*, *B*, *C*.

Next, let *EF* not be prime;
therefore it is measured by some prime number. [VII. 31]

Let it be measured by the prime number *G*.

I say that *G* is not the same with any of the numbers *A*, *B*, *C*.

For, if possible, let it be so.

Now *A*, *B*, *C* measure *DE*;
therefore *G* also will measure *DE*.

But it also measures *EF*.

Therefore *G*, being a number, will measure the remainder, the unit *DF*:
which is absurd.

Therefore *G* is not the same with any one of the numbers *A*, *B*, *C*.

And by hypothesis it is prime.

Therefore the prime numbers *A*, *B*, *C*, *G* have been found which are more than the assigned multitude of *A*, *B*, *C*.

Q. E. D.

Figure 1.3. Proposition 20 of the *Elements* Book IX (by courtesy of Thomas L. Heath [73])

This is the famous Bertrand's postulate, conjectured by Joseph Bertrand (1822–1900) in 1845, and proved by Chebyshev in 1850. The proof of this result is rather lengthy; interested readers are advised to consult Hardy and Wright's book [100]. However, there do exist long sequences of consecutive integers which are barren of primes, as the next result shows.

Proposition 1.2.1. If n is an integer ≥ 2, then there are no primes between $n! + 2$ and $n! + n$.

Proof. Since if $n!$ is a product of all integers between 1 and n, then $2 \mid n!+2$, $3 \mid n! + 3, \cdots, n \mid n! + n$. □

Theorem 1.2.6. If n is a composite, then n has a prime divisor p such that $p \leq \sqrt{n}$.

Proof. Let p be the smallest prime divisor of n. If $n = rs$, then $p \leq r$ and $p \leq s$. Hence, $p^2 \leq rs = n$. That is, $p \leq \sqrt{n}$. □

Theorem 1.2.6 can be used to find all the prime numbers up to a given positive integer x; this procedure is called the Sieve of Eratosthenes, attributed to the ancient Greek astronomer and mathematician Eratosthenes of Cyrene[12], assuming that x is relatively small. To apply the sieve, list all the integers from 2 up to x in order:

$$2, 3, 4, 5, 6, 7, 8, 9, 10, 11, 12, 13, 14, 15, \cdots, x.$$

Starting from 2, delete all the multiples $2m$ of 2 such that $2 < 2m \leq x$:

$$2, 3, 5, 7, 9, 11, 13, 15, \cdots, x.$$

Starting from 3, delete all the multiples $3m$ of 3 such that $3 < 3m \leq x$:

$$2, 3, 5, 7, 11, 13, \cdots, x.$$

In general, if the resulting sequence at the kth stage is

$$2, 3, 5, 7, 11, 13, \cdots, p, \cdots, x.$$

then delete all the multiples pm of p such that $p < pm \leq x$. Continue this exhaustive computation, until $p \leq \lfloor \sqrt{x} \rfloor$. The remaining integers are all the primes between $\lfloor \sqrt{x} \rfloor$ and x and if we take care not to delete $2, 3, 5, \cdots, p \leq \lfloor \sqrt{x} \rfloor$, the sieve then gives all the primes less than or equal to x. For example, let $x = 36$, then $\sqrt{x} = 6$, there are only three primes 2, 3 and 5 below 6, and all the positive integers from 2 to 36 are as follows.

	2	3	4	5	6	7	8	9	10	11	12
13	14	15	16	17	18	19	20	21	22	23	24
25	26	27	28	29	30	31	32	33	34	35	36

First of all, we delete (marked with the symbol "_") all the multiples of 2 with $2 < 2m \leq 36$, for $m = 1, 2, \cdots, 18$, and get:

[12] 
Eratosthenes of Cyrene (274–194 B.C.) was born in Cyrene which is now in Libya in North Africa, was one of the great men in the ancient world. He was the first to calculate the size of the Earth by making measurements of the angle of the Sun at two different places a known distance apart. His other achievements include measuring the tilt of the Earth's axis. Eratosthenes also worked on prime numbers. He is best remembered by generations of number theorists for his prime number sieve, the "Sieve of Eratosthenes" which, in modified form, is still an important tool in number theory research.

	2	3	-	5	-	7	-	9	-	11	-
13	-	15	-	17	-	19	-	21	-	23	-
25	-	27	-	29	-	31	-	33	-	35	-

Then we delete (marked with the symbol "$_*$") all the multiples of 3 with $3 < 3m \leq 36$, for $m = 1, 2, \cdots, 11$, and get:

	2	3	-	5	-	7	-	*	-	11	-
13	-	*	-	17	-	19	-	*	-	23	-
25	-	*	-	29	-	31	-	*	-	35	-

Finally, we delete (marked with the symbol "$_\times$") all the multiples of 5 with $5 < 5m \leq 35$, for $m = 1, 2, \cdots, 7$, and get:

	2	3	-	5	-	7	-	*	-	11	-
13	-	*	-	17	-	19	-	*	-	23	-
×	-	*	-	29	-	31	-	*	-	×	-

The remaining numbers $2, 3, 5, 7, 11, 13, 17, 19, 23, 29, 31$ are then the primes up to 36.

According to the above analysis, we can get the following algorithm for the Sieve of Eratosthenes:

Algorithm 1.2.1 (The Sieve of Eratosthenes). Given a positive integer $n > 1$, this algorithm will find all prime numbers up to n.

[1] Create a list of integers from 2 to n;

[2] For prime numbers p_i $(i = 1, 2, \cdots)$ from $2, 3, 5$ up to $\lfloor\sqrt{n}\rfloor$, delete all the multiples $p_i < p_i m \leq n$ from the list;

[3] Print the integers remaining in the list.

1.2.2 Fundamental Theorem of Arithmetic

First, let us investigate a simple but important property of composite numbers.

Theorem 1.2.7. Every composite number has a prime factor.

Proof. Let n be a composite number. Then

$$n = n_1 n_2$$

where n_1 and n_2 are positive integers with $n_1, n_2 < n$. If either n_1 or n_2 is a prime, then the theorem is proved. If n_1 and n_2 are not prime, then

$$n_1 = n_3 n_4$$

where n_3 and n_4 are positive integers with $n_3, n_4 < n_1$. Again if n_3 or n_4 is a prime, then the theorem is proved. If n_3 and n_4 are not prime, then we can write

$$n_3 = n_5 n_6$$

where n_5 and n_6 are positive integers with $n_5, n_6 < n_3$. In general, after k steps we write

$$n_{2k-1} = n_{2k+1} n_{2k+2}$$

where n_{2k+1} and n_{2k+2} are positive integers with $n_{2k+1}, n_{2k+1} < n_{2k-1}$. Since

$$n > n_1 > n_3 > n_5 > \cdots n_{2k-1} > 0$$

for any value k, the process must terminate. So there must exist an n_{2k-1} for some value of k, that is prime. Hence, every composite has a prime factor. □

Prime numbers are the building blocks of positive integers, as the following theorem shows:

Theorem 1.2.8 (Fundamental Theorem of Arithmetic). Every positive integer n greater than 1 can be written uniquely as the product of primes:

$$n = p_1^{\alpha_1} p_2^{\alpha_2} \cdots p_k^{\alpha_k} = \prod_{i=1}^{k} p_i^{\alpha_i} \tag{1.43}$$

where $p_1, p_2, \cdots, p_k$ are distinct primes, and $\alpha_1, \alpha_2, \cdots, \alpha_k$ are natural numbers. The equation (1.43) is often called the prime power decomposition of n, or the standard prime factorization of n.

Proof. We shall first show that a factorization exists. Starting from $n > 1$, if n is a prime, then it stands as a *product* with a single factor. Otherwise, n can be factored into, say, ab, where $a > 1$ and $b > 1$. Apply the same argument to a and b: each is either a prime or a product of two numbers both > 1. The numbers other than primes involved in the expression for n are greater than 1 and decrease at every step; hence eventually all the numbers must be prime.

Now we come to uniqueness. Suppose that the theorem is false and let $n > 1$ be the smallest number having more than one expression as the product of primes, say

$$n = p_1 p_2 \cdots p_r = q_1 q_2 \cdots q_s$$

where each p_i $(i = 1, 2, \cdots, r)$ and each q_j $(j = 1, 2, \cdots, s)$ is prime. Clearly both r and s must be greater than 1 (otherwise n is prime, or a prime is equal to a composite). If for example p_1 were one of the q_j $(j = 1, 2, \cdots, s)$, then n/p_1 would have two expressions as a product of primes, but $n/p_1 < n$ so this would contradict the definition of n. Hence p_1 is not equal to any of the q_j $(j = 1, 2, \cdots, s)$, and similarly none of the p_i $(i = 1, 2, \cdots, r)$ equals any of the q_j $(j = 1, 2, \cdots, s)$. Next, there is no loss of generality in presuming that $p_1 < q_1$, and we define the positive integer N as

$$N = (q_1 - p_1)q_2q_3 \cdots q_s = p_1(p_2p_3 \cdots p_r - q_2q_3 \cdots q_s).$$

Certainly $1 < N < n$, so N is uniquely factorable into primes. However, $p_1 \nmid (q_1 - p_1)$, since $p_1 < q_1$ and q_1 is prime. Hence one of the above expressions for N contains p_1 and the other does not. This contradiction proves the result: there cannot be any exceptions to the theorem. □

Note that if n is prime, then the product is, of course, n itself.

Example 1.2.5. The following are some sample prime factorizations:

$$\begin{array}{ll} 643 = 643 & 2^{31} - 1 = 2147483647 \\ 644 = 2^2 \cdot 7 \cdot 23 & 2^{31} + 1 = 3 \cdot 715827883 \\ 645 = 3 \cdot 5 \cdot 43 & 2^{32} - 1 = 3 \cdot 5 \cdot 17 \cdot 257 \cdot 65537 \\ 646 = 2 \cdot 17 \cdot 19 & 2^{32} + 1 = 641 \cdot 6700417 \\ 647 = 647 & 2^{31} + 2 = 2 \cdot 5^2 \cdot 13 \cdot 41 \cdot 61 \cdot 1321 \end{array}$$

Definition 1.2.5. Let a and b be integers, not both zero. The largest divisor d such that $d \mid a$ and $d \mid b$ is called the *greatest common divisor* (gcd) of a and b. The greatest common divisor of a and b is denoted by $\gcd(a, b)$.

Example 1.2.6. The sets of positive divisors of 111 and 333 are as follows:

$$1, 3, 37, 111,$$
$$1, 3, 9, 37, 111, 333,$$

so $\gcd(111, 333) = 111$. But $\gcd(91, 111) = 1$, since 91 and 111 have no common divisors other than 1.

The next theorem indicates that $\gcd(a, b)$ can be represented as a linear combination of a and b.

Theorem 1.2.9. Let a and b be integers, not both zero. Then there exists integers x and y such that

$$d = \gcd(a, b) = ax + by. \tag{1.44}$$

Proof. Consider the set of all linear combinations $au + bv$, where u and v range over all integers. Clearly this set of integers $\{au + bv\}$ includes positive, negative as well as 0. Choose x and y such that $m = ax + by$ is the smallest integer in the set. Use the Division algorithm, to write $a = mq + r$, where $0 \leq r < m$. Then

$$r = a - mq = a - q(ax + by) = (1 - qx)a + (-qy)b$$

and hence r is also a linear combination of a and b. But $r < m$, so it follows from the definition of m that $r = 0$. Thus $a = mq$, that is, $m \mid a$; similarly, $m \mid b$. Therefore, m is a common divisor of a and b. Since $d \mid a$ and $d \mid b$, $d \leq m$. Since $d = \gcd(a, b)$, we must have $d = m$. □

Remark 1.2.4. The greatest common divisor of a and b can also be characterized as follows:

(1) $d \mid a$ and $d \mid b$,

(2) if $c \mid a$ and $c \mid b$, then $c \mid d$.

Corollary 1.2.1. If a and b are integers, not both zero, then the set

$$S = \{ax + by : \ x, y \in \mathbb{Z}\}$$

is precisely the set of all multiples of $d = \gcd(a, b)$.

Proof. It follows from Theorem 1.2.9, because d is the smallest positive values of $ax + by$ where x and y range over all integers. □

Definition 1.2.6. Two integers a and b are called *relatively prime* if $\gcd(a, b) = 1$. We say that integers $n_1, n_2, ..., n_k$ are *pairwise relatively prime* if, whenever $i \neq j$, we have $\gcd(n_i, n_j) = 1$.

Example 1.2.7. 91 and 111 are relatively prime, since $\gcd(91, 111) = 1$.

The following theorem charaterizes relatively primes in terms of linear combinations.

Theorem 1.2.10. Let a and b be integers, not both zero, then a and b are relatively prime if and only if there exsit integers x and y such that $ax + by = 1$.

Proof. If a and b are relatively prime, so that $\gcd(a, b) = 1$, then Theorem 1.2.9 guarantees the existence of integers x and y satisfying $ax + by = 1$. As for the converse, suppose that $ax + by = 1$ and that $d = \gcd(a, b)$. Since $d \mid a$ and $d \mid b$, $d \mid (ax + by)$, that is, $d \mid 1$. Thus $d = 1$. The results follows. □

Theorem 1.2.11. If $a \mid bc$ and $\gcd(a, b) = 1$, then $a \mid c$.

Proof. By Theorem 1.2.9, we can write $ax + by = 1$ for some choice of integers x and y. Multiplying this equation by c we get

$$acx + bcy = c.$$

Since $a \mid ac$ and $a \mid bc$, it follows that $a \mid (acx + bcy)$. The result thus follows. □

For the greatest common divisor of more than two integers, we have the following result.

Theorem 1.2.12. Let $a_1, a_2, \cdots, a_n$ be n integers. Let also

$$\left.\begin{array}{l} \gcd(a_1, a_2) = d_2, \\ \gcd(d_2, a_3) = d_3, \\ \quad\cdots \\ \quad\cdots \\ \gcd(d_{n-1}, a_n) = d_n. \end{array}\right\} \tag{1.45}$$

Then

$$\gcd(a_1, a_2, \cdots, a_n) = d_n. \tag{1.46}$$

Proof. By (1.45), we have $d_n \mid a_n$ and $d_n \mid d_{n-1}$. But $d_{n-1} \mid a_{n-1}$ and $d_{n-1} \mid d_{n-2}$, so $d_n \mid a_{n-1}$ and $d_n \mid d_{n-2}$. Continuing in this way, we finally have $d_n \mid a_n$, $d_n \mid a_{n-1}$, $\cdots$, $d_n \mid a_1$, so d_n is a common divisor of $a_1, a_2, \cdots, a_n$. Now suppose that d is any common divisor of $a_1, a_2, \cdots, a_n$, then $d \mid a_1$ and $d \mid d_2$. Observe the fact the common divisor of a and b and the divisor of $\gcd(a, b)$ are the same, so $d \mid d_2$. Similarly, we have $d \mid d_3, \cdots, d \mid d_n$. Therefore, $d \le |d| \le d_n$. So, d_n is the greatest common divisor of $a_1, a_2, \cdots, a_n$. □

Definition 1.2.7. If d is a multiple of a and also a multiple of b, then d is a common multiple of a and b. The *least common multiple* (lcm) of two integers a and b, is the smallest of the common multiples of a and b. The least common multiple of a and b is denoted by $\text{lcm}(a, b)$.

Theorem 1.2.13. Suppose a and b are not both zero (i.e., one of the a and b can be zero, but not both zero), and that $m = \text{lcm}(a, b)$. If x is a common multiple of a and b, then $m \mid x$. That is, every common multiple of a and b is a multiple of the least common multiple.

Proof. If any one of a and b is zero, then all common multiples of a and b are zero, so the statement is trivial. Now we assume that both a and b are not zero. Dividing x by m, we get

$$x = mq + r, \quad \text{where } 0 \le r < m.$$

Now $a \mid x$ and $b \mid x$ and also $a \mid m$ and $b \mid m$; so by Theorem 1.2.1, $a \mid r$ and $b \mid r$. That is, r is a common multiple of a and b. But m is the least common multiple of a and b, so $r = 0$. Therefore, $x = mq$, the result follows. □

For the lest common multiple of more than two integers, we have the following result.

Theorem 1.2.14. Let $a_1, a_2, \cdots, a_n$ be n integers. Let also

$$\left.\begin{array}{l} \text{lcm}(a_1, a_2) = m_2, \\ \text{lcm}(m_2, a_3) = m_3, \\ \quad\cdots \\ \quad\cdots \\ \text{lcm}(m_{n-1}, a_n) = m_n. \end{array}\right\} \tag{1.47}$$

Then

$$\operatorname{lcm}(a_1, a_2, \cdots, a_n) = m_n. \tag{1.48}$$

Proof. By (1.47), we have $m_i \mid m_{i+1}$, $i = 2, 3, \cdots, n-1$, and $a_1 \mid m_2$, $a_i \mid m_i$, $i = 2, 3, \cdots, n$. So, m_n is a common multiple of $a_1, a_2, \cdots, a_n$. Now let m is any common multiple of $a_1, a_2, \cdots, a_n$, then $a_1 \mid m$, $a_2 \mid m$. Observe the result that all the common multiples of a and b are the multiples of $\operatorname{lcm}(a, b)$ $d_n \mid a_n$ and $d_n \mid d_{n-1}$. So $m_1 \mid m$ and $a_3 \mid m$. Continuing the process in this way, we finally have $m_n \mid m$. Thus, $m_n \leq |m|$. Therefore, $m_n = \operatorname{lcm}(a_1, a_2, \cdots, a_n)$. □

One way to calculate the $\gcd(a, b)$ or the $\operatorname{lcm}(a, b)$ is to use the standard prime factorizations of a and b. That is:

Theorem 1.2.15. If

$$a = \prod_{i=1}^{k} p_i^{\alpha_i}, \quad \alpha_i \geq 0,$$

and

$$b = \prod_{i=1}^{k} p_i^{\beta_i}, \quad \beta_i \geq 0,$$

then

$$\gcd(a, b) = \prod_{i=1}^{k} p_i^{\gamma_i} \tag{1.49}$$

$$\operatorname{lcm}(a, b) = \prod_{i=1}^{k} p_i^{\delta_i} \tag{1.50}$$

where $\gamma_i = \min(\alpha_i, \beta_i)$ and $\delta_i = \max(\alpha_i, \beta_i)$ for $i = 1, 2, \cdots, k$.

Proof. It is easy to see that

$$\gcd(a, b) = \prod_{i=1}^{k} p_i^{\gamma_i}, \text{ where } \gamma_i \text{ is the lesser of } \alpha_i \text{ and } \beta_i,$$

$$\operatorname{lcm}(a, b) = \prod_{i=1}^{k} p_i^{\delta_i}, \text{ where } \delta_i \text{ is the greater of } \alpha_i \text{ and } \beta_i.$$

The result thus follows. □

Of course, if we know any one of the $\gcd(a, b)$ or $\operatorname{lcm}(a, b)$, we can easily find the other via the following corollary which follows immediately from Theorem 1.2.15:

Corollary 1.2.2. Suppose a and b are positive integers, then

$$\operatorname{lcm}(a, b) = \frac{ab}{\gcd(a, b)}. \tag{1.51}$$

Proof. Since $\gamma_i + \delta_i = \alpha_i + \beta_i$, it is now obvious that

$$\gcd(a,b) \cdot \operatorname{lcm}(a,b) = ab.$$

The result thus follows. □

Example 1.2.8. Find gcd(240, 560) and lcm(240, 560).
Since the prime factorizations of 240 and 560 are

$$\begin{aligned} 240 &= 2^4 \cdot 3 \cdot 5 = 2^4 \cdot 3^1 \cdot 5^1 \cdot 7^0 \\ 560 &= 2^4 \cdot 5 \cdot 7 = 2^4 \cdot 3^0 \cdot 5^1 \cdot 7^1, \end{aligned}$$

then

$$\begin{aligned} \gcd(240, 560) &= 2^{\min(4,4)} \cdot 3^{\min(1,0)} \cdot 5^{\min(1,1)} \cdot 7^{\min(0,1)} \\ &= 2^4 \cdot 3^0 \cdot 5^1 \cdot 7^0 \\ &= 80. \\ \operatorname{lcm}(240, 560) &= 2^{\max(4,4)} \cdot 3^{\max(1,0)} \cdot 5^{\max(1,1)} \cdot 7^{\max(0,1)} \\ &= 2^4 \cdot 3^1 \cdot 5^1 \cdot 7^1 \\ &= 1680. \end{aligned}$$

Of course, if we know gcd(240, 560) = 80, then we can find lcm(240, 560) by

$$\operatorname{lcm}(240, 560) = 240 \cdot 560/80 = 1680.$$

Similarly, if we know lcm(240, 560), we can find gcd(240, 560) by

$$\gcd(240, 560) = 240 \cdot 560/1680 = 80.$$

1.2.3 Mersenne Primes and Fermat Numbers

In this section, we shall introduce some basic concepts and results on Mersenne primes and perfect numbers.

Definition 1.2.8. A number is called a Mersenne[13] number if it is in the form of

[13]

Marin Mersenne (1588–1648) was a French monk, philosopher and mathematician who provided a valuable channel of communication between such contemporaries as Descartes, Fermat, Galileo and Pascal: "to inform Mersenne of a discovery is to publish it throughout the whole of Europe". Mersenne stated in *Cognitata Physico-Mathematica* but without proof that M_p is prime for $p = 2, 3, 5, 7, 13, 17, 19, 31, 67, 127, 257$ and for no other primes p with $p < 257$. Of course, Mersenne's list is not quite correct. It took over 300 years to totally settle this claim made by Mersenne, and finally in 1947, it was shown that Mersenne made five errors in his work; namely, M_{67} and M_{257} are composite and hence should be deleted from the list, whereas M_{61}, M_{89}, M_{107} are all primes and hence should be added to the list.

$$M_p = 2^p - 1, \tag{1.52}$$

where p is a prime. If a Mersenne number $M_p = 2^p - 1$ is a prime, then it is called a *Mersenne prime*.

Example 1.2.9. The following numbers

$$\begin{array}{ll} 2^2 - 1 = 3, & 2^3 - 1 = 7, \\ 2^5 - 1 = 31, & 2^7 - 1 = 127, \\ 2^{13} - 1 = 8191 & 2^{17} - 1 = 131071 \end{array}$$

are all Mersenne numbers as well as Mersenne primes, but $2^{11} - 1$ is only a Mersenne number, not a Mersenne prime, since $2^{11} - 1 = 2047 = 23 \times 89$ is a composite.

In Table 1.2, we list all thirty-nine Mersenne primes known to date (where GIMPS is the short for the Great Internet Mersenne Prime Search). There seems to be an astounding amount of interest in the world's largest known prime. When Curt Noll and Laura Nickel, two 18-year-old American high-school students in California, discovered the 25th Mersenne prime in October 1987, the announcement was carried by every major wire service in the United States and even announced by Walter Cronkite on the CBS Evening News. Currently the largest known prime is the 37th Mersenne prime $2^{3021377} - 1$, a 909526 digit number. In fact, since 1876, when Lucas determined the primality of $2^{127} - 1$ (confirmed later in 1914) the largest known prime has always been a Mersenne prime, except for a brief interregnum between June 1951 and January 1952. In this period Miller and Wheeler found the prime $934(2^{127} - 1) + 1$ and later $180(2^{127} - 1) + 1$. Also Ferrier in 1952 found, by hand calculation, that $(2^{148} + 1)/17$ is a prime. This is probably the largest prime that will ever be identified without using a computer (Williams [255]). It is amusing to note that after the 23rd Mersenne prime was found at the University of Illinois, the mathematics department there was so proud that they had their postage meter changed to stamp "$2^{11213} - 1$ IS PRIME" on each envelope (see Figure 1.4), at no profit to the U.S. Post Office, considering the zero value of the stamp.

Figure 1.4. A stamp of the 23rd Mersenne prime (by courtesy of Schroeder [222])

Table 1.2. The thirty-nine known Mersenne primes $M_p = 2^p - 1$

No.	p	**digits in** M_p	**discoverer(s) and time**
1	2	1	——
2	3	1	——
3	5	2	——
4	7	3	——
5	13	4	anonymous, 1456
6	17	6	Cataldi, 1588
7	19	6	Cataldi, 1588
8	31	10	Euler, 1772,
9	61	19	Pervushin, 1883
10	89	27	Powers, 1911
11	107	33	Powers, 1914
12	127	39	Lucas, 1876
13	521	157	Robinson, 1952
14	607	183	Robinson, 1952
15	1279	386	Robinson, 1952
16	2203	664	Robinson, 1952
17	2281	687	Robinson, 1952
18	3217	969	Riesel, 1957
19	4253	1281	Hurwitz, 1961
20	4423	1332	Hurwitz, 1961
21	9689	2917	Gillies, 1963
22	9941	2993	Gillies, 1963
23	11213	3376	Gillies, 1963
24	19937	6002	Tuckerman, 1971
25	21701	6533	Noll & Nickel, 1978
26	23209	6987	Noll, 1979
27	44497	13395	Nelson & Slowinski, 1979
28	86243	25962	Slowinski, 1982
29	110503	33265	Colquitt & Welsh, 1988
30	132049	39751	Slowinski, 1983
31	216091	65050	Slowinski, 1985
32	756839	227832	Slowinski & Gage, 1992
33	859433	258716	Slowinski & Gage, 1994
34	1257787	378632	Slowinski & Gage, 1996
35	1398269	420921	Armengaud & Woltman et al. (GIMPS), 1996
36	2976221	895932	Spence & Woltman et al. (GIMPS), 1997
37	3021377	909526	Clarkson, Woltman & Kurowski et al. (GIMPS, PrimeNet), 1998
38	6972593	2098960	Hajratwala, Woltman & Kurowski et al. (GIMPS, PrimeNet), 1999
39	13466917	4053946	Cameron, Woltman & Kurowski et al. (GIMPS, PrimeNet), 2001

There are some probabilistic estimates for the distribution of Mersenne primes; for example, in 1983, Wagstaff proposed the following conjecture:

Conjecture 1.2.1. (1) Let the number of Mersenne primes less than x be $\pi_M(x)$, then

$$\pi_M(x) \approx \frac{e^\gamma}{\ln 2} \log\log x = (2.5695\cdots)\ln\ln x, \tag{1.53}$$

where $\gamma = 0.5772$ is Euler's constant.

(2) The expected number of Mersenne primes M_q with $x < q < 2x$ is about $e^\gamma = 1.7806\cdots$.

(3) The probability that M_q is a prime is about

$$\frac{e^\gamma}{\ln 2} \cdot \frac{\ln aq}{\ln 2} = (2.5695\cdots)\frac{\ln aq}{q}, \tag{1.54}$$

where

$$a = \begin{cases} 2 & \text{if } q \equiv 3 \pmod 4 \\ 6 & \text{if } q \equiv 1 \pmod 4). \end{cases}$$

Schroeder [222] also refers to a conjecture of Eberhart, namely:

Conjecture 1.2.2. Let q_n be the nth prime such that M_{q_n} is a Mersenne prime. Then

$$q_n \approx \left(\frac{3}{2}\right)^n. \tag{1.55}$$

Definition 1.2.9. Numbers of the form $F_n = 2^{2^n} + 1$, whether prime or composite, are called *Fermat numbers*. A Fermat number is called a *prime Fermat number* if it is prime. A Fermat number is called a *composite Fermat number* if it is composite.

These special numbers obey the simple recursion:

$$F_{n+1} = (F_n - 1)^2 + 1 \tag{1.56}$$

or

$$F_{n+1} - 2 = F_n(F_n - 2) \tag{1.57}$$

which leads to the interesting product:

$$F_{n+1} - 2 = F_0 F_1 \cdots F_n. \tag{1.58}$$

In other words, $F_{n+1} - 2$ is divisible by all lower Fermat numbers:

$$F_{n-k} \mid (F_{n+1} - 2), \qquad 1 \le k \le n. \tag{1.59}$$

Fermat in 1640 conjectured, in a letter to Mersenne, that all numbers of the form $F_n = 2^{2^n} + 1$ were primes after he had verified it up to $n = 4$;

but Euler in 1732 found that the fifth Fermat number is not a prime, since $F_5 = 2^{2^5} + 1$ is the product of two primes 641 and 6700417. Later, it was found that F_6, F_7, and many others are not primes. Fermat was wrong! To date, the Fermat numbers $F_5, F_6, \cdots, F_{11}$ have been completely factored:

(1) F_5 was factored by Euler in 1732:

$$2^{2^5} + 1 = 2^{32} + 1 = 641 \cdot 6700417$$

(2) F_6 was factored by Landry and Lasseur in 1880:

$$2^{2^6} + 1 = 2^{64} + 1 = 274177 \cdot 67280421310721$$

(3) F_7 was factored by Morrison and Brillhart in 1970 using the Continued FRACtion (CFRAC) method:

$$2^{2^7} + 1 = 2^{128} + 1 = 59649589127497217 \cdot 5704689200685129054721$$

(4) F_8 was factored by Brent and Pollard in 1980 by using Brent and Pollard's "rho" (Monte Carlo) method:

$$2^{2^8} + 1 = 2^{256} + 1 = 1238926361552897 \cdot p_{63}$$

(5) F_9 was factored by Lenstra et al. in 1990 by using the Number Field Sieve (NFS) method:

$$\begin{aligned} 2^{2^9} + 1 &= 2^{512} + 1 = 2424833 \cdot \\ &\quad 7455602825647884208337395736200454918783366342657 \cdot p_{99} \end{aligned}$$

(6) F_{10} was factored by Brent in 1995 by using the Elliptic Curve Method (ECM):

$$\begin{aligned} 2^{2^{10}} + 1 &= 2^{1024} + 1 = 45592577 \cdot 6487031809 \cdot \\ &\quad 4659775785220018543264560743076778192897 \cdot p_{252} \end{aligned}$$

(7) F_{11} was factored by Brent in 1989 by using again the Elliptic Curve Method (ECM):

$$\begin{aligned} 2^{2^{11}} + 1 &= 2^{2048} + 1 = 319489 \cdot 974849 \cdot \\ &\quad 167988556341760475137 \cdot 3560841906445833920513 \cdot p_{564} \end{aligned}$$

In the above list, p_{63}, p_{99}, p_{252} and p_{564} are primes with 40, 49, 63, 99, 252 and 564 decimal digits, respectively. As a summary, we give the factorization status for the Fermat numbers F_n with $0 \le n \le 24$ in Table 1.3 (where p denotes a proven prime, and c a proven composite; ? means that the primality/compositeness of the number is not known). Four Fermat numbers in Table 1.3, namely, F_{14}, F_{20}, F_{22} and F_{24} are known to be composite, though

Table 1.3. The factorization status for Fermat numbers

n	F_n
0, 1, 2, 3, 4	p
5	$641 \cdot 6700417$
6	$274177 \cdot 67280421310721$
7	$59649589127497217 \cdot 5704689200685129054721$
8	$1238926361552897 \cdot p$
9	$2424833 \cdot$ $7455602825647884208337395736200454918783366342657 \cdot p$
10	$45592577 \cdot 6487031809 \cdot$ $4659775785220018543264560743076778192897 \cdot p$
11	$319489 \cdot 974849 \cdot 167988556341760475137$ $\cdot 3560841906445833920513 \cdot p$
12	$114689 \cdot 26017793 \cdot 63766529 \cdot$ $190274191361 \cdot 1256132134125569 \cdot c$
13	$2710954639361 \cdot 2663848877152141313 \cdot 3603109844522919$9 $\cdot 3603109844542291969 \cdot c$
14	c
15	$1214251009 \cdot 2327042503868417 \cdot c$
16	$825753601 \cdot 188981757975021318420037633 \cdot c$
17	$31065037602817 \cdot c$
18	$13631489 \cdot c$
19	$70525124609 \cdot 646730219521 \cdot c$
20	c
21	$4485296422913 \cdot c$
22	c
23	$167772161 \cdot ?$
24	c

no factors have yet been found (see Crandall, Doenias, et al [55], Crandall and Pomerance [56]). Table 1.3 also shows that the smallest not completely factored Fermat number is F_{12}, thus, it is the *most* wanted number at present. The smallest Fermat numbers which are not known to be prime or composite are F_{24} and F_{28}. Riesel [207] lists 99 prime factors of the form $k \cdot 2^m + 1$ in Fermat numbers, the largest being $5 \cdot 2^{23473} + 1$ of F_{23471}. Combining Riesel [207] and Young [263], we give in Table 1.4 the known prime factors of the form $k \cdot 2^m + 1$ for Fermat numbers F_n with $23 \leq n \leq 303088$.

There are still many open problems related to the Fermat numbers; some of them are the following:

(1) Are there infinitely many *prime* Fermat numbers?

(2) Are there infinitely many *composite* Fermat numbers?

(3) Is every Fermat number square-free?

Table 1.4. Prime factors of the form $k \cdot 2^m + 1$ in $F_n = 2^{2^n} + 1$ for $23 \leq n \leq 303088$

F_n	Prime Factor of F_n	F_n	Prime Factor of F_n
F_{23}	$5 \cdot 2^{25} + 1$	F_{25}	$48413 \cdot 2^{29} + 1$
F_{25}	$1522849979 \cdot 2^{27} + 1$	F_{26}	$143165 \cdot 2^{29} + 1$
F_{27}	$141015 \cdot 2^{30} + 1$	F_{27}	$430816215 \cdot 2^{29} + 1$
F_{29}	$1120049 \cdot 2^{31} + 1$	F_{30}	$149041 \cdot 2^{32} + 1$
F_{30}	$127589 \cdot 2^{33} + 1$	F_{32}	$1479 \cdot 2^{34} + 1$
F_{36}	$5 \cdot 2^{39} + 1$	F_{36}	$3759613 \cdot 2^{38} + 1$
F_{38}	$3 \cdot 2^{41} + 1$	F_{38}	$2653 \cdot 2^{40} + 1$
F_{39}	$21 \cdot 2^{41} + 1$	F_{42}	$43485 \cdot 2^{45} + 1$
F_{52}	$4119 \cdot 2^{54} + 1$	F_{52}	$21626655 \cdot 2^{54} + 1$
F_{55}	$29 \cdot 2^{57} + 1$	F_{58}	$95 \cdot 2^{61} + 1$
F_{61}	$54985063 \cdot 2^{66} + 1$	F_{62}	$697 \cdot 2^{64} + 1$
F_{63}	$9 \cdot 2^{67} + 1$	F_{64}	$17853639 \cdot 2^{67} + 1$
F_{66}	$7551 \cdot 2^{69} + 1$	F_{71}	$683 \cdot 2^{73} + 1$
F_{73}	$5 \cdot 2^{75} + 1$	F_{75}	$3447431 \cdot 2^{77} + 1$
F_{77}	$425 \cdot 2^{79} + 1$	F_{81}	$271 \cdot 2^{84} + 1$
F_{91}	$1421 \cdot 2^{93} + 1$	F_{93}	$92341 \cdot 2^{96} + 1$
F_{99}	$16233 \cdot 2^{104} + 1$	F_{117}	$7 \cdot 2^{120} + 1$
F_{122}	$5234775 \cdot 2^{124} + 1$	F_{125}	$5 \cdot 2^{127} + 1$
F_{142}	$8152599 \cdot 2^{145} + 1$	F_{144}	$17 \cdot 2^{147} + 1$
F_{147}	$3125 \cdot 2^{149} + 1$	F_{150}	$1575 \cdot 2^{157} + 1$
F_{150}	$5439 \cdot 2^{154} + 1$	F_{201}	$4585 \cdot 2^{204} + 1$
F_{205}	$232905 \cdot 2^{207} + 1$	F_{207}	$3 \cdot 2^{209} + 1$
F_{215}	$32111 \cdot 2^{217} + 1$	F_{226}	$15 \cdot 2^{229} + 1$
F_{228}	$29 \cdot 2^{231} + 1$	F_{250}	$403 \cdot 2^{252} + 1$
F_{255}	$629 \cdot 2^{257} + 1$	F_{267}	$177 \cdot 2^{271} + 1$
F_{268}	$21 \cdot 2^{276} + 1$	F_{275}	$22347 \cdot 2^{279} + 1$
F_{284}	$7 \cdot 2^{290} + 1$	F_{287}	$5915 \cdot 2^{289} + 1$
F_{298}	$247 \cdot 2^{302} + 1$	F_{316}	$7 \cdot 2^{320} + 1$
F_{329}	$1211 \cdot 2^{333} + 1$	F_{334}	$27609 \cdot 2^{341} + 1$
F_{398}	$120845 \cdot 2^{401} + 1$	F_{416}	$8619 \cdot 2^{418} + 1$
F_{416}	$38039 \cdot 2^{419} + 1$	F_{452}	$27 \cdot 2^{455} + 1$
F_{544}	$225 \cdot 2^{547} + 1$	F_{556}	$127 \cdot 2^{558} + 1$
F_{637}	$11969 \cdot 2^{643} + 1$	F_{692}	$717 \cdot 2^{695} + 1$
F_{744}	$17 \cdot 2^{747} + 1$	F_{906}	$57063 \cdot 2^{908} + 1$
F_{931}	$1985 \cdot 2^{993} + 1$	F_{1551}	$291 \cdot 2^{1553} + 1$
F_{1945}	$5 \cdot 2^{1947} + 1$	F_{2023}	$29 \cdot 2^{2027} + 1$
F_{2089}	$431 \cdot 2^{2099} + 1$	F_{2456}	$85 \cdot 2^{2458} + 1$
F_{3310}	$5 \cdot 2^{3313} + 1$	F_{4724}	$29 \cdot 2^{4727} + 1$
F_{6537}	$17 \cdot 2^{6539} + 1$	F_{6835}	$19 \cdot 2^{6838} + 1$
F_{9428}	$9 \cdot 2^{9431} + 1$	F_{9448}	$19 \cdot 2^{9450} + 1$
F_{23471}	$5 \cdot 2^{23473} + 1$	F_{25006}	$57 \cdot 2^{25010} + 1$
F_{94798}	$21 \cdot 2^{94801} + 1$	F_{95328}	$7 \cdot 2^{95330} + 1$
F_{114293}	$13 \cdot 2^{114296} + 1$	F_{125410}	$5 \cdot 2^{125413} + 1$
F_{157167}	$3 \cdot 2^{157169} + 1$	F_{213319}	$3 \cdot 2^{213321} + 1$
F_{303088}	$3 \cdot 2^{303093} + 1$		

1.2.4 Euclid's Algorithm

We might call Euclid's method the granddaddy of all algorithms, because it is the oldest nontrivial algorithm that has survived to the present day.

DONALD E. KNUTH
The Art of Computer Programming: Seminumerical Algorithms [123]

Euclid's algorithm for finding the greatest common divisor of two integers is perhaps the oldest nontrivial algorithm that has survived to the present day. It is based on the division theorem (Theorem 1.2.2). In fact, it is based on the following fact.

Theorem 1.2.16. Let a, b, q, r be integers with $b > 0$ and $0 \le r < b$ such that $a = bq + r$. Then $\gcd(a, b) = \gcd(b, r)$.

Proof. Let $X = \gcd(a, b)$ and $Y = \gcd(b, r)$, it suffices to show that $X = Y$. If integer c is a divisor of a and b, it follows from the equation $a = bq + r$ and the divisibility properties that c is a divisor of r also. By the same argument, every common divisor of b and r is a divisor of a. □

Theorem 1.2.16 can be used to reduce the problem of finding $\gcd(a, b)$ to the simpler problem of finding $\gcd(b, r)$. The problem is simpler because the numbers are smaller, but it has the same answer as the original one. The process of finding $\gcd(a, b)$ by repeated application of Theorem 1.2.16 is called Euclid's algorithm which proceeds as follows.

$$
\begin{array}{lll}
a = bq_0 + r_1, & 0 \le r_1 < b & \text{(dividing } b \text{ into } a\text{)}, \\
b = r_1 q_1 + r_2, & 0 \le r_2 < r_1 & \text{(dividing } r_1 \text{ into } b\text{)}, \\
r_1 = r_2 q_2 + r_3, & 0 \le r_3 < r_2 & \text{(dividing } r_2 \text{ into } r_1\text{)}, \\
 & & \\
r_2 = r_3 q_3 + r_4, & 0 \le r_4 < r_3 & \text{(dividing } r_3 \text{ into } r_2\text{)}, \\
\cdots\cdots & \cdots\cdots & \cdots\cdots \\
\cdots\cdots & \cdots\cdots & \cdots\cdots \\
r_{n-2} = r_{n-1} q_{n-1} + r_n, & 0 \le r_n < r_{n-1} & \text{(dividing } r_{n-1} \text{ into } r_{n-2}\text{)}, \\
r_{n-1} = r_n q_n + 0, & r_{n+1} = 0 & \text{(arriving at a zero-remainder)},
\end{array}
$$

or, diagrammatically,

a		
$- bq_0$	q_0	b
r_1	q_1	$- r_1q_1$
$- r_2q_2$	q_2	r_2
r_3	q_3	r_3q_3
$\vdots$	$\vdots$	$\vdots$
r_{n-1}	q_{n-1}	$- r_{n-1}q_{n-1}$
$- r_nq_n$	q_n	$\boxed{r_n}$
$r_{n+1} = 0$		

Then the greatest common divisor *gcd* of a and b is r_n. That is,

$$d = \gcd(a, b) = r_n. \tag{1.60}$$

We now restate it in a theorem form.

Theorem 1.2.17 (Euclid's algorithm). Let a and b be positive integers with $a > b$. If $b \mid a$, then $\gcd(a, b) = b$. If $b \nmid a$, then apply the division algorithm repeatedly as follows:

$$\left.\begin{array}{ll} a = bq_0 + r_1, & 0 < r_1 < b, \\ b = r_1q_1 + r_2, & 0 < r_2 < r_1, \\ r_1 = r_2q_2 + r_3, & 0 < r_3 < r_2, \\ r_2 = r_3q_3 + r_4, & 0 < r_4 < r_3, \\ \cdots\cdots & \cdots\cdots \\ r_{n-2} = r_{n-1}q_{n-1} + r_n, & 0 < r_n < r_{n-1}, \\ r_{n-1} = r_nq_n + 0. & \end{array}\right\} \tag{1.61}$$

Then r_n, the last nonzero remainder, is the *greatest common divisor* of a and b. That is,

$$\gcd(a, b) = r_n. \tag{1.62}$$

Values of x and y in

$$\gcd(a, b) = ax + by \tag{1.63}$$

can be obtained by writing each r_i as a linear combination of a and b.

Proof. The chain of equations is obtained by dividing b into a, r_1 into b, r_2 into r_1, $\cdots$, r_{n-1} into r_n. (Note that we have written the inequalities for the remainder without an equality sign.) The process stops when the division is exact, that is, whenever $r_i = 0$ for $i = 1, 2, \cdots, j$.

We now prove that r_j is the greatest common divisor of a and b. by Theorem 1.2.16, we have

$$\begin{aligned}\gcd(a, b) &= \gcd(a - bq_0, b) \\ &= \gcd(r_1, b) \\ &= \gcd(r_1, b - r_1q_1) \\ &= \gcd(r_1, r_2) \\ &= \gcd(r_1 - r_2q_2, r_2) \\ &= \gcd(r_3, r_2)\end{aligned}$$

Continuing by mathematical induction, we have

$$\gcd(a, b) = \gcd(r_{j-1}, r_j) = \gcd(r_j, 0) = r_j.$$

To see that r_j is a linear combination of a and b, we argue by induction that each r_i is a a linear combination of a and b. Clearly, r_1 is a linear combination of a and b, since $r_1 = a - bq_0$, so does r_2. In general, r_i is a linear combination of r_{i-1} and r_{i-2}. By the inductive hypothesis we may suppose that these latter two numbers are linear combinations of a and b, and it follows that r_i is also a linear combination of a and b. □

Remark 1.2.5. Euclid's algorithm is found in Book VII, Proposition 1 and 2 of his *Elements*, but it probably wasn't his own invention. Scholars believe that the method was known up to 200 years earlier. However, it first appeared in Euclid's *Elements*, and more importantly, it is the first nontrivial algorithm that has survived to this day.

Remark 1.2.6. It is evident that the algorithm cannot recur indefinitely, since the second argument strictly decreases in each recursive call. Therefore, the algorithm always terminates with the correct answer. More importantly, it can be performed in polynomial time. That is, if Euclid's algorithm is applied to two positive integers a and b with $a \geq b$, then the number of divisions required to find $\gcd(a, b)$ is $\mathcal{O}(\log b)$, a polynomial-time complexity (the big-$\mathcal{O}$ notation is used to denote the upper bound of a complexity function, i.e., $f(n) = \mathcal{O}(g(n))$ if there exists some constant $c > 0$ such that $f(n) \leq c \cdot g(n)$; see Subsection 2.1.3 in Chapter 2 for more information).

Example 1.2.10. Use Euclid's algorithm to find the gcd of 1281 and 243. Since

1281		
− 1215	5	243
66	3	− 198
− 45	1	45
21	2	− 42
− 21	7	$\boxed{3}$
0		

we have $\gcd(1281, 243) = 3$.

Exercise 1.2.2. Calculate $\gcd(1403, 549)$ using Euclid's algorithm.

Theorem 1.2.18. If a and b are any two integers, then

$$Q_k a - P_k b = (-1)^{k-1} r_k, \quad k = 1, 2, \cdots, n \tag{1.64}$$

where

$$\left.\begin{array}{l} P_0 = 1, \ P_1 = q_0, \ P_k = q_{k-1}P_{k-1} + P_{k-2} \\ Q_0 = 0, \ Q_1 = 1, \ Q_k = q_{k-1}Q_{k-1} + Q_{k-2} \end{array}\right\} \tag{1.65}$$

for $k = 2, 3, \cdots, n$.

Proof. When $k = 1$, (1.64) is clearly true, since $Q_1 a - P_1 b = (-1)^{1-1} r_1$ implies $a - q_0 b = r_1$. When $k = 2$, $r_2 = -(aq_1 - b(1 + q_1 q_1))$. But $1 + q_1 q_1 = q_2 P_1 + P_0$, $q_1 = q_1 \cdot 1 + 0 = q_1 Q_1 + Q_0$, therefore, $Q_2 a - P_2 b = (-1)^{2-1} r_2$, $P_2 = q_1 P_1 + P_0$, $Q_2 = q_1 Q_1 + Q_0$. Assume (1.64) and (1.65) hold for all positive integers $\leq k$, then

$$\begin{aligned} (-1)^k r_{k+1} &= (-1)^k (r_{k-1} - q_k r_k) \\ &= (Q_{k-1} a - P_k b) + q_k (Q_k a - P_k b) \\ &= (q_k Q_k + Q_{k-1}) a - (q_{k+1} P_k + P_{k+1}) b. \end{aligned}$$

Thus, $Q_{k+1} a - P_{k+1} a = (-1)^k r_{k+1}$, where $P = k + 1 = q_k P_k + P_{k-1}$, $Q_{k+1} = q_{k+1} Q_k + Q_{k-1}$. By induction, the reult is true for all positive integers. □

The $\gcd(a, b)$ will be equal to unity for more than 60 percent of the time for random inputs; this is a consequence of the following well-known result of number theory (Knuth [123]):

Theorem 1.2.19. If a and b are integers chosen at random, the probability that $\gcd(a, b) = 1$ is $6/\pi^2 = 0.60793$. That is,

$$\text{Prob}[\gcd(a, b) = 1] = 0.6. \tag{1.66}$$

This result was first proved by the Italian mathematician Ernesto Cesàro (1859–1906) in 1881. The idea of the proof is as follows. Let p be the probability

$$p = \text{Prob}[\gcd(a, b) = 1].$$

Then, for any positive integer d, consider the probability

$$p = \text{Prob}[\gcd(a, b) = d].$$

This happens when a is a multiple of d, b is a multiple of d, and $\gcd(a/d,\ b/d) = 1$. The probability that $d \mid a$ is $1/d$.

1.2.5 Continued Fractions

Euclid's algorithm for computing the greatest common divisor of two integers is intimately connected with continued fractions.

Definition 1.2.10. Let a and b be integers and let Euclid's algorithm run as

$$\begin{aligned}
a &= bq_0 + r_1, \\
b &= r_1 q_1 + r_2 \\
r_1 &= r_2 q_2 + r_3, \\
r_2 &= r_3 q_3 + r_4, \\
&\cdots\cdots \\
&\cdots\cdots \\
r_{n-2} &= r_{n-1} q_{n-1} + r_n, \\
r_{n-1} &= r_n q_n + 0.
\end{aligned}$$

That is,

$$\begin{array}{r|c|r}
a & & \\
-\ bq_0 & q_0 & b \\ \hline
r_1 & q_1 & -\ r_1 q_1 \\ \hline
-\ r_2 q_2 & q_2 & r_2 \\ \hline
r_3 & q_3 & r_3 q_3 \\
\vdots & \vdots & \vdots \\
r_{n-1} & q_{n-1} & -\ r_{n-1} q_{n-1} \\ \hline
-\ r_n q_n & q_n & r_n \\ \hline
r_{n+1} = 0 & &
\end{array}$$

Then the fraction $\frac{a}{b}$ can be expressed as a simple continued fraction:

$$\frac{a}{b} = q_0 + \cfrac{1}{q_1 + \cfrac{1}{q_2 + \cfrac{1}{\ddots\, q_{n-1} + \cfrac{1}{q_n}}}} \tag{1.67}$$

where $q_0, q_1, \cdots, q_{n-1}, q_n$ are taken directly from Euclid's algorithm expressed in (1.61), and are called the *partial quotients* of the continued fraction. For simplicity, the continued fraction expansion (1.67) of $\frac{a}{b}$ is usually written as

$$\frac{a}{b} = q_0 + \frac{1}{q_1+} \frac{1}{q_2+} \cdots \frac{1}{q_{n-1}+} \frac{1}{q_n} \tag{1.68}$$

or even more briefly as

$$\frac{a}{b} = [q_0, q_1, q_2, \cdots q_{n-1}, q_n]. \tag{1.69}$$

If each q_i is an integer, the continued fraction is called *simple*; a simple continued fraction can either be *finite* or *infinite*. A continued fraction formed from $[q_0, q_1, q_2, \cdots q_{n-1}, q_n]$ by neglecting all of the terms after a given term is called a *convergent* of the original continued fraction. If we denote the k-th convergent by $C_k = \frac{P_k}{Q_k}$, then

$$(1) \begin{cases} C_0 = \dfrac{P_0}{Q_0} = \dfrac{q_0}{1}; \\ C_1 = \dfrac{P_1}{Q_1} = \dfrac{q_0 q_1 + 1}{q_1}; \\ \cdots\cdots \\ \cdots\cdots \\ C_k = \dfrac{P_k}{Q_k} = \dfrac{q_k P_{k-1} + P_{k-2}}{q_k Q_{k-1} + Q_{k-2}}, \text{ for } k \geq 2. \end{cases}$$

(2) If $P_k = q_k Q_{k-1} + Q_{k-2}$ and $Q_k = q_k P_{k-1} + P_{k-2}$, then $\gcd(P_k, Q_k) = 1$.

(3) $P_k Q_{k-1} - P_{k-1} Q_k = (-1)^{k-1}$, for $k \geq 1$.

The following example shows how to use Euclid's algorithm to express a rational number as a finite simple continued fraction.

Example 1.2.11. Expand the rational number $\frac{1281}{243}$ as a simple continued fraction. First let $a = 1281$ and $b = 243$, and then let Euclid's algorithm run as follows:

$$
\begin{array}{r|c|r}
1281 & & \\
-\ 1215 & 5 & 243 \\
\hline
66 & 3 & -\ 198 \\
\hline
-\ 45 & 1 & 45 \\
\hline
21 & 2 & -\ 42 \\
\hline
-\ 21 & 7 & 3 \\
\hline
0 & &
\end{array}
$$

So $\dfrac{1281}{243} = [5, 3, 1, 2, 7]$. Thus

$$\frac{1281}{243} = 5 + \cfrac{1}{3 + \cfrac{1}{1 + \cfrac{1}{2 + \cfrac{1}{7}}}}.$$

Of course, as a by-product, we also find that $\gcd(1281, 243) = 3$.

Exercise 1.2.3. Expand the rational numbers $\dfrac{239}{51}$ and $\dfrac{51}{239}$ as simple continued fractions.

The above discussion tells us that any rational number $\frac{a}{b}$ with $b \neq 0$ can be expressed as a simple *finite* continued fraction.

Theorem 1.2.20. Any finite simple continued fraction represents a rational number. Conversely, any rational number can be expressed as a finite simple continued fraction, in exactly two ways, one with an odd number of terms and one with an even number of terms.

Proof. The first assertion is proved by induction. When $n = 1$, we have

$$[q_0, q_1] = q_0 + \frac{1}{q_1} = \frac{q_0 q_1 + 1}{q_1}$$

which is rational. Now we assume for $n = k$ the simple continued fraction $[q_0, q_1, \cdots, q_k]$ is rational whenever $q_0, q_1, \cdots, q_k$ are integers with $q_1, \cdots, q_k$ positive. Let $q_0, q_1, \cdots, q_{k+1}$ are integers with $q_1, \cdots, q_{k+1}$ positive. Note that

$$[q_0, q_1, \cdots, q_k, q_{k+1}] = a_0 + \frac{1}{[q_1, \cdots, q_k, q_{k+1}]}.$$

By the induction hypothesis, $[q_1, q_2, \cdots, q_k, q_{k+1}]$ is rational. That is, there exist two integers r and s with $s \neq 0$ such that

$$[q_1, q_2, \cdots, q_k, q_{k+1}] = \frac{r}{s}.$$

Thus,

$$[q_0, q_1, \cdots, q_k, q_{k+1}] = a_0 + \frac{1}{r/s} = \frac{a_0 r + s}{r}$$

which is rational.

Now we use Euclid's algorithm to show that every rational number can be written as a finite simple continued fraction. Let a and b be a rational number with $b > 0$. Euclid's algorithm tells us that

$$\begin{array}{ll}
a = bq_0 + r_1, & 0 < r_1 < b, \\
b = r_1 q_1 + r_2, & 0 < r_2 < r_1, \\
r_1 = r_2 q_2 + r_3, & 0 < r_3 < r_2, \\
r_2 = r_3 q_3 + r_4, & 0 < r_4 < r_3, \\
\quad \cdots\cdots & \quad \cdots\cdots \\
r_{n-2} = r_{n-1} q_{n-1} + r_n, & 0 < r_n < r_{n-1}, \\
r_{n-1} = r_n q_n + 0. &
\end{array}$$

In these equations, $q_1, q_2, \cdots, q_n$ are positive integers. Rewriting these equations, we obtain

$$\begin{array}{rcl}
\frac{a}{b} & = & q_0 + \frac{r_1}{b} \\
\frac{b}{r_1} & = & q_1 + \frac{r_2}{r_1} \\
\frac{r_1}{r_2} & = & q_2 + \frac{r_3}{r_2} \\
\cdots & & \cdots \\
\cdots & & \cdots \\
\frac{r_{n-1}}{r_n} & = & q_n
\end{array}$$

By successive substitution, we have

$$
\begin{aligned}
\frac{a}{b} &= q_0 + \cfrac{1}{\cfrac{b}{r_1}} \\
&= q_0 + \cfrac{1}{q_1 + \cfrac{1}{\cfrac{r_1}{r_2}}} \\
&\cdots \\
&\cdots \\
&= q_0 + \cfrac{1}{q_1 + \cfrac{1}{q_2 + \cfrac{1}{\ddots q_{n-1} + \cfrac{1}{q_n}}}}
\end{aligned}
$$

This shows that every rational number can be written as a finite simple continued fraction.

Further, it can be shown that any rational number can be expressed as a finite simple continued fraction in exactly two ways, one with an odd number of terms and one with an even number of terms; we leave this as an exercise. □

In what follows, we shall show that any *irrational* number can be expressed as an *infinite* simple continued fraction.

Definition 1.2.11. Let $q_0, q_1, q_2, \cdots$ be a sequence of integers, all positive except possibly q_0. Then the expression $[q_0, q_1, q_2, \cdots]$ is called an *infinite* simple continued fraction and is defined to be equal to the number $\lim_{n \to \infty} [q_0, q_1, q_2, \cdots, q_{n-1}, q_n]$.

Theorem 1.2.21. Any *irrational* number can be written uniquely as an *infinite* simple continued fraction. Conversely, if α is an infinite simple continued fraction, then α is irrational.

Proof. Let α be an irrational number. We write

$$\alpha = [\alpha] + \{\alpha\} = [\alpha] + \cfrac{1}{\cfrac{1}{\{\alpha\}}}$$

where $[\alpha]$ is the integral part and $\{\alpha\}$ the fractional part of α, respectively. Because α is irrational, $1/\{\alpha\}$ is irrational and greater than 1. Let

$$q_0 = [\alpha], \quad \text{and} \quad \alpha_1 = \frac{1}{\{\alpha\}}.$$

We now write

$$\alpha_1 = [\alpha_1] + \{\alpha_1\} = [\alpha_1] + \cfrac{1}{\cfrac{1}{\{\alpha_1\}}}$$

where $1/\{\alpha_1\}$ is irrational and greater than 1. Let

$$q_1 = [\alpha_1], \quad \text{and} \quad \alpha_2 = \frac{1}{\{\alpha_1\}}.$$

We continue inductively

$$\begin{aligned} q_2 &= [\alpha_2], \quad \text{and} \quad \alpha_3 = \frac{1}{\{\alpha_2\}} > 1 \quad (\alpha_3 \text{ irrational}) \\ q_3 &= [\alpha_3], \quad \text{and} \quad \alpha_4 = \frac{1}{\{\alpha_3\}} > 1 \quad (\alpha_3 \text{ irrational}) \\ & \quad \cdots \\ & \quad \cdots \\ q_n &= [\alpha_n], \quad \text{and} \quad \alpha_n = \frac{1}{\{\alpha_{n-1}\}} > 1 \quad (\alpha_3 \text{ irrational}) \\ & \quad \cdots \\ & \quad \cdots \end{aligned}$$

Since each α_n, $i = 2, 3, \cdots$ is greater than 1, then $q_{n-1} \geq 1$, $n = 2, 3, \cdots$. If we substitute successively, we obtain

$$\begin{aligned} \alpha &= [q_0, \alpha_1] \\ &= [q_0, q_1, \alpha_2] \\ &= [q_0, q_1, q_2, \alpha_3] \\ & \quad \cdots \\ & \quad \cdots \\ &= [q_0, q_1, q_2, \cdots, q_n, \alpha_{n+1}] \\ & \quad \cdots \\ & \quad \cdots \end{aligned}$$

Next we shall show that $\alpha = [q_0, q_1, q_2, \cdots]$. Note that C_n, the nth convergent to $[q_0, q_1, q_2, \cdots]$ is also the nth convergent to $[q_0, q_1, q_2, \cdots, q_n, \alpha_{n+1}]$. If we denote the $(n+1)$st convergent to this finite continued fraction by $P'_{n+1}/Q'_{n+1} = \alpha$, then

$$\alpha - C_n = \frac{P'_{n+1}}{Q'_{n+1} - \frac{P_n}{Q_n}} = \frac{(-1)^{n+1}}{Q'_{n+1}Q_n}.$$

Since Q_n and Q'_{n+1} become infinite as $n \to \infty$, then

$$\lim_{n\to\infty} (\alpha - C_n) = \lim_{n\to\infty} \frac{(-1)^{n+1}}{Q'_{n+1}Q_n} = 0$$

and

$$\alpha = \lim_{n \to \infty} C_n = [q_0, q_1, \cdots].$$

The uniqueness of the representation, as well as the second assertion are left as an exercise. □

Definition 1.2.12. A real irrational number which is the root of a quadratic equation $ax^2+bx+c=0$ with integer coefficients is called *quadratic irrational.*

For example, $\sqrt{3}$, $\sqrt{5}$, $\sqrt{7}$ are quadratic irrationals. For convenience, we shall denote $\sqrt{N}$, with N not a perfect square, as a quadratic irrational. Quadratic irrationals are the simplest possible irrationals.

Definition 1.2.13. An infinite simple continued fraction is said to be *periodic* if there exists integers k and m such that $q_{i+m} = q_i$ for all $i \geq k$. The periodic simple continued fraction is usually denoted by $[q_0, q_1, \cdots, q_k, \overline{q_{k+1}, q_{k+2}, \cdots, q_{k+m}}]$. If it is of the form $[\overline{q_0, q_1, \cdots, q_{m-1}}]$, then it is called *purely periodic*. The smallest positive integer m satisfying the above relationship is called the *period* of the expansion.

Theorem 1.2.22. Any *periodic* simple continued fraction is a quadratic irrational. Conversely, any *quadratic irrational* has a periodic expansion as a simple continued fraction.

Proof. The proof is rather lengthy and left as an exercise; a complete proof can be found on pages 224–226 in [197]. □

We are now in a position to present an algorithm for finding the simple continued fraction expansion of a *real number*.

Theorem 1.2.23 (Continued fraction algorithm). Let $x = x_0$ be a real number. Then α can be expressed as a simple continued fraction

$$[q_0, q_1, q_2, \cdots, q_n, q_{n+1}, \cdots]$$

by the following process:

$$\left.\begin{array}{ll} q_0 = \lfloor x_0 \rfloor, & x_1 = \dfrac{1}{x_0 - q_0} \\ q_1 = \lfloor x_1 \rfloor, & x_2 = \dfrac{1}{x_1 - q_1} \\ \vdots & \vdots \\ q_n = \lfloor x_n \rfloor, & x_{n+1} = \dfrac{1}{x_n - q_n} \\ q_{n+1} = \lfloor x_{n+1} \rfloor, & x_{n+2} = \dfrac{1}{x_{n+1} - q_{n+1}} \\ \vdots & \vdots \end{array}\right\} \quad (1.70)$$

Proof. Follows from Theorem 1.2.21. □

Note that just as the numbers $q_0, q_1, \cdots$ are called the partial quotients of the continued fraction, the numbers $x_0, x_1, \cdots$ are called the *complete quotients* of the continued fraction. For quadratic irrational numbers, of course, we do not need to calculate the infinitely many q_i's, since according to Theorem 1.2.22, any quadratic irrational number is periodic and can be written as an infinite simple continued fraction of the form $[q_0, q_1, q_2, \cdots, q_k, \overline{q_{k+1}, \cdots, q_{k+m}}]$.

Now we can use the algorithm given in Theorem 1.2.23 to represent any real number as a simple continued fraction.

Example 1.2.12. Expand $\sqrt{3}$ as a periodic simple continued fraction. Let $x_0 = \sqrt{3}$. Then we have:

$$q_0 = \lfloor x_0 \rfloor = \lfloor \sqrt{3} \rfloor = 1$$

$$x_1 = \frac{1}{x_0 - q_0} = \frac{1}{\sqrt{3}-1} = \frac{\sqrt{3}+1}{2}$$

$$q_1 = \lfloor x_1 \rfloor = \lfloor \frac{\sqrt{3}+1}{2} \rfloor = \lfloor 1 + \frac{\sqrt{3}-1}{2} \rfloor = 1$$

$$x_2 = \frac{1}{x_1 - q_1} = \frac{1}{\frac{\sqrt{3}+1}{2} - 1} = \frac{1}{\frac{\sqrt{3}-1}{2}} = \frac{2(\sqrt{3}+1)}{(\sqrt{3}-1)(\sqrt{3}+1)} = \sqrt{3}+1$$

$$q_2 = \lfloor x_2 \rfloor = \lfloor \sqrt{3}+1 \rfloor = 2$$

$$x_3 = \frac{1}{x_2 - q_2} = \frac{1}{\sqrt{3}+1-2} = \frac{1}{\sqrt{3}-1} = \frac{\sqrt{3}+1}{2} = x_1$$

$$q_3 = \lfloor x_3 \rfloor = \lfloor \frac{\sqrt{3}+1}{2} \rfloor = \lfloor 1 + \frac{\sqrt{3}-1}{2} \rfloor = 1 = q_1$$

$$x_4 = \frac{1}{x_3 - q_3} = \frac{1}{\frac{\sqrt{3}+1}{2} - 1} = \frac{1}{\frac{\sqrt{3}-1}{2}} = \frac{2(\sqrt{3}+1)}{(\sqrt{3}-1)(\sqrt{3}+1)} = \sqrt{3}+1 = x_2$$

$$q_4 = \lfloor x_3 \rfloor = \lfloor \sqrt{3}+1 \rfloor = 2 = q_2$$

$$x_5 = \frac{1}{x_4 - q_4} = \frac{1}{\sqrt{3}+1-2} = \frac{1}{\sqrt{3}-1} = \frac{\sqrt{3}+1}{2} = x_3 = x_1$$

$$q_5 = \lfloor x_5 \rfloor = \lfloor x_3 \rfloor = 1 = q_3 = q_1$$

$$\cdots\cdots\cdots\cdots$$

$$\cdots\cdots\cdots\cdots$$

So, for $n = 1, 2, 3, \cdots$, we have $q_{2n-1} = 1$ and $q_{2n} = 2$. Thus, the *period* of the continued fraction expansion of $\sqrt{3}$ is 2. Therefore, we finally get

$$\sqrt{3} = 1 + \cfrac{1}{1 + \cfrac{1}{2 + \cfrac{1}{1 + \cfrac{1}{2 + \cfrac{1}{\ddots}}}}} = [1, \overline{1, 2}].$$

Exercise 1.2.4. Find the continued fraction expansions of $\sqrt{5}$ and $\sqrt{7}$.

1.3 Diophantine Equations

I consider that I understand an equation when I can predict the properties of its solutions, without actually solving it.

PAUL A. M. DIRAC (1902–1984)

In this section, we shall introduce some basic concepts of Diophantine equations and study some solutions of certain types of Diophantine equations.

1.3.1 Basic Concepts of Diophantine Equations

The word "Diophantine" is derived from the name of Diophantus[14] of Alexandria who was one of the first to make a study of equations in integers. The simplest form of problem involved is the determination of whether or not a polynomial equation $f(x, y, z, \cdots) = 0$ in variables $x, y, z, \cdots$, with integral coefficients, has integral solutions, or in some cases rational solutions.

[14] Diophantus (about 200–284), the father of algebra, lived in the great city of Alexandria about 1700 years ago. He is perhaps best known as the writer of the book *Arithmetica*, of which only six of the original thirteen volumes of the book have been preserved; the photograph in Figure 1.5 shows the title page of the Latin translation of the book. About 130 problems in Arithmetic and Algebra are considered in the book, some of which are surprisingly hard. The work of Diophantus was forgotten until a copy of the book was discovered in 1570. Italian mathematicians in the 16th century introduced his works into Europe where they were read with great interest and where they stimulated the study of Algebra, more specifically, Diophantine Analysis. Very little knowledge about his personal life has survived except his epitaph which contains clues to his age: One sixth of his life was spent as a child; after one twelfth more he grew a beard; when one seventh more had passed, he married. Five years later a son was born; the son lived to half his father's age; four years after the son's death, he also died.

DIOPHANTI
ALEXANDRINI
ARITHMETICORVM
LIBRI SEX,
ET DE NVMERIS MVLTANGVLIS
LIBER VNVS.
CVM COMMENTARIIS C. G. BACHETI V. C.
& obseruationibus D. P. de FERMAT *Senatoris Tolosani.*
Accessit Doctrinæ Analyticæ inuentum nouum, collectum
ex varijs eiusdem D. de FERMAT Epistolis.

TOLOSÆ,
Excudebat BERNARDVS BOSC, è Regione Collegij Societatis Iesu.
M. DC. LXX.

Figure 1.5. The title page of Diophantus' book *Arithmetica*

A Diophantine equation may have no solution, a finite number of solutions or an infinite number of solutions, and in the infinite case, the solutions may be given in terms of one or more integral parameters.

From a geometrical point of view, the integral solutions of a Diophantine equation $f(x, y) = 0$ represents the points with integral coordinates on the curve $f(x, y) = 0$. For example, in the case of equation $x^2 - 2y^2 = 0$, the only integral solution is $(x, y) = (0, 0)$, which shows that the point $(0, 0)$ is the only point on the line $x^2 - 2y^2 = 0$ with integral coordinates, whilst the equation $x^2 + y^2 = z^2$ has an infinite number of solutions. There are corresponding geometrical interpretations in higher dimensions.

1.3.2 Linear Diophantine Equations

Definition 1.3.1. The algebraic equation with two variables

$$ax + by = c \tag{1.71}$$

is called a *linear Diophantine equation*, for which we wish to find integer solutions in x and y.

A linear Diophantine equation is a type of algebraic equation with two linear variables. For this reason, it is sometimes also called a *bilinear Diophantine equation.* In this type of equation $ax+by=c$, we are only interested in the integer solutions in x and y.

Theorem 1.3.1. Let a, b, c be integers with not both a and b equal to 0, and let $d = \gcd(a, b)$. If $d \nmid c$, then the linear Diophantine equation

$$ax + by = c$$

has no integer solution. The equation has an integer solution in x and y if and only if $d \mid c$. Moreover, if (x_0, y_0) is a solution of the equation, then the general solution of the equation is

$$(x, y) = \left(x_0 + \frac{b}{d} \cdot t,\ y_0 - \frac{a}{d} \cdot t\right) \quad t \in \mathbb{Z}. \tag{1.72}$$

Proof. Assume that x and y are integers such that $ax + by = c$. Since $d \mid a$ and $d \mid b$, $d \mid c$. Hence, if $d \nmid c$, there is no integer solutions of the equation.

Now suppose $d \mid c$. There is an integer k such that $c = kd$. Since d is a sum of multiples of a and b, we may write

$$am + bn = d.$$

Multiplying this equation by k, we get

$$a(mk) + b(nk) = dk = c$$

so that $x = mk$ and $y = nk$ is a solution.

For the "only if" part, suppose x_0 and y_0 is a solution of the equation. Then

$$ax_0 + by_0 = c.$$

Since $d \mid a$ and $d \mid b$, then $d \mid c$. □

Observe that the proof of Theorem 1.3.1, together with Euclid's algorithm, provides us with a practical method to obtain one solution of the equation. In what follows, however, we shall show how to find x and y by using the continued fraction method.

Suppose that a and b are two integers whose *gcd* is d and we wish to solve

$$ax - by = d. \tag{1.73}$$

We expand a/b as a finite continued fraction with convergents

$$\left[\frac{P_0}{Q_0}, \frac{P_1}{Q_1}, \ldots, \frac{P_{n-1}}{Q_{n-1}}, \frac{P_n}{Q_n}\right] = \frac{a}{b}. \tag{1.74}$$

Since $d = \gcd(a, b)$ we must have $a = da'$, $b = db'$ and $\gcd(a', b') = 1$. Then $P_n/Q_n = a'/b'$ and both fractions are in their lowest terms, giving $P_n = a'$, $Q_n = b'$. So equation (1.73) gives

$$P_nQ_{n-1} - Q_nP_{n-1} = a'Q_{n-1} - b'P_{n-1} = (-1)^{n-1} \tag{1.75}$$

Hence

$$aQ_{n-1} - bP_{n-1} = da'Q_{n-1} - db'P_{n-1} = (-1)^{n-1}d \tag{1.76}$$

or

$$(-1)^{n-1}aQ_{n-1} - (-1)^{n-1}bP_{n-1} = d \tag{1.77}$$

A solution to the equation $ax - by = d$ is therefore given by

$$\left.\begin{aligned} x &= (-1)^{n-1}Q_{n-1}, \\ y &= (-1)^{n-1}P_{n-1}. \end{aligned}\right\} \tag{1.78}$$

To conclude the above analysis, we have the following theorem for solving the linear Diophantine equation $ax - by = d$:

Theorem 1.3.2. Let the convergents of the finite continued fraction of a/b be as follows:

$$\left[\frac{P_0}{Q_0}, \frac{P_1}{Q_1}, \ldots, \frac{P_{n-1}}{Q_{n-1}}, \frac{P_n}{Q_n}\right] = \frac{a}{b}. \tag{1.79}$$

Then the integer solution in x and y of the equation $ax - by = d$ is

$$\left.\begin{aligned} x &= (-1)^{n-1}Q_{n-1}, \\ y &= (-1)^{n-1}P_{n-1}. \end{aligned}\right\} \tag{1.80}$$

Remark 1.3.1. We have already known a way of solving equations like 1.73 by applying Euclid's algorithm to a and b and working backwards through the resulting equations (the so-called extended Euclid's algorithm). Our new method here turns out to be equivalent to this since the continued fraction for a/b is derived from Euclid's algorithm. However, it is quicker to generate the convergents P_i/Q_i using the recurrence relations than to work backwards through the equations in Euclid's algorithm.

Example 1.3.1. Use the continued fraction method to solve the following linear Diophantine equation:

$$364x - 227y = 1.$$

Since $364/227$ can be expanded as a finite continued fraction with convergents

$$\left[1,\ 2,\ \frac{3}{2},\ \frac{5}{3},\ \frac{8}{5},\ \frac{85}{53},\ \frac{93}{58},\ \frac{364}{227}\right]$$

we have

$$\begin{aligned} x &= (-1)^{n-1}q_{n-1} = (-1)^{7-1}58 = 58, \\ y &= (-1)^{n-1}p_{n-1} = (-1)^{7-1}93 = 93. \end{aligned}$$

That is,

$$364 \cdot 58 - 227 \cdot 93 = 1.$$

Example 1.3.2. Use the continued fraction method to solve the following linear Diophantine equation:

$$20719x + 13871y = 1.$$

Note first that

$$20719x + 13871y = 1 \Longleftrightarrow 20719x - (-13871y) = 1.$$

Now since $20719/13871$ can be expanded as a finite simple continued fraction with convergents

$$\left[1,\ \frac{3}{2},\ \frac{118}{79},\ \frac{829}{555},\ \frac{947}{634},\ \frac{1776}{1189},\ \frac{2723}{1823},\ \frac{4499}{3012},\ \frac{20719}{13871}\right],$$

we have

$$\begin{aligned} x &= (-1)^{n-1}q_{n-1} = (-1)^{8-1}3012 = -3012, \\ y &= (-1)^{n-1}p_{n-1} = (-1)^{8-1}4499 = -4499. \end{aligned}$$

That is,

$$20719 \cdot (-3012) - 13871 \cdot (-4499) = 1.$$

The linear Diophantine equation $ax + by = d$ can also be interpreted geometrically. If we allow (x, y) to be any real values, then the graph of this equation is a straight line L in the xy-plane. The points (x, y) in the plane with integer coordinates (x, y) are the integer lattice-points. Pairs of integers (x, y) satisfying the equation correspond to integer lattice-points (x, y) on L. Thus, Theorem 1.3.1 tells us that L passes through such a lattice-point if and only if $\gcd(a, b) \mid d$, in which case it passes through infinitely many of them.

Remark 1.3.2. In some areas of number theory (see e.g., Yan [261]), it may be necessary to solve the following more general form of linear Diophantine equation:

$$axy + bx + cy = d. \tag{1.81}$$

Note first that this type of equation can be reduced to a factorization: multiplying (1.81) by a, adding bc to both sides and factoring results in

$$(ax + c)(ay + b) = ad + bc. \tag{1.82}$$

If mn is a factorization of $ad + bc$ and a divides $n - c$ and $m - b$, an integer solution of (1.81) is

$$\left.\begin{aligned} x &= \frac{n-c}{a}, \\ y &= \frac{m-b}{a}. \end{aligned}\right\} \tag{1.83}$$

1.3.3 Pell's Equations

In this subsection, we shall study the elementary theory of Pell's equations, a type of quadratic Diophantine equation.

Definition 1.3.2. A *Pell's equation* is a quadratic Diophantine equation in any one of the following three forms:

$$x^2 - Ny^2 = 1, \tag{1.84}$$

$$x^2 - Ny^2 = -1, \tag{1.85}$$

$$x^2 - Ny^2 = n, \tag{1.86}$$

where N is a positive integer other than a perfect square, and n a positive integer greater than 1.

Remark 1.3.3. Pell's equations are named after the 17th century British mathematician John Pell (1611–1685). It is often said that Euler mistakenly attributed these types of equations to Pell. They probably should be called Fermat's equations since Fermat initiated the comparatively recent study of the topic. But because Euler is so famous, everybody adopts Euler's convention.

The solutions to Pell's equations or its more general forms can be easily obtained in terms of the continued fraction of $\sqrt{N}$. In this subsection, we shall use the continued fraction method to solve Pell's equations.

Theorem 1.3.3. Let α be an irrational number. If a/b is a rational number in lowest terms, where a and b are integers $b > 0$, such that

$$\left|\alpha - \frac{a}{b}\right| < \frac{1}{2s^2}, \tag{1.87}$$

then a/b is a convergent of the simple continued fraction expansion of α.

Theorem 1.3.4. Let α be an irrational number greater than 1. The $(k+1)$th convergent to $1/\alpha$ is the reciprocal of the kth convergent to α, for $k = 1, 2, \cdots$.

Theorem 1.3.5. Let N be a positive integer other than a perfect square, and let n be an integer with $|n| < \sqrt{N}$. If x_0 and y_0 is a positive integer solution of

$$x^2 - Ny^2 = n, \tag{1.88}$$

then x_0/y_0 is one of the convergents of $\sqrt{N}$.

Proof. Suppose $n > 0$. Since x_0 and y_0 is a positive integer solution of $x^2 - Ny^2 = n$, then

$$(x_0 - y_0\sqrt{N})((x_0 + y_0\sqrt{N}) = n,$$

which implies that

$$x_0 > y_0\sqrt{N}.$$

Therefore,

$$\begin{aligned} 0 &< \frac{x_0}{y_0} - \sqrt{N} \\ &= \frac{n}{y_0(x_0 + y_0\sqrt{N})} \\ &< \frac{n}{y_0(y_0 + y_0\sqrt{N})} \\ &< \frac{\sqrt{N}}{2y_0^2\sqrt{N}} \end{aligned}$$

It follows from Theorem 1.3.3 that x_0/y_0 is a convergent to $\sqrt{N}$. Similarly, if $n < 0$, we find that y_0/x_0 is a convergent to $1/\sqrt{N}$. Using Theorem 1.3.4, we conclude that x_0/y_0 is a convergent to $\sqrt{N}$. □

Corollary 1.3.1. Let (x_0, y_0) be a positive integer solution of

$$x^2 - Ny^2 = \pm 1, \tag{1.89}$$

then

$$x_0 = P_n, \quad y_0 = Q_n, \tag{1.90}$$

where P_n/Q_n is a convergent to $\sqrt{N}$.

Proof. By Theorem 1.3.5 we know that $x_0/y_0 = P_n/Q_n$. Since the fractions are reduced to lowest terms, then $x_0 = P_n$, $y_0 = Q_n$. □

Theorem 1.3.6. Let N be a positive integer other than a perfect square, and m the period of the expansion of $\sqrt{N}$ as a simple continued fraction. Then we have:

(1) m is even

(i) The positive integer solutions of $x^2 - Ny^2 = 1$ are

$$\left.\begin{aligned} x &= P_{km-1}, \\ y &= Q_{km-1}, \end{aligned}\right\} \tag{1.91}$$

for $k = 1, 2, 3, \cdots$, with

$$\left.\begin{aligned} x &= P_{m-1}, \\ y &= Q_{m-1}, \end{aligned}\right\} \tag{1.92}$$

as the *smallest* positive integer solution.

(ii) The equation $x^2 - Ny^2 = -1$ has no integer solution.

(2) m is odd

(i) The positive integer solutions of $x^2 - Ny^2 = 1$ are

$$\left.\begin{aligned} x &= P_{km-1}, \\ y &= Q_{km-1}, \end{aligned}\right\} \tag{1.93}$$

for $k = 2, 4, 6, \cdots$, with

$$\left.\begin{aligned} x &= P_{2m-1}, \\ y &= Q_{2m-1}, \end{aligned}\right\} \tag{1.94}$$

as the *smallest* positive integer solution.

(ii) The positive integer solutions of $x^2 - Ny^2 = -1$ are

$$\left.\begin{aligned} x &= P_{km-1}, \\ y &= Q_{km-1}, \end{aligned}\right\} \tag{1.95}$$

for $k = 1, 3, 5, \cdots$, with

$$\left.\begin{aligned} x &= P_{m-1}, \\ y &= Q_{m-1}, \end{aligned}\right\} \tag{1.96}$$

as the *smallest* positive integer solution.

Proof. Left as an exercise. □

Example 1.3.3. Find the integer solutions of $x^2 - 73y^2 = \pm 1$. Note first that

$$\sqrt{73} = [8, \overline{1, 1, 5, 5, 1, 1, 16}].$$

So the period $m = 7$ and of course m is odd. Thus, both equations are soluble and their solutions are as follows:

(1) The smallest positive integral solution of $x^2 - 73y^2 = 1$ is

$$\left.\begin{array}{l} x = P_{km-1} = P_{2\cdot 7-1} = P_{13} = 2281249, \\ y = Q_{km-1} = Q_{2\cdot 7-1} = Q_{13} = 267000. \end{array}\right\} \tag{1.97}$$

That is, $2281249^2 - 73 \cdot 267000^2 = 1$.

(2) The smallest positive integer solution of $x^2 - 73y^2 = -1$ is

$$\left.\begin{array}{l} x = P_{km-1} = P_{1\cdot 7-1} = P_6 = 1068, \\ y = Q_{km-1} = Q_{1\cdot 7-1} = Q_6 = 125. \end{array}\right\} \tag{1.98}$$

That is, $1068^2 - 73 \cdot 125^2 = -1$.

Example 1.3.4. Find the integer solutions of $x^2 - 97y^2 = \pm 1$. Note first that

$$\sqrt{97} = [9, \overline{1, 5, 1, 1, 1, 1, 1, 1, 5, 1, 18}].$$

So the period $m = 11$ and of course m is odd. Thus, both equations are soluble and their solutions are as follows:

(1) The smallest positive integral solution of $x^2 - 97y^2 = 1$ is

$$\left.\begin{array}{l} x = P_{2m-1} = P_{2\cdot 11-1} = P_{21} = 62809633, \\ y = Q_{2m-1} = Q_{2\cdot 11-1} = Q_{21} = 6377352. \end{array}\right\} \tag{1.99}$$

That is, $62809633^2 - 97 \cdot 6377352^2 = 1$.

(2) The smallest positive integer solution of $x^2 - 97y^2 = -1$ is

$$\left.\begin{array}{l} x = P_{m-1} = P_{1\cdot 11-1} = P_{10} = 5604, \\ y = Q_{m-1} = Q_{1\cdot 11-1} = Q_{10} = 569. \end{array}\right\} \tag{1.100}$$

That is, $5604^2 - 97 \cdot 569^2 = 1$.

Remark 1.3.4. Incidentally, the continued fraction for $\sqrt{N}$, with N not a perfect square, always has the form

$$\sqrt{N} = [q_0, \overline{q_1, q_2, q_3, \cdots, q_3, q_2, q_1, 2q_0}],$$

as can be seen in Table 1.5.

Table 1.6 and Table 1.7 show the smallest positive integer solutions (x, y) to Pell's equations $x^2 - Ny^2 = 1$ and $x^2 - Ny^2 = -1$ for $1 < N < 100$ (except the perfect squares), respectively.

The following is actually a corollary of Theorem 1.3.6.

Corollary 1.3.2. Let N be a positive integer other than a perfect square, m the period of the expansion of $\sqrt{N}$ as a simple continued fraction, and $\dfrac{P_n}{Q_n}$, $n = 1, 2, \cdots$ the convergents to $\sqrt{N}$. Then the complete set of all solutions, including positive and negative (if any) of Pell's equation are:

Table 1.5. Continued fractions for $\sqrt{N}$ with $N \leq 50$ and not perfect square

$\sqrt{2} = [1, \overline{2}]$	$\sqrt{3} = [1, \overline{1, 2}]$
$\sqrt{5} = [2, \overline{4}]$	$\sqrt{6} = [2, \overline{2, 4}]$
$\sqrt{7} = [2, \overline{1, 1, 1, 4}]$	$\sqrt{8} = [2, \overline{1, 4}]$
$\sqrt{10} = [3, \overline{6}]$	$\sqrt{11} = [3, \overline{3, 6}]$
$\sqrt{12} = [3, \overline{2, 6}]$	$\sqrt{13} = [3, \overline{1, 1, 1, 1, 6}]$
$\sqrt{14} = [3, \overline{1, 2, 1, 6}]$	$\sqrt{15} = [3, \overline{1, 6}]$
$\sqrt{17} = [4, \overline{8}]$	$\sqrt{18} = [4, \overline{4, 8}]$
$\sqrt{19} = [4, \overline{2, 1, 3, 1, 2, 8}]$	$\sqrt{20} = [4, \overline{2, 8}]$
$\sqrt{21} = [4, \overline{1, 1, 2, 1, 1, 8}]$	$\sqrt{22} = [4, \overline{1, 2, 4, 2, 1, 8}]$
$\sqrt{23} = [4, \overline{1, 3, 1, 8}]$	$\sqrt{24} = [4, \overline{1, 8}]$
$\sqrt{26} = [5, \overline{10}]$	$\sqrt{27} = [5, \overline{5, 10}]$
$\sqrt{28} = [5, \overline{3, 2, 3, 10}]$	$\sqrt{29} = [5, \overline{2, 1, 1, 2, 10}]$
$\sqrt{30} = [5, \overline{2, 10}]$	$\sqrt{31} = [5, \overline{1, 1, 3, 5, 3, 1, 1, 10}]$
$\sqrt{32} = [5, \overline{1, 1, 1, 10}]$	$\sqrt{33} = [5, \overline{1, 2, 1, 10}]$
$\sqrt{34} = [5, \overline{1, 4, 1, 10}]$	$\sqrt{35} = [5, \overline{1, 10}]$
$\sqrt{37} = [6, \overline{12}]$	$\sqrt{38} = [6, \overline{6, 12}]$
$\sqrt{39} = [6, \overline{4, 12}]$	$\sqrt{40} = [6, \overline{3, 12}]$
$\sqrt{41} = [6, \overline{2, 2, 12}]$	$\sqrt{42} = [6, \overline{2, 12}]$
$\sqrt{43} = [6, \overline{1, 1, 3, 1, 5, 1, 3, 1, 1, 12}]$	$\sqrt{44} = [6, \overline{1, 1, 1, 2, 1, 1, 1, 12}]$
$\sqrt{45} = [6, \overline{1, 2, 2, 2, 1, 12}]$	$\sqrt{46} = [6, \overline{1, 3, 1, 1, 2, 6, 2, 1, 1, 3, 1, 12}]$
$\sqrt{47} = [6, \overline{1, 5, 1, 12}]$	$\sqrt{48} = [6, \overline{1, 12}]$
$\sqrt{50} = [7, \overline{14}]$	

(1) m even

(i) $x^2 - Ny^2 = 1$: For $i = 0, 1, 2, 3, \cdots$,

$$x + y\sqrt{N} = \pm(P_{m-1} \pm y\sqrt{N}Q_{m-1})^i. \tag{1.101}$$

(ii) $x^2 - Ny^2 = -1$: No solutions.

(2) m odd

(i) $x^2 - Ny^2 = 1$: For $i = 1, 3, 5, \cdots$,

$$x + y\sqrt{N} = \pm(P_{m-1} \pm y\sqrt{N}Q_{m-1})^i. \tag{1.102}$$

(ii) $x^2 - Ny^2 = -1$: For $i = 0, 2, 4, \cdots$,

$$x + y\sqrt{N} = \pm(P_{m-1} \pm y\sqrt{N}Q_{m-1})^i. \tag{1.103}$$

Proof. Left as an exercise. □

If N is not a perfect square, Pell's equation $x^2 - Ny^2 = 1$ always has infinitely many integer solutions. For the more general form of Pell's equation

$$x^2 - Ny^2 = n,$$

we have the following result:

Table 1.6. The smallest solution to $x^2 - Ny^2 = 1$ for $N \leq 100$

N	x	y	N	x	y
2	3	2	3	2	1
5	9	4	6	5	2
7	8	3	8	3	1
10	19	6	11	10	3
12	7	2	13	649	180
14	15	4	15	4	1
17	33	8	18	17	4
19	170	39	20	9	2
21	55	12	22	197	42
23	24	5	24	5	1
26	51	10	27	26	5
28	127	24	29	9801	1820
30	11	2	31	1520	273
32	17	3	33	23	4
34	35	6	35	6	1
37	73	12	38	37	6
39	25	4	40	19	3
41	2049	320	42	13	2
43	3482	531	44	199	30
45	161	24	46	24335	3588
47	48	7	48	7	1
50	99	14	51	50	7
52	649	90	53	66249	9100
54	485	66	55	89	12
56	15	2	57	151	20
58	19603	2574	59	530	69
60	31	4	61	1766319049	226153980
62	63	8	63	8	1
65	129	16	66	65	8
67	48842	5967	68	33	4
69	7775	936	70	251	30
71	3480	413	72	17	2
73	2281249	267000	74	3699	430
75	26	3	76	57799	6630
77	351	40	78	53	6
79	80	9	80	9	1
82	163	18	83	82	9
84	55	6	85	285769	30996
86	10405	1122	87	28	3
88	197	21	89	500001	53000
90	19	2	91	1574	165
92	1151	120	93	12151	1260
94	2143295	221064	95	39	4
96	49	5	97	62809633	6377352
98	99	10	99	10	1

Table 1.7. The smallest solution to $x^2 - Ny^2 = -1$ for $N \leq 100$

N	x	y	N	x	y	N	x	y
2	1	1	5	2	1	10	3	1
13	18	5	17	4	1	26	5	1
29	70	13	37	6	1	41	32	5
50	7	1	53	182	25	58	99	13
61	29718	3805	65	8	1	73	1068	125
74	43	5	82	9	1	85	378	41
89	500	53	97	5604	569			

Theorem 1.3.7. If N is not a perfect square and n an integer, then the equation

$$x^2 - Ny^2 = n \tag{1.104}$$

has a finite set T of solutions such that for any solution (x, y), we have

$$(x \pm y\sqrt{N}) = (x_0 \pm y_0\sqrt{N})(u \pm v\sqrt{N}) \tag{1.105}$$

for some $(x_0, y_0) \in T$ and some (u, v) with $u^2 - Nv^2 = 1$.

Proof. Left as an exercise. □

1.4 Arithmetic Functions

It is true that a mathematician who is not also somewhat of a poet will never be a perfect mathematician.

KARL WEIERSTRASS (1815–1897)

Arithmetic (or number-theoretic) functions are the most fundamental functions in mathematics and computer science; for example, the computable functions studied in mathematical logic and computer science are actually arithmetic functions. In this section, we shall study some basic arithmetic functions that are useful in number theory.

1.4.1 Multiplicative Functions

Definition 1.4.1. A function f is called an *arithmetic function* or a *number-theoretic function* if it assigns to each positive integer n a unique real or complex number $f(n)$. Typically, an arithmetic function is a real-valued function whose domain is the set of positive integers.

Example 1.4.1. The equation

$$f(n) = \sqrt{n}, \quad n \in \mathbb{N} \tag{1.106}$$

defines an arithmetic function f which assigns the real number $\sqrt{n}$ to each positive integer n.

Definition 1.4.2. A real function f defined on the positive integers is said to be *multiplicative* if

$$f(m)f(n) = f(mn), \quad \forall m, n \in \mathbb{N}, \text{ with } \gcd(m, n) = 1. \tag{1.107}$$

If

$$f(m)f(n) = f(mn), \quad \forall m, n \in \mathbb{N}, \tag{1.108}$$

then f is *completely multiplicative*. Every completely multiplicative function is multiplicative.

Theorem 1.4.1. If f is completely multiplicative and not the zero function, then $f(1) = 1$.

Proof. If f is not zero function, then there exists a positive integer k such that $f(k) \neq 0$. Hence, $f(k) = f(k \cdot 1) = f(k)f(1)$. Dividing both sides by $f(k)$, we get $f(1) = 1$. □

Theorem 1.4.2. Let

$$n = \prod_{i=1}^{k} p_i^{\alpha_i}$$

be the prime factorization of n and let f be a multiplicative function, then

$$f(n) = \prod_{i=1}^{k} f(p_i^{\alpha_i}).$$

Proof. Clearly, if $k = 1$, we have the identity, $f(p_i^{\alpha_i}) = f(p_i^{\alpha_i})$. Assume that the representation is valid whenever n has r or fewer distinct prime factors, and consider $n = \prod_{i=1}^{r+1} f(p_i^{\alpha_i})$. Since $\gcd\left(\prod_{i=1}^{r} p_i^{\alpha_i}, p_{r+1}^{r+1}\right) = 1$ and f is multiplicative, we have

$$\begin{aligned}
f(n) &= f\left(\prod_{i=1}^{r+1} p_i^{\alpha_i}\right) \\
&= f\left(\prod_{i=1}^{r} p_i^{\alpha_i} \cdot p_{r+1}^{\alpha_{r+1}}\right) \\
&= f\left(\prod_{i=1}^{r} p_i^{\alpha_i}\right) \cdot f\left(p_i^{\alpha_{r+1}}\right) \\
&= \prod_{i=1}^{r}(f(p_i^{\alpha_i})) \cdot f(p_{r+1}^{\alpha_{r+1}}) \\
&= \prod_{i=1}^{r+1} f(p_i^{\alpha_i}).
\end{aligned}$$

□

Theorem 1.4.3. If f is multiplicative and if

$$g(n) = \sum_{d|n} f(d) \tag{1.109}$$

where the sum is over all divisors d of n, then g is also multiplicative.

Proof. Since f is multiplicative, if $\gcd(m, n) = 1$, then

$$\begin{aligned}
g(mn) &= \sum_{d|m}\sum_{d'|n} f(dd') \\
&= \sum_{d|m} f(d) \sum_{d'|n} f(d') \\
&= g(m)g(n).
\end{aligned}$$

□

Theorem 1.4.4. If f and g are multiplicative, then so is

$$F(n) = \sum_{d|m} f(d)g\left(\frac{n}{d}\right).$$

Proof. If $\gcd(m, n) = 1$, then $d \mid mn$ if and only if $d = d_1 d_2$, where $d_1 \mid m$ and $d_2 \mid n$, $\gcd(d_1, d_2) = 1$ and $\gcd(m/d_1, n/d_2) = 1$. Thus,

$$\begin{aligned} F(mn) &= \sum_{d|mn} f(d)g\left(\frac{mn}{d}\right) \\ &= \sum_{d_1|m}\sum_{d_2|n} f(d_1d_2)g\left(\frac{mn}{d_1d_2}\right) \\ &= \sum_{d_1|m}\sum_{d_2|n} f(d_1)f(d_2)g\left(\frac{m}{d_1}\right)\left(\frac{n}{d_2}\right) \\ &= \left[\sum_{d_1|m} f(d_1)g\left(\frac{m}{d_1}\right)\right]\left[\sum_{d_2|m} f(d_2)g\left(\frac{n}{d_2}\right)\right] \\ &= F(m)F(n). \end{aligned}$$

□

1.4.2 Functions $\tau(n)$, $\sigma(n)$ and $s(n)$

Definition 1.4.3. Let n be a positive integer. Then the arithmetic functions $\tau(n)$ and $\sigma(n)$ are defined as follows:

$$\tau(n) = \sum_{d|n} 1, \qquad \sigma(n) = \sum_{d|n} d. \tag{1.110}$$

That is, $\tau(n)$ designates the number of all positive divisors of n, and $\sigma(n)$ designates the sum of all positive divisors of n.

It is sometimes also convenient to use the function $s(n)$ rather than $\sigma(n)$. The function $s(n)$ is defined as follows:

Definition 1.4.4. Let n be a positive integer. Then

$$s(n) = \sigma(n) - n. \tag{1.111}$$

Example 1.4.2. By Definitions 1.4.3 and 1.4.4, we have:

n	1	2	3	4	5	6	7	8	9	10	100	101	220	284
$\tau(n)$	1	2	2	3	2	4	2	4	3	4	9	2	12	6
$\sigma(n)$	1	3	4	7	6	12	8	15	13	18	217	102	504	504
$s(n)$	0	1	1	3	1	6	1	7	4	8	117	1	284	220

Lemma 1.4.1. If n be a positive integer greater than 1 and

$$n = \prod_{i=1}^{k} p_i^{\alpha_i},$$

then the positive divisors of n are precisely those integers d of the form

$$d = \prod_{i=1}^{k} p_i^{\beta_i},$$

where $0 \leq \beta_i \leq \alpha_i$.

Proof. If $d \mid n$, then $n = dq$. By the Fundamental theorem of arithmetic, the prime factorization of n is unique, so the prime numbers in the prime factorization of d must occur in p_j, $(j = 1, 2, \cdots, k)$. Furthermore, the power β_j of p_j occurring in the prime factorization of d cannot be greater than α_j, that is, $\beta_j \leq \alpha_j$. Conversely, when $\beta_j \leq \alpha_j$, d clearly divides n. □

Theorem 1.4.5. Let n be a positive integer. Then

(1) $\tau(n)$ is multiplicative, i.e.,

$$\tau(mn) = \tau(m)\tau(n). \tag{1.112}$$

(2) if n is a prime, say p, then $\tau(p) = 2$. More generally, if n is a prime power p^α, then

$$\tau(p^\alpha) = \alpha + 1. \tag{1.113}$$

(3) if n is a composite and has the standard prime factorization form, then

$$\begin{aligned} \tau(n) &= (\alpha_1 + 1)(\alpha_2 + 1) \cdots (\alpha_k + 1) \\ &= \prod_{i=1}^{k} (\alpha_i + 1). \end{aligned} \tag{1.114}$$

Proof.

(1) Since the constant function $f(n) = 1$ is multiplicative and $\tau(n) = \sum_{d|n} 1$, the result follows immediately from Theorem 1.4.3.

(2) Clearly, if n is a prime, there are only two divisors, namely, 1 and n itself. If $n = p^\alpha$, then by Lemma 1.4.1, the positive divisors of n are precisely those integers $d = p^\beta$, with $0 \leq \beta \leq \alpha$. Since there are $\alpha + 1$ choices for the exponent β, there are $\alpha + 1$ possible positive divisors of n.

(3) By Lemma 1.4.1 and Part (2) of this theorem, there are $\alpha_1 + 1$ choices for the exponent β_1, $\alpha_2 + 1$ choices for the exponent β_2, $\cdots$, $\alpha_k + 1$ choices for the exponent β_k. From the multiplication principle it follows that there are $(\alpha_1 + 1)(\alpha_2 + 1) \cdots (\alpha_k + 1)$ different choices for the $\beta_1, \beta_2, \cdots, \beta_k$, thus that many divisors of n. Therefore, $\tau(n) = (\alpha_1 + 1)(\alpha_2 + 1) \cdots (\alpha_k + 1)$.

Theorem 1.4.6. The product of all divisors of a number n is

$$\prod_{d|n} d = n^{\tau(n)/2}. \tag{1.115}$$

Proof. Let d denote an arbitrary positive divisor of n, so that

$$n = dd'$$

for some d'. As d ranges over all $\tau(n)$ positive divisors of n, there are $\tau(n)$ such equations. Multiplying these together, we get

$$n^{\tau(n)} = \prod_{d|n} d \prod_{d'|n} d'.$$

But as d runs through the divisors of n, so does d', hence

$$\prod_{d|n} d = \prod_{d'|n} d'.$$

So,

$$n^{\tau(n)} = \left(\prod_{d|n} d \cdot d'\right)^2,$$

or equivalently

$$n^{\tau(n)/2} = \prod_{d|n} d \cdot d'.$$

□

Example 1.4.3. Let $n = 1371$, then

$$\tau(1371) = 4.$$

Therefore

$$\prod d = 1371^{4/2} = 1879641.$$

It is of course true, since

$$d(1371) = \{1, 3, 457, 1371\}$$

implies that

$$\prod d = 1 \cdot 3 \cdot 457 \cdot 1371 = 1879641.$$

The result in Theorem 1.4.6 can be expressed in a different manner. Let $\{x_1, x_2, \cdots, x_k\}$ be a set of k positive integers. The *geometric mean* of these k numbers is defined by

$$G = (x_1 x_2 \cdots x_k)^{1/k}. \tag{1.116}$$

When this applies to the product of $\tau(n)$ divisors of n, we have:

Theorem 1.4.7. *The geometric mean of the divisors of n is*

$$G(n) = \sqrt{n}. \tag{1.117}$$

Example 1.4.4. Let again $n = 1371$, then

$$G(1371) = (1 \cdot 3 \cdot 457 \cdot 1371)^{1/4} = 37.02701716.$$

It is of course true since

$$\sqrt{1371} = 37.02701716.$$

Theorem 1.4.8. Let n be a positive integer. Then

(1) $\sigma(n)$ is multiplicative, i.e.,

$$\sigma(mn) = \sigma(m)\sigma(n). \tag{1.118}$$

(2) if n is a prime, say p, then $\sigma(p) = p + 1$. More generally, if n is a prime power p^α, then

$$\sigma(p^\alpha) = \frac{p^{\alpha+1} - 1}{p - 1}. \tag{1.119}$$

(3) if n is a composite and has the standard prime factorization form, then

$$\begin{aligned} \sigma(n) &= \frac{p_1^{\alpha_1+1} - 1}{p_1 - 1} \cdot \frac{p_2^{\alpha_2+1} - 1}{p_2 - 1} \cdots \frac{p_k^{\alpha_k+1} - 1}{p_k - 1} \\ &= \prod_{i=1}^{k} \frac{p_i^{\alpha_i+1} - 1}{p_i - 1}. \end{aligned} \tag{1.120}$$

Proof.

(1) The result follows immediately from Theorem 1.4.3 since the identity function $f(n) = n$ and $\sigma(n)$ can be represented in the form $\sigma(n) = \sum_{d|n} d$.

(2) Left as an exercise; we prove the most general case in Part (3).

(3) The sum of the divisors of the positive integer

$$n = p_1^{\alpha_1} p_2^{\alpha_2} \cdots p_k^{\alpha_k}$$

can be expressed by the product

$$\left(1 + p_1 + p_1^2 + \cdots + p_1^{\alpha_1}\right)\left(1 + p_2 + p_2^2 + \cdots + p_2^{\alpha_2}\right) \cdots \left(1 + p_k + p_k^2 + \cdots + p_k^{\alpha_k}\right).$$

Using the finite geometric series

$$1 + x + x^2 + \cdots + x^n = \frac{x^{n+1} - 1}{x - 1},$$

we simplify each of the k sums in the above product to find that the sum of the divisors can be expressed as

$$\begin{aligned}\sigma(n) &= \frac{p_1^{\alpha_1+1}-1}{p_1-1}\cdot\frac{p_2^{\alpha_2+1}-1}{p_2-1}\cdots\frac{p_k^{\alpha_k+1}-1}{p_k-1} \\ &= \prod_{i=1}^{k}\frac{p_i^{\alpha_i+1}-1}{p_i-1}. \end{aligned} \quad (1.121)$$

Just as the geometric mean $G(n)$ of the divisors of a number n, we can define the *arithmetic mean* as follows:

$$A(n) = \frac{\sigma(n)}{\tau(n)}. \quad (1.122)$$

Similarly, we can also define the *harmonic mean* $H(n)$ of the divisors of a number n in terms of the arithmetic mean as follows:

$$\frac{1}{H(n)} = \frac{A(n)}{n}. \quad (1.123)$$

Note that the harmonic mean $H(n)$ of a set of numbers $\{x_1, x_2, \cdots, x_n\}$ is defined by

$$\frac{1}{H} = \frac{1}{n}\cdot\left(\frac{1}{x_1}+\frac{1}{x_2}+\cdots\frac{1}{x_n}\right). \quad (1.124)$$

The following theorem gives the relationships between the number n and the harmonic and arithmetic means of the divisors of n.

Theorem 1.4.9. Let $A(n)$, $G(n)$ and $H(n)$ be arithmetic, geometric and harmonic means, respectively. Then

(1) The product of the harmonic and arithmetic means of the divisors of n is equal to n

$$n = A(n)\cdot H(n), \quad (1.125)$$

(2)

$$H(n) \le G(n) = \sqrt{n} \le A(n). \quad (1.126)$$

1.4.3 Perfect, Amicable and Sociable Numbers

"Perfect numbers" certainly never did any good, but then they never did any particular harm.

J. E. LITTLEWOOD (1885–1977)

Perfect and amicable numbers have been studied since ancient times; however, many problems concerning them still remain unsolved. This subsection introduces some basic concepts and results on perfect and amicable numbers based on the arithmetic functions studied previously.

Definition 1.4.5. Let $(m_1, m_2, \cdots, m_k)$ be k positive integers all greater than 1, satisfying:

$$\left.\begin{array}{l} \sigma(m_1) = m_1 + m_2 \\ \sigma(m_2) = m_2 + m_3 \\ \cdots\cdots \\ \cdots\cdots \\ \sigma(m_k) = m_k + m_1 \end{array}\right\} \tag{1.127}$$

then the k positive integers form a *sociable group* with order k (or an aliquot k-cycle). If $k = 1$, that is

$$\sigma(m_1) = m_1 + m_1 = 2m_1, \tag{1.128}$$

then m_1 is called a *perfect number*. If $k = 2$, that is

$$\sigma(m_1) = m_1 + m_2 = \sigma(m_2), \tag{1.129}$$

then (m_1, m_2) is called an *amicable pair*. The k integers $m_1, m_2, \cdots, m_k$ are called an amicable k-tuple if

$$\sigma(m_1) = \sigma(m_2) = \cdots = \sigma(m_k) = m_1 + m_2 + \cdots + m_k. \tag{1.130}$$

(In case $k = 3$, we call them amicable triples.)

Example 1.4.5. The following are some examples of perfect, amicable and sociable numbers:

(1) 6, 28, 496 and 8128 are the first four perfect numbers, whereas $2^{4053946}(2^{4053946} - 1)$ is the largest known perfect number at present. Since once we found a Mersenne prime of the form $2^p - 1$, we found an (even) perfect number of the form $2^{p-1}(2^p - 1)$. As there are 39 known Mersenne primes at present (see Table 1.2), there are 39 known perfect numbers.

(2) (220, 284), (1184, 1210), (2620, 2924) and (5020, 5564) are the first four amicable pairs. The following is a large amicable pair: $(2^9 \cdot p^{65} \cdot m \cdot q_1,\ \ 2^9 \cdot p^{65} \cdot q \cdot q_2)$ with

$$\begin{aligned} p &= 37669773212168992472511541, \\ q &= 609610904872320606430695102719, \\ m &= 569 \cdot 5023 \cdot 22866511 \cdot 287905188653, \\ q_1 &= (p+q) \cdot p^{65} - 1, \\ q_2 &= (p-m) \cdot p^{65} - 1, \end{aligned}$$

Both numbers in the pair have 3383 digits; it was found by M. Garcia in 1997. But it is still not the largest known amicable pair; the largest known amicable pair at present has 5577 digits in both its numbers. To date, there are in total 2494343 amicable pairs are known. Table 1.8 gives the frequency of these known amicable pairs distributed over the number of digits in the smaller number (the list exhaustive up to 10^{12}).

(3) (1980, 2016, 2556), (9180, 9504, 11556) and (21668, 22200, 27312) are the first three amicable triples with $m_1 \neq m_2 \neq m_3$; the last two triples were found by Te Riele in 1994 whereas the first one was known a long time ago. $(am_1, am_2, am_3, am_4, am_5)$ is an amicable 5-tuple, with $a = 2^{19} \cdot 3^5 \cdot 5 \cdot 7^3 \cdot 13 \cdot 31 \cdot 41$, $m_1 = 11 \cdot 359$, $m_2 = 23 \cdot 179$, $m_3 = 47 \cdot 89$, $m_4 = 53 \cdot 79$, $m_4 = 59 \cdot 71$; it was found by C. Krishnamurthy in 1980.

(4) (1236402232, 1369801928, 1603118392, 1412336648) is an aliquot 4-cycle. The longest aliquot known cycle is the aliquot 28-cycle with $m_1 = 14316 = 2^2 \cdot 3 \cdot 1193$; it was found by P. Poulet in 1918. About 119 aliquot k-cycles for $4 \leq k \leq 28$ have been found to date (with $k = 28$ the longest, generated by 14316):

k	4	5	6	8	9	28
Number of k-cycles	112	1	2	2	1	1

For perfect numbers, we have the very convenient necessary and sufficient condition for an even number to be perfect:

Theorem 1.4.10 (The Euclid–Euler Theorem). n is an even perfect number if and only if $n = 2^{p-1}(2^p - 1)$, where $2^p - 1$ is a Mersenne prime.

Proof. We first prove that this is a necessary condition for n to be perfect. Let $n = 2^{p-1}(2^p - 1)$. Then

$$\begin{aligned} \sigma(n) &= \sigma(2^{p-1})\sigma(2^p - 1) \\ &= (2^p - 1)2^p \qquad \text{(since } 2^p - 1 \text{ is prime)} \\ &= 2 \cdot 2^{p-1}(2^p - 1) \\ &= 2n. \end{aligned}$$

Table 1.8. Number of Known Amicable Pairs (By courtesy of Mr. Jan Munch Pedersen)

Digits	0	1	2	3	4	5	6	7	8	9
0-9		0	0	1	4	8	29	66	128	350
10-19	841	1913	4302	9867	15367	30604	5881	1991	1851	1976
20-29	1750	1916	1936	2225	2405	2817	2914	3306	3977	4699
30-39	5240	5565	6276	6503	6899	7433	8029	8661	8804	9145
40-49	12013	12876	13078	12972	12343	12383	15085	17050	17022	16933
50-59	18409	18477	20555	18142	15734	16068	16576	16564	13678	12697
60-69	11470	11302	11220	12045	10961	12099	45779	48368	40170	34601
70-79	31817	27639	75099	57453	48401	41159	46813	44160	50008	39017
80-89	41982	46845	51611	47552	55896	49069	49221	41510	39944	41246
90-99	46649	39511	36427	32406	33921	31181	29169	25124	25986	28029
100-109	27840	23753	20766	18801	18288	18267	16257	14274	12668	11713
110-119	11189	18642	16929	15070	13570	12468	11744	10517	9557	8892
120-129	8358	7684	6792	8733	16396	15748	14108	13417	12695	11986
130-139	11348	10522	10271	9498	9103	8434	7704	7141	6468	6177
140-149	5546	5217	4449	4042	3620	3297	2999	2651	2281	2240
150-159	2352	2065	1746	1484	1344	1184	1101	979	833	773
160-169	757	814	754	672	882	1445	1158	1158	1154	1100
170-179	1001	968	939	852	754	773	718	674	666	646
180-189	667	606	566	533	517	517	453	412	439	387
190-199	358	379	362	341	325	289	190	288	257	251
200-209	229	232	185	152	161	174	131	150	96	119
210-219	123	122	112	95	87	66	112	68	74	72
220-229	60	70	70	71	55	69	66	48	56	57
230-239	55	66	62	50	53	41	49	42	32	46
240-249	52	55	54	52	40	33	41	98	84	90
250-259	79	66	70	74	85	80	67	64	63	57
260-269	51	51	50	99	78	75	62	63	60	45
270-279	53	56	53	49	34	49	35	53	39	35
280-289	35	36	37	41	29	33	27	24	28	26
290-299	21	21	20	18	19	15	14	12	17	13
613 pairs with 300-5577 digits										
There are 2574378 pairs in total										

Therefore, by Definition (1.4.5), n is a perfect number. Next, we prove that even perfect numbers must be of the given form. Let n be an even perfect number and write it as

$$n = 2^{p-1}q \qquad \text{with } q \text{ odd.}$$

Since $\gcd(2^{p-1}, q) = 1$, then

$$\sigma(n) = \sigma(2^{p-1})\sigma(q) = (2^p - 1)\sigma(q). \tag{1.131}$$

By Definition 1.4.5, we must have

$$\sigma(n) = 2n = 2^p q. \tag{1.132}$$

Combining (1.131) and (1.132), we get

$$\begin{aligned} 2^p q &= (2^p - 1)\sigma(q) \\ &= (2^p - 1)(s(q) + q) \qquad \text{(since } s(q) = \sigma(q) - q) \end{aligned}$$

Therefore,

$$q = s(q)(2^p - 1). \tag{1.133}$$

Clearly, (1.133) implies that $d = s(q)$ is a proper divisor of q. On the other hand, $s(q)$ is the sum of all proper divisors of q, including d, so that there cannot be any other proper divisors besides d. But a number q with a single proper divisor d must be a prime and $d = 1$. So from (1.133), we can conclude that

$$q = 2^p - 1$$

is a Mersenne prime. Thus each even perfect number is of the form

$$2^{p-1}(2^p - 1)$$

where $2^p - 1$ is a Mersenne prime. □

The sufficient condition of the above theorem was established in Euclid's Elements (Book IX, Proposition 36) 2000 years ago, but the fact that it is also necessary was established by Euler in work published posthumously. Thus we have an example of a theorem in Number Theory that took about 2000 years to prove. However, we still do not know if there are infinitely many perfect numbers and we also do not know if there exists an odd perfect number; we know that there are no odd perfect numbers up to 10^{300} (Brent, Cohen and Te Riele, [39]) and if there is an odd perfect number it should be divisible by at least eight distinct prime numbers. Compared with perfect numbers, unfortunately, we not only do not know whether or not there exist finitely many amicable pairs, but also do not have necessary and sufficient conditions for amicable numbers (i.e., we do not have a general rule for generating all amicable pairs).

The first (algebraic) rule for amicable numbers was invented by the Arab mathematician Abu-l-Hasan Thabit ibn Qurra[15] and appeared in his book in the ninth century:

[15] Thabit ibn Qurra (824–901), a famous Arab mathematician of the 9th century, lived in Baghdad as a money changer, but he was highly esteemed for his writings on medicine, philosophy, mathematics, astronomy and astrology. He wrote a *Book on the Determination of Amicable Numbers* (Figure 1.6 shows the front cover of the book), in which he proposed his famous rule for amicable numbers: "if $p = 3 \cdot 2^{n-1}$, $q = 3 \cdot 2^n - 1$ and $r = 9 \cdot 2^{2n-1} - 1$ are primes, then $M = 2^n \cdot p \cdot q$ and $N = 2^n \cdot r$ are amicable numbers". In his remarkable treatise entitled "On the Verification of the Problems of Algebra by Geometrical Proofs", he showed that the three types of quadratic equations: $x^2 - ax + c = 0$, $x^2 - ax - c = 0$, and $x^2 + ax - c = 0$ can be solved by means of Propositions 5 and 6 in Book II of Euclid's *Elements*. Thabit was also a most competent translator from Greek and Syriac to Arabic; he translated works of Euclid, Archimedes, Apollonios, Autolykos, Ptolemaios, Nikomachos, Proklos and others.

Figure 1.6. The cover of Thabit's book on amicable numbers (by courtesy of Guedj [95])

Theorem 1.4.11 (Thabit's rule for amicable pairs). If

$$\left.\begin{aligned} p &= 3 \cdot 2^{n-1} - 1 \\ q &= 3 \cdot 2^{n} - 1 \\ r &= 9 \cdot 2^{2n-1} - 1 \end{aligned}\right\} \tag{1.134}$$

are all primes, then

$$(M,\ N) = (2^n \cdot p \cdot q,\ 2^n \cdot r) \tag{1.135}$$

is an amicable pair.

Proof. First, we have

$$\begin{aligned} \sigma(M) &= \sigma(2^n \cdot p \cdot q) \\ &= \sigma(2^n)\sigma(p)\sigma(q) \\ &= \sigma(2^n)\sigma(3 \cdot 2^{n-1} - 1)\sigma(3 \cdot 2^n - 1) \\ &= (2^{n+1} - 1)(3 \cdot 2^{n-1})(3 \cdot 2^n) \end{aligned}$$

$$\begin{aligned}
&= 9 \cdot 2^{2n-1}(2^{n+1} - 1) \\
\sigma(N) &= \sigma(2^n \cdot r) \\
&= \sigma(2^n)\sigma(9 \cdot 2^{2n-1} - 1) \\
&= 9 \cdot 2^{2n-1}(2^{n+1} - 1) \\
M + N &= 2^n(p \cdot q + r) \\
&= 2^n[(3 \cdot 2^{n-1} - 1)(3 \cdot 2^n - 1) + (9 \cdot 2^{2n-1} - 1)] \\
&= 2^n(9 \cdot 2^{2n-1} - 3 \cdot 2^n - 3 \cdot 2^{n-1} + 9 \cdot 2^{2n-1}) \\
&= 2^n(9 \cdot 2^{2n} - 9 \cdot 2^{n-1}) \\
&= 2^n[(9 \cdot 2^{n-1}(2^{n+1} - 1)] \\
&= 9 \cdot 2^{2n-1}(2^{n+1} - 1).
\end{aligned}$$

So $(M,\ N) = (2^n \cdot p \cdot q,\ 2^n \cdot r)$ is an amicable pair. □

For $n = 2$ Thabit's rule gives the first and also the smallest amicable pair

$$(M,\ N) = (2^2 \cdot 5 \cdot 11,\ 2^2 \cdot 71) = (220,\ 284)$$

attributed to the legendary Pythagoras[16]. Two further pairs obtained by Thabit's rule are for $n = 4$ and $n = 7$ (see Borho and Hoffmann [32]); in the early 14th century Ibn al-Banna in Marakesh and also Kamaladdin Farisi in Baghdad discovered the pair for $n = 4$:

$$(M,\ N) = (2^4 \cdot 23 \cdot 47,\ 2^4 \cdot 1151) = (17296,\ 18416)$$

and in the 17th century Muhammad Baqir Yazdi in Iran discovered the pair for $n = 7$

$$(M,\ N) = (2^7 \cdot 191 \cdot 383,\ 2^7 \cdot 73727) = (9363584,\ 9437056).$$

However, after $n = 7$, Thabit's method seems to dry up and has not produced any other amicable pairs.

16

Pythagoras (died about 500 B.C.) was born on the Greek island of Samos. He founded his famous school at the Greek port of Crotona (now in southern Italy) and discovered the *Pythagoras Theorem*, namely that $a^2 + b^2 = c^2$ where a, b and c are the lengths of the two legs and of the hypotenuse of a right-angled triangle, respectively. The Pythagoreans believed that *Everything is Number*. Because of their fascination with natural numbers, the Pythagoreans made many discoveries in number theory, and in particular, they studied perfect numbers and amicable pairs for the mystical properties they felt these numbers possessed.

Euler[17] was the first to study amicable numbers *systematically.* Based on Thabit's work, he developed several new methods for generating amicable numbers and found 59 new amicable pairs. Since Euler's time, many more amicable pairs have been found, most of them with the help of variations of Euler's methods. The following rule developed by Euler is directly based on Thabit's rule:

Theorem 1.4.12 (Euler's rule for amicable pairs). Let n be a positive number, and choose $0 < x < n$ such that $g = 2^{n-x} + 1$. If

$$\left.\begin{aligned} p &= 2^x \cdot g - 1 \\ q &= 2^n \cdot g - 1 \\ s &= 2^{n+x} \cdot g^2 - 1 \end{aligned}\right\} \tag{1.136}$$

are all primes, then

$$(M,\ N) = (2^n \cdot p \cdot q,\ \ 2^n \cdot s) \tag{1.137}$$

is an amicable pair.

It is clear that Euler's rule is a generalization of Thabit's rule. That is, when $n - x = 1$, it reduces to Thabit's rule. There are many rules (although none of them are general) for generating amicable pairs; interested readers may wish to verify that if

$$\left.\begin{aligned} f &= 2^k + 1 \\ g &= 2^{m-k} \cdot f^2 \\ r_1 &= f \cdot 2^{m-k} - 1 \\ r_2 &= f \cdot 2^m - 1 \\ p &= g\left(2^{m+1} - 1\right) + 1 \\ q_1 &= p^n\left[g \cdot (2^m - 1) + 2\right] - 1 \\ q_2 &= 2^m \cdot p^n \cdot g\left[(2^m - 1)g + 2\right] - 1 \end{aligned}\right\} \tag{1.138}$$

17

Leonhard Euler (1707–1783), a key figure in 18th century mathematics, was the son of a minister from the vicinity of Basel, Switzerland, who, besides theology, also studied mathematics. He spent most of his life in the Imperial Academy in St. Petersburg,. Russia (1727–1741 and 1766–1783). "Prolific" is the word most often applied to Euler, from whom gushed forth a steady flow of work from the age of 19 on, even though he was blind for the last 17 years of his life. (He also had 13 children.) Mainly known for his work in analysis, Euler wrote a calculus textbook and introduced the present-day symbols for e, ϕ and i. Among Euler's discoveries in number theory is the law of quadratic reciprocity, which connects the solvability of the congruences $x^2 \equiv p \pmod q$ and $y^2 \equiv q \pmod p$, where p and q are distinct primes, although it remained for Gauss to provide the first proof. Euler also gave a marvellous proof of the existence of infinitely many primes based on the divergence of the harmonic series $\sum n^{-1}$.

are all primes (where $k, m, n \in \mathbb{N}$ and $m > k$) , then

$$(M,\ N) = (2^m \cdot p^n \cdot r_1 \cdot r_2 \cdot q_1,\ \ 2^m \cdot p^n \cdot q_2) \tag{1.139}$$

is an amicable pair.

It is interesting to note that although we do not know whether or not there exist infinitely many amicable pairs, we do have some methods which can be used to generate new amicable pairs from old ones; the following is one of the very successful methods invented by Te Riele[18] [203] in 1983:

Theorem 1.4.13. Let $(M',\ N') = (a \cdot u,\ a \cdot p)$ be a given amicable pair (called a breeder pair) with $\gcd(a, u) = \gcd(a, p) = 1$, where p is a prime. If a pair of primes (r, s), with $p < r < s$ and $\gcd(a,\ r \cdot s) = 1$, exists, satisfying the following bilinear Diophantine equation

$$(r - p)(s - p) = \frac{\sigma(a)}{a} \cdot \sigma(u)^2 \tag{1.140}$$

and if a third prime q exists, with $\gcd(a \cdot u,\ q) = 1$ and

$$q = r + s + u \tag{1.141}$$

then $(M,\ N) = (a \cdot u \cdot q,\ a \cdot r \cdot s)$ is also an amicable pair.

Proof. See pages 170–172 in [261]. □

Very surprisingly we are in trouble as soon as $k = 3$, for no one has yet come up with an example, and this in spite of the fact that an algorithm (Borho [31]) exists which purports to produce them! This algorithm generates the following four numbers:

$$\left.\begin{aligned} p &= 2^v - 1 \\ p_1 &= \frac{(2^{v+1} - 1)(2^u - 1) + 2^{u-v}(2^{u+1} - 1)}{p} \\ p_2 &= 2^v(p_1 + 1) - 1 \\ p_3 &= 2^v(2^{v+1} - 1) + 2^{u+1} - 1 \end{aligned}\right\} \tag{1.142}$$

where $v, u \in \mathbb{N}$, $u \geq v$ and $2u + 1 \equiv 0 \pmod{v}$. If p, p_1, p_2, p_3 are all primes, then

[18] Herman J. J. te Riele, a leading computational number theorist, is a senior scientist at the Centre for Mathematics and Computer Science (CWI) in Amsterdam, the Netherlands. Te Riele works in several central areas of computational number theory and has made significant contributions to the field; he jointly with A. M. Odlyzko at AT&T, showed in 1985 that Mertens's conjecture was false. (Mertens conjectured that $|M(x)| < \sqrt{x}$ for all $x > 1$, where $M(x) = \sum_{n \leq x} \mu(n)$.) This question, which was in the minds of many classical number theorists, including Stieltjes and Hadamard, was very important to settle. Together with the German mathematician W. Borho, he has discovered more amicable pairs and rules that generate amicable pairs than anyone else.

$$(m_1,\ m_2,\ m_3) = (2^v \cdot p \cdot p_1,\ 2^v \cdot p_2,\ 2^u \cdot p_3) \tag{1.143}$$

is an aliquot 3-cycle. Unfortunately, these four numbers don't seem to want to play! Nevertheless it is conjectured that aliquot 3-cycles exist. Readers who are interested in perfect, amicable and sociable numbers are invited to consult Yan [261] for more information.

1.4.4 Functions $\phi(n)$, $\lambda(n)$ and $\mu(n)$

Let us first introduce Euler's (totient) ϕ-function, attributed to Euler.

Definition 1.4.6. Let n be a positive integer. *Euler's (totient) ϕ-function*, $\phi(n)$, is defined to be the number of positive integers k less than n which are relatively prime to n:

$$\phi(n) = \sum_{\substack{1 \le k < n \\ \gcd(k,n)=1}} 1. \tag{1.144}$$

Example 1.4.6. By Definition 1.4.6, we have:

n	1	2	3	4	5	6	7	8	9	10	100	101	102	103
$\phi(n)$	1	1	2	2	4	2	6	4	6	4	40	100	32	102

Lemma 1.4.2. For any positive integer n,

$$\sum_{d|n} \phi(d) = n. \tag{1.145}$$

Proof. Let n_d denote the number of elements in the set $\{1, 2, \cdots, n\}$ having a greatest common divisor of d and n. Then

$$n = \sum_{d|n} n_d = \sum_{d|n} \phi\left(\frac{n}{d}\right) = \sum_{d|n} \phi(d).$$

□

Theorem 1.4.14. Let n be a positive integer. Then

(1) Euler's ϕ-function is multiplicative, that is, if $\gcd(m, n) = 1$, then

$$\phi(mn) = \phi(m)\phi(n). \tag{1.146}$$

(2) If n is a prime, say p, then

$$\phi(p) = p - 1. \tag{1.147}$$

(Conversely, if p is a positive integer with $\phi(p) = p - 1$, then p is prime.)

(3) If n is a prime power p^α with $\alpha > 1$, then

$$\phi(p^\alpha) = p^\alpha - p^{\alpha-1}. \tag{1.148}$$

(4) if n is a composite and has the standard prime factorization form, then

$$\begin{aligned}\phi(n) &= p_1^{\alpha_1}\left(1-\frac{1}{p_1}\right)p_2^{\alpha_2}\left(1-\frac{1}{p_2}\right)\cdots p_k^{\alpha_k}\left(1-\frac{1}{p_k}\right)\\ &= n\prod_{p|n}\left(1-\frac{1}{p}\right).\end{aligned} \tag{1.149}$$

Proof.

(1) Since $g(n) = n$ is multiplicative and $n = \sum_{d|n} \phi(n)$, it follows from Theorem 1.4.3 that the ϕ-function is multiplicative.

(2) If n is prime, then $1, 2, \cdots, n-1$ are relatively prime to n, so it follows from the definition of Euler's ϕ-function that $\phi(n) = n - 1$. Conversely, if n is not prime, n has a divisor d such that $\gcd(d, n) \neq 1$. Thus, there is at least one positive integer less than n that is not relatively prime to n, and hence $\phi(n) \leq n - 2$.

(3) Note that $\gcd(n, p^\alpha) = 1$ if and only if $p \nmid n$. There are exactly $p^{\alpha-1}$ integers between 1 and p^α divisible by p, namely,

$$p,\ 2p,\ 3p, \cdots,\ (p^{\alpha-1})p.$$

Thus, the set $\{1, 2, \cdots, p^\alpha\}$ contains exactly $p^\alpha - p^{\alpha-1}$ integers that are relatively prime to p^α, and so by the definition of the ϕ-function, $\phi(p^\alpha) = p^\alpha - p^{\alpha-1}$.

(4) We use mathematical induction on k, the number of distinct prime factors. By Part (3) of this theorem, the result is true for $k = 1$. Suppose that the result is true for $k = i$. Since

$$\gcd\left(p_1^{\alpha-1}p_2^{\alpha_2}\cdots p_i^{\alpha_i},\ p_{i+1}^{\alpha_{i+1}}\right) = 1,$$

the definition of multiplicative function gives

$$\begin{aligned}&\phi\left(\left(p_1^{\alpha-1}p_2^{\alpha_2}\cdots p_i^{\alpha_i}\right)p_{i+1}^{\alpha_{i+1}}\right)\\ =\ &\phi\left(p_1^{\alpha-1}p_2^{\alpha_2}\cdots p_i^{\alpha_i}\right)\phi\left(p_{i+1}^{\alpha_{i+1}}\right)\\ =\ &\phi\left(p_1^{\alpha-1}p_2^{\alpha_2}\cdots p_i^{\alpha_i}\right)\left(p_{i+1}^{\alpha_{i+1}} - p_{i+1}^{\alpha_{i+1-1}}\right)\\ =\ &\left(p_1^{\alpha_1} - p_1^{\alpha_1-1}\right)\left(p_2^{\alpha_2} - p_2^{\alpha_2-1}\right)\cdots\left(p_i^{\alpha_i} - p_i^{\alpha_i-1}\right)\left(p_{i+1}^{\alpha_{i+1}} - p_{i+1}^{\alpha_{i+1-1}}\right)\\ =\ &p_1^{\alpha_1}\left(1-\frac{1}{p_1}\right)p_2^{\alpha_2}\left(1-\frac{1}{p_2}\right)\cdots p_{i+1}^{\alpha_{i+1}}\left(1-\frac{1}{p_{i+1}}\right)\\ =\ &n\left(1-\frac{1}{p_1}\right)\left(1-\frac{1}{p_2}\right)\cdots\left(1-\frac{1}{p_{i+1}}\right).\end{aligned}$$

So, the result holds for all positive integer k. □

Remark 1.4.1. Suppose that n is known to be the product of two distinct primes p and q. Then knowledge of p and q is equivalent to knowledge of $\phi(n)$, since $\phi(n) = (p-1)(q-1)$. However, there is no known efficient method to compute $\phi(n)$ if the prime factorization of n is not known. More precisely, one can compute $\phi(n)$ from p and q in $\mathcal{O}(\log n)$ bit operations, and one can compute p and q from n and $\phi(n)$ in $\mathcal{O}(\log n)^3$ bit operations (see Koblitz [128]). This interesting fact is useful in the RSA public-key cryptography, which will be studied in detail in Chapter 3.

The following function, first proposed by the American mathematician Carmichael[19], is a very useful number theoretic function.

Definition 1.4.7. Carmichael's λ-function, $\lambda(n)$, is defined as follows:

$$\left.\begin{array}{ll}
\lambda(p) = \phi(p) = p-1 & \text{for prime } p, \\
\lambda(p^\alpha) = \phi(p^\alpha) & \text{for } p=2 \text{ and } \alpha \le 2, \\
 & \text{and for } p \ge 3 \\
\lambda(2^\alpha) = \dfrac{1}{2}\phi(2^\alpha) & \text{for } \alpha \ge 3 \\
\lambda(n) = \operatorname{lcm}\left(\lambda(p_1^{\alpha_1})\lambda(p_2^{\alpha_2})\cdots\lambda(p_k^{\alpha_k})\right) & \text{if } n = \prod_{i=1}^{k} p_i^{\alpha_i}.
\end{array}\right\} \quad (1.150)$$

Example 1.4.7. By Definition 1.4.7, we have:

n	1	2	3	4	5	6	7	8	9	10	100	101	102	103
$\lambda(n)$	1	1	2	2	4	2	6	2	6	4	20	100	16	102

Example 1.4.8. Let $n = 65520 = 2^4 \cdot 3^2 \cdot 5 \cdot 7 \cdot 13$, and $a = 11$. Then $\gcd(65520, 11) = 1$ and we have

$$\begin{aligned}
\phi(65520) &= 8 \cdot 6 \cdot 4 \cdot 6 \cdot 12 = 13824, \\
\lambda(65520) &= \operatorname{lcm}(4, 6, 4, 6, 12) = 12.
\end{aligned}$$

Euler's ϕ-function and Carmichael's λ-function are two very useful arithmetic functions particularly in public-key cryptography which we shall discuss in Chapter 3 of this book; some important properties about Euler's ϕ-function and Carmichael's λ-function will be discussed in Subsection 1.6.2.

[19] Robert D. Carmichael (1879–1967) was born in Goodwater, Alabama. He received his BA from Lineville College in 1898 and his PhD in 1911 from Princeton University. His thesis, written under G. D. Birkhoff, was considered the first significant American contribution to differential equations. Perhaps best known in number theory for his *Carmichael numbers*, *Carmichael's function*, and *Carmichael's theorem*, Carmichael worked in a wide range of areas, including real analysis, differential equations, mathematical physics, group theory and number theory. It is also worthwhile mentioning that Carmichael published two very readable little books about number theory: *Theory of Numbers* in 1914 and *Diophantine Analysis* in 1915, both published by John Wiley & Sons, New York.

Now we move on to another important arithmetic function, the Möbius function, named after A. F. Möbius[20].

Definition 1.4.8. Let n be a positive integer. Then the *Möbius μ-function*, $\mu(n)$, is defined as follows:

$$\mu(n) = \begin{cases} 1, & \text{if } n = 1, \\ 0, & \text{if } n \text{ contains a squared factor}, \\ (-1)^k, & \text{if } n = p_1 p_2 \cdots p_k \text{ is the product of} \\ & \quad k \text{ distinct primes.} \end{cases} \tag{1.151}$$

Example 1.4.9. By Definition 1.151, we have:

n	1	2	3	4	5	6	7	8	9	10	100	101	102
$\mu(n)$	1	−1	−1	0	−1	1	−1	0	0	1	0	−1	−1

Theorem 1.4.15. Let $\mu(n)$ be the Möbius function. Then

(1) $\mu(n)$ is multiplicative, i.e., for $\gcd(m, n) = 1$,

$$\mu(mn) = \mu(m)\mu(n). \tag{1.152}$$

(2) Let

$$\nu(n) = \sum_{d|n} \mu(d). \tag{1.153}$$

Then

$$\nu(n) = \begin{cases} 1, & \text{if } n = 1, \\ 0, & \text{if } n > 1. \end{cases} \tag{1.154}$$

Proof.

(1) If either $p^2 \mid m$ or $p^2 \mid n$, p is a prime, then $p^2 \mid mn$. Hence, $\mu(mn) = 0 = \mu(m)\mu(n)$. If both m and n are square-free integers, say, $m = p_1 p_2 \cdots p_s$ and $n = q_1 q_2 \cdots q_t$, then

[20]

Augustus Ferdinand Möbius (1790–1868) was born in Schilpforta in Prussia. Möbius studied mathematics at Leipzig, Halle and finally at Göttingen with Gauss. He became a lecturer at Leipzig in 1815 and Professor in 1844; he held the post there until his death. Möbius is perhaps best known for his work in topology, especially for his conception of the Möbius strip, a two dimensional surface with only one side. He is also well-known for proposing the colouring of maps in 1840, which led to the famous *four-colouring problem.*

$$\begin{aligned}\mu(mn) &= \mu(p_1 p_2 \cdots p_s q_1 q_2 \cdots q_t)\\ &= (-1)^{s+t}\\ &= (-1)^s(-1)^t\\ &= \mu(m)\mu(n)\end{aligned}$$

(2) If $n = 1$, then $\nu(1) = \sum_{d|n} \nu(d) = \mu(1) = 1$. If $n > 1$, since $\nu(n)$ is multiplicative, we need only evaluate ν on prime to powers. In addition, if p is prime,

$$\begin{aligned}\nu(p^\alpha) &= \sum_{d|p^\alpha}\\ &= \mu(1) + \mu(p) + \mu(p^2) + \cdots + \mu(p^\alpha)\\ &= 1 + (-1) + 0 + \cdots + 0\\ &= 0.\end{aligned}$$

Thus, $\nu(n) = 0$ for any positive integer n greater than 1. □

The importance of the Möbius function lies in the fact that it plays an important role in the inversion formula given in the following theorem. The formula involves a general arithmetic function f which is not necessarily multiplicative.

Theorem 1.4.16 (The Möbius inversion formula). If f is any arithmetic function and if

$$g(n) = \sum_{d|n} f(d), \tag{1.155}$$

then

$$f(n) = \sum_{d|n} \mu\left(\frac{n}{d}\right) g(d) = \sum_{d|n} \mu(d)\, g\left(\frac{n}{d}\right). \tag{1.156}$$

Proof. If f is an arithmetic function and $g(n) = \sum_{d|n} f(d)$. Then

$$\begin{aligned}\sum_{d|n} \mu(d)\, g\left(\frac{n}{d}\right) &= \sum_{d|n} \mu(d) \sum_{a|(n/d)} f(a)\\ &= \sum_{d|n} \sum_{a|(n/d)} \mu(d) f(a)\\ &= \sum_{d|n} \sum_{a|(n/d)} f(a)\mu(d)\\ &= \sum_{d|n} f(a) \sum_{a|(n/d)} \mu(d)\\ &= f(n) \cdot 1\\ &= f(n).\end{aligned}$$

The converse of Theorem 1.4.16 is also true and can be stated as follows:

Theorem 1.4.17 (The converse of the Möbius inversion formula). If

$$f(n) = \sum_{d|n} \mu\left(\frac{n}{d}\right) g(d), \tag{1.157}$$

then

$$g(n) = \sum_{d|n} f(d). \tag{1.158}$$

Note that the functions τ and σ

$$\tau(n) = \sum_{d|n} 1 \quad \text{and} \quad \sigma(n) = \sum_{d|n} d$$

may be inverted to give

$$1 = \sum_{d|n} \mu\left(\frac{n}{d}\right) \tau(d) \quad \text{and} \quad n = \sum_{d|n} \mu\left(\frac{n}{d}\right) \sigma(d)$$

for all $n \geq 1$. The relationship between Euler's *phi*-function and Möbius' μ-function is given by the following theorem.

Theorem 1.4.18. For any positive integer n,

$$\phi(n) = n \sum_{d|n} \frac{\mu(d)}{d}. \tag{1.159}$$

Proof. By applying Möbius inversion formula to

$$g(n) = n = \sum_{d|n} \phi(d)$$

we get

$$\begin{aligned} \phi(n) &= \sum_{d|n} \mu(d)\, g\left(\frac{n}{d}\right) \\ &= \frac{\sum_{d|n} \mu(d)\, n}{d}. \end{aligned}$$

□

1.5 Distribution of Prime Numbers

It will be another million years, at least, before we understand the primes.

PAUL ERDÖS (1913–1996)

As mentioned earlier, prime numbers are building blocks of positive integers. In fact, the theory of numbers is essentially the theory of prime numbers. In this section, we shall introduce some important results about the distribution of prime numbers. More specifically, we shall study some functions of a *real* or a *complex* variable that are related to the distribution of prime numbers.

1.5.1 Prime Distribution Function $\pi(x)$

Let us first investigate the occurrence of the prime numbers among the positive integers. The following are some counting results of the number of primes in each hundred positive integers:

(1) Each 100 from 1 to 1000 contains respectively the following number of primes:

$$25,\ 21,\ 16,\ 16,\ 17,\ 14,\ 16,\ 14,\ 15,\ 14.$$

(2) For each 100 from 10^6 to $10^6 + 1000$, the corresponding sequences are:

$$6,\ 10,\ 8,\ 8,\ 7,\ 7,\ 10,\ 5,\ 6,\ 8.$$

(3) For each 100 from 10^7 to $10^7 + 1000$, the corresponding sequences are:

$$2,\ 6,\ 6,\ 6,\ 5,\ 4,\ 7,\ 10,\ 9,\ 6.$$

(4) For each 100 from 10^{12} to $10^{12} + 1000$, the corresponding sequences are:

$$4,\ 6,\ 2,\ 4,\ 2,\ 4,\ 3,\ 5,\ 1,\ 6.$$

Except 2 and 3, any two consecutive primes must have a distance that is at least equal to 2. Pairs of primes with this shortest distance are called twin primes. Of the positive integers ≤ 100, there are eight twin primes, namely,

$$(3,5),\ (5,7),\ (11,13),\ (17,19),\ (29,31),\ (41,43),\ (59,61),\ (71,73).$$

In spite of the seemingly frequent occurrence of twin primes, there are however arbitrarily long distances between two consecutive primes, that is, there are arbitrarily long sequences of consecutive composite numbers. To prove this, one needs only to observe that for an arbitrary positive integer $n > 1$, the following $n - 1$ numbers

$$n! + 2, \ n! + 3, \ n! + 4, \ \cdots, \ n! + n.$$

are all composite numbers. The above investigations show that the occurrence of primes among positive integers is very irregular. However, when the large-scale distribution of primes is considered, it appears in many ways quite regular and obeys simple laws. In the study of these laws, a central question is: "How many primes are there less than or equal to x"? The answer to this question leads to a famous expression, $\pi(x)$, which is defined as follows.

Definition 1.5.1. Let x be a positive real number > 1. Then $\pi(x)$, is defined as follows:

$$\pi(x) = \sum_{\substack{p \leq x \\ p \text{ prime}}} 1. \tag{1.160}$$

That is, $\pi(x)$ is the number of primes less than or equal to x; it is also called the *prime counting function* (or the *prime distribution function*).

Example 1.5.1. The prime numbers up to 100 are:

$$2, 3, 5, 7, 11, 13, 17, 19, 23, 29, 31, 37, 41, 43,$$
$$47, 53, 59, 61, 67, 71, 73, 79, 83, 89, 97.$$

Thus we have

$$\begin{array}{lllll} \pi(1) = 0, & \pi(2) = 1, & \pi(3) = 2, & \pi(10) = 4, & \pi(20) = 8, \\ \pi(30) = 10, & \pi(40) = 12, & \pi(50) = 15, & \pi(75) = 21, & \pi(100) = 25. \end{array}$$

A longer table of values of $\pi(x)$ can be found in Table 1.9.

The numerical values of the ratio of $\pi(x)/x$ in Table 1.9 suggest (in fact it is not difficult to prove) that

$$\lim_{x \to \infty} \frac{\pi(x)}{x} = 0. \tag{1.161}$$

That is, almost all the positive integers are composite numbers. It must be, however, pointed out that even though almost all positive integers are composites, there are infinitely many prime numbers, as proved by Euclid 2000 years ago. So, in terms of $\pi(x)$, Euclid's theorem on the infinitude of prime numbers can then be re-formulated as follows:

$$\lim_{x \to \infty} \pi(x) = \infty. \tag{1.162}$$

The asymptotic behaviour of $\pi(x)$ has been studied extensively by many of the world's greatest mathematicians beginning with Legendre in 1798 and culminating in 1899 when de la Valleé-Poussin proved that for some constant $c > 0$,

$$\pi(x) = \int_2^x \frac{\mathrm{d}t}{\ln t} + \mathcal{O}\left(x \exp\left\{-c\sqrt{\ln x}\right\}\right). \tag{1.163}$$

Table 1.9. Table of values of $\pi(x)$

x	$\pi(x)$	$\pi(x)/x$
10	4	0.4
10^2	25	0.25
10^3	168	0.168
10^4	1229	0.1229
10^5	9592	0.09592
10^6	78498	0.078498
10^7	664579	0.0664579
10^8	5761455	0.05761455
10^9	50847534	0.050847534
10^{10}	455052511	0.04550525110
10^{11}	4118054813	0.04118054813
10^{12}	37607912018	0.037607912018
10^{13}	346065536839	0.0346065536839
10^{14}	3204941750802	0.03204941750802
10^{15}	29844570422669	0.029844570422669
10^{16}	279238341033925	0.0279238341033925
10^{17}	2625557157654233	0.02625557157654233
10^{18}	24739954287740860	0.02473995428774086
10^{19}	234057667276344607	0.0234057667276344607
10^{20}	2220819602560918840	0.02220819602560918840
10^{21}	21127269486018731928	0.021127269486018731928
10^{22}	201467286689315906290	0.0201467286689315906290

Note that the big-$\mathcal{O}$ notation used above was first introduced by German mathematician Edmund Landau. Intuitively, f is $\mathcal{O}(g)$ if there is a real positive constant k such that $f(x) < k \cdot g(x)$ for all sufficiently large x. The big-$\mathcal{O}$ notation is very useful in computational complexity, and we shall use it throughout the book.

In the next few subsections, we shall study the asymptotic behaviour of $\pi(x)$. More specifically, we shall study the approximations of $\pi(x)$ by the functions $\frac{x}{\ln x}$, $\text{Li}(x)$ and $R(x)$.

1.5.2 Approximations of $\pi(x)$ by $x/\ln x$

Although the distribution of primes among the integers is very irregular, the prime distribution function $\pi(x)$ is surprisingly well behaved. Let us first study the approximation $\frac{x}{\ln x}$ to $\pi(x)$. Table 1.10 gives the values of $\pi(x)$, $\frac{x}{\ln x}$ and $\frac{\pi(x)}{x/\ln x}$, for $x = 10, 10^2, 10^2, \cdots, 10^{20}$. It can be easily seen from Table 1.10 that the approximation $x/\ln x$ gives reasonably accurate estimates of

Table 1.10. Approximations to $\pi(x)$ by $x/\ln x$

x	$\pi(x)$	$\frac{x}{\ln x}$	$\frac{\pi(x)}{x/\ln x}$
10^1	4	$4.3\cdots$	$0.93\cdots$
10^2	25	$21.7\cdots$	$1.152\cdots$
10^3	168	$144.8\cdots$	$1.16\cdots$
10^4	1229	$1085.7\cdots$	$1.13\cdots$
10^5	9592	$8685.8\cdots$	$1.131\cdots$
10^6	78498	$72382.5\cdots$	$1.084\cdots$
10^7	664579	$620420.5\cdots$	$1.071\cdots$
10^8	5761455	$5428680.9\cdots$	$1.061\cdots$
10^9	50847534	$48254942.5\cdots$	$1.053\cdots$
10^{10}	455052511	$434294481.9\cdots$	$1.047\cdots$
10^{11}	4118054813	$3948131653.7\cdots$	$1.043\cdots$
10^{12}	37607912018	$36191206825.3\cdots$	$1.039\cdots$
10^{13}	346065536839	$334072678387.1\cdots$	$1.035\cdots$
10^{14}	3204941750802	$3102103442166.0\cdots$	$1.033\cdots$
10^{15}	29844570422669	$28952965460216.8\cdots$	$1.030\cdots$
10^{16}	279238341033925	$271434051189532.4\cdots$	$1.028\cdots$
10^{17}	2625557157654233	$2554673422960304.8\cdots$	$1.027\cdots$
10^{18}	24739954287740860	$24127471216847323.8\cdots$	$1.025\cdots$
10^{19}	234057667276344607	$228576043106974646.1\cdots$	$1.023\cdots$
10^{20}	2220819602560918840	$2171472409516259138.2\cdots$	$1.022\cdots$
10^{21}	21127269486018731928	$20680689614440563221.4\cdots$	$1.021\cdots$
10^{22}	201467286689315906290	$197406582683296285295.9\cdots$	$1.020\cdots$
$4\cdot 10^{22}$	783964159852157952242	$768592742555118350978.9\cdots$	$1.019\cdots$

$\pi(x)$. In fact, the study of this approximation leads to the following *famous* theorem of number theory, and indeed of all mathematics.

Theorem 1.5.1 (Prime Number Theorem). $\pi(x)$ is asymptotic to $\frac{x}{\ln x}$. That is,

$$\lim_{x\to\infty} \frac{\pi(x)}{x/\ln x} = 1. \tag{1.164}$$

The Prime Number Theorem (PNT) was postulated by Gauss[21] in 1792 on numerical evidence. It is known that Gauss constructed by hand a table of all primes up to three million, and investigated the number of primes occurring in each group of 1000. Note that it was also conjectured by Legendre[22] before Gauss, in a different form, but of course both Legendre and Gauss were unable to prove the PNT.

The first serious attempt (after Gauss) to study the function $\pi(x)$ was due to Legendre, who used the sieve of Eratosthenes and proved in 1808 that

$$\pi(n) = \pi(\sqrt{n}) - 1 + \sum \mu(d) \left\lfloor \frac{n}{d} \right\rfloor \tag{1.165}$$

where the sum is over all divisors d of the product of all primes $p \leq n$, and $\mu(d)$ is the Möbius function. Legendre also conjectured in 1798 and again in 1808 that

$$\pi(x) = \frac{x}{\ln x - A(x)}, \tag{1.166}$$

21

Carl Friedrich Gauss (1777–1855), the greatest mathematician of all time (Prince of Mathematicians), was the son of a German bricklayer. It was quickly apparent that he was a child prodigy. In fact, at the age of three he corrected an error in his father's payroll, and at the age of seven, he can quickly calculate $1 + 2 + 3 + \cdots + 100 = 5050$ because $50(1 + 100) = 5050$. Gauss made fundamental contributions to astronomy including calculating the orbit of the asteroid Ceres. On the basis of this calculation, Gauss was appointed Director of the Göttingen Observatory. He laid the foundations of modern number theory with his book *Disquisitiones Arithmeticae* in 1801. Gauss conceived most of his discoveries before the age of 20, but spent the rest of his life polishing and refining them.

22

Adrien-Marie Legendre (1752–1833), a French mathematician who, with Lagrange and Laplace, formed a trio associated with the period of the French Revolution. Legendre was educated at Collège Mazarin in Paris and was Professor of Mathematics at École Militaire Mazarin in Paris for five years. He resigned to devote more time to his research. In 1782, he won a prize offered by the Berlin Academy with a paper in ballistics. Legendre gave the first proof that every prime has a primitive root. He was also the first to determine the number of representations of an integer as a sum of two squares and proved that every odd positive integer which is not of the form $8k+7$ is a sum of three squares. Legendre conjectured the Prime Number Theorem and the Law of Quadratic Reciprocity but of course unable to prove them. In his later years, Legendre's investigations focussed on elliptic integrals. At the age of 75, Legendre proved the Fermat Last Theorem for $n = 5$. It was unfortunate that Legendre lived in the era of Lagrange and Gauss and received less recognition than he deserved.

where $\lim_{x\to\infty} A(x) = 1.08366\cdots$. It was shown 40 years later by Chebyshev[23] that if $\lim_{x\to\infty} A(x)$ exists, it must be equal to 1 (see Ribenboim [200]). It is also interesting to note that around 1850 (about 50 years before the Prime Number Theorem was proved), Chebyshev showed that

$$0.92129\frac{x}{\ln x} < \pi(x) < 1.1056\frac{x}{\ln x} \tag{1.167}$$

for large x. Chebyshev's result was further refined by Sylvester in 1892 to

$$0.95695\frac{x}{\ln x} < \pi(x) < 1.04423\frac{x}{\ln x} \tag{1.168}$$

for every sufficiently large x. Chebyshev also worked with the function $\theta(x)$, defined by

$$\theta(x) = \sum_{p\leq x} \ln p \tag{1.169}$$

now called Chebyshev's function, which is closely related to $\pi(x)$. That is,

Theorem 1.5.2.

$$\lim_{x\to\infty} \frac{\theta(x)}{x} = 1. \tag{1.170}$$

Note that the summatory function of $\Lambda(n)$ defined in (1.177), denoted by $\psi(x)$, is easily expressible in terms of Chebyshev's θ-function

$$\psi(x) = \theta(x) + \theta(x^{1/2}) + \theta(x^{1/2}) + \cdots. \tag{1.171}$$

The Prime Number Theorem may then be rephrased as follows:

Theorem 1.5.3.

$$\lim_{x\to\infty} \frac{\psi(x)}{x} = 1. \tag{1.172}$$

It can be seen that Chebyshev came rather close to the Prime Number Theorem; however, the complete proof of the PNT had to wait for about 50

---

[23]

Pafnuty Lvovich Chebyshev (1821–1894), was a Russian mathematician and founder of a notable school of mathematicians in St Petersburg. He made St Petersburg for the second time, after Euler, a world centre of mathematics. He contributed to several branches of mathematics and his name is remembered in results in algebra, analysis and mathematical probability. In number theory, he proved, among many other things, Bertrand's postulate that, if $n \in \mathbb{N}$, then there is at least one prime p such that $n < p \leq 2n$. Chebyshev was appointed in 1847 to the University of St Petersburg, became a foreign associate of the Institut de France in 1874 and also a foreign Fellow of the Royal Society, London.

years more. During this time, Riemann[24] had the idea of defining the zeta function for complex numbers s having real part greater than 1, namely,

$$\zeta(s) = \sum_{n=1}^{\infty} \frac{1}{n^s} \tag{1.173}$$

(we shall return to the zeta function soon), and attempted to give a proof of the prime number Theorem using the zeta function. Although Riemann's proof was not adequate but contained the ideas essential for a complete proof. The theorem was established in 1896 independently by two eminent mathematical analysts: Jacques Hadamard[25] and the Belgian mathematician De la Vallée-Poussin[26] independently proved the theorem. Since Euclid discovered 2000 years ago that *"there are infinitely many prime numbers"*, thousands of

---

[24]

Georg Friedrich Bernhard Riemann (1826–1866), the son of a minister, was born in Breselenz, Germany. Riemann was a major figure in 19th century mathematics, somewhat the father of modern *analytic number theory*, and the last of the famous trilogy at Göttingen (the other two were Gauss and Dirichlet). In many ways, Riemann was the intellectual successor of Gauss (Riemann did his PhD at Göttingen under Gauss). In geometry, he started the development of those tools which Einstein would eventually use to describe the universe and which in the 20th century would be turned into the theory of manifolds. He also made fundamental contributions to analysis, in which his name is preserved in the Riemann integral, the Riemann sum, the Cauchy–Riemann equations and Riemann surfaces. Riemann only wrote one paper on number theory, but this paper had tremendous impact on the development of the Prime Number Theorem; it was in this paper that Riemann provided a foundation of modern *analytic number theory*. Riemann died of tuberculosis at the early age of 40.

[25]

Jacques Hadamard (1865–1963) was born in Versailles, France. He was good at all subjects at school except mathematics; he wrote in 1936 that "in arithmetic, until the seventh grade, I was last or nearly last". A good mathematics teacher happened to turn him towards mathematics and changed his life. Hadamard made important contributions to complex analysis, functional analysis and partial differential equations of mathematical physics. His proof of the *Prime Number Theorem* was based on his work in complex analysis. Hadamard was also a famous teacher; he taught at a Paris secondary school and wrote numerous articles on elementary mathematics for schools.

[26]

Charles-Jean de la Vallée-Poussin (1866–1962) was born in Louvain, Belgium. He proved the *Prime Number Theorem* independently of Hadamard in 1896. He also extended this work and established results about the distribution of arithmetic progressions of prime numbers, and refined the *Prime Number Theorem* to include error estimates. Notice that both Hadamard and De la Vallée-Poussin lived well into their 90's (Hadamard 98, and De la Vallée-Poussin 96); it is a common belief among mathematicians

theorems about prime numbers have been discovered; many are significant, some are beautiful, but only *this* serious theorem is called the *Prime Number Theorem (PNT).*

The mathematicians of the 19th century were somewhat disturbed by the use of complex analysis to prove the PNT; for example, in their proofs of the PNT, both Hadamard and De la Vallée-Poussin used very complicated analytical methods. Mathematicians attempted for a long time to give an *elementary* proof of the PNT. This was first achieved by Atle Selberg[27] in 1949, whose proof used only elementary estimates of arithmetic functions such as

$$\sum_{p\leq x}(\ln p)^2 + \sum_{pq\leq x}\ln p \ln q = 2x\ln x + \mathcal{O}(x), \tag{1.174}$$

where p and q are primes (the above estimate was given by Selberg in 1949). Soon after, using also a variant of Selberg's estimate

$$\frac{\psi(x)}{x} + \frac{1}{\ln x}\sum_{n\leq x}\frac{\psi(x/n)}{x/n}\frac{\Lambda(n)}{n} = 2 + \mathcal{O}\left(\frac{1}{\ln x}\right), \tag{1.175}$$

where $\Lambda(n)$ is the von Mangoldt function defined by

$$\Lambda(n) = \begin{cases} \ln p & \text{if } n = p^k \text{ is prime power} \\ 0 & \text{otherwise,} \end{cases} \tag{1.176}$$

and $\psi(x)$ is the summatory function of $\Lambda(n)$

$$\psi(x) = \sum_{n\leq x}\frac{\Lambda(n)}{\ln n}, \tag{1.177}$$

that anyone who produces a proof of the *Prime Number Theorem* is guaranteed longevity!

27

Atle Selberg (1917–), is a Norwegian mathematician and the 1950 Fields Medal recipient. Selberg's interest in mathematics began when he was a schoolboy. By reading about Ramanujan and Ramanujan's collected papers, Selberg was not only greatly impressed by the mathematics he read but also intrigued by Ramanujan's personality. Inspired by Ramanujan's work, Selberg began to make his own mathematical explorations and made significant contributions to the theory of numbers, particularly the Riemann zeta function. Selberg is perhaps best known for his elementary proof of the prime number theorem. He has been a permanent member of the Institute for Advanced Study at Princeton since 1949 and is currently Professor Emeritus in the Institute.

Paul Erdös[28] gave, with a different elementary method, his proof of the prime number theorem. (It was planned to write a joint paper between Selberg and Erdös, but for some reason this did not happen.) These *elementary* proofs of the PNT were considered so important that Selberg got a Fields medal in 1950 and Erdös received the American Mathematical Society's Cole Prize in 1951 and the Wolf prize in 1984.

The PNT is not only an important theoretical result about prime numbers, but also a very applicable result in mathematics and computing science. For example, we can use the PNT to:

(1) Estimate the probability that a randomly chosen integer n will turn out to be prime as $1/\ln n$. Thus we would need to examine approximately $\ln n$ integers chosen randomly near n in order to find a prime that is of the same size as n; for example, to find a 1000-digit prime might require testing approximately $\ln 10^{1000} \approx 2303$ randomly chosen 1000-digit numbers for primality. Of course, this figure can be cut in half if only the odd numbers are chosen.

(2) Estimate the number of computation steps required for primality testing by trial divisions. The maximum number of divisions in the trial division test for primality of n is $\pi(\sqrt{n})$; for large n we have $\pi(\sqrt{n}) \approx \dfrac{\sqrt{n}}{\ln \sqrt{n}} = \dfrac{2\sqrt{n}}{\ln n}$. A computer which takes $(\ln n)/10^6$ seconds to perform one such division would take approximately $\dfrac{2\sqrt{n}}{\ln n} \cdot \dfrac{\ln n}{10^6} = \dfrac{2\sqrt{n}}{10^6}$ seconds to check that n was prime, provided that all the primes up to $\sqrt{n}$ were known. Using this direct method it would take more than 63 years to verify that a 30-digit number was prime. Later on, we shall introduce more efficient methods for primality testing.

28

The legendary Paul Erdös was born in Budapest, Hungary, on 26 March 1913 and died on 20 September 1996 while attending a minisemester at the Banach Mathematical Centre in Warsaw, Poland. A mathematician with no home, no wife, no job, and no permanent address, Erdös was the most versatile and prolific mathematician of our time, and indeed probably of all times. He traveled a lot around the world to meet mathematicians, to deliver lectures, and to discuss mathematical problems. He wrote about 1500 papers, about five times as many as other prolific mathematicians, co-authored with over 250 people. These people are said to have Erdős number 1. People who do not have Erdős number 1, but who have written a paper with someone who does, are said to have Erdős number 2, and so on inductively. Erdős's papers cover a broad range of topics, but the majority are in number theory, combinatorics and probability theory. (Photo by courtesy of the Mathematical Institute of the Hungarian Academy of Sciences.)

1.5.3 Approximations of $\pi(x)$ by $\mathrm{Li}(x)$

Although the expression $x/\ln x$ is a fairly simple approximation to $\pi(x)$, it is not *terribly* close (i.e., it is good, but not very good), and mathematicians have been interested in improving it. Of course, one does this at the price of complicating the approximation. For example, one can use the following much better approximation $\mathrm{Li}(x)$ to $\pi(x)$. $\mathrm{Li}(x)$ is called the logarithmic integral of x; the formal definition of the logarithmic integral $\mathrm{Li}(x)$ is as follows.

Definition 1.5.2. Let x be a positive real number greater than 1. Then

$$\mathrm{Li}(x) = \int_0^x \frac{\mathrm{d}t}{\ln t}, \tag{1.178}$$

the integral is usually interpreted as

$$\int_0^x \frac{\mathrm{d}t}{\ln t} = \lim_{\eta \to 0} \left(\int_0^{1-\eta} + \int_{1+\eta}^x \right) \frac{\mathrm{d}t}{\ln t}. \tag{1.179}$$

As illustrated in Table 1.11 (compared also with Table 1.10), the logarithmic integral $\mathrm{Li}(x)$ is indeed a much better approximation to $\pi(x)$, although for large values of x the two approximations behave asymptotically alike. Riemann and Gauss believed that $\mathrm{Li}(x) > \pi(x)$ for every $x > 3$. It is of course true in the present range of Table 1.11. However, Littlewood showed in 1914 that the difference $\mathrm{Li}(x) - \pi(x)$ changes sign infinitely often, whilst Te Riele showed in 1986 that between $6.62 \cdot 10^{370}$ and $6.69 \cdot 10^{370}$ there are more than 10^{180} successive integers x for which $\mathrm{Li}(x) < \pi(x)$.

The study of the approximation of $\pi(x)$ by $\mathrm{Li}(x)$ leads naturally to an equivalent form of the Prime Number Theorem, since

$$\lim_{x \to \infty} \frac{\mathrm{Li}(x)}{x/\ln x} = \lim_{x \to \infty} \frac{(\mathrm{Li}(x))'}{(x/\ln x)'} = \lim_{x \to \infty} \frac{1/\ln x}{1/\ln x - 1/\ln^2 x} = 1.$$

Theorem 1.5.4. $\pi(x)$ is asymptotic to $\mathrm{Li}(x)$. That is,

$$\lim_{x \to \infty} \frac{\pi(x)}{\mathrm{Li}(x)} = 1. \tag{1.180}$$

Remark 1.5.1. At the age of 15, in 1792, Gauss conjectured that

$$\pi(x) \sim \mathrm{Li}(x), \tag{1.181}$$

but Gauss used the following definition for $\mathrm{Li}(x)$

$$\mathrm{Li}(x) = \int_2^x \frac{\mathrm{d}t}{\ln t}, \tag{1.182}$$

which differs by a constant $\mathrm{Li}(2)$ from (1.178).

Table 1.11. Approximations to $\pi(x)$ by $\mathrm{Li}(x)$

x	$\pi(x)$	$\mathrm{Li}(x)$	$\frac{\pi(x)}{\mathrm{Li}(x)}$
10^3	168	178	$0.943820224719\cdots$
10^4	1229	1246	$0.986356340288\cdots$
10^5	9592	9630	$0.996053997923\cdots$
10^6	78498	78628	$0.998346644961\cdots$
10^7	664579	664918	$0.999490162696\cdots$
10^8	5761455	5762209	$0.999869147405\cdots$
10^9	50847534	50849235	$0.999966548169\cdots$
10^{10}	455052511	455055615	$0.999993178855\cdots$
10^{11}	4118054813	4118066401	$0.999997186058\cdots$
10^{12}	37607912018	37607950281	$0.999998982582\cdots$
10^{13}	346065536839	346065645810	$0.999999685114\cdots$
10^{14}	3204941750802	3204942065692	$0.999999901748\cdots$
10^{15}	29844570422669	29844571475288	$0.999999964729\cdots$
10^{16}	279238341033925	279238344248557	$0.999999988487\cdots$
10^{17}	2625557157654233	2625557165610822	$0.999999996969\cdots$
10^{18}	24739954287740860	24739954309690415	$0.999999999112\cdots$
10^{19}	234057667276344607	234057667376222382	$0.999999999573\cdots$

1.5.4 The Riemann ζ-Function $\zeta(s)$

In 1859, Bernhard Riemann astounded the mathematical world by writing an eight-page memoir on $\pi(x)$ entitled *Über die Anzahl der Primzahlen unter einer gegebenen Grösse* (*On the Number of Primes Less Than a Given Magnitude*) which is now regarded as one of the greatest classics of mathematics. In this remarkable paper, which was incidentally the only paper he ever wrote on Number Theory, Riemann related the study of prime numbers to the properties of various functions of a *complex* number. In particular, he studied the ζ-function (now widely known as the Riemann ζ-function) as a function of a complex variable, and made various conjectures about its behaviour. We shall first give the definition of the Riemann ζ-function as follows.

Definition 1.5.3. Let s be a complex variable (we write $s = \sigma + it$ with σ and t real; here $\sigma = \mathrm{Re}(s)$ is the real part of s, whereas $t = \mathrm{Im}(s)$ is the imaginary part of s). Then the Riemann ζ-function, $\zeta(s)$, is defined to be the sum of the following series

$$\zeta(s) = \sum_{n=1}^{\infty} \frac{1}{n^s}. \tag{1.183}$$

In particular,

$$\zeta(2) = \sum_{n=1}^{\infty} \frac{1}{n^2} = \frac{\pi^2}{6}, \quad (1.184)$$

$$\zeta(4) = \frac{\pi^4}{90}, \quad (1.185)$$

and more generally,

$$\zeta(2n) = \frac{2^{2n-1}B_n}{(2n)!}\pi^{2n}, \quad (1.186)$$

where B_n is the Bernoulli number, named after Jacob Bernoulli (1654–1705). Bernoulli numbers are defined as follows:

$$B_0 = 1, \quad B_1 = -\frac{1}{2}, \quad B_2 = \frac{1}{6}, \quad B_3 = 0, \quad B_4 = -\frac{1}{30}, \quad B_5 = 0, \quad B_6 = \frac{1}{42}, \cdots,$$

B_k being recursively defined by the relation

$$\binom{k+1}{1}B_k + \binom{k+1}{2}B_{k-1} + \cdots + \binom{k+1}{k}B_1 + B_0 = 0. \quad (1.187)$$

It is clear that the series $\zeta(s)$ converges absolutely for $\sigma > 1$, and indeed that it converges uniformly for $\sigma > 1 + \delta$ for any $\delta > 0$. Euler actually studied the zeta function earlier, but only considered it for real values of s. The famous Euler's product formula expresses the unique factorization of integers as product of primes:

Theorem 1.5.5 (Euler's product). If $\sigma > 1$, then

$$\zeta(s) = \prod_p \left(\frac{1}{1 - p^{-s}}\right), \quad (1.188)$$

where the product runs over all prime numbers.

In particular, this implies that $\zeta(s) \neq 0$ for $\sigma > 1$. Euler's product formula is very important in the theory of prime numbers; it is, in fact, this formula that allows one to use analytic methods in the study of prime numbers. (Note that Euler's product formula may also be regarded as an analytic version of the Fundamental Theorem of Arithmetic.) Riemann's great insight was to study the ζ-function for *complex* values of s and to use the powerful methods of complex analysis. This enabled him to discover a remarkable connection between the zeros of the ζ-function and prime numbers; he showed that $\zeta(s)$ is analytic for $\sigma > 1$ and can be continued across the line $\sigma = 1$ (see Figure 1.7). More precisely, the difference

$$\zeta(s) - \frac{1}{s-1}$$

can be continued analytically to the half-plane $\sigma > 0$ and in fact to all of $\mathbb{C}$.

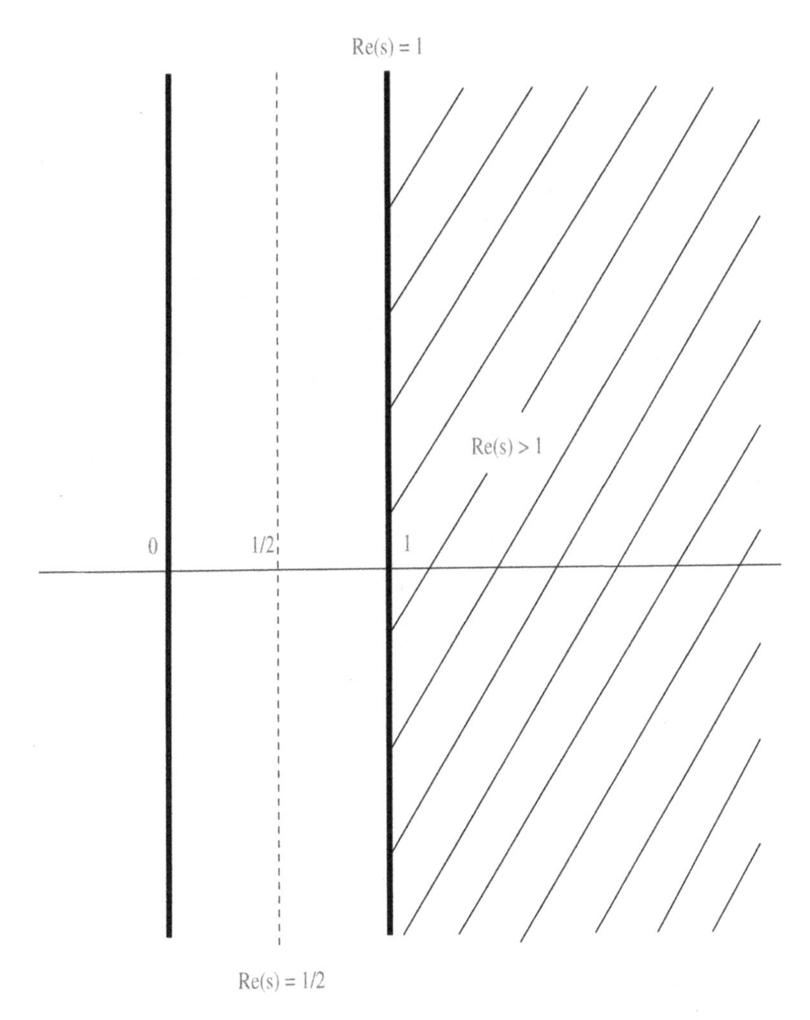

Figure 1.7. The complex plane of the Riemann ζ-function

The most interesting thing about the Riemann ζ-function is the *distribution* of the zeros of the ζ-function, since it is intimately connected with the distribution of the prime numbers. Now let us investigate the distribution of the zeros of the Riemann ζ-function (see Figure 1.7). It is known that

(1) The ζ-function has no zeros in the half-plane $\text{Re}(s) > 1$. (Since by Euler's product, if $\text{Re}(s) > 1$, then $\zeta(s) \neq 0$.)

(2) The ζ-function has no zeros on the line $\text{Re}(s) = 1$. (Since for any real value of t, $\zeta(1 + it) \neq 0$.)

Therefore, there are only three possible types of zeros of $\zeta(s)$:

(1) Zeros lying outside the critical strip $0 < \text{Re}(s) < 1$: These are the zeros at the points

$$-2,\ -4,\ -6,\ -8,\ -10,\ \cdots.$$

These zeros are the only zeros of $\zeta(s)$ outside the critical strip and are called *trivial zeros* of $\zeta(s)$. They are also called *real zeros* of $\zeta(s)$, since the zeros $-2, -4, \cdots$ are certainly real, and no other zeros are real.

(2) Zeros lying in the critical strip $0 < \text{Re}(s) < 1$: These zeros are called *nontrivial zeros* of $\zeta(s)$; there are infinitely many such nontrivial zeros. Note that the nontrivial zeros are *not* real, and hence they are sometimes called *complex zeros*. Note also that these zeros are symmetric about the real axis (so that if s_0 is a zero, so is $\overline{s}_0$, where the bar denotes the complex conjugate) and the critical line $\text{Re}(s) = \frac{1}{2}$ so that if $\frac{3}{4} + it$ were a zero, then $\frac{1}{4} + it$ would also be a zero).

(3) Zeros lying on the critical line $\text{Re}(s) = \frac{1}{2}$: These are the zeros at $\frac{1}{2} + it$. These zeros are, of course, nontrivial (complex) zeros (because they all lie in the critical strip). There are infinitely many such nontrivial zeros lying on the critical line.

Riemann made the somewhat startling conjecture about the distribution of the nontrivial zeros of $\zeta(s)$ in his famous memoir, namely that

Conjecture 1.5.1 (Riemann Hypothesis (RH)). All the nontrivial (complex) zeros ρ of $\zeta(s)$ lying in the critical strip $0 < \text{Re}(s) < 1$ must lie on the critical line $\text{Re}(s) = \frac{1}{2}$, that is, $\rho = \frac{1}{2} + it$, where ρ denotes a nontrivial zero of $\zeta(s)$.

Remark 1.5.2. The Riemann Hypothesis may be true; if it is true, then it can be diagrammatically shown as in the left picture of Figure 1.8. The Riemann Hypothesis may also be false; if it is false, then it can be diagrammatically shown as in the right picture of Figure 1.8. At present, no one knows whether or not the Riemann Hypothesis is true.

Remark 1.5.3. The Riemann Hypothesis has never been proved or disproved; in fact, finding a proof or a counter-example is generally regarded as one of most difficult and important unsolved problems in all of mathematics. There is, however, a lot of numerical evidence to support the conjecture; as we move away from the real axis, the first thirty nontrivial zeros ρ_n (where ρ_n denotes the nth nontrivial zero) of $\zeta(s)$ are given in Table 1.12 (all figures here are given to six decimal digits). In fact, as we move further and further away from the real axis, the first 1500000001 nontrivial zeros of $\zeta(s)$ in the critical strip have been calculated; all these zeros lie on the critical line $\text{Re}(s) = \frac{1}{2}$ and have imaginary part with $0 < t < 545439823.215$. That is,

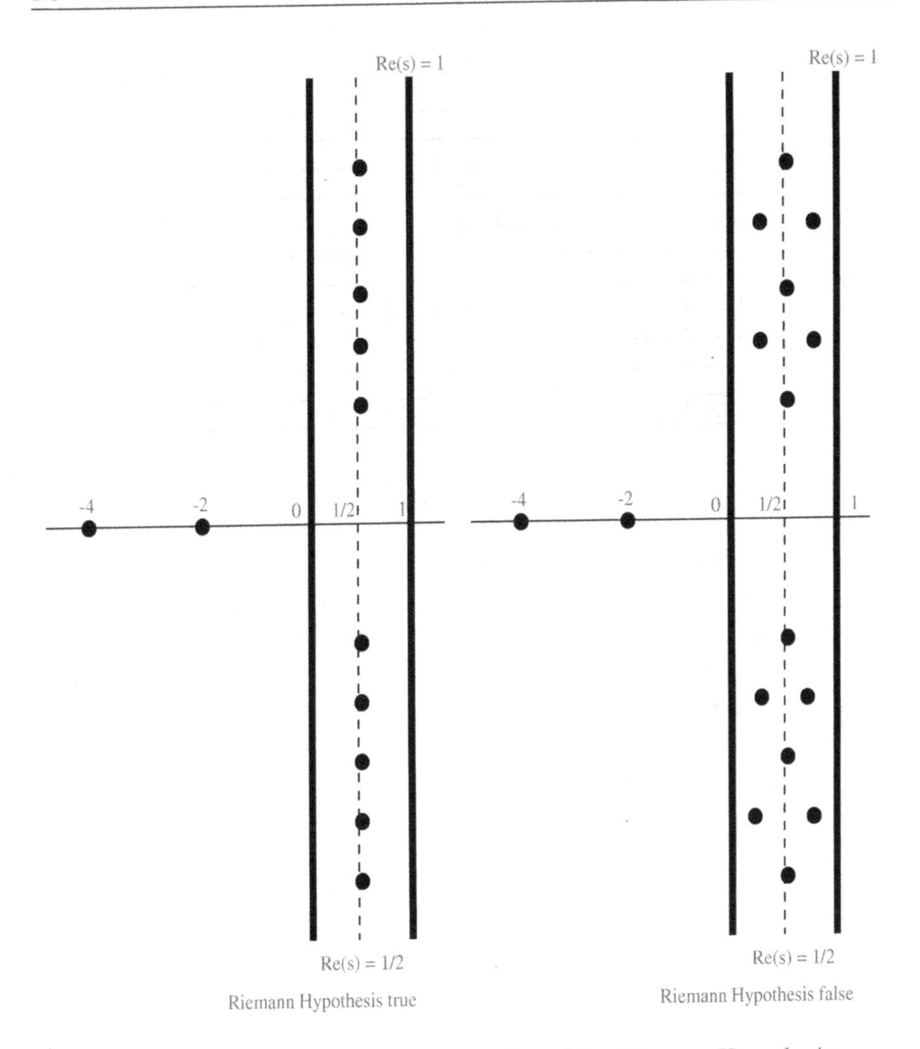

Figure 1.8. The diagrammatical representation of the Riemann Hypothesis

$\rho_n = \dfrac{1}{2} + it_n$ with $n = 1, 2, \cdots, 1500000001$ and $0 < t_n < 545439823.215$. In spite of this, there are several distinguished number theorists who believe the Riemann Hypothesis to be false, and that the presence of the first 1500000001 nontrivial zeros of $\zeta(s)$ on the critical line $\mathrm{Re}(s) = \dfrac{1}{2}$ does not indicate the behaviour of $\zeta(s)$ for every large t. The current status of knowledge of this conjecture is:

(1) The ζ-function has infinitely many zeros lying on the critical line $\mathrm{Re}(s) = \dfrac{1}{2}$.

Table 1.12. The first thirty nontrivial zeros of $\zeta(s)$

n	t_n	n	t_n	n	t_n
1	14.134725	2	21.022040	3	25.010857
4	30.424876	5	32.935062	6	37.586178
7	40.918719	8	43.327073	9	48.005151
10	49.773832	11	52.970321	12	56.446248
13	59.347044	14	60.831779	15	65.112544
16	67.079811	17	69.546402	18	72.067158
19	75.704691	20	77.144840	21	79.337375
22	82.910381	23	84.735479	24	87.425275
25	88.809111	26	92.491899	27	94.651344
28	95.874634	29	98.831194	30	101.317851

(2) A positive proportion of the zeroes of $\zeta(s)$ in the critical strip $0 < \mathrm{Re}(s) < 1$ lie on the critical line $\mathrm{Re}(s) = \frac{1}{2}$ (thanks to Selberg).

(3) It is not known whether there are any nontrivial zeros which are not simple; certainly, none has ever been found.

Remark 1.5.4. The Riemann Hypothesis (RH) is fundamental to the Prime Number Theorem (PNT). For example, if this conjecture is true, then there is a refinement of the Prime Number Theorem

$$\pi(x) = \int_2^x \frac{dt}{\ln t} + \mathcal{O}\left(xe^{-c\sqrt{\ln x}}\right) \tag{1.189}$$

to the effect that

$$\pi(x) = \int_2^x \frac{dt}{\ln t} + \mathcal{O}\left(\sqrt{x}\ln x\right). \tag{1.190}$$

Remark 1.5.5. The knowledge of a large zero-free region for $\zeta(s)$ is important in the proof of the PNT and better estimates of the various functions connected with the distribution of prime numbers; the larger the region, the better the estimates of differences $|\pi(x) - \mathrm{Li}(x)|$ and $|\psi(x) - x|$, appearing in the PNT. If we assume RH, we then immediately have a good zero-free region and hence the proof of PNT becomes considerably easier (see picture on the right in Figure 1.9). De la Vallée-Poussin constructed in 1896 a zero-free region in the critical strip (see the picture on the left in Figure 1.9). This zero-free region is not as good as that given by the RH, but it turns out to be good enough for the purpose of proving the PNT.

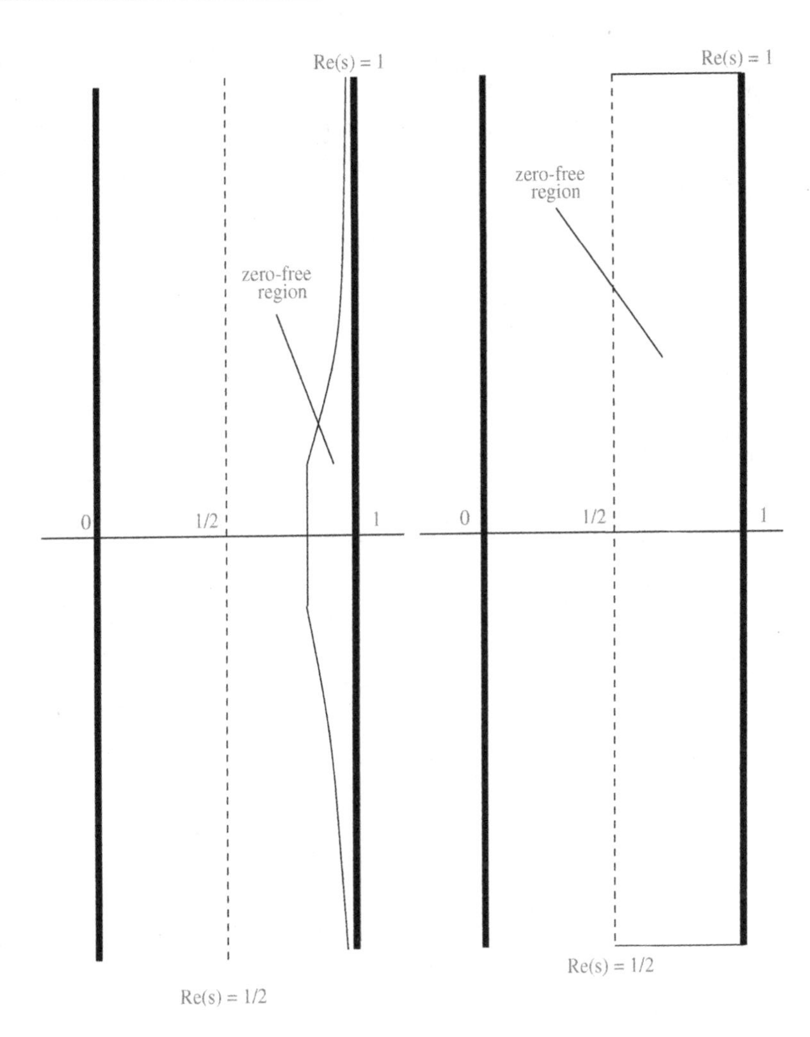

Figure 1.9. Zero-free regions for $\zeta(s)$

In a celebrated memoir published in 1837, when studying the arithmetic progression $kn + h$, Dirichlet[29] introduced new arithmetic functions $\chi(n)$,

---

29

Johann Peter Gustav Lejeune Dirichlet (1805–1859) was born into a French family in the vicinity of Cologne, Germany. He studied at the University of Paris, and held positions at the Universities of Breslau and Berlin and, in 1855, he was chosen to succeed Gauss at the University of Göttingen. Dirichlet is said to be the first person to master Gauss's *Disquisitiones Arithmeticae*. He is said to have kept a copy of this book at his side even when he traveled. His own book on number theory *Vorlesungen über Zahlentheorie*, helped make Gauss's discoveries

now called *Dirichlet characters* modulo k. These are multiplicative functions that have period k and vanish only on numbers not relatively prime to k. Clearly, there are $\phi(k)$ Dirichlet characters modulo k. In terms of Dirichlet characters, Dirichlet also introduced functions analogous to the Riemann ζ-function $\zeta(s)$. These functions, now called *Dirichlet L-functions*, are defined by infinite series of the form:

$$L(s, \chi) = \sum_{n=1}^{\infty} \frac{\chi(n)}{n^s}, \tag{1.191}$$

where $\chi(n)$ is a Dirichlet character modulo k and s is a real number greater than 1 (or a complex number with real part greater than 1). Dirichlet's work on L-functions led naturally to the description of a more general class of functions defined by infinite series of the form:

$$F(s) = \sum_{n=1}^{\infty} \frac{f(n)}{n^s}, \tag{1.192}$$

where $f(n)$ is a given arithmetic function. This is called a *Dirichlet series* with coefficients $f(n)$, and the function $F(s)$ is called a *generating function* of $f(n)$. For example, the simplest possible Dirichlet series is the Riemann ζ-function $\zeta(s)$, which generates the constant function $f(n) = 1$ for all n,

$$\zeta(s) = \sum_{n=1}^{\infty} \frac{1}{n^s}. \tag{1.193}$$

The square of the ζ-function generates the divisor function $\tau(n)$,

$$\zeta(s)^2 = \sum_{n=1}^{\infty} \frac{\tau(n)}{n^s}, \tag{1.194}$$

and the reciprocal of the ζ-function generates the Möbius function $\mu(n)$,

$$\zeta(s)^{-1} = \sum_{n=1}^{\infty} \frac{\mu(n)}{n^s}. \tag{1.195}$$

The study of L-functions is an active area of contemporary mathematical research, but it is not our purpose to explain here the theory and applications of Dirichlet L-functions in detail; we shall only use the basic concepts of Dirichlet L-functions to formulate the following *Generalized Riemann Hypothesis.*

---

accessible to other mathematicians. Dirichlet made many important contributions to several branches of mathematics. He proved that in any arithmetic progression $a, a+d, a+2d, \cdots$, where $\gcd(a, d) = 1$, there are infinitely many primes. His famous *Pigeonhole Principle* is used extensively in combinatorics and in number theory.

Conjecture 1.5.2 (Generalized Riemann Hypothesis (GRH)). All the nontrivial zeros of the Dirichlet L-functions in the critical strip $0 < \mathrm{Re}(s) < 1$ must lie on the critical line $\mathrm{Re}(s) = \dfrac{1}{2}$.

Clearly, the Generalized Riemann Hypothesis generalizes the (plain) Riemann hypothesis to Dirichlet L-functions. There are again many consequences of the generalized Riemann hypothesis. For example, if this conjecture is true, then the primality testing problem is in $\mathcal{P}$. ($\mathcal{P}$ stands for a class of problems solvable in polynomial time on a deterministic Turing machine; see Section 2.1.3 of Chapter 2 for more information.)

Having introduced the Riemann ζ-function and Dirichlet L-functions, let us introduce one more function named also after Riemann (but this time we just call it the "plain" Riemann function) and its relationship to $\pi(x)$.

Definition 1.5.4. Let x be a positive real number. Then the Riemann function, $R(x)$, is defined as follows

$$R(x) = \sum_{n=1}^{\infty} \frac{\mu(n)}{n} \mathrm{Li}(x^{1/n}). \tag{1.196}$$

Remark 1.5.6. The Riemann function $R(x)$ is computable by the following quickly converging power series

$$R(x) = 1 + \sum_{n=1}^{\infty} \frac{1}{n\zeta(n+1)} \cdot \frac{(\ln x)^n}{n!}. \tag{1.197}$$

In terms of the Riemann function $R(x)$, Riemann gave the following exact formula for $\pi(x)$

$$\pi(x) = R(x) - \sum_{\rho} R(x^{\rho}) \tag{1.198}$$

where the sum is extended over all the zeros ρ of the Riemann ζ-function, each counted with its own multiplicity (Ribenboim [200]).

The Riemann function $R(x)$ provides a *very* good approximation to $\pi(x)$. Table 1.13 shows what a remarkably good approximation $R(x)$ is to $\pi(x)$. Table 1.14 shows the differences between $\pi(x)$ and $\dfrac{x}{\ln x}$, $\mathrm{Li}(x)$ and $R(x)$.

Theorem 1.5.6. $\pi(x)$ is asymptotic to $R(x)$. That is,

$$\lim_{x \to \infty} \frac{\pi(x)}{R(x)} = 1. \tag{1.199}$$

Table 1.13. Approximations to $\pi(x)$ by $R(x)$

x	$\pi(x)$	$R(x)$	$\pi(x)/R(x)$
10^8	5761455	5761552	0.999983164258
10^9	50847534	50847455	1.000001553666
10^{10}	455052511	455050683	1.000004017134
10^{11}	4118054813	4118052495	1.000000562887
10^{12}	37607912018	37607910542	1.000000039247
10^{13}	346065536839	346065531066	1.000000016681
10^{14}	3204941750802	3204941731602	1.000000005990
10^{15}	29844570422669	29844570495887	0.999999997546
10^{16}	279238341033925	279238341360977	0.999999998828
10^{17}	2625557157654233	2625557157055978	1.000000000227
10^{18}	24739954287740860	24739954284239494	1.000000000141
10^{19}	234057667276344607	234057667300228940	0.999999999897

Table 1.14. Differences between $\pi(x)$ and $x/\ln x$, $\text{Li}(x)$ and $R(x)$

x	$x/\ln x - \pi(x)$	$\text{Li}(x) - \pi(x)$	$R(x) - \pi(x)$
10^8	-332774	754	97
10^9	-2592592	1701	-79
10^{10}	-20758030	3104	-1828
10^{11}	-169923160	11588	-2318
10^{12}	-1416705193	38263	-1476
10^{13}	-11992858452	108971	-5773
10^{14}	-102838308636	314890	-19200
10^{15}	-891604962453	1052619	73218
10^{16}	-7804289844393	3214632	327052
10^{17}	-70883734693929	7956589	-598255
10^{18}	-612483070893537	21949555	-3501366
10^{19}	-5481624169369961	99877775	23884333

1.5.5 The nth Prime

We have seen several equivalent forms of the prime number theorem, for example

$$\lim_{n\to\infty} \frac{\pi(n)}{n/\ln n} = 1 \iff \lim_{n\to\infty} \frac{\psi(x)}{x} = 1.$$

In this subsection, we shall study one more equivalent form of the prime number theorem. More specifically, we shall show that the following two forms of the prime number theorem are also equivalent.

$$\frac{\pi(n)}{n/\ln n} = 1 \iff \frac{p_n}{n \ln n} = 1.$$

Let p_n be the nth prime. Then we have:

$$\begin{array}{lll}
p_1 = 2, & p_2 = 3, & p_3 = 5, \\
p_4 = 7, & p_5 = 11, & p_6 = 13, \\
p_7 = 17, & p_8 = 19, & p_9 = 23, \\
\cdots & \cdots & \cdots \\
\cdots & \cdots & \cdots \\
p_{100} = 541, & p_{101} = 547, & p_{102} = 557, \\
p_{103} = 563, & p_{104} = 569, & p_{105} = 571, \\
p_{106} = 577, & p_{107} = 587, & p_{108} = 593, \\
p_{109} = 599, & p_{110} = 601, & p_{111} = 607, \\
\cdots & \cdots & \cdots \\
\cdots & \cdots & \cdots
\end{array}$$

Now we wish to show that

$$\lim_{n\to\infty} \frac{\pi(n)}{n/\ln n} = 1 \tag{1.200}$$

is equivalent to

$$\lim_{n\to\infty} \frac{p_n}{n \ln n} = 1. \tag{1.201}$$

By taking logarithms of both side sides of equation (1.200) and then removing a factor $\ln n$, we get

$$\lim_{n\to\infty} \left(\ln n \left(\frac{\ln \pi(n)}{\ln n} + \frac{\ln \ln n}{\ln n} - 1 \right) \right) = 0. \tag{1.202}$$

Since

$$\lim_{n\to\infty} \frac{\ln \ln n}{\ln n} = 0, \tag{1.203}$$

we have

$$\lim_{n\to\infty} \frac{\ln \pi(n)}{\ln n} = 1. \tag{1.204}$$

Multiplying (1.204) by (1.200), we get

$$\lim_{n\to\infty} \frac{\pi(n) \ln \pi(n)}{n} = 1. \tag{1.205}$$

Now replace n by the nth prime p_n. Then $\pi(p_n) = n$, and (1.205) becomes

$$\lim_{n\to\infty} \frac{n \ln n}{p_n} = 1, \tag{1.206}$$

which implies (1.201). Equation (1.200) can also be deduced from (1.201); we leave it as an exercise. So the two forms are equivalent. It is worthwhile pointing out that each of the statements (involving the Möbius function):

$$\lim_{n\to\infty} \frac{1}{n}\sum_{k=1}^{n} \mu(k) = 0 \tag{1.207}$$

and

$$\lim_{n\to\infty} \sum_{k=1}^{n} \frac{\mu(k)}{k} = 0 \tag{1.208}$$

is also equivalent to the prime number theorem.

It is known in fact that $p_n > n \ln n$ for all n. The error $p_n - n\ln n$ can be very large, but if n is large, the error is much smaller than $n \ln n$. In other words, for large n, the nth prime is about the size of $n \ln n$. Felgner showed in 1990 a weaker estimate that

$$0.91 n \ln n < p_n < 1.7 n \ln n. \tag{1.209}$$

Example 1.5.2. Table 1.15 gives some comparisons of p_n with $n \ln n$, $0.91n \ln n$, $1.7n \ln n$, and $P_n - n \ln n$. For example, let $n = 664999$. Then we have

$$\begin{aligned} 10006721 \sim 8916001.238 &\implies p_n \sim n \ln n \\ 10006721 > 8113561.127 &\implies p_n > 0.91 n \ln \\ 10006721 < 15157202.10 &\implies p_n < 1.7 n \ln \end{aligned}$$

These agree well with (1.209).

Table 1.15. Some comparisons about p_n

n	p_n	$n \ln n$	$0.9n \ln n$	$1.7n \ln n$	$\frac{p_n}{n \ln n} - 1$
10	29	23.02585093	20.95352435	39.14394658	.259453998
100	541	460.5170186	419.0704869	782.8789316	.174766574
1000	7919	6907.755279	6286.057304	11743.18397	.146392667
10000	104729	92103.40372	83814.09739	156575.7863	.137080670
664999	10006721	8916001.238	8113561.127	15157202.10	.122332841

1.5.6 Distribution of Twin Primes

Compared with the distribution of prime numbers, little is known about the distribution of the twin primes; for example, it was known 2000 years ago that *there are infinitely many prime numbers*, but it is not known *whether or not there are infinitely many twin primes*. In spite of this, remarkable progress has been made on the distribution of twin primes. Let $\pi_2(x)$ be the number of primes p such that $p \leq x$ and $p+2$ is also a prime. Then we have $\pi_2(10) = 2$ and $\pi_2(100) = 12$. A larger table of values of $\pi_2(x)$, together with some other information ($\mathrm{L}_2(x)$ is defined in (1.215) in the same way as $\mathrm{Li}(x)$), can be found in Table 1.16. Note that in Brent's paper [34], some interesting tables and graphs are given; they show, in particular, the difference between the behaviour of $\pi_2(x)$ (which has slow oscillations) and $\pi(x)$ (which has much faster oscillations).

Table 1.16. Some results for twin primes up to 10^{14}

x	$\pi_2(x)$	$\mathrm{L}_2(x)$	$\frac{\pi_2(x)}{\mathrm{L}_2(x)}$	$\pi_2(x) - \mathrm{L}_2(x)$
10	2	5	0.4	−3
10^2	8	14	0.5714285	−6
10^3	35	46	0.7608695	−11
10^4	205	214	0.9579439	−9
10^5	1224	1249	0.9799839	−25
10^6	8169	8248	0.9904219	−79
10^7	58980	58754	1.0038465	226
10^8	440312	440368	0.9998728	−56
10^9	3424506	3425308	0.9997658	−802
10^{10}	27412679	27411417	1.0000460	1262
10^{11}	224376048	224368865	1.0000320	7183
10^{12}	1870585220	1870559867	1.0000135	25353
10^{13}	15834664872	15834598305	1.0000042	66567
10^{14}	135780321665	135780264894	1.0000004	56771

There is also keen competition to find the largest pair of twin primes; we list in Table 1.17 twenty-nine large pairs of twin primes. (Note that the multifactorial notation $n!!!!$ in the 27th pair of the twin primes denotes the quadruple factorial function, i.e., $n!!!! = n(n-4)(n-8)(n-12)(n-16)\cdots$.)

Clemant in 1949 gave a necessary and sufficient condition for twin primes, although it has no practical value in the determination of twin primes.

Theorem 1.5.7. Let $n \geq 2$. The pair of integers $(n, n+2)$ form a pair of twin primes if and only if

$$4((n-1)!+1) + n \equiv 0 \pmod{n(n+2)}. \tag{1.210}$$

V. Brun announced in 1919 and proved in 1920 that there exists an effectively computable integer x_0 such that if $x \geq x_0$ then

$$\pi_2(x) < \frac{100x}{\ln^2 x}. \tag{1.211}$$

Table 1.17. Twenty-nine large twin primes

Twin Primes	Digits	Year	Discoverer(s)
$318032361 \cdot 2107001 \pm 1$	32220	2001	Underbakke et al.
$1807318575 \cdot 298305 \pm 1$	29603	2001	Underbakke et al.
$665551035 \cdot 280025 \pm 1$	24099	2000	Underbakke et al.
$781134345 \cdot 266445 \pm 1$	20011	2001	Underbakke et al.
$1693965 \cdot 266443 \pm 1$	20008	2000	LaBarbera et al.
$83475759 \cdot 264955 \pm 1$	19562	2000	Underbakke et al.
$291889803 \cdot 260090 \pm 1$	18098	2001	Boivin & Gallot
$4648619711505 \cdot 260000 \pm 1$	18075	2000	Indlekofer et al.
$2409110779845 \cdot 260000 \pm 1$	18075	2000	Indlekofer et al.
$2230907354445 \cdot 248000 \pm 1$	14462	1999	Indlekofer et al.
$871892617365 \cdot 2^{48000} \pm 1$	14462	1999	Indlekofer et al.
$361700055 \cdot 2^{39020} \pm 1$	11755	1999	Lifchitz
$835335 \cdot 2^{39014} \pm 1$	11751	1999	Ballinger & Gallot
$242206083 \cdot 2^{38880} \pm 1$	11713	1995	Indlekofer & Jarai
$570918348 \cdot 10^{5120} \pm 1$	5129	1995	Dubner
$697053813 \cdot 2^{16352} \pm 1$	4932	1994	Indlekofer & Jarai
$6797727 \cdot 2^{15328} \pm 1$	4622	1995	Forbes
$1692923232 \cdot 10^{4020} \pm 1$	4030	1993	Dubner
$4655478828 \cdot 10^{3429} \pm 1$	3439	1993	Dubner
$1706595 \cdot 2^{11235} \pm 1$	3389	1989	Parady et al.
$459 \cdot 2^{8529} \pm 1$	2571	1993	Dubner
$1171452282 \cdot 10^{2490} \pm 1$	2500	1991	Dubner
$571305 \cdot 2^{7701} \pm 1$	2324	1989	Paradi et al.
$75188117004 \cdot 10^{2298} \pm 1$	2309	1989	Dubner
$663777 \cdot 2^{7650} \pm 1$	2309	1989	Paradi et al.
$107570463 \cdot 10^{2259} \pm 1$	2259	1985	Dubner
$2846!!!! \pm 1$	2151	1992	Dubner
$43690485351513 \cdot 10^{1995} \pm 1$	2009	1985	Dubner
$260497545 \cdot 2^{6625} \pm 1$	2003	1984	Atkin & Rickert

Brun proved that the sum of $\dfrac{1}{p+1}$ for all primes p such that $p+2$ is also a prime converges. The sum is now called Brun's constant, B, and it has now in fact been calculated:

$$
\begin{aligned}
B &= \sum\left(\frac{1}{p}+\frac{1}{p+2}\right) \\
&= \left(\frac{1}{3}+\frac{1}{5}\right)+\left(\frac{1}{5}+\frac{1}{7}\right)+\cdots+\left(\frac{1}{p}+\frac{1}{p+2}\right)+\cdots \\
&= 1.902160577783278\cdots \qquad (1.212)
\end{aligned}
$$

It also has been proved by Bombieri and Davenport [97] in 1966 that

$$
\pi_2(x) \le 8 \prod_{p>2}\left(1-\frac{1}{(p-1)^2}\right)\frac{x}{\ln^2 x}\left(1+\mathcal{O}\left(\frac{\ln\ln x}{\ln x}\right)\right). \qquad (1.213)
$$

The constant 8 has been improved to $6.26+\epsilon$, however it was conjectured by Hardy and Littlewood that the constant should be 2 rather than 8.

The famous Twin Prime Conjecture states that

Conjecture 1.5.3. Let $\pi_2(x)$ be the number of primes p such that $p \le x$ and $p+2$ is also a prime. Then

(1) (A weak form) There are infinitely many twin primes. That is,

$$
\lim_{x\to\infty} \pi_2(x) = \infty. \qquad (1.214)
$$

(2) (A strong form) Let

$$
\begin{aligned}
\mathrm{L}_2(x) &= 2\prod_{p\ge 3}\frac{p(p-2)}{(p-1)^2}\int_2^x \frac{dt}{\ln^2 t} \\
&\approx 1.320323632\int_2^x \frac{dt}{\ln^2 t} \qquad (1.215)
\end{aligned}
$$

then

$$
\lim_{x\to\infty}\frac{\pi_2(x)}{\mathrm{L}_2(x)} = 1. \qquad (1.216)
$$

Using very complicated arguments based on sieve methods, the Chinese mathematician J. R. Chen showed that *there are infinitely many pairs of integers* $(p,\ p+2)$, *with* p *prime and* $p+2$ *a product of at most two primes.*

If we write $d_n = p_{n+1} - p_n$ so that $d_1 = 1$ and all other d_n are even. Then an equivalent form of the Twin Prime Conjecture is that $d_n = 2$ *occurs infinitely often*. How large can d_n be? Rankin has shown that

$$
d_n > \frac{c\ln n\ln\ln n\ln\ln\ln\ln n}{(\ln\ln\ln n)^2} \qquad (1.217)
$$

for infinitely many n. Erdös offered \$5000 for a proof or a disproof that the constant c can be taken arbitrarily large.

1.5.7 The Arithmetic Progression of Primes

In this subsection we shall move on to the study of the arithmetic progression of primes.

An *arithmetic progression of primes* is defined to be the sequence of primes

$$p,\ p+d,\ p+2d,\ \cdots,\ p+(n-1)d, \tag{1.218}$$

where p is the first term, d the common difference, and $p+(n-1)d$ the last term of the progression, respectively. For example,

$$5,\ 11,\ 17,\ 23,\ 29$$

is an arithmetic progression of primes with $p=5$, $d=6$ and $n=5$. Table 1.18 contains fifteen long arithmetic progressions of primes, discovered by James Fry, Andrew Moran, Paul Pritchard, S. C. Root, S. Weintraub and Jeff Young; see Table 1 of Guy [94] and Table 32 of Ribenboim [200] for more information (note that there are some printing errors in Table 32 of Ribenboim [200], which have been corrected here in Table 1.18).

Table 1.18. Fifteen long arithmetic progressions of primes

n	p	d	$p+(n-1)d$	Year
22	11410337850553	4609098694200	108201410428753	PMT, 1993
21	5749146449311	26004868890	6269243827111	P, 1992
21	142072321123	1419763024680	28537332814723	MP, 1990
20	24845147147111	19855265430	25222397190281	MP, 1990
20	1845449006227	1140004565700	23505535754527	MP, 1990
20	214861583621	18846497670	572945039351	YF, 1987
20	1140997291211	7643355720	1286221049891	F, 1987
20	803467381001	2007835830	841616261771	F, 1987
19	244290205469	13608665070	489246176729	F, 1987
19	8297644387	4180566390	83547839407	P, 1984
18	107928278317	9922782870	276615587107	P, 1982
18	4808316343	717777060	17010526363	P, 1983
17	3430751869	87297210	4827507229	W, 1977
16	2236133941	223092870	5582526991	R, 1969
16	53297929	9699690	198793279	R, 1969

It is conjectured that n can be as large as you like (but it has not been proven yet):

Conjecture 1.5.4. There are arbitrarily long arithmetic progressions of primes.

Erdös conjectured that if $\{a_i\}$ is any infinite sequence of integers for which $\sum \frac{1}{a_i}$ is divergent, then the sequence contains arbitrarily long arithmetic progressions; he offered \$3000 for a proof or disproof of this conjecture (before his death, Erdös signed some blank cheques and left them in the care of Ronald L. Graham to present to future problem solvers). A related but even more difficult problem is the following:

Conjecture 1.5.5. There are arbitrarily long arithmetic progressions of consecutive primes.

1.6 Theory of Congruences

As with everything else, so with a mathematical theory: beauty can be perceived, but not explained.

ARTHUR CAYLEY (1821–1895)

1.6.1 Basic Concepts and Properties of Congruences

The notion of congruences was first introduced by Gauss, who gave their definition in his celebrated *Disquisitiones Arithmeticae* in 1801, though the ancient Greeks and Chinese had already had the idea.

Definition 1.6.1. Let a be an integer and n a positive integer greater than 1. We define "a mod n" to be the remainder r when a is divided by n, that is

$$r = a \bmod n = a - \lfloor a/n \rfloor n. \tag{1.219}$$

We may also say that "r is equal to a reduced modulo n".

Remark 1.6.1. It follows from the above definition that a mod n is the integer r such that $a = \lfloor a/n \rfloor n + r$ and $0 \le r < n$, which was known to the ancient Greeks and Chinese some 2000 years ago.

Example 1.6.1. The following are some examples of a mod n:

$$\begin{aligned}
&35 \bmod 12 = 11,\\
&-129 \bmod 7 = 4,\\
&3210 \bmod 101 = 79,\\
&1412^{13115} \bmod 12349 = 1275.
\end{aligned}$$

Given the well-defined notion of the remainder of one integer when divided by another, it is convenient to provide a special notion to indicate equality of remainders.

Definition 1.6.2. Let a and b be integers and n a positive integer. We say that "a is *congruent* to b modulo n", denoted by

$$a \equiv b \pmod{n} \tag{1.220}$$

if n is a divisor of $a - b$, or equivalently, if $n \mid (a - b)$. Similarly, we write

$$a \not\equiv b \pmod{n} \tag{1.221}$$

if a is not congruent (or incongruent) to b modulo n, or equivalently, if $n \nmid (a - b)$. Clearly, for $a \equiv b \pmod{n}$ (resp. $a \not\equiv b \pmod{n}$), we can write $a = kn - b$ (resp. $a \neq kn - b$) for some integer k. The integer n is called the *modulus*.

Clearly,

$$\begin{aligned} a \equiv b \pmod{n} &\iff n \mid (a - b) \\ &\iff a = kn + b, \quad k \in \mathbb{Z} \end{aligned}$$

and

$$\begin{aligned} a \not\equiv b \pmod{n} &\iff n \nmid (a - b) \\ &\iff a \neq kn + b, \quad k \in \mathbb{Z} \end{aligned}$$

So, the above definition of congruences, introduced by Gauss in his *Disquisitiones Arithmeticae*, does not offer any new idea than the divisibility relation, since "$a \equiv b \pmod{n}$" and "$n \mid (a - b)$" (resp. "$a \not\equiv b \pmod{n}$" and "$n \nmid (a - b)$") have the same meaning, although each of them has its own advantages. However, Gauss did present a *new* way (i.e., congruences) of looking at the old things (i.e., divisibility); this is exactly what we are interested in. It is interesting to note that the ancient Chinese mathematician Ch'in Chiu-Shao[30] already had the idea and theory of congruences in his famous book *Mathematical Treatise in Nine Chapters* appeared in 1247.

Definition 1.6.3. If $a \equiv b \pmod{n}$, then b is called a *residue* of a modulo n. If $0 \le b \le m - 1$, b is called the *least nonnegative residue* of a modulo n.

Remark 1.6.2. It is common, particularly in computer programs, to denote the least nonnegative residue of a modulo n by $a \bmod n$. Thus, $a \equiv b \pmod{n}$ if and only if $a \bmod n = b \bmod n$, and, of course, $a \not\equiv b \pmod{n}$ if and only if $a \bmod n \neq b \bmod n$.

[30] Ch'in Chiu-Shao (1202–1261) was born in the southwest Chinese province of Sichuan, but studied astronomy in Hangzhou, the capital of the Song dynasty, now the capital of the Chinese southeast province Zhejiang. Ch'in was a genius in mathematics and was also accomplished in poetry, fencing, archery, riding, music and architecture. He wrote *Mathematical Treatise in Nine Chapters* which appeared in 1247. It contains simultaneous integer congruences, the Chinese Remainder Theorem, and considers algebraic equations, areas of geometrical figures and linear simultaneous equations. This work on *congruences* was rediscovered by Gauss, Lebesgue and Stieltjes.

Example 1.6.2. The following are some examples of congruences or incongruences.

$$\begin{aligned} 35 &\equiv 11 \pmod{12} && \text{since} && 12 \mid (35-11) \\ &\not\equiv 12 \pmod{11} && \text{since} && 11 \nmid (35-12) \\ &\equiv 2 \pmod{11} && \text{since} && 11 \mid (35-2) \end{aligned}$$

The congruence relation has many properties in common with the equality relation. For example, we know from high-school mathematics that equality is

(1) reflexive: $a = a$, $\forall a \in \mathbb{Z}$;
(2) symmetric: if $a = b$, then $b = a$, $\forall a, b \in \mathbb{Z}$;
(3) transitive: if $a = b$ and $b = c$, then $a = c$, $\forall a, b, c \in \mathbb{Z}$.

We shall see that congruence modulo n has the same properties:

Theorem 1.6.1. Let n be a positive integer. Then the congruence modulo n is

(1) reflexive: $a \equiv a \pmod{n}$, $\forall a \in \mathbb{Z}$;
(2) symmetric: if $a \equiv b \pmod{n}$, then $b \equiv a \pmod{n}$, $\forall a, b \in \mathbb{Z}$;
(3) transitive: if $a \equiv b \pmod{n}$ and $b \equiv c \pmod{n}$, then $a \equiv c \pmod{n}$, $\forall a, b, c \in \mathbb{Z}$.

Proof.

(1) For any integer a, we have $a = 0 \cdot n + a$, hence $a \equiv a \pmod{n}$.
(2) For any integers a and b, if $a \equiv b \pmod{n}$, then $a = kn + b$ for some integer k. Hence $b = a - kn = (-k)n + a$, which implies $b \equiv a \pmod{n}$, since $-k$ is an integer.
(3) If $a \equiv b \pmod{n}$ and $b \equiv c \pmod{n}$, then $a = k_1 n + b$ and $b = k_2 n + c$. Thus, we can get

$$a = k_1 n + k_2 n + c = (k_1 + k_2)n + c = k'n + c$$

which implies $a \equiv c \pmod{n}$, since k' is an integer. □

Theorem 1.6.1 shows that the congruence modulo n is an equivalence relation on the set of integers $\mathbb{Z}$. But note that the divisibility relation $a \mid b$ is reflexive, and transitive but not symmetric; in fact if $a \mid b$ and $b \mid a$ then $a = b$, so it is not an equivalence relation. The congruence relation modulo n partitions $\mathbb{Z}$ into n *equivalence classes*. In number theory, we call these classes *congruence classes*, or *residue classes*. More formally, we have:

Definition 1.6.4. If $x \equiv a \pmod{n}$, then a is called a *residue* of x modulo n. The *residue class* of a modulo n, denoted by $[a]_n$ (or just $[a]$ if no confusion will be caused), is the set of all those integers that are congruent to a modulo n. That is,

$$\begin{aligned} [a]_n &= \{x : x \in \mathbb{Z} \text{ and } x \equiv a \pmod{n}\} \\ &= \{a + kn : k \in \mathbb{Z}\}. \end{aligned} \tag{1.222}$$

Note that writing $a \in [b]_n$ is the same as writing $a \equiv b \pmod{n}$.

Example 1.6.3. Let $n = 5$. Then there are five residue classes, modulo 5, namely the sets:

$$\begin{aligned} [0]_5 &= \{\cdots, -15, -10, -5, 0, 5, 10, 15, 20, \cdots\}, \\ [1]_5 &= \{\cdots, -14, \ -9, -4, 1, 6, 11, 16, 21, \cdots\}, \\ [2]_5 &= \{\cdots, -13, \ -8, -3, 2, 7, 12, 17, 22, \cdots\}, \\ [3]_5 &= \{\cdots, -12, \ -7, -2, 3, 8, 13, 18, 23, \cdots\}, \\ [4]_5 &= \{\cdots, -11, \ -6, -1, 4, 9, 14, 19, 24, \cdots\}. \end{aligned}$$

The first set contains all those integers congruent to 0 modulo 5, the second set contains all those congruent to 1 modulo 5, $\cdots$, and the fifth (i.e., the last) set contains all those congruent to 4 modulo 5. So, for example, the residue class $[2]_5$ can be represented by any one of the elements in the set

$$\{\cdots, -13, \ -8, -3, 2, 7, 12, 17, 22, \cdots\}.$$

Clearly, there are infinitely many elements in the set $[2]_5$.

Example 1.6.4. In residue classes modulo 2, $[0]_2$ is the set of all even integers, and $[1]_2$ is the set of all odd integers:

$$\begin{aligned} [0]_2 &= \{\cdots, -6, -4, -2, 0, 2, 4, 6, 8, \cdots\}, \\ [1]_2 &= \{\cdots, -5, -3, -1, 1, 3, 5, 7, 9, \cdots\}. \end{aligned}$$

Example 1.6.5. In congruence modulo 5, we have

$$\begin{aligned} [9]_5 &= \{9 + 5k : k \in \mathbb{Z}\} = \{9, 9 \pm 5, 9 \pm 10, 9 \pm 15, \cdots\} \\ &= \{\cdots, -11, -6, -1, 4, 9, 14, 19, 24, \cdots\}. \end{aligned}$$

We also have

$$\begin{aligned} [4]_5 &= \{4 + 5k : k \in \mathbb{Z}\} = \{4, 4 \pm 5, 4 \pm 10, 4 \pm 15, \cdots\} \\ &= \{\cdots, -11, -6, -1, 4, 9, 14, 19, 24, \cdots\}. \end{aligned}$$

So, clearly, $[4]_5 = [9]_5$.

Definition 1.6.5. If $x \equiv a \pmod{n}$ and $0 \leq a \leq n - 1$, then a is called the *least (nonnegative) residue* of x modulo n.

Example 1.6.6. Let $n = 7$. There are seven residue classes, modulo 7. In each of these seven residue classes, there is exactly one least residue of x modulo 7. So, the complete set of all least residues x modulo 7 is $\{0, 1, 2, 3, 4, 5, 6\}$.

Definition 1.6.6. The set of all residue classes modulo n, often denoted by $\mathbb{Z}/n\mathbb{Z}$ or $\mathbb{Z}_n$, is

$$\mathbb{Z}/n\mathbb{Z} = \{[a]_n : 0 \le a \le n-1\}. \tag{1.223}$$

Remark 1.6.3. One often sees the definition

$$\mathbb{Z}/n\mathbb{Z} = \{0, 1, 2, \cdots, n-1\}, \tag{1.224}$$

which should be read as equivalent to (1.223) with the understanding that 0 represents $[0]_n$, 1 represents $[1]_n$, 2 represents $[2]_n$, and so on; each class is represented by its least nonnegative residue, but the underlying residue classes must be kept in mind. For example, a reference to $-a$ as a member of $\mathbb{Z}/n\mathbb{Z}$ is a reference to $[n-a]_n$, provided $n \ge a$, since $-a \equiv n - a \pmod{n}$.

The following theorem gives some elementary properties of residue classes:

Theorem 1.6.2. Let n be a positive integer. Then we have:

(1) $[a]_n = [b]_n$ if and only if $a \equiv b \pmod{n}$.

(2) Two residue classes modulo n are either disjoint or identical.

(3) There are exactly n distinct residue classes modulo n, namely, $[0]_n, [1]_n, [2]_n, [3]_n, \cdots, [n-1]_n$, and they contain all of the integers.

Proof.

(1) If $a \equiv b \pmod{n}$, it follows from the transitive property of congruence that an integer is congruent to a modulo n if and only if it is congruent to b modulo n. Thus, $[a]_n = [b]_n$. To prove the converse, suppose $[a]_n = [b]_n$. Because $a \in [a]_n$ and $a \in [b]_n$, Thus, $a \equiv b \pmod{n}$.

(2) Suppose that $[a]_n$ and $[b]_n$ have a common element c. Then $c \equiv a \pmod{n}$ and $c \equiv b \pmod{n}$. From the symmetric and transitive properties of congruence, it follows that $a \equiv b \pmod{n}$. From part (1) of this theorem, it follows that $[a]_n$ and $[b]_n$. Thus, either $[a]_n$ and $[b]_n$ are disjoint or identical.

(3) If a is an integer, we can divide a by n to get

$$a = kq + r, \quad 0 \le r < k.$$

Thus, $a \equiv r \pmod{n}$ and so $[a]_n = [r]_n$. This implies that a is one of the residue classes $[0]_n, [1]_n, [2]_n, \cdots, [n-1]_n$, Because the integers $0, 1, 2, \cdots, n-1$ are incongruent modulo n, it follows that there are exactly n residue classes modulo n.

□

Definition 1.6.7. Let n be a positive integer. A set of integers $a_1, a_2, \cdots, a_n$ is called a *complete system of residues* modulo n, if the set contains exactly one element from each residue class modulo n.

Example 1.6.7. Let $n = 4$. Then $\{-12, 9, -6, -1\}$ is a complete system of residues modulo 4, since $-12 \in [0]$, $9 \in [1]$, $-6 \in [2]$ and $-1 \in [3]$. Of course, it can be easily verified that $\{12, -7, 18, -9\}$ is another complete system of residues modulo 4. It is clear that the simplest complete system of residues modulo 4 is $\{0, 1, 2, 3\}$, the set of all nonnegative least residues modulo 4.

Example 1.6.8. Let $n = 7$. Then $\{x,\ x+3,\ x+3^2,\ x+3^3,\ x+3^4,\ x+3^5,\ x+3^6\}$ is a complete system of residues modulo 7, for any $x \in \mathbb{Z}$. To see this let us first evaluate the powers of 3 modulo 7:

$$
\begin{array}{lll}
3 & 3^2 \equiv 2 \pmod 7 & 3^3 \equiv 6 \pmod 7 \\
3^4 \equiv 4 \pmod 7 & 3^5 \equiv 5 \pmod 7 & 3^6 \equiv 1 \pmod 7
\end{array}
$$

hence, the result follows from $x = 0$. Now the general result follows immediately, since $(x + 3^i) - (x + 3^j) = 3^i - 3^j$.

Theorem 1.6.3. Let n be a positive integer and S a set of integers. S is a complete system of residues modulo n if and only if

(1) S contains n elements, and

(2) no two elements of S are congruent, modulo n.

Proof. If S is a complete system of residues, then the two conditions are satisfied. To prove the converse, we note that if no two elements of S are congruent, the elements of S are in different residue classes modulo n. Since S has n elements, all the residue classes must be represented among the elements of S. Thus, S is a complete system of residues modulo n □

We now introduce one more type of systems of residues, the *reduced* systems of residues modulo n.

Definition 1.6.8. Let $[a]_n$ be a residue class modulo n. We say that $[a]_n$ is relatively prime to n if each element in $[a]_n$ is relatively prime to n.

Example 1.6.9. Let $n = 10$. Then the ten residue classes, modulo 10, are as follows:

$$
\begin{aligned}
[0]_{10} &= \{\cdots, -30, -20, -10, 0, 10, 20, 30, \cdots\} \\
[1]_{10} &= \{\cdots, -29, -19, \ -9, 1, 11, 21, 31, \cdots\} \\
[2]_{10} &= \{\cdots, -28, -18, \ -8, 2, 12, 22, 32, \cdots\} \\
[3]_{10} &= \{\cdots, -27, -17, \ -7, 3, 13, 23, 33, \cdots\} \\
[4]_{10} &= \{\cdots, -26, -16, \ -6, 4, 14, 24, 34, \cdots\} \\
[5]_{10} &= \{\cdots, -25, -15, \ -5, 5, 15, 25, 35, \cdots\} \\
[6]_{10} &= \{\cdots, -24, -14, \ -4, 6, 16, 26, 36, \cdots\} \\
[7]_{10} &= \{\cdots, -23, -13, \ -3, 7, 17, 27, 37, \cdots\} \\
[8]_{10} &= \{\cdots, -22, -12, \ -2, 8, 18, 28, 38, \cdots\} \\
[9]_{10} &= \{\cdots, -21, -11, \ -1, 9, 19, 29, 39, \cdots\}.
\end{aligned}
$$

Clearly, $[1]_{10}$, $[3]_{10}$, $[7]_{10}$, and $[9]_{10}$ are residue classes that are relatively prime to 10.

Proposition 1.6.1. If a residue class modulo n has *one* element which is relatively prime to n, then every element in that residue class is relatively prime to n.

Proposition 1.6.2. If n is prime, then every residue class modulo n (except $[0]_n$) is relatively prime to n.

Definition 1.6.9. Let n be a positive integer, then $\phi(n)$ is the number of residue classes modulo n, which is relatively prime to n. A set of integers $\{a_1, a_2, \cdots, a_{\phi(n)}\}$ is called a *reduced system of residues*, if the set contains exactly one element from each residue class modulo n which is relatively prime to n.

Example 1.6.10. In Example 1.6.9, we know that $[1]_{10}$, $[3]_{10}$, $[7]_{10}$ and $[9]_{10}$ are residue classes that are relatively prime to 10, so by choosing -29 from $[1]_{10}$, -17 from $[3]_{10}$, 17 from $[7]_{10}$ and 39 from $[9]_{10}$, we get a reduced system of residues modulo 10: $\{-29, -17, 17, 39\}$. Similarly, $\{31, 3, -23, -1\}$ is another reduced system of residues modulo 10.

One method to obtain a reduced system of residues is to start with a complete system of residues and delete those elements that are not relatively prime to the modulus m. Thus, the simplest reduced system of residues (mod m) is just the collections of all integers in the set $\{0, 1, 2, \cdots, m-1\}$ that are relatively prime to m.

Theorem 1.6.4. Let m be a positive integer, and S a set of integers. Then S is a reduced system of residues (mod$n$) if and only if

(1) S contains exactly $\phi(n)$ elements,
(2) no two elements of S are congruent (mod$n$),
(3) each element of S is relatively prime to n.

Proof. It is obvious that a reduced system of residues satisfies the three conditions. To prove the converse, we suppose that S is a set of integers having the three properties. Because no two elements of S are congruent, the elements are in different residues modulo n. Since the elements of S are relatively prime n, there are in residue classes that are relatively prime n. Thus, the $\phi(n)$ elements of S are distributed among the $\phi(n)$ residue classes that are relatively prime n, one in each residue class. Therefore, S is a reduced system of residues modulo m. □

Corollary 1.6.1. Let $\{a_1, a_2, \cdots, a_{\phi(m)}\}$ be a reduced system of residues modulo m, and suppose that $\gcd(k, m) = 1$. Then $\{ka_1, ka_2, \cdots, ka_{\phi(m)}\}$ is also a reduced system of residues modulo m.

Proof. Left as exercise. □

1.6.2 Modular Arithmetic

The finite set $\mathbb{Z}/n\mathbb{Z}$ is closely related to the infinite set $\mathbb{Z}$. So, it is natural to ask if it is possible to define addition and multiplication in $\mathbb{Z}/n\mathbb{Z}$ and do some reasonable kind of arithmetic there. Surprisingly, addition, subtraction and multiplication in $\mathbb{Z}/n\mathbb{Z}$ will be much the same as that in $\mathbb{Z}$. Let us first investigate some elementary arithmetic properties of congruences.

Theorem 1.6.5. For all $a, b, c, d \in \mathbb{Z}$ and $n \in \mathbb{Z}_{>1}$, if $a \equiv b \pmod{n}$ and $c \equiv d \pmod{n}$, then

(1) $a \pm b \equiv c \pm d \pmod{n}$,
(2) $a \cdot b \equiv c \cdot d \pmod{n}$,
(3) $a^m \equiv b^m \pmod{n}, \quad \forall m \in \mathbb{N}$.

Proof.

(1) Write $a = kn + b$ and $c = ln + d$ for some $k, l \in \mathbb{Z}$. Then $a + c = (k+l)n + b + d$. Therefore, $a + c = b + d + tn$, $t = k + l \in \mathbb{Z}$. Consequently, $a + c \equiv b + d \pmod{n}$, which is what we wished to show. The case of subtraction is left as an exercise.
(2) Similarly,
$$\begin{aligned} ac &= bd + bln + knd + kln^2 \\ &= bd + n(bl + k(d + ln)) \\ &= bd + n(bl + kc) \\ &= bd + sn \end{aligned}$$
where $s = bl + kc \in \mathbb{Z}$. Thus, $a \cdot b \equiv c \cdot d \pmod{n}$.
(3) We prove Part (3) by induction. We have $a \equiv b \pmod{n}$ (base step) and $a^m \equiv b^m \pmod{n}$ (inductive hypothesis). Then by Part (2) we have $a^{m+1} \equiv aa^m \equiv bb^m \equiv b^{m+1} \pmod{n}$. □

Theorem 1.6.5 is equivalent to the following theorem, since

$$\begin{aligned} a \equiv b \pmod{n} &\iff a \bmod n = b \bmod n, \\ a \bmod n &\iff [a]_n, \\ b \bmod n &\iff [b]_n. \end{aligned}$$

Theorem 1.6.6. For all $a, b, c, d \in \mathbb{Z}$, if $[a]_n = [b]_n$, $[c]_n = [d]_n$, then

(1) $[a \pm b]_n = [c \pm d]_n$,
(2) $[a \cdot b]_n = [c \cdot d]_n$,
(3) $[a^m]_n = [b^m]_n, \quad \forall m \in \mathbb{N}$.

The fact that the congruence relation modulo n is stable for addition (subtraction) and multiplication means that we can define binary operations, again called addition (subtraction) and multiplication on the set of $\mathbb{Z}/n\mathbb{Z}$ of equivalence classes modulo n as follows (in case only one n is being discussed, we can simply write $[x]$ for the class $[x]_n$):

$$[a]_n + [b]_n = [a+b]_n \tag{1.225}$$
$$[a]_n - [b]_n = [a-b]_n \tag{1.226}$$
$$[a]_n \cdot [b]_n = [a \cdot b]_n \tag{1.227}$$

Example 1.6.11. Let $n = 12$, then

$$[7]_{12} \ +_{12} \ [8]_{12} = [7+8]_{12} = [15]_{12} = [3]_{12},$$
$$[7]_{12} \ -_{12} \ [8]_{12} = [7-8]_{12} = [-1]_{12} = [11]_{12},$$
$$[7]_{12} \ \cdot_{12} \ [8]_{12} = [7 \cdot 8]_{12} = [56]_{12} = [8]_{12}.$$

In many cases, we may still prefer to write the above operations as follows:

$$7 + 8 = 15 \equiv 3 \pmod{12},$$
$$7 - 8 = -1 \equiv 11 \pmod{12},$$
$$7 \cdot 8 = 56 \equiv 8 \pmod{12}$$

We summarise the properties of addition and multiplication modulo n in the following two theorems.

Theorem 1.6.7. The set $\mathbb{Z}/n\mathbb{Z}$ of integers modulo n has the following properties with respect to addition:

(1) Closure: $[x] + [y] \in \mathbb{Z}/n\mathbb{Z}$, for all $[x], [y] \in \mathbb{Z}/n\mathbb{Z}$.
(2) Associative: $([x] + [y]) + [z] = [x] + ([y] + [z])$, for all $[x], [y], [z] \in \mathbb{Z}/n\mathbb{Z}$.
(3) Commutative: $[x] + [y] = [y] + [x]$, for all $[x], [y] \in \mathbb{Z}/n\mathbb{Z}$.
(4) Identity, namely, $[0]$.
(5) Additive inverse: $-[x] = [-x]$, for all $[x] \in \mathbb{Z}/n\mathbb{Z}$.

Proof. These properties follow directly from the stability and the definition of the operation in $\mathbb{Z}/n\mathbb{Z}$. □

Theorem 1.6.8. The set $\mathbb{Z}/n\mathbb{Z}$ of integers modulo n has the following properties with respect to multiplication:

(1) Closure: $[x] \cdot [y] \in \mathbb{Z}/n\mathbb{Z}$, for all $[x], [y] \in \mathbb{Z}/n\mathbb{Z}$.
(2) Associative: $([x] \cdot [y]) \cdot [z] = [x] \cdot ([y] \cdot [z])$, for all $[x], [y], [z] \in \mathbb{Z}/n\mathbb{Z}$.
(3) Commutative: $[x] \cdot [y] = [y] \cdot [x]$, for all $[x], [y] \in \mathbb{Z}/n\mathbb{Z}$.
(4) Identity, namely, $[1]$.
(5) Distributivity of multiplication over addition: $[x] \cdot ([y]) + [z]) = ([x] \cdot [y]) + ([x] \cdot [z])$, for all $[x], [y], [z] \in \mathbb{Z}/n\mathbb{Z}$.

Proof. These properties follow directly from the stability of the operation in $\mathbb{Z}/n\mathbb{Z}$ and and the corresponding properties of $\mathbb{Z}$. □

The division a/b (we assume a/b is in lowest terms and $b \not\equiv 0 \pmod{n}$) in $\mathbb{Z}/n\mathbb{Z}$, however, will be more of a problem; sometimes you can divide, sometimes you cannot. For example, let $n = 12$ again, then

$$\begin{aligned} 3/7 &\equiv 9 \pmod{12} \qquad &&\text{(no problem)},\\ 3/4 &\equiv \perp \pmod{12} \qquad &&\text{(impossible)}. \end{aligned}$$

Why is division sometimes possible (e.g., $3/7 \equiv 9 \pmod{12}$) and sometimes impossible (e.g., $3/4 \equiv \perp \pmod{12}$)? The problem is with the modulus n; if n is a prime number, then the division $a/b \pmod{n}$ is always possible and unique, whilst if n is a composite then the division $a/b \pmod{n}$ may be not possible or the result may be not unique. Let us observe two more examples, one with $n = 13$ and the other with $n = 14$. First note that $a/b \equiv a \cdot 1/b \pmod{n}$ if and only if $1/b \pmod{n}$ is possible, since multiplication modulo n is always possible. We call $1/b \pmod{n}$ the *multiplicative inverse* (or *modular inverse*) of b modulo n. More generally, we have:

Definition 1.6.10. Two integers x and y are said to be multiplicative inverses if

$$xy \equiv 1 \pmod{n}, \tag{1.228}$$

where n is a positive integer greater than 1.

It is now clear that given (x, n), y does not always exist. Let $n = 13$ be a prime, then the following table gives all the possible values of the multiplicative inverses $y = 1/x \pmod{13}$ for $x = 1, 2, \cdots, 12$:

x	1	2	3	4	5	6	7	8	9	10	11	12
y	1	7	9	10	8	11	2	5	3	4	6	12

This means that divisions in $\mathbb{Z}/13\mathbb{Z}$ are always possible and unique (i.e., the multiplicative inverses y of x in $\mathbb{Z}/13\mathbb{Z}$ do always exist and are unique). On the other hand, if $n = 14$ (n now is a composite), then for $x = 1, 2, \cdots, 13$, some values for $y = 1/x \pmod{14}$ exist, whereas others do not:

x	1	2	3	4	5	6	7	8	9	10	11	12	13
y	1	$\perp$	5	$\perp$	3	$\perp$	$\perp$	$\perp$	11	$\perp$	9	$\perp$	13

This means that only the numbers $1, 3, 5, 9, 11$ and 13 have multiplicative inverses modulo 14, or equivalently only those divisions by $1, 3, 5, 9, 11$ and 13 modulo 14 are possible. This observation leads to the following important results:

Theorem 1.6.9. The multiplicative inverse $1/b$ modulo n exists if and only if $\gcd(b, n) = 1$.

But how many b's are there that satisfy $\gcd(b, n) = 1$? The following result answers this question.

Corollary 1.6.2. There are $\phi(n)$ numbers b for which $1/b \pmod{n}$ exists.

Example 1.6.12. Let $n = 21$. Since $\phi(21) = 12$, there are twelve b for which $1/b \pmod{21}$ exists. In fact, the multiplicative inverse modulo 21 only exists for each of the following b:

b	1	2	4	5	8	10	11	13	16	17	19	20
$1/b \pmod{21}$	1	11	16	17	8	19	2	13	4	5	10	20

Corollary 1.6.3. The division a/b modulo n (assume that a/b is in lowest terms) is possible if and only if $1/b \pmod{n}$ exists, i.e., if and only if $\gcd(b, n) = 1$.

Example 1.6.13. Compute $6/b \pmod{21}$ whenever it is possible. By the multiplicative inverses of $1/b \pmod{21}$ in the previous table, we just need to calculate $6 \cdot 1/b \pmod{21}$:

b	1	2	4	5	8	10	11	13	16	17	19	20
$6/b \pmod{21}$	6	3	12	18	6	9	12	15	3	9	18	15

As it can be seen, addition (subtraction) and multiplication are always possible in $\mathbb{Z}/n\mathbb{Z}$, with $n > 1$, since $\mathbb{Z}/n\mathbb{Z}$ is a ring. Note also that $\mathbb{Z}/n\mathbb{Z}$ with n prime is an Abelian group with respect to addition, and all the non-zero elements in $\mathbb{Z}/n\mathbb{Z}$ form an Abelian group with respect to multiplication (i.e., a division is always possible for any two non-zero elements in $\mathbb{Z}/n\mathbb{Z}$ if n is prime); hence $\mathbb{Z}/n\mathbb{Z}$ with n prime is a field. That is,

Theorem 1.6.10. $\mathbb{Z}/n\mathbb{Z}$ is a field if and only if n is prime.

The above results only tell us when the multiplicative inverse $1/a$ modulo n is possible, without mentioning how to find the inverse. To actually find the multiplicative inverse, we let

$$1/a \pmod{n} = x, \tag{1.229}$$

which is equivalent to

$$ax \equiv 1 \pmod{n}. \tag{1.230}$$

Since

$$ax \equiv 1 \pmod{n} \Longleftrightarrow ax - ny = 1. \tag{1.231}$$

So the finding of the multiplicative inverse becomes to find the solution of the linear Diophantine equation $ax - ny = 1$, which, as we know in Section 1.3, can be solved by using the continued fraction expansion of a/n, and can, of course, be solved by using Euclid's algorithm.

Example 1.6.14. Find

(1) $1/154 \pmod{801}$,

(2) $4/154 \pmod{801}$.

Solution:

(1) Since

$$1/a \pmod n = x \iff ax \equiv 1 \pmod n \iff ax - ny = 1, \quad (1.232)$$

we only need to find x and y in

$$154x - 801y = 1.$$

To do so, we first use the Euclid's algorithm to find $\gcd(154, 801)$ as follows.

$$\begin{aligned} 801 &= 154 \cdot 5 + 31 \\ 154 &= 31 \cdot 4 + 30 \\ 31 &= 30 \cdot 1 + 1 \\ 30 &= 10 \cdot 3 + 0. \end{aligned}$$

Since $\gcd(154, 801) = 1$, by Theorem 1.6.9, the equation $154x - 801y = 1$ is soluble. We now rewrite the above resulting equations

$$\begin{aligned} 31 &= 801 - 154 \cdot 5 \\ 30 &= 154 - 31 \cdot 4 \\ 1 &= 31 - 30 \cdot 1 \end{aligned}$$

and work backwards on the above new equations

$$\begin{aligned} 1 &= 31 - 30 \cdot 1 \\ &= 31 - (154 - 31 \cdot 4) \cdot 1 \\ &= 31 - 154 + 4 \cdot 31 \\ &= 5 \cdot 31 - 154 \\ &= 5 \cdot (801 - 154 \cdot 5) - 154 \\ &= 5 \cdot 801 - 26 \cdot 154 \\ &= 801 \cdot 5 - 154 \cdot 26 \end{aligned}$$

So, $x \equiv -26 \equiv 775 \pmod{801}$. That is, $1/154 \bmod 801 = 775$.

(2) Since $4/154 \equiv 4 \cdot 1/154 \pmod{801}$, then $4/154 \equiv 4 \cdot 775 \equiv 697 \pmod{801}$.

The above procedure used to find the x and y in $ax + by = 1$ can be generalized to find the x and y in $ax + by = c$; this procedure usually called the *extended Euclid's algorithm*. We shall discuss the solution of the general equation $ax + by = c$ in the next subsection.

1.6.3 Linear Congruences

Congruences have much in common with equations. In fact, the linear congruence $ax \equiv b \pmod{n}$ is equivalent to the linear Diophantine equation $ax - ny = b$. That is,

$$ax \equiv b \pmod{n} \Longleftrightarrow ax - ny = b. \tag{1.233}$$

Thus, linear congruences can be solved by using a continued fraction method just as for linear Diophantine equations. In this section, however, we shall use some theoretical properties of congruences to solve linear congruences (the continued fraction approach to linear congruences is left as an exercise for readers). The basic theory of linear congruences is described in the next three theorems.

Theorem 1.6.11. Let $\gcd(a, n) = d$. If $d \nmid b$, then the linear congruence

$$ax \equiv b \pmod{n} \tag{1.234}$$

has no solutions.

Proof. We will prove the contrapositive of the assertion: if $ax \equiv b \pmod{n}$ has a solution, then $\gcd(a, n) \mid b$. Suppose that s is a solution. Then $as \equiv b \pmod{n}$, and from the definition of the congruence, $n \mid (as - b)$. or from the definition of divisibility, $as - b = kn$ for some integer k. Since $\gcd(a, m) \mid a$ and $\gcd(a, m) \mid kn$, it follows that $\gcd(a, m) \mid b$. □

Theorem 1.6.12. Let $\gcd(a, n) = d$. Then the linear congruence $ax \equiv b \pmod{n}$ has solutions if and only if $d \mid b$.

Proof. Follows from Theorem 1.6.11. □

Theorem 1.6.13. Let $\gcd(a, n) = 1$. Then the linear congruence $ax \equiv b \pmod{n}$ has exactly one solution.

Proof. If $\gcd(a, n) = 1$, then there exist x and y such that $ax + ny = 1$. Multiplying by b gives

$$a(xb) + n(yb) = b.$$

As $a(xb) - b$ is a multiple of n, or $a(xb) \equiv b \pmod{n}$. The least residue of xb modulo n is then a solution of the linear congruence. The uniqueness of the solution is left as an exercise. □

Theorem 1.6.14. Let $\gcd(a, n) = d$ and suppose that $d \mid b$. Then the linear congruence

$$ax \equiv b \pmod{n}. \tag{1.235}$$

has exactly d solutions modulo n. These are given by

$$t, \; t+\frac{n}{d}, \; t+\frac{2n}{d}, \; \cdots, \; t+\frac{(d-1)n}{d} \tag{1.236}$$

where t is the solution, unique modulo n/d, of the linear congruence

$$\frac{a}{d}x \equiv \frac{b}{d} \left(\text{mod } \frac{n}{d}\right). \tag{1.237}$$

Proof. By Theorem 1.6.12, the linear congruence has solutions since $d \mid b$. Now let t be be such a solution, then $t + k(n/d)$ for $k = 1, 2, \cdots, d-1$ are also solutions, since $a(t + k(n/d)) \equiv at + kn(t/d) \equiv at \equiv b \pmod{n}$. □

Together with the above theorems and the extended Euclid's algorithm discussed in the previous subsection (or the continued fraction method discussed in Subsection 1.3), we can find the solutions of $ax \equiv b \pmod{n}$, provided they exist.

Example 1.6.15. Find $154x \equiv 22 \pmod{803}$. Notice first that

$$154x \equiv 22 \pmod{803} \Longleftrightarrow 154x - 803y = 22.$$

Now we use the Euclid's algorithm to find $\gcd(154, 803)$ as follows.

$$\begin{aligned} 803 &= 154 \cdot 5 + 33 \\ 154 &= 33 \cdot 4 + 22 \\ 33 &= 22 \cdot 1 + 11 \\ 22 &= 11 \cdot 2. \end{aligned}$$

Since $\gcd(154, 803) = 11$ and $11 \mid 22$, by Theorem 1.6.12, the equation $154x - 801y = 22$ is soluble. Now we rewrite the above resulting equations

$$\begin{aligned} 33 &= 803 - 154 \cdot 5 \\ 22 &= 154 - 33 \cdot 4 \\ 11 &= 33 - 22 \cdot 1 \end{aligned}$$

and work backwards on the above new equations

$$\begin{aligned} 11 &= 33 - 22 \cdot 1 \\ &= 33 - (154 - 33 \cdot 4) \cdot 1 \\ &= 33 - 154 + 4 \cdot 33 \\ &= 5 \cdot 33 - 154 \\ &= 5 \cdot (803 - 154 \cdot 5) - 154 \\ &= 5 \cdot 803 - 26 \cdot 154 \\ &= 803 \cdot 5 - 154 \cdot 26. \end{aligned}$$

So, $x \equiv -26 \equiv 777 \pmod{803}$. By Theorem 1.6.13, there are, in total, 11 solutions of $154x - 801y = 22$; we list all of them as follows (we also write the verifications of the results on right):

$$\begin{array}{rl}
777, & 154 \cdot 777 \equiv 11 \pmod{803} \\
777 + 803/11 \equiv 47, & 154 \cdot 47 \equiv 11 \pmod{803} \\
777 + 2 \cdot 803/11 \equiv 120, & 154 \cdot 120 \equiv 11 \pmod{803} \\
777 + 3 \cdot 803/11 \equiv 193, & 154 \cdot 193 \equiv 11 \pmod{803} \\
777 + 4 \cdot 803/11 \equiv 266, & 154 \cdot 266 \equiv 11 \pmod{803} \\
777 + 5 \cdot 803/11 \equiv 339, & 154 \cdot 339 \equiv 11 \pmod{803} \\
777 + 6 \cdot 803/11 \equiv 412, & 154 \cdot 412 \equiv 11 \pmod{803} \\
777 + 7 \cdot 803/11 \equiv 485, & 154 \cdot 485 \equiv 11 \pmod{803} \\
777 + 8 \cdot 803/11 \equiv 558, & 154 \cdot 558 \equiv 11 \pmod{803} \\
777 + 9 \cdot 803/11 \equiv 631, & 154 \cdot 631 \equiv 11 \pmod{803} \\
777 + 10 \cdot 803/11 \equiv 704, & 154 \cdot 704 \equiv 11 \pmod{803}.
\end{array}$$

Remark 1.6.4. To find the solution for the linear Diophantine equation

$$ax \equiv b \pmod{n} \tag{1.238}$$

is equivalent to find the quotient of the modular division

$$x \equiv \frac{b}{a} \pmod{n} \tag{1.239}$$

which is, again, equivalent to find the multiplicative inverse

$$x \equiv \frac{1}{a} \pmod{n} \tag{1.240}$$

because, if $\frac{1}{a}$ modulo n exists, the multiplication $b \cdot \frac{1}{a}$ is always possible.

In what follows, we shall introduce some important results on linear congruences. Our first result will be Fermat's little theorem (or just Fermat's theorem, for short), due to Fermat.

Theorem 1.6.15 (Fermat's little theorem). Let a be a positive integer, and p prime. If $\gcd(a,p) = 1$, then

$$a^{p-1} \equiv 1 \pmod{p}. \tag{1.241}$$

Proof. First notice that the residues modulo p of $a,\ 2a,\ \cdots, (p-1)a$ are $1, 2, \cdots, (p-1)$ in some order, because no two of them can be equal. So, if we multiply them together, we get

$$\begin{aligned}
a \cdot 2a \cdots (p-1)a &\equiv [(a \bmod p) \cdot (2a \bmod p) \cdots (p-1)a \bmod p)] \pmod{p} \\
&\equiv (p-1)! \pmod{p}.
\end{aligned}$$

This means that

$$(p-1)!a^{p-1} \equiv (p-1)! \pmod{p}.$$

Now we can cancel the $(p-1)!$ since $p \nmid (p-1)!$, and the result thus follows. □

There is a more convenient and more general form of Fermat's little theorem:

$$a^p \equiv a \pmod{p}, \tag{1.242}$$

for $a \in \mathbb{N}$. The proof is easy: if $\gcd(a,p)=1$, we simply multiply (1.241) by a. If not, then $p \mid a$. So $a^p \equiv 0 \equiv a \pmod{p}$.

Fermat's theorem has several important consequences which are very useful in compositeness; one of the these consequences is as follows:

Corollary 1.6.4 (Converse of Fermat's little theorem, 1640). Let n be an odd positive integer. If $\gcd(a,n)=1$ and

$$a^{n-1} \not\equiv 1 \pmod{n}, \tag{1.243}$$

then n is composite.

Remark 1.6.5. As mentioned in Subsection 1.2.3, Fermat, in 1640, made a false conjecture that all the numbers of the form $F_n = 2^{2^n}+1$ were prime. Fermat really should not have made such a "stupid" conjecture, since $F_5 = 2^{32}+1$ can be relatively easily verified to be composite, by just using his own recently discovered theorem – Fermat's little theorem:

$$\begin{array}{rcll}
3^{2^2} & \equiv & 81 & \pmod{2^{32}+1} \\
3^{2^3} & \equiv & 6561 & \pmod{2^{32}+1} \\
3^{2^4} & \equiv & 43046721 & \pmod{2^{32}+1} \\
3^{2^5} & \equiv & 3793201458 & \pmod{2^{32}+1} \\
& \cdots\cdots & & \\
& \cdots\cdots & & \\
3^{2^{32}} & \equiv & 3029026160 & \pmod{2^{32}+1} \\
& \not\equiv & 1 & \pmod{2^{32}+1}.
\end{array}$$

Thus, by Fermat's little theorem, $F_5 = 2^{32}+1$ is not prime!

Based on Fermat's little theorem, Euler established a more general result in 1760:

Theorem 1.6.16 (Euler's theorem). Let a and n be positive integers with $\gcd(a,n)=1$. Then

$$a^{\phi(n)} \equiv 1 \pmod{n}. \tag{1.244}$$

Proof. Let $r_1, r_2, \cdots, r_{\phi(n)}$ be a reduced residue system modulo n. Then $ar_1, ar_2, \cdots, ar_{\phi(n)}$ is also a residue system modulo n. Thus we have

$$(ar_1)(ar_2)\cdots(ar_{\phi(n)}) \equiv r_1 r_2 \cdots r_{\phi(n)} \pmod{n},$$

since $ar_1, ar_2, \cdots, ar_{\phi(n)}$, being a reduced residue system, must be congruent in some order to $r_1, r_2, \cdots, r_{\phi(n)}$. Hence,

$$a^{\phi(n)} r_1 a r_2 \cdots r_{\phi(n)} \equiv r_1 r_2 \cdots r_{\phi(n)} \pmod{n},$$

which implies that $a^{\phi(n)} \equiv 1 \pmod{n}$. □

It can be difficult to find the order[31] of an element a modulo n but sometimes it is possible to improve (1.244) by proving that every integer a modulo n must have an order smaller than the number $\phi(n)$ – this order is actually the number $\lambda(n)$.

Theorem 1.6.17 (Carmichael's theorem). Let a and n be positive integers with $\gcd(a, n) = 1$. Then

$$a^{\lambda(n)} \equiv 1 \pmod{n}, \tag{1.245}$$

where $\lambda(n)$ is Carmichael's function.

Proof. Let $n = p_1^{\alpha_1} p_2^{\alpha_2} \cdots p_k^{\alpha_k}$. We shall show that

$$a^{\lambda(n)} \equiv 1 \pmod{p_i^{\alpha_i}}$$

for $1 \le i \le k$, since this implies that $a^{\lambda(n)} \equiv 1 \pmod{n}$. If $p_k^{\alpha_k} = 2, 4$ or a power of an odd prime, then by Definition 1.4.7, $\lambda(\alpha_k) = \phi(\alpha_k)$, so $a^{\lambda(p_i^{\alpha_i})} \equiv 1 \pmod{p_i^{\alpha_i}}$. Since $\lambda(p_i^{\alpha_i}) \mid \lambda(n)$, $a^{\lambda(n)} \equiv 1 \pmod{p_i^{\alpha_i}}$. The case that $p_i^{\alpha_i}$ is a power of 2 greater than 4 is left as an exercise. □

Note that $\lambda(n)$ will never exceed $\phi(n)$ and is often much smaller than $\phi(n)$; it is the value of the largest order it is possible to have.

Example 1.6.16. Let $a = 11$ and $n = 24$. Then $\phi(24) = 8$, $\lambda(24) = 2$. So,

$$11^{\phi(24)} = 11^8 \equiv 1 \pmod{24},$$

$$11^{\lambda(24)} = 11^2 \equiv 1 \pmod{24}.$$

That is, $\mathrm{ord}_{24}(11) = 2$.

[31] The order of an element a modulo n is the smallest integer r such that $a^r \equiv 1 \pmod{n}$; we shall discuss this later in Subsection 1.6.7.

In 1770 Edward Waring (1734–1793) published the following result, which is attributed to John Wilson[32].

Theorem 1.6.18 (Wilson's theorem). If p is a prime, then

$$(p-1)! \equiv -1 \pmod{p}. \tag{1.246}$$

Proof. It suffices to assume that p is odd. Now to every integer a with $0 < a < p$ there is a unique integer a' with $0 < a' < p$ such that $aa' \equiv 1 \pmod{p}$. Further if $a = a'$ then $a^2 \equiv 1 \pmod{p}$ whence $a = 1$ or $a = p-1$. Thus the set $2, 3, \cdots, p-2$ can be divided into $(p-3)/2$ pairs a, a' with $aa' \equiv 1 \pmod{p}$. Hence we have $2 \cdot 3 \cdots (p-2) \equiv 1 \pmod{p}$, and so $(p-1)! \equiv -1 \pmod{p}$, as required. □

Theorem 1.6.19 (Converse of Wilson's theorem). If n is an odd positive integer greater than 1 and

$$(n-1)! \equiv -1 \pmod{n}, \tag{1.247}$$

then n is a prime.

Remark 1.6.6. Prime p is called a *Wilson prime* if

$$W(p) \equiv 0 \pmod{p}, \tag{1.248}$$

where

$$W(p) = \frac{(p-1)! + 1}{p}$$

is an integer, or equivalently if

$$(n-1)! \equiv -1 \pmod{p^2}. \tag{1.249}$$

For example, $p = 5, 13, 563$ are Wilson primes, but 599 is not since

$$\frac{(599-1)! + 1}{599} \bmod 599 = 382 \neq 0.$$

It is not known whether there are infinitely many Wilson primes; to date, the only known Wilson primes for $p < 5 \cdot 10^8$ are $p = 5, 13, 563$. A prime p is called a *Wieferich prime*, named after A. Wieferich, if

$$2^{p-1} \equiv 1 \pmod{p^2}. \tag{1.250}$$

To date, the only known Wieferich primes for $p < 4 \cdot 10^{12}$ are $p = 1093$ and 3511.

[32] The English mathematician John Wilson (1741–1793) is best known for Wilson's theorem. This result was first published by Waring. Almost certainly Wilson's theorem was a guess and Waring didn't know how to prove it. It was first proved by Joseph-Louis Lagrange (1736–1813) in 1773 who showed that the converse is true. Wilson's theorem has a direct application in primality testing, although the test is not very efficient.

In what follows, we shall show how to use Euler's theorem to calculate the multiplicative inverse modulo n, and hence the solutions of a linear congruence.

Theorem 1.6.20. Let x be the multiplicative inverse $1/a$ modulo n. If $\gcd(a, n) = 1$, then

$$x \equiv \frac{1}{a} \pmod{n} \tag{1.251}$$

is given by

$$x \equiv a^{\phi(n)-1} \pmod{n}. \tag{1.252}$$

Proof. By Euler's theorem, we have $a^{\phi(n)} \equiv 1 \pmod{n}$. Hence

$$aa^{\phi(n)-1} \equiv 1 \pmod{n},$$

and $a^{\phi(n)-1}$ is the multiplicative inverse of a modulo n, as desired. □

Corollary 1.6.5. Let x be the division b/a modulo n (b/a is assumed to be in lowest terms). If $\gcd(a, n) = 1$, then

$$x \equiv \frac{b}{a} \pmod{n} \tag{1.253}$$

is given by

$$x \equiv b \cdot a^{\phi(n)-1} \pmod{n}. \tag{1.254}$$

Corollary 1.6.6. If $\gcd(a, n) = 1$, then the solution of the linear congruence

$$ax \equiv b \pmod{n} \tag{1.255}$$

is given by

$$x \equiv ba^{\phi(n)-1} \pmod{n}. \tag{1.256}$$

Example 1.6.17. Solve the congruence $5x \equiv 14 \pmod{24}$. First note that because $\gcd(5, 24) = 1$, the congruence has exactly one solution. Using (1.256) we get

$$x \equiv 14 \cdot 5^{\phi(24)-1} \pmod{24} = 22.$$

Example 1.6.18. Solve the congruence $20x \equiv 15 \pmod{135}$. First note that as $d = \gcd(20, 135) = 5$ and $d \mid 15$, the congruence has exactly five solutions modulo 135. To find these five solutions, we divide by 5 and get a new congruence

$$4x' \equiv 3 \pmod{27}.$$

To solve this new congruence, we get

$$x' \equiv 3 \cdot 4^{\phi(27)-1} \equiv 21 \pmod{27}.$$

Therefore, the five solutions are as follows:

$$
\begin{aligned}
(x_0, x_1, x_2, x_3, x_4) &\equiv \left(x', \; x' + \frac{n}{d}, \; x' + \frac{2n}{d}, \; x' + \frac{3n}{d}, \; x' + \frac{4n}{d}\right) \\
&\equiv (21, \; 21 + 27, \; 21 + 2 \cdot 27, \; 21 + 3 \cdot 27, \; 21 + 4 \cdot 27) \\
&\equiv (21, 48, 75, 102, 129) \pmod{135}.
\end{aligned}
$$

1.6.4 The Chinese Remainder Theorem

In this subsection, we introduce a method for solving systems of linear congruences. The method, widely known as the Chinese Remainder Theorem (or just CRT, for short), was discovered by the ancient Chinese mathematician Sun Tsu[33].

Theorem 1.6.21 (The Chinese Remainder Theorem CRT). If m_1, $m_2, \cdots, m_n$ are pairwise relatively prime and greater than 1, and $a_1, a_2, \cdots,$ a_n are any integers, then there is a solution x to the following simultaneous congruences:

$$
\left.
\begin{array}{l}
x \equiv a_1 \pmod{m_1}, \\
x \equiv a_2 \pmod{m_2}, \\
\quad \cdots\cdots \\
\quad \cdots\cdots \\
x \equiv a_n \pmod{m_n}.
\end{array}
\right\}
\tag{1.257}
$$

If x and x' are two solutions, then $x \equiv x' \pmod{M}$, where $M = m_1 m_2 \cdots m_n$.

Proof. Existence: Let us first solve a special case of the simultaneous congruences (1.257), where i is some fixed subscript,

$$a_i = 1, \; a_1 = a_2 = \cdots = a_{i-1} = a_{i+1} = \cdots = a_n = 0.$$

[33] Sun Zi (known as Sun Tsu in the West), a Chinese mathematician, lived sometime between 200 B.C. and 200 A.D. He is perhaps best known for his discovery of the Chinese Remainder Theorem which may be found in Problem 26 in Volume 3 of his classic three-volume mathematics book *Mathematical Manual*: find a number that leaves a remainder of 2 when divided by 3, a remainder of 3 when divided by 5, and a remainder of 2 when divided by 7; in modern algebraic language, to find the smallest positive integer satisfying the following systems of congruences:

$$
\begin{array}{l}
x \equiv 2 \pmod{3}, \\
x \equiv 3 \pmod{5}, \\
x \equiv 2 \pmod{7}.
\end{array}
$$

Sun Zi gave a rule called "tai-yen" ("great generalisation") to find the solution. Sun Zi's rule was generalized in today's "theorem-form" by the great Chinese mathematician Ch'in Chiu-Shao in his book *Mathematical Treatise in Nine Chapters* in 1247; Ch'in also rediscovered Euclid's algorithm, and gave a complete procedure for solving numerically polynomial equations of any degree, which is very similar to, or almost the same as, what is now called the *Horner method* published by William Horner in 1819.

Let $k_i = m_1 m_2 \cdots m_{i-1} m_{i+1} \cdots m_n$. Then k_i and m_i are relatively prime, so we can find integers r and s such that $rk_i + sm_i = 1$. This gives the congruences:

$$\begin{aligned} rk_i &\equiv 0 \pmod{k_i}, \\ rk_i &\equiv 1 \pmod{m_i}. \end{aligned}$$

Since $m_1, m_2, \cdots, m_{i-1}, m_{i+1}, \cdots m_n$ all divide k_i, it follows that $x_i = rk_i$ satisfies the simultaneous congruences:

$$\begin{aligned} x_i &\equiv 0 \pmod{m_1}, \\ x_i &\equiv 0 \pmod{m_2}, \\ &\cdots\cdots \\ x_i &\equiv 0 \pmod{m_{i-1}}. \\ x_i &\equiv 1 \pmod{m_i}. \\ x_i &\equiv 0 \pmod{m_{i+1}}. \\ &\cdots\cdots \\ x_i &\equiv 0 \pmod{m_n}. \end{aligned}$$

For each subscript i, $1 \le i \le n$, we find such an x_i. Now, to solve the system of the simultaneous congruences (1.257), set $x = a_1 x_1 + a_2 x_2 + \cdots + a_n x_n$. Then $x \equiv a_i x_i \equiv a_i \pmod{m_i}$ for each i, $1 \le i \le n$, therefore x is a solution of the simultaneous congruences.

Uniqueness: Let x' be another solution to the simultaneous congruences (1.257), but different from the solution x, so that $x' \equiv x \pmod{m_i}$ for each x_i. Then, $x - x' \equiv 0 \pmod{m_i}$ for each i. So, m_i divides $x - x'$ for each i; hence the least common multiple of all the m_j's divides $x - x'$. But since the m_i are pairwise relatively prime, this least common multiple is the product M. So, $x \equiv x' \pmod{M}$. □

The above proof of the CRT is constructive, providing an efficient method for finding all solutions of systems of simultaneous congruences (1.257). There are, of course, many other different proofs of the CRT; there is even a very short proof, due to Mozzochi [171]; it makes use of the following lemma:

Lemma 1.6.1. Suppose that $m_1, m_2, \cdots, m_n$ are pairwise relatively prime. Then $x \equiv y \pmod{m_i}$, $i = 1, 2, \cdots, n$ if and only if $x \equiv y \pmod{M}$, where $M = m_1 m_2 \cdots m_n$.

Now we are in a position to present Mozzochi's short proof of the CRT.

Proof. Let $a \in \mathbb{Z}$, $[x]_a = \{\text{y: } x \equiv y \pmod{a}\}$, and $\mathbb{Z}/a\mathbb{Z}$ the set of all residue classes modulo a. Define

$$\alpha : \mathbb{Z}/M\mathbb{Z} \to \mathbb{Z}/m_1\mathbb{Z} \times \mathbb{Z}/m_2\mathbb{Z} \times \cdots \times \mathbb{Z}/m_n\mathbb{Z}$$

by

$$\alpha\left([x]_M\right) = \left([x]_{m_1} [x]_{m_2} \cdots [x]_{m_n}\right)$$

for each $x \in \mathbb{Z}$. By Lemma 1.6.1, α is a well-defined, one-to-one mapping of $\mathbb{Z}/M\mathbb{Z}$ into $\mathbb{Z}_{m_1} \times \mathbb{Z}_{m_2} \times \cdots \times \mathbb{Z}_{m_n}$. Since

$$|\mathbb{Z}/M\mathbb{Z}| = M = |\mathbb{Z}/m_1\mathbb{Z} \times \mathbb{Z}/m_2\mathbb{Z} \times \cdots \times \mathbb{Z}/m_n\mathbb{Z}|,$$

α is onto. But then, given integers $a_1, a_2, \cdots, a_n$, there is an integer x such that

$$\alpha([x]_{m_1 m_2 \cdots m_n}) = ([a_1]_{m_1}[a_2]_{m_2} \cdots [a_n]_{m_n})$$

and therefore, $x \equiv a_i \pmod{m_i}$, for $i = 1, 2, \cdots, n$. By Lemma 1.6.1, any two solutions are congruent modulo M. □

Remark 1.6.7. If the system of the linear congruences (1.257) is soluble, then its solution can be conveniently described as follows:

$$x \equiv \sum_{i=1}^{n} a_i M_i M_i' \pmod{m} \tag{1.258}$$

where

$$\begin{aligned} m &= m_1 m_2 \cdots m_n, \\ M_i &= m/m_i, \\ M_i' &= M_i^{-1} \pmod{m_i}, \end{aligned}$$

for $i = 1, 2, \cdots, n$.

Example 1.6.19. Consider the Sun Zi problem:

$$\begin{aligned} x &\equiv 2 \pmod{3}, \\ x &\equiv 3 \pmod{5}, \\ x &\equiv 2 \pmod{7}. \end{aligned}$$

By (1.258), we have

$$\begin{aligned} m &= m_1 m_2 m_3 = 3 \cdot 5 \cdot 7 = 105, \\ M_1 &= m/m_1 = 105/3 = 35, \\ M_1' &= M_1^{-1} \pmod{m_1} = 35^{-1} \pmod{3} = 2, \\ M_2 &= m/m_2 = 105/5 = 21, \\ M_2' &= M_2^{-1} \pmod{m_2} = 21^{-1} \pmod{5} = 1, \\ M_3 &= m/m_3 = 105/7 = 15, \\ M_3' &= M_3^{-1} \pmod{m_3} = 15^{-1} \pmod{7} = 1. \end{aligned}$$

Hence,

$$\begin{aligned} x &= a_1 M_1 M_1' + a_2 M_2 M_2' + a_3 M_3 M_3' \pmod{m} \\ &= 2 \cdot 35 \cdot 2 + 3 \cdot 21 \cdot 1 + 2 \cdot 15 \cdot 1 \pmod{105} \\ &= 23. \end{aligned}$$

Exercise 1.6.1. Solve the following simultaneous congruences:

$$\begin{aligned} x &\equiv 2 \pmod{7}, \\ x &\equiv 7 \pmod{9}, \\ x &\equiv 3 \pmod{4}. \end{aligned}$$

The Chinese Remainder Theorem is very applicable in several central areas of mathematics and computer science, including algebra, number theory, computer arithmetic, fast computation, cryptography, computer security, and hash functions. We shall discuss some of these applications later.

1.6.5 High-Order Congruences

The congruences $ax \equiv b \pmod{m}$ we have studied so far are a special type of high-order congruence; that is, they are all linear congruences. In this subsection, we shall study the higher degree congruences, particularly the quadratic congruences.

Definition 1.6.11. Let m be a positive integer, and let

$$f(x) = a_0 + a_1 x + a_2 x^2 + \cdots + a_n x^n$$

be any polynomial with integer coefficients. Then a *high-order congruence* or a *polynomial congruence* is a congruence of the form

$$f(x) \equiv 0 \pmod{n}. \tag{1.259}$$

A polynomial congruence is also called a *polynomial congruential equation.*

Let us consider the polynomial congruence

$$f(x) = x^3 + 5x - 4 \equiv 0 \pmod{7}.$$

This congruence holds when $x = 2$, since

$$f(2) = 2^3 + 5 \cdot 2 - 4 \equiv 0 \pmod{7}.$$

Just as for algebraic equations, we say that $x = 2$ is a root or a solution of the congruence. In fact, any value of x which satisfies the following condition

$$x \equiv 2 \pmod{7}$$

is also a solution of the congruence. In general, as in linear congruence, when a solution x_0 has been found, all values x for which

$$x \equiv x_0 \pmod{n}$$

are also solutions. But, by convention, we still consider them as a *single* solution. Thus, our problem is to find all incongruent (different) solutions of $f(x) \equiv 0 \pmod{n}$. In general, this problem is very difficult, and many techniques for solution depend partially on trial-and-error methods. For example, to find all solutions of the congruence $f(x) \equiv 0 \pmod{n}$, we could certainly try all values $0, 1, 2, \cdots, n-1$ (or the numbers in the complete residue system modulo n), and determine which of them satisfy the congruence; this would give us the total number of *incongruent* solutions modulo n.

Theorem 1.6.22. Let $M = m_1 m_2 \cdots m_n$, where $m_1, m_2, \cdots, m_n$ are pairwise relatively prime. Then the integer x_0 is a solution of

$$f(x) \equiv 0 \pmod{M} \tag{1.260}$$

if and only if x_0 is a solution of the system of polynomial congruences:

$$\left.\begin{array}{l} f(x) \equiv 0 \pmod{m_1}, \\ f(x) \equiv 0 \pmod{m_2}, \\ \quad \cdots\cdots \\ \quad \cdots\cdots \\ f(x) \equiv 0 \pmod{m_n}. \end{array}\right\} \tag{1.261}$$

If x and x' are two solutions, then $x \equiv x' \pmod{M}$, where $M = m_1 m_2 \cdots m_n$.

Proof. If $f(a) \equiv 0 \pmod{M}$, then obviously $f(a) \equiv 0 \pmod{m_i}$, for $i = 1, 2, \cdots, n$. Conversely, suppose a is a solution of the system

$$f(x) \equiv 0 \pmod{m_i}, \quad \text{for } i = 1, 2, \cdots, n.$$

Then $f(a)$ is a solution of the system

$$\left.\begin{array}{l} y \equiv 0 \pmod{m_1}, \\ y \equiv 0 \pmod{m_2}, \\ \quad \cdots\cdots \\ \quad \cdots\cdots \\ y \equiv 0 \pmod{m_n} \end{array}\right\}$$

and it follows from the Chinese Remainder Theorem that $f(a) \equiv 0 \pmod{m_1 m_2 \cdots m_n}$. Thus, a is a solution of $f(x) \equiv 0 \pmod{M}$. □

We now restrict ourselves to quadratic congruences, the simplest possible nonlinear polynomial congruences.

Definition 1.6.12. A quadratic congruence is a congruence of the form:

$$x^2 \equiv a \pmod{n} \tag{1.262}$$

where $\gcd(a, n) = 1$. To solve the congruence is to find an integral solution for x which satisfies the congruence.

In most cases, it is sufficient to study the above congruence rather than the following more general quadratic congruence

$$ax^2 + bx + c \equiv 0 \pmod{n} \tag{1.263}$$

since if $\gcd(a, n) = 1$ and b is even or n is odd, then the congruence (1.263) can be reduced to a congruence of type (1.262). The problem can even be further reduced to solving a congruence of the type (if $n = p_1^{\alpha_1} p_2^{\alpha_2} \cdots p_k^{\alpha_k}$, where $p_1, p_2, \cdots p_k$ are primes, and $\alpha_1, \alpha_2, \cdots, \alpha_k$ are positive integers):

$$x^2 \equiv a \pmod{p_1^{\alpha_1} p_2^{\alpha_2} \cdots p_k^{\alpha_k}} \tag{1.264}$$

because solving the congruence (1.264) is equivalent to solving the following system of congruences:

$$\left.\begin{array}{l} x^2 \equiv a \pmod{p_1^{\alpha_1}} \\ x^2 \equiv a \pmod{p_2^{\alpha_2}} \\ \quad \cdots\cdots \\ \quad \cdots\cdots \\ x^2 \equiv a \pmod{p_k^{\alpha_k}}. \end{array}\right\} \tag{1.265}$$

In what follows, we shall be only interested in quadratic congruences of the form

$$x^2 \equiv a \pmod{p} \tag{1.266}$$

where p is an odd prime and $a \not\equiv 0 \pmod{p}$.

Definition 1.6.13. Let a be any integer and n a natural number, and suppose that $\gcd(a, n) = 1$. Then a is called a quadratic residue modulo n if the congruence

$$x^2 \equiv a \pmod{n}$$

is soluble. Otherwise, it is called a quadratic nonresidue modulo n.

Remark 1.6.8. Similarly, we can define the cubic residues, and fourth-power residues, etc. For example, a is a kth power residue modulo n if the congruence

$$x^k \equiv a \pmod{n} \tag{1.267}$$

is soluble. Otherwise, it is a kth power nonresidue modulo n.

Theorem 1.6.23. Let p be an odd prime and a an integer not divisible by p. Then the congruence

$$x^2 \equiv a \pmod{p} \tag{1.268}$$

has either no solution, or exactly two congruence solutions modulo p.

Proof. If x and y are solutions to $x^2 \equiv a \pmod{p}$, then $x^2 \equiv y^2 \pmod{p}$, that is, $p \mid (x^2 - y^2)$. Since $x^2 - y^2 = (x+y)(x-y)$, we must have $p \mid (x-y)$ or $p \mid (x+y)$, that is, $x \equiv \pm y \pmod{p}$. Hence, any two distinct solutions modulo p differ only by a factor of -1. □

Example 1.6.20. Find the quadratic residues and quadratic nonresidues for moduli $5, 7, 11$, respectively.

(1) Modulo 5, the integers $1, 4$ are quadratic residues, whilst $2, 3$ are quadratic nonresidues, since

$$1^2 \equiv 4^2 \equiv 1, \qquad 2^2 \equiv 3^2 \equiv 4.$$

(2) Modulo 7, the integers $1, 2, 4$ are quadratic residues, whilst $3, 5, 6$ are quadratic nonresidues, since

$$\begin{aligned} 1^2 &\equiv 6^2 \equiv 1, & 2^2 &\equiv 5^2 \equiv 4, \\ 3^2 &\equiv 4^2 \equiv 2. \end{aligned}$$

(3) Modulo 11, the integers $1, 3, 4, 5, 9$ are quadratic residues, whilst $2, 6, 7, 8, 10$ are quadratic nonresidues, since

$$\begin{aligned} 1^2 &\equiv 10^2 \equiv 1, & 2^2 &\equiv 9^2 \equiv 4, \\ 3^2 &\equiv 8^2 \equiv 9, & 4^2 &\equiv 7^2 \equiv 5, \\ 5^2 &\equiv 6^2 \equiv 3. \end{aligned}$$

(4) Modulo 15, only the integers 1 and 4 are quadratic residues, whilst $2, 3, 5, 6, 7, 8, 9, 10, 11, 12, 13, 14$ are all quadratic nonresidues, since

$$\begin{aligned} 1^2 \equiv 4^2 \equiv 11^2 \equiv 14^2 \equiv 1, \\ 2^2 \equiv 7^2 \equiv 8^2 \equiv 13^2 \equiv 4. \end{aligned}$$

(5) Modulo 23, the integers $1, 2, 3, 4, 6, 8, 9, 12, 13, 16, 18$ are quadratic residues, whilst $5, 7, 10, 11, 14, 15, 17, 19, 20, 21, 22$ are quadratic nonresidues, since

$$\begin{aligned} 1^2 &\equiv 22^2 \equiv 1, & 5^2 &\equiv 18^2 \equiv 2, \\ 7^2 &\equiv 16^2 \equiv 3, & 2^2 &\equiv 21^2 \equiv 4, \\ 11^2 &\equiv 12^2 \equiv 6, & 10^2 &\equiv 13^2 \equiv 8, \\ 3^2 &\equiv 20^2 \equiv 9, & 9^2 &\equiv 14^2 \equiv 12, \\ 6^2 &\equiv 17^2 \equiv 13, & 4^2 &\equiv 19^2 \equiv 16, \\ 8^2 &\equiv 15^2 \equiv 18. \end{aligned}$$

The above example illustrates the following two theorems:

Theorem 1.6.24. Let p be an odd prime and $N(p)$ the number of consecutive pairs of quadratic residues modulo p in the interval $[1, p-1]$. Then

$$N(p) = \frac{1}{4}\left(p - 4 - (-1)^{(p-1)/2}\right). \tag{1.269}$$

Proof. (Sketch) The complete proof of this theorem can be found in [10]; here we only give the sketch of the proof. Let (**RR**), (**RN**), (**NR**) and (**NN**) denote the number of pairs of two quadratic residues, of a quadratic residue followed by a quadratic nonresidue, of a quadratic nonresidue followed by a quadratic residue, of two quadratic nonresidues, among pairs of consecutive positive integers less than p, respectively. Then from [10], we have:

$$\begin{aligned}
(\mathbf{RR}) + (\mathbf{RN}) &= \frac{1}{2}\left(p - 2 - (-1)^{(p-1)/2}\right) \\
(\mathbf{NR}) + (\mathbf{NN}) &= \frac{1}{2}\left(p - 2 + (-1)^{(p-1)/2}\right) \\
(\mathbf{RR}) + (\mathbf{NR}) &= \frac{1}{2}(p-1)) - 1 \\
(\mathbf{RN}) + (\mathbf{NN}) &= \frac{1}{2}(p-1)) \\
(\mathbf{RR}) + (\mathbf{NN}) - (\mathbf{RN}) - (\mathbf{NR}) &= -1 \\
(\mathbf{RR}) + (\mathbf{NN}) &= \frac{1}{2}(p-3) \\
(\mathbf{RR}) - (\mathbf{NN}) &= -\frac{1}{2}\left(1 + (-1)^{(p-1)/2}\right)
\end{aligned}$$

Hence $(\mathbf{RR}) = \frac{1}{4}\left(p - 4 - (-1)^{(p-1)/2}\right)$. □

Remark 1.6.9. Similarly, Let $\nu(p)$ denote the number of consecutive triples of quadratic residues in the interval $[1, p-1]$, where p is odd prime. Then

$$\nu(p) = \frac{1}{8}p + E_p, \tag{1.270}$$

where $|E_p| < \frac{1}{8}\sqrt{p} + 2$.

Example 1.6.21. For $p = 23$, there are five consecutive pairs of quadratic residues, namely, $(1, 2)$, $(2, 3)$, $(3, 4)$, $(8, 9)$ and $(12, 13)$, modulo 23; there are also one consecutive triple of quadratic residues, namely, $(1, 2, 3)$, modulo 23.

Theorem 1.6.25. Let p be an odd prime. Then there are exactly $(p-1)/2$ quadratic residues and exactly $(p-1)/2$ quadratic nonresidues modulo p.

Proof. Consider the $p - 1$ congruences:

$$\begin{aligned}
x^2 &\equiv 1 \pmod{p} \\
x^2 &\equiv 2 \pmod{p} \\
&\cdots\cdots \\
&\cdots\cdots \\
x^2 &\equiv p - 1 \pmod{p}.
\end{aligned}$$

Since each of the above congruences has either no solution or exactly two congruence solutions modulo p, there must be exactly $(p-1)/2$ quadratic residues modulo p among the integers $1, 2, \cdots, p-1$. The remaining $p - 1 - (p-1)/2 = (p-1)/2$ positive integers less than $p-1$ are quadratic nonresidues modulo p. □

Example 1.6.22. Again for $p = 23$, there are eleven quadratic residues, and eleven quadratic nonresidues modulo 23.

Remark 1.6.10. Note that here $15 = 3 \cdot 5$ is a composite number. Let Q_n be the quadratic residues modulo n with n composite. Then for $n = p \cdot q$ with p, q prime, we have

$$Q_n = Q_p \cup Q_q.$$

This fact suggests that the quadratic residues modulo a composite n can be determined quickly if the prime factorization of n is known. For example, let $n = 15$, we have

$$Q_{15} = Q_3 \cup Q_5 = \{1\} \cup \{1, 4\} = \{1, 4\}.$$

Euler devised a simple criterion for deciding whether an integer a is a quadratic residue modulo a prime number p.

Theorem 1.6.26 (Euler's criterion). Let p be an odd prime and $\gcd(a, p) = 1$. Then a is a quadratic residue modulo p if and only if

$$a^{(p-1)/2} \equiv 1 \pmod{p}.$$

Proof. Using Fermat's little theorem, we find that

$$(a^{(p-1)/2} - 1)(a^{(p-1)/2} + 1) \equiv a^{p-1} - 1 \equiv 0 \pmod{p}$$

thus

$$a^{(p-1)/2} \equiv 1 \pmod{p}.$$

If a is a quadratic residue modulo p, then there exists an integer x_0 such that $x_0^2 \equiv a \pmod{p}$. By Fermat's little theorem, we have

$$a^{(p-1)/2} \equiv (x_0^2)^{(p-1)/2} \equiv x_0^{p-1} \equiv 1 \pmod{p}.$$

To prove the converse, we assume that $a^{(p-1)/2} \equiv 1 \pmod{p}$. If g is a primitive root modulo p (g is a primitive root modulo p if $\text{order}(g, p) = \phi(p)$; we shall formally define primitive roots in Subsection 1.6.7), then there exists a positive integer t such that $g^t \equiv a \pmod{p}$. Then

$$g^{t(p-1)/2} \equiv a^{(p-1)/2} \equiv 1 \pmod{p}$$

which implies that

$$t(p-1)/2 \equiv 0 \pmod{p-1}.$$

Thus, t is even, and so

$$(g^{t/2})^2 \equiv g^t \equiv a \pmod{p}$$

which implies that a is a quadratic residue modulo p. □

Euler's criterion is not very useful as a practical test for deciding whether or not an integer is a quadratic residue, unless the modulus is small. Euler's studies on quadratic residues were further developed by Legendre, who introduced, in his own honour, the Legendre symbol, which will be the subject matter of our next subsection.

1.6.6 Legendre and Jacobi Symbols

Definition 1.6.14. Let p be an odd prime and a an integer. Suppose that $\gcd(a,p)=1$. Then the *Legendre symbol*, $\left(\dfrac{a}{p}\right)$, is defined by

$$\left(\frac{a}{p}\right) = \begin{cases} = 1, & \text{if } a \text{ is a quadratic residue modulo } p, \\ = -1, & \text{if } a \text{ is a quadratic non-residue modulo } p. \end{cases} \tag{1.271}$$

We shall use the notation $a \in Q_p$ to denote that a is a quadratic residue modulo p; similarly, $a \in \overline{Q}_p$ will be used to denote that a is a quadratic nonresidue modulo p.

Example 1.6.23. Let $p = 7$ and

$$\begin{array}{lll} 1^2 \equiv 1 \pmod 7, & 2^2 \equiv 4 \pmod 7, & 3^2 \equiv 2 \pmod 7, \\ 4^2 \equiv 2 \pmod 7, & 5^2 \equiv 4 \pmod 7, & 6^2 \equiv 1 \pmod 7. \end{array}$$

Then

$$\left(\frac{1}{7}\right) = \left(\frac{2}{7}\right) = \left(\frac{4}{7}\right) = 1, \quad \left(\frac{3}{7}\right) = \left(\frac{5}{7}\right) = \left(\frac{6}{7}\right) = -1.$$

Some elementary properties of the Legendre symbol, which can be used to evaluate it, are given in the following theorem.

Theorem 1.6.27. Let p be an odd prime, and a and b integers that are relatively prime to p. Then:

(1) If $a \equiv b \pmod p$, then $\left(\dfrac{a}{p}\right) = \left(\dfrac{b}{p}\right)$.

(2) $\left(\dfrac{a^2}{p}\right) = 1$, and so $\left(\dfrac{1}{p}\right) = 1$.

(3) $\left(\dfrac{a}{p}\right) \equiv a^{(p-1)/2} \pmod p$.

(4) $\left(\dfrac{ab}{p}\right) = \left(\dfrac{a}{p}\right)\left(\dfrac{b}{p}\right)$.

(5) $\left(\dfrac{-1}{p}\right) = (-1)^{(p-1)/2}$.

Proof. Assume p is an odd prime and $\gcd(p, a) = \gcd(p, b) = 1$.

(1) If $a \equiv b \pmod{p}$, then $x^2 \equiv a \pmod{p}$ has solution if and only if $x^2 \equiv b \pmod{p}$ has a solution. Hence $\left(\dfrac{a}{p}\right) = \left(\dfrac{b}{p}\right)$.

(2) The quadratic congruence $x^2 \equiv a^2 \pmod{p}$ clearly has a solution, namely a, so $\left(\dfrac{a^2}{p}\right) = 1$.

(3) This is Euler's criterion in terms of Legendre's symbol.

(4) We have

$$\begin{aligned} \left(\frac{ab}{p}\right) &\equiv (ab)^{(p-1)/2} \pmod{p} \quad \text{(by Euler's criterion)} && (1.272) \\ &\equiv a^{(p-1)/2}b^{(p-1)/2} \pmod{p} && (1.273) \\ &\equiv \left(\frac{a}{p}\right)\left(\frac{b}{p}\right) && (1.274) \end{aligned}$$

(5) By Euler's criterion, we have

$$\left(\frac{-1}{p}\right) = (-1)^{(p-1)/2}.$$

This completes the proof. □

Corollary 1.6.7. Let p be an odd prime. Then

$$\left(\frac{-1}{p}\right) = \begin{cases} 1 & \text{if } p \equiv 1 \pmod{4} \\ -1 & \text{if } p \equiv 3 \pmod{4}. \end{cases} \qquad (1.275)$$

Proof. If $p \equiv 1 \pmod{4}$, then $p = 4k + 1$ for some integer k. Thus,

$$(-1)^{(p-1)/2} = (-1)^{((4k+1)-1)/2} = (-1)^{2k} = 1,$$

so that $\left(\dfrac{-1}{p}\right) = 1$. The proof for $p \equiv 3 \pmod{4}$ is similar. □

Example 1.6.24. Does $x^2 \equiv 63 \pmod{11}$ have a solution? We first evaluate the Legendre symbol $\left(\frac{63}{11}\right)$ corresponding to the quadratic congruence as follows:

$$\begin{aligned}
\left(\frac{63}{11}\right) &= \left(\frac{8}{11}\right) && \text{by (1) of Theorem 1.6.27} \\
&= \left(\frac{2}{11}\right)\left(\frac{2^2}{11}\right) && \text{by (2) of Theorem 1.6.27} \\
&= \left(\frac{2}{11}\right)\cdot 1 && \text{by (2) of Theorem 1.6.27} \\
&= -1 && \text{by ``trial and error''.}
\end{aligned}$$

Therefore, the quadratic congruence $x^2 \equiv 63 \pmod{11}$ has no solution.

To avoid the "trial and error" in the above and similar examples, we introduce in the following Gauss's lemma for evaluating the Legendre symbol.

Definition 1.6.15. Let $a \in \mathbb{Z}$ and $n \in \mathbb{N}$. Then the *least residue* of a modulo n is the integer a' in the interval $(-n/2, n/2]$ such that $a \equiv a' \pmod{n}$. We denote the least residue of a modulo n by $\mathrm{LR}_n(a)$.

Example 1.6.25. The set $\{-5, -4, -3, -2, -1, 0, 1, 2, 3, 4, 5\}$ is a complete set of of the least residues modulo 11. Thus, $\mathrm{LR}_{11}(21) = -1$ since $21 \equiv 10 \equiv -1 \pmod{11}$; similarly, $\mathrm{LR}_{11}(99) = 0$ and $\mathrm{LR}_{11}(70) = 4$.

Lemma 1.6.2 (Gauss's lemma). Let p be an odd prime number and suppose that $\gcd(a, b) = 1$. Further, let ω be the number of integers in the set

$$\left\{1a,\ 2a,\ 3a,\ \cdots,\ \left(\frac{p-1}{2}\right)a\right\}$$

whose least residues modulo p are negative (or greater than $p/2$), then

$$\left(\frac{a}{p}\right) = (-1)^{\omega}. \tag{1.276}$$

Proof. When we reduce the following numbers (modulo p)

$$\left\{a,\ 2a,\ 3a,\ \cdots,\ \left(\frac{p-1}{2}\right)a\right\}$$

to lie in set

$$\left\{\pm 1, \pm 2, \cdots,\ \pm\left(\frac{p-1}{2}\right)a\right\},$$

then no two different numbers ma and na can go to the same numbers. Further, it cannot happen that ma goes to k and na goes to $-k$, because then $ma + na \equiv k + (-k) \equiv 0 \pmod{p}$, and hence (multiplying by the inverse of a), $m + n \equiv 0 \pmod{p}$, which is impossible. Hence, when reduced the numbers

$$\left\{a,\ 2a,\ 3a,\ \cdots,\ \left(\frac{p-1}{2}\right)a\right\}$$

we get exactly one of -1 and 1, exactly one of -2 and 2, $\cdots$, exactly one of $-(p-1)/2$ and $(p-1)/2$. Hence, modulo p, we get

$$a \cdot 2a \cdots \left(\frac{p-1}{2}\right) a \equiv 1 \cdot 2 \cdots \left(\frac{p-1}{2}\right) (-1)^{\omega} \pmod{p}.$$

Cancelling the numbers $1, 2, \cdots, (p-1)/2$, we have

$$a^{(p-1)/2} \equiv (-1)^{\omega} \pmod{p}.$$

By Euler's criterion, we have $\left(\frac{a}{p}\right) \equiv (-1)^{\omega} \pmod{p}$. Since $\left(\frac{a}{p}\right) \equiv \pm 1$, we must have $\left(\frac{a}{p}\right) \equiv (-1)^{\omega}$. □

Example 1.6.26. Use Gauss's lemma to evaluate the Legendre symbol $\left(\frac{6}{11}\right)$. By Gauss's lemma, $\left(\frac{6}{11}\right) = (-1)^{\omega}$, where ω is the number of integers in the set

$$\{1 \cdot 6,\ 2 \cdot 6,\ 3 \cdot 6,\ 4 \cdot 6,\ 5 \cdot 6\}$$

whose least residues modulo 11 are negative (or greater than 11/2). Clearly,

$$(6, 12, 18, 24, 30) \bmod 11 \equiv (6, 1, 7, 2, 8) \equiv (-5, 1, -4, 2, -3) \pmod{11}$$

So, there are 3 least residues that are negative (or greater than 11/2). Thus, $\omega = 3$. Therefore, $\left(\frac{6}{11}\right) = (-1)^3 = -1$. Consequently, the quadratic congruence $x^2 \equiv 6 \pmod{11}$ is not solvable.

Remark 1.6.11. Gauss's lemma is similar to Euler's criterion in the following ways:

(1) Gauss's lemma provides a method for direct evaluation of the Legendre symbol;

(2) It has more significance as a theoretical tool than as a computational tool.

Gauss's lemma provides, among many others, a means for deciding whether or not 2 is a quadratic residue modulo an odd prime p.

Theorem 1.6.28. If p is an odd prime, then

$$\left(\frac{2}{p}\right) = (-1)^{(p^2-1)/8} = \begin{cases} 1, & \text{if } p \equiv \pm 1 \pmod{8} \\ -1, & \text{if } p \equiv \pm 3 \pmod{8}. \end{cases} \tag{1.277}$$

Proof. By Gauss's lemma, we know that if ω is the number of least positive residues of the integers

$$1 \cdot 2,\ 2 \cdot 2, \cdots, \frac{p-1}{2} \cdot 2$$

that are greater than $p/2$, then $\left(\frac{2}{p}\right) = (-1)^{\omega}$. Let $k \in \mathbb{Z}$ with $1 \leq k \leq (p-1)/2$. Then $2k < p/2$ if and only if $k < p/4$; so $[p/4]$ of the integers $1 \cdot 2,\ 2 \cdot 2, \cdots, \frac{p-1}{2} \cdot 2$ are less than $p/2$. So, there are $\omega = (p-1)/2 - [p/4]$ integers greater than $p/2$. Therefore, by Gauss's lemma, we have

$$\left(\frac{2}{p}\right) = (-1)^{\frac{p-1}{2} - \left[\frac{p}{4}\right]}.$$

For the first equality, it suffices to show that

$$\frac{p-1}{2} - \left[\frac{p}{4}\right] \equiv \frac{p^2-1}{8} \pmod{2}.$$

If $p \equiv 1 \pmod{8}$, then $p = 8k+1$ for some $k \in \mathbb{Z}$, from which

$$\frac{p-1}{2} - \left[\frac{p}{4}\right] = \frac{(8k+1)-1}{2} - \left[\frac{8k+1}{4}\right] = 4k - 2k = 2k \equiv 0 \pmod{2},$$

and

$$\frac{p^2-1}{8} = \frac{(8k+1)^2-1}{8} = \frac{64k^2+16k}{8} = 8k^2 + 2k \equiv 0 \pmod{2},$$

so the desired congruence holds for $p \equiv 1 \pmod{8}$. The cases for $p \equiv -1, \pm 3 \pmod{8}$ are similar. This completes the proof for the first equality of the theorem. Note that the cases above yield

$$\frac{p^2-1}{8} = \begin{cases} 1, & \text{if } p \equiv \pm 1 \pmod{8} \\ -1, & \text{if } p \equiv \pm 3 \pmod{8} \end{cases}$$

which implies

$$(-1)^{(p^2-1)/8} = \begin{cases} 1, & \text{if } p \equiv \pm 1 \pmod{8} \\ -1, & \text{if } p \equiv \pm 3 \pmod{8} \end{cases}$$

This completes the second equality of the theorem. □

Example 1.6.27. Evaluate $\left(\frac{2}{7}\right)$ and $\left(\frac{2}{53}\right)$.

(1) By Theorem 1.6.28, we have $\left(\frac{2}{7}\right) = 1$, since $7 \equiv 7 \pmod{8}$. Consequently, the quadratic congruence $x^2 \equiv 2 \pmod{7}$ is solvable.

(2) By Theorem 1.6.28, we have $\left(\frac{2}{53}\right) = -1$, since $53 \equiv 5 \pmod{8}$. Consequently, the quadratic congruence $x^2 \equiv 2 \pmod{53}$ is not solvable.

Using Lemma 1.6.2, Gauss proved the following theorem, one of the great results of mathematics:

Theorem 1.6.29 (Quadratic reciprocity law). If p and q are distinct odd primes, then

(1) $\left(\frac{p}{q}\right) = \left(\frac{q}{p}\right)$ if one of $p, q \equiv 1 \pmod 4$.

(2) $\left(\frac{p}{q}\right) = -\left(\frac{q}{p}\right)$ if both $p, q \equiv 3 \pmod 4$.

Remark 1.6.12. This theorem may be stated equivalently in the form

$$\left(\frac{p}{q}\right)\left(\frac{q}{p}\right) = (-1)^{(p-1)(q-1)/4}. \tag{1.278}$$

Proof. We first observe that, by Gauss's lemma, $\left(\frac{p}{q}\right) = 1^{\omega}$, where ω is the number of lattice points (x, y) (that is, pairs of integers) satisfying $0 < x < q/2$ and $-q/2 < px - qy < 0$. These inequalities give $y < (px/q) + 1/2 < (p+1)/2$. Hence, since y is an integer, we see ω is the number of lattice points in the rectangle R defined by $0 < x < q/2$, $0 < y < p/2$, satisfying $-q/2 < px - qy < 0$ (see Figure 1.10). Similarly, $\left(\frac{q}{p}\right) = 1^{\mu}$, where μ is the number of lattice points in R satisfying $-p/2 < qx - py < 0$. Now, it suffices to prove that $(p-1)(q-1)/4 - (\omega + \mu)$ is even. But $(p-1)(q-1)/4$ is just the number of lattice points in R satisfying that $px - qy \leq q/2$ or $qy - px \leq -p/2$. The regions in R defined by these inequalities are disjoint and they contain the same number of lattice points, since the substitution

$$x = (q+1)/2 - x', \quad y = (p+1)/2 - y'$$

furnishes a one-to-one correspondence between them. The theorem follows. □

Remark 1.6.13. The Quadratic Reciprocity Law was one of Gauss's major contributions. For those who consider number theory "the Queen of Mathematics", this is one of the jewels in her crown. Since Gauss's time, over 150 proofs of it have been published; Gauss himself published not less than six different proofs. Among the eminent mathematicians who contributed to the proofs are Cauchy, Jacobi, Dirichlet, Eisenstein, Kronecker and Dedekind.

Combining all the above results for Legendre symbols, we get the following set of formulas for evaluating Legendre symbols:

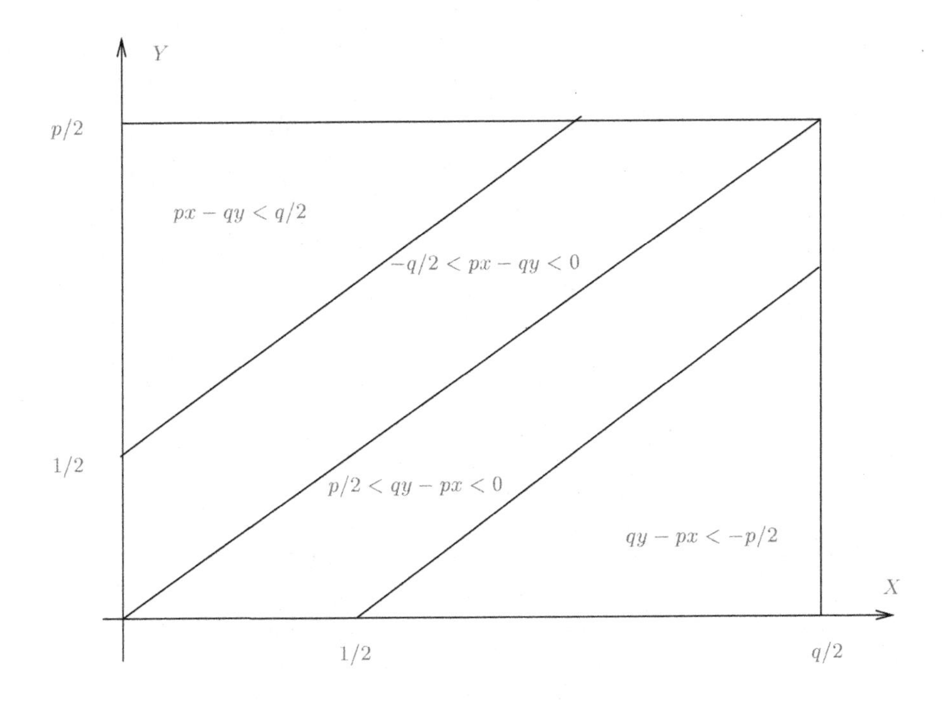

Figure 1.10. Proof of the quadratic reciprocity law

$$\left(\frac{a}{p}\right) = a^{(p-1)/2} \pmod{p} \tag{1.279}$$

$$\left(\frac{1}{p}\right) = 1 \tag{1.280}$$

$$\left(\frac{-1}{p}\right) = (-1)^{(p-1)/2} \tag{1.281}$$

$$a \equiv b \pmod{p} \Longrightarrow \left(\frac{a}{p}\right) = \left(\frac{b}{p}\right) \tag{1.282}$$

(1.283)

$$\left(\frac{a_1 a_2 \cdots a_k}{p}\right) = \left(\frac{a_1}{p}\right)\left(\frac{a_2}{p}\right) \cdots\cdots \left(\frac{a_k}{p}\right) \tag{1.284}$$

$$\left(\frac{ab^2}{p}\right) = \left(\frac{a}{p}\right), \text{ for } p \nmid b \tag{1.285}$$

$$\left(\frac{2}{p}\right) = (-1)^{(p^2-1)/8} \tag{1.286}$$

$$\left(\frac{p}{q}\right) = (-1)^{(p-1)(q-1)/4} \left(\frac{q}{p}\right). \tag{1.287}$$

Example 1.6.28. Evaluate the Legendre symbol $\left(\frac{33}{83}\right)$.

$$\begin{aligned}
\left(\frac{33}{83}\right) &= \left(\frac{-50}{83}\right) && \text{by (1.282)} \\
&= \left(\frac{-2}{83}\right)\left(\frac{5^2}{83}\right) && \text{by (1.284)} \\
&= \left(\frac{-2}{83}\right) && \text{by (1.285)} \\
&= -\left(\frac{2}{83}\right) && \text{by (1.281)} \\
&= 1 && \text{by (1.286).}
\end{aligned}$$

It follows that the quadratic congruence $33 \equiv x^2 \pmod{83}$ is soluble.

Example 1.6.29. Evaluate the Legendre symbol $\left(\frac{46}{997}\right)$.

$$\begin{aligned}
\left(\frac{46}{997}\right) &= \left(\frac{2}{997}\right)\left(\frac{23}{997}\right) && \text{by (1.284)} \\
&= -\left(\frac{23}{997}\right) && \text{by (1.286)} \\
&= -\left(\frac{997}{23}\right) && \text{by (1.287)} \\
&= -\left(\frac{8}{23}\right) && \text{by (1.282)} \\
&= -\left(\frac{2^2 \cdot 2}{23}\right) && \text{by (1.284)} \\
&= -\left(\frac{2}{23}\right) && \text{by (1.285)} \\
&= -1 && \text{by (1.286).}
\end{aligned}$$

It follows that the quadratic congruence $46 \equiv x^2 \pmod{997}$ is not soluble.

Gauss's quadratic reciprocity law enables us to evaluate the values of Legendre symbols $\left(\frac{a}{p}\right)$ very quickly, provided a is a prime or a product of primes, and p is an odd prime. However, when a is a composite, we must factor it into its prime decomposition form in order to use Gauss's quadratic reciprocity law. Unfortunately, there is no efficient algorithm so far for prime decomposition (see Chapter 2 for more information). One way to overcome the difficulty of factoring a is to introduce the following Jacobi symbol (in honour of the German mathematician Jacobi[34]), which is a natural generalization of the Legendre symbol:

Definition 1.6.16. Let a be an integer and $n > 1$ an odd positive integer. If $n = p_1^{\alpha_1} p_2^{\alpha_2} \cdots p_k^{\alpha_k}$, then the *Jacobi symbol*, $\left(\frac{a}{n}\right)$, is defined by

$$\left(\frac{a}{n}\right) = \left(\frac{a}{p_1}\right)^{\alpha_1} \left(\frac{a}{p_2}\right)^{\alpha_2} \cdots \left(\frac{a}{p_k}\right)^{\alpha_k}, \tag{1.288}$$

where $\left(\frac{a}{p_i}\right)$ for $i = 1, 2, \cdots, k$ is the Legendre symbol for the odd prime p_i. If n is an odd prime, the Jacobi symbol is *just* the Legendre symbol.

The Jacobi symbol has some similar properties to the Legendre symbol, as shown in the following theorem.

Theorem 1.6.30. Let m and n be any positive odd composites, and $\gcd(a, n) = \gcd(b, n) = 1$. Then

(1) If $a \equiv b \pmod{n}$, then $\left(\frac{a}{n}\right) = \left(\frac{b}{n}\right)$.

(2) $\left(\frac{a}{n}\right)\left(\frac{b}{n}\right) = \left(\frac{ab}{n}\right)$.

(3) If $\gcd(m, n) = 1$, then $\left(\frac{a}{mn}\right)\left(\frac{a}{m}\right) = \left(\frac{a}{n}\right)$.

(4) $\left(\frac{-1}{n}\right) = (-1)^{(n-1)/2}$.

(5) $\left(\frac{2}{n}\right) = (-1)^{(n^2-1)/8}$.

(6) If $\gcd(m, n) = 1$, then $\left(\frac{m}{n}\right)\left(\frac{n}{m}\right) = (-1)^{(m-1)(n-1)/4}$.

Remark 1.6.14. It should be noted that the Jacobi symbol $\left(\frac{a}{n}\right) = 1$ does not imply that a is a quadratic residue modulo n. Indeed, a is a quadratic

---

34

Carl Gustav Jacobi (1804–1851) was largely self-taught, learning his mathematics from the works of Euler and Lagrange. He entered the University of Berlin in 1821 and obtained his PhD in 1825, with a thesis on continued fractions. In 1826 he became a lecturer at the University of Königsberg and was appointed professor there in 1831. Jacobi is mainly known for his work in the theory of elliptic functions and was not primarily a number theorist; nevertheless, he made important contributions to number theory.

residue modulo n if and only if a is a quadratic residue modulo p for each prime divisor p of n. For example, the Jacobi symbol $\left(\frac{2}{3599}\right) = 1$, but the quadratic congruence $x^2 \equiv 2 \pmod{3599}$ is actually not soluble. This is the significant difference between the Legendre symbol and the Jacobi symbol. However, $\left(\frac{a}{n}\right) = -1$ does imply that a is a quadratic nonresidue modulo n. For example, the Jacobi symbol

$$\left(\frac{6}{35}\right) = \left(\frac{6}{5}\right)\left(\frac{6}{7}\right) = \left(\frac{1}{5}\right)\left(\frac{-1}{7}\right) = -1,$$

and so we can conclude that 6 is a quadratic nonresidue modulo 35. In short, we have

$$\left.\begin{array}{l} \left(\frac{a}{p}\right) = \begin{cases} 1, & a \equiv x^2 \pmod{p} \text{ is soluble} \\ -1, & a \equiv x^2 \pmod{p} \text{ is not soluble.} \end{cases} \\ \\ \left(\frac{a}{n}\right) = \begin{cases} 1, & a \equiv x^2 \pmod{n} \text{ may or may not be soluble} \\ -1, & a \equiv x^2 \pmod{n} \text{ is not soluble.} \end{cases} \end{array}\right\} \qquad (1.289)$$

Combining all the above results for Jacobi symbols, we get the following set of formulas for evaluating Jacobi symbols:

$$\left(\frac{1}{n}\right) = 1 \qquad (1.290)$$

$$\left(\frac{-1}{n}\right) = (-1)^{(n-1)/2} \qquad (1.291)$$

$$a \equiv b \pmod{n} \Longrightarrow \left(\frac{a}{n}\right) = \left(\frac{b}{n}\right) \qquad (1.292)$$

$$\left(\frac{a_1 a_2 \cdots a_k}{n}\right) = \left(\frac{a_1}{n}\right)\left(\frac{a_2}{n}\right) \cdots\cdots \left(\frac{a_k}{n}\right) \qquad (1.293)$$

$$\left(\frac{ab^2}{n}\right) = \left(\frac{a}{n}\right), \text{ for } \gcd(b, n) = 1 \qquad (1.294)$$

$$\left(\frac{2}{n}\right) = (-1)^{(n^2-1)/8} \qquad (1.295)$$

$$\left(\frac{m}{n}\right) = (-1)^{(m-1)(n-1)/4} \left(\frac{n}{m}\right). \qquad (1.296)$$

Example 1.6.30. Evaluate the Jacobi symbol $\left(\frac{286}{563}\right)$.

$$\begin{aligned}
\left(\frac{286}{563}\right) &= \left(\frac{2}{563}\right)\left(\frac{143}{563}\right) && \text{by} \quad (1.293) \\
&= -\left(\frac{143}{563}\right) && \text{by} \quad (1.295) \\
&= \left(\frac{563}{143}\right) && \text{by} \quad (1.296) \\
&= \left(\frac{-3^2}{143}\right) && \text{by} \quad (1.292) \\
&= -\left(\frac{3^2}{143}\right) && \text{by} \quad (1.291) \\
&= -1 && \text{by} \quad (1.294).
\end{aligned}$$

It follows that the quadratic congruence $286 \equiv x^2 \pmod{563}$ is not soluble.

Example 1.6.31. Evaluate the Jacobi symbol $\left(\frac{1009}{2307}\right)$.

$$\begin{aligned}
\left(\frac{1009}{2307}\right) &= \left(\frac{2307}{1009}\right) && \text{by} \quad (1.296) \\
&= \left(\frac{289}{1009}\right) && \text{by} \quad (1.292) \\
&= \left(\frac{17^2}{1009}\right) && \text{by} \quad (1.293) \\
&= 1 && \text{by} \quad (1.294).
\end{aligned}$$

Although the Jacobi symbol $\left(\frac{1009}{2307}\right) = 1$, we still cannot determine whether or not the quadratic congruence $1009 \equiv x^2 \pmod{2307}$ is soluble.

Remark 1.6.15. Jacobi symbols can be used to facilitate the calculation of Legendre symbols. In fact, Legendre symbols can be eventually calculated by Jacobi symbols [17]. That is, the Legendre symbol can be calculated as if it were a Jacobi symbol. For example, consider the Legendre symbol $\left(\frac{335}{2999}\right)$, where $335 = 5 \cdot 67$ is not a prime (of course, 2999 is prime, otherwise, it is not a Legendre symbol). To evaluate this Legendre symbol, we first regard it as a Jacobi symbol and evaluate it as if it were a Jacobi symbol (note that once it is regarded as a Jacobi symbol, it does not matter whether or not 335 is prime; it even does not matter whether or not 2999 is prime, but anyway, it is a Legendre symbol).

$$\left(\frac{335}{2999}\right) = -\left(\frac{2999}{335}\right) = -\left(\frac{-16}{335}\right) = -\left(\frac{-1 \cdot 4^2}{335}\right) = -\left(\frac{-1}{335}\right) = 1.$$

Since 2999 is prime, $\left(\frac{335}{2999}\right)$ is a Legendre symbol, and so 355 is a quadratic residue modulo 2999.

Example 1.6.32. In Table 1.19, we list the elements in $(\mathbb{Z}/21\mathbb{Z})^*$ and their Jacobi symbols. Incidentally, exactly half of the Legendre and Jacobi symbols

Table 1.19. Jacobi Symbols for $a \in (\mathbb{Z}/21\mathbb{Z})^*$

$a \in (\mathbb{Z}/21\mathbb{Z})^*$	1	2	4	5	8	10	11	13	16	17	19	20
$a^2 \bmod 21$	1	4	16	4	1	16	16	1	4	16	4	1
$\left(\frac{a}{3}\right)$	1	−1	1	−1	−1	1	−1	1	1	−1	1	−1
$\left(\frac{a}{7}\right)$	1	1	1	−1	1	−1	1	−1	1	−1	−1	−1
$\left(\frac{a}{21}\right)$	1	−1	1	1	−1	−1	−1	−1	1	1	−1	1

$\left(\frac{a}{3}\right)$, $\left(\frac{a}{7}\right)$ and $\left(\frac{a}{21}\right)$ are equal to 1 and half equal to −1. Also for those Jacobi symbols $\left(\frac{a}{21}\right) = 1$, exactly half of the a's are indeed quadratic residues, whereas the other half are not. (Note that a is a quadratic residue of 21 if and only if it is a quadratic residue of both 3 and 7.) That is,

$$\left(\frac{a}{3}\right) = \begin{cases} 1, & \text{for } a \in \{1,4,10,13,16,19\} = Q_3 \\ -1, & \text{for } a \in \{2,5,8,11,17,20\} = \overline{Q}_3 \end{cases}$$

$$\left(\frac{a}{7}\right) = \begin{cases} 1, & \text{for } a \in \{1,2,4,8,11,16\} = Q_7 \\ -1, & \text{for } a \in \{5,10,13,17,19,20\} = \overline{Q}_7 \end{cases}$$

$$\left(\frac{a}{21}\right) = \begin{cases} 1, & \text{for } a \in \{1,4,5,16,17,20\} \quad \begin{cases} a \in \{1,4,16\} = Q_{21} \\ a \in \{5,17,20\} \subset \overline{Q}_{21} \end{cases} \\ -1, & \text{for } a \in \{2,8,10,11,13,19\} \subset \overline{Q}_{21}. \end{cases}$$

1.6.7 Orders and Primitive Roots

In this subsection, we introduce two very important and useful concepts in elementary number theory: orders and primitive roots. First let us give the definition of the order of an integer modulo n.

Definition 1.6.17. Let n be a positive integer and a an integer such that $\gcd(a, n) = 1$. Then the *order* of a modulo n, denoted by $\text{ord}_n(a)$ or by $\text{ord}(a, n)$, is the smallest integer r such that $a^r \equiv 1 \pmod{n}$.

Remark 1.6.16. The terminology "the order of a modulo n" is the modern algebraic term from group theory. The older terminology "a belongs to the exponent r" is the classical term from number theory used by Gauss.

Example 1.6.33. In Table 1.20, values of a^i mod 11 for $i = 1, 2, \cdots, 10$ are given. By Table 1.20, we get, e.g.,

$$\begin{aligned}
&\text{ord}_{11}(1) = 1 \\
&\text{ord}_{11}(2) = \text{ord}_{11}(6) = \text{ord}_{11}(7) = \text{ord}_{11}(8) = 10 \\
&\text{ord}_{11}(3) = \text{ord}_{11}(4) = \text{ord}_{11}(5) = \text{ord}_{11}(9) = 5 \\
&\text{ord}_{11}(10) = 2.
\end{aligned}$$

Table 1.20. Values of a^i mod 11, for $1 \le i < 11$

a	a^2	a^3	a^4	a^5	a^6	a^7	a^8	a^9	a^{10}
1	1	1	1	1	1	1	1	1	1
2	4	8	5	10	9	7	3	6	1
3	9	5	4	1	3	9	5	4	1
4	5	9	3	1	4	5	9	3	1
5	3	4	9	1	5	3	4	9	1
6	3	7	9	10	5	8	4	2	1
7	5	2	3	10	4	6	9	8	1
8	9	6	4	10	3	2	5	7	1
9	4	3	5	1	9	4	3	5	1
10	1	10	1	10	1	10	1	10	1

Exercise 1.6.2. What are the orders of 3, 5 and 7 modulo 8?

We list in the following theorem some useful properties of the order of an integer a modulo n.

Theorem 1.6.31. Let n be a positive integer, $\gcd(a, n) = 1$, and $r = \text{ord}_n(a)$. Then

(1) If $a^m \equiv 1 \pmod{n}$, where m is a positive integer, then $r \mid m$.
(2) $r \mid \phi(n)$.
(3) For integers s and t, $a^s \equiv a^t \pmod{n}$ if and only if $s \equiv t \pmod{r}$.
(4) No two of the integers $a, a^2, a^3, \cdots, a^r$ are congruent modulo r.

(5) If m is a positive integer, then the order of a^m modulo n is $\dfrac{r}{\gcd(r,m)}$.

(6) The order of a^m modulo n is r if and only if $\gcd(m,r) = 1$.

The following theorem shows an unexpected relationship between group theory and number theory.

Theorem 1.6.32. If x is an element of a group $\mathcal{G}$, then the order of x divides the order of $\mathcal{G}$.

Example 1.6.34. Let $\mathcal{G} = (\mathbb{Z}/91\mathbb{Z})^*$ and $x = 17$. Then the order of $\mathcal{G}$ is $|\mathcal{G}| = \phi(91) = 72$, and the order of 17 modulo 91 is 6. It is clear that $6 \mid 72$.

Definition 1.6.18. Let n be a positive integer and a an integer such that $\gcd(a,n) = 1$. If the order of an integer a modulo n is $\phi(n)$, that is, $\text{order}(a,n) = \phi(n)$, then a is called a *primitive root* of n.

Example 1.6.35. Determine whether or not 7 is a primitive root of 45. First note that $\gcd(7,45) = 1$. Now observe that

$$
\begin{array}{ll}
7^1 \equiv 7 \pmod{45} & 7^2 \equiv 4 \pmod{45} \\
7^3 \equiv 28 \pmod{45} & 7^4 \equiv 16 \pmod{45} \\
7^5 \equiv 22 \pmod{45} & 7^6 \equiv 19 \pmod{45} \\
7^7 \equiv 43 \pmod{45} & 7^8 \equiv 31 \pmod{45} \\
7^9 \equiv 37 \pmod{45} & 7^{10} \equiv 34 \pmod{45} \\
7^{11} \equiv 13 \pmod{45} & 7^{12} \equiv 1 \pmod{45}.
\end{array}
$$

Thus, $\text{ord}_{45}(7) = 12$. However, $\phi(45) = 24$. That is, $\text{ord}_{45}(7) \neq \phi(45)$. Therefore, 7 is not a primitive root of 45.

Example 1.6.36. Determine whether or not 7 is a primitive root of 46. First note that $\gcd(7,46) = 1$. Now observe that

$$
\begin{array}{ll}
7^1 \equiv 7 \pmod{46} & 7^2 \equiv 3 \pmod{46} \\
7^3 \equiv 21 \pmod{46} & 7^4 \equiv 9 \pmod{46} \\
7^5 \equiv 17 \pmod{46} & 7^6 \equiv 27 \pmod{46} \\
7^7 \equiv 5 \pmod{46} & 7^8 \equiv 35 \pmod{46} \\
7^9 \equiv 15 \pmod{46} & 7^{10} \equiv 13 \pmod{46} \\
7^{11} \equiv 45 \pmod{46} & 7^{12} \equiv 39 \pmod{46} \\
7^{13} \equiv 43 \pmod{46} & 7^{14} \equiv 25 \pmod{46} \\
7^{15} \equiv 37 \pmod{46} & 7^{16} \equiv 29 \pmod{46} \\
7^{17} \equiv 19 \pmod{46} & 7^{18} \equiv 41 \pmod{46} \\
7^{19} \equiv 11 \pmod{46} & 7^{20} \equiv 31 \pmod{46} \\
7^{21} \equiv 33 \pmod{46} & 7^{22} \equiv 1 \pmod{46}.
\end{array}
$$

Thus, $\text{ord}_{46}(7) = 22$. Note also that $\phi(46) = 22$. That is, $\text{ord}_{46}(7) = \phi(46) = 22$. Therefore 7 is a primitive root of 46.

Exercise 1.6.3. Show that 11 is a primitive root of 31.

Exercise 1.6.4. Find, by trial, the second smallest primitive root of 106.

Theorem 1.6.33 (Primitive roots as residue system). Suppose $\gcd(g,n) = 1$. If g is a primitive root modulo n, then the set of integers $\{g, g^2, g^3, \cdots, g^{n-1}\}$ is a reduced system of residues modulo n.

Example 1.6.37. Let $n = 34$. Then there are $\phi(\phi(34)) = 8$ primitive roots of 34, namely, $3, 5, 7, 11, 23, 27, 29, 31$. Now let $g = 5$ such that $\gcd(g,n) = \gcd(5,34) = 1$. Then

$$\begin{aligned}
&\{g, g^2, \cdots, g^{\phi(n)}\}\\
&\quad = \{5, 5^2, 5^3, 5^4, 5^5, 5^6, 5^7, 5^8, 5^9, 5^{10}, 5^{11}, 5^{12}, 5^{13}, 5^{14}, 5^{15}, 5^{16}\} \bmod 34\\
&\quad = \{5, 25, 23, 13, 31, 19, 27, 33, 29, 9, 11, 21, 3, 15, 7, 1\}\\
&\quad = \{1, 3, 5, 7, 9, 11, 13, 15, 19, 21, 23, 25, 27, 29, 33, 31\}
\end{aligned}$$

which forms a reduced system of residues modulo 34. We can, of course, choose $g = 23$ such that $\gcd(g,n) = \gcd(23,34) = 1$. Then we have

$$\begin{aligned}
&\{g, g^2, \cdots, g^{\phi(n)}\}\\
&\quad = \{23, 23^2, 23^3, 23^4, 23^5, 23^6, 23^7, 23^8, 23^9, 23^{10}, 23^{11}, 23^{12}, 23^{13}, 23^{14},\\
&\qquad\quad 23^{15}, 23^{16}\} \bmod 34\\
&\quad = \{23, 19, 29, 21, 7, 25, 31, 33, 11, 15, 5, 13, 27, 9, 3, 1\}\\
&\quad = \{1, 3, 5, 7, 9, 11, 13, 15, 19, 21, 23, 25, 27, 29, 33, 31\}
\end{aligned}$$

which again forms a reduced system of residues modulo 34.

Theorem 1.6.34. If p is a prime number, then there exist $\phi(p-1)$ (incongruent) primitive roots modulo p.

Example 1.6.38. Let $p = 47$, then there are $\phi(47-1) = 22$ primitive roots modulo 47, namely,

5	10	11	13	15	19	20	22	23	26	29
30	31	33	35	38	39	40	41	43	44	45

Note that no method is known for predicting what will be the smallest primitive root of a given prime p, nor is there much known about the distribution of the $\phi(p-1)$ primitive roots among the least residues modulo p.

Corollary 1.6.8. If n has a primitive root, then there are $\phi(\phi(n))$ (incongruent) primitive roots modulo n.

Example 1.6.39. Let $n = 46$, then there are $\phi(\phi(46)) = 10$ primitive roots modulo 46, namely,

$$5 \quad 7 \quad 11 \quad 15 \quad 17 \quad 19 \quad 21 \quad 33 \quad 37 \quad 43$$

Note that not all moduli n have primitive roots; in Table 1.21 we give the smallest primitive root g for $2 \le n \le 1017$ that has primitive roots.

The following theorem establishes conditions for moduli to have primitive roots:

Table 1.21. Primitive roots g modulo n (if any) for $1 \leq n \leq 1017$

n	g	n	g	n	g	n	g	n	g	n	g
2	1	3	2	4	3	5	2	6	5	7	3
9	2	10	3	11	2	13	2	14	3	17	3
18	5	19	2	22	7	23	5	25	2	26	7
27	2	29	2	31	3	34	3	37	2	38	3
41	6	43	3	46	5	47	5	49	3	50	3
53	2	54	5	58	3	59	2	61	2	62	3
67	2	71	7	73	5	74	5	79	3	81	2
82	7	83	2	86	3	89	3	94	5	97	5
98	3	101	2	103	5	106	3	107	2	109	6
113	3	118	11	121	2	122	7	125	2	127	3
131	2	134	7	137	3	139	2	142	7	146	5
149	2	151	6	157	5	158	3	162	5	163	2
166	5	167	5	169	2	173	2	178	3	179	2
181	2	193	5	194	5	197	2	199	3	202	3
206	5	211	2	214	5	218	11	223	3	226	3
227	2	229	6	233	3	239	7	241	7	242	7
243	2	250	3	251	6	254	3	257	3	262	17
263	5	269	2	271	6	274	3	277	5	278	3
281	3	283	3	289	3	293	2	298	3	302	7
307	5	311	17	313	10	314	5	317	2	326	3
331	3	334	5	337	10	338	7	343	3	346	3
347	2	349	2	353	3	358	7	359	7	361	2
362	21	367	6	373	2	379	2	382	19	383	5
386	5	389	2	394	3	397	5	398	3	401	3
409	21	419	2	421	2	422	3	431	7	433	5
439	15	443	2	446	3	449	3	454	5	457	13
458	7	461	2	463	3	466	3	467	2	478	7
479	13	482	7	486	5	487	3	491	2	499	7
502	11	503	5	509	2	514	3	521	3	523	2
526	5	529	5	538	3	541	2	542	15	547	2
554	5	557	2	562	3	563	2	566	3	569	3
571	3	577	5	578	3	586	3	587	2	593	3
599	7	601	7	607	3	613	2	614	5	617	3
619	2	622	17	625	2	626	15	631	3	634	3
641	3	643	11	647	5	653	2	659	2	661	2
662	3	673	5	674	15	677	2	683	5	686	3
691	3	694	5	698	7	701	2	706	3	709	2
718	7	719	11	722	3	727	5	729	2	733	6
734	11	739	3	743	5	746	5	751	3	757	2
758	3	761	6	766	5	769	11	773	2	778	3
787	2	794	5	797	2	802	3	809	3	811	3
818	21	821	2	823	3	827	2	829	2	838	11
839	11	841	2	842	23	853	2	857	3	859	2
862	7	863	5	866	5	877	2	878	15	881	3
883	2	886	5	887	5	898	3	907	2	911	17
914	13	919	7	922	3	926	3	929	3	934	5
937	5	941	2	947	2	953	3	958	13	961	3
967	5	971	6	974	3	977	3	982	7	983	5
991	6	997	7	998	7	1006	5	1009	11	1013	3

Theorem 1.6.35. An integer $n > 1$ has a primitive root modulo n if and only if

$$n = 2, 4, p^\alpha, \text{ or } 2p^\alpha, \tag{1.297}$$

where p is an odd prime and α is a positive integer.

Corollary 1.6.9. If $n = 2^\alpha$ with $\alpha \geq 3$, or $n = 2^\alpha p_1^{\alpha_1} \cdots p_k^{\alpha_k}$ with $\alpha \geq 2$ or $k \geq 2$, then there are no primitive roots modulo n.

Example 1.6.40. For $n = 16 = 2^4$, since it is of the form $n = 2^\alpha$ with $\alpha \geq 3$, there are no primitive roots modulo 16.

Although we know which numbers possess primitive roots, it is not a simple matter to find these roots. Except for trial and error methods, very few general techniques are known. Artin in 1927 made the following conjecture (Rose [210]):

Conjecture 1.6.1. Let $N_a(x)$ be the number of primes less than x of which a is a primitive root, and suppose a is not a square and is not equal to -1, 0 or 1. Then

$$N_a(x) \sim A \frac{x}{\ln x}, \tag{1.298}$$

where A depends only on a.

Hooley in 1967 showed that if the extended Riemann hypothesis is true then so is Artin's conjecture. It is also interesting to note that before the age of computers Jacobi in 1839 listed all solutions $\{a, b\}$ of the congruences $g^a \equiv b \pmod{p}$ where $1 \leq a < p$, $1 \leq b < p$, g is the least positive primitive root of p and $p < 1000$.

1.6.8 Indices and kth Power Residues

We shall now move on to the study of the theory of index, and the kth power residues.

The concept of *index* of an integer modulo n was first introduced by Gauss in his *Disquisitiones Arithmeticae*. Given an integer n, if n has primitive root g, then the set

$$\{g, g^2, g^3, \cdots, g^{\phi(n)}\} \tag{1.299}$$

forms a reduced system of residues modulo n; g is a generator of the cyclic group of the reduced residues modulo n. (Clearly, the group $(\mathbb{Z}/n\mathbb{Z})^*$ is cyclic if $n = 2, 4, p^\alpha$, or $2p^\alpha$, for p odd prime and α positive integer.) Hence, if $\gcd(a, n) = 1$, then a can be expressed in the form:

$$a \equiv g^k \pmod{n} \tag{1.300}$$

for a suitable k with $1 \leq k \leq \phi(n)$. This motivates our following definition, which is an analogue of the real base logarithm function.

Definition 1.6.19. Let g be a primitive root of n. If $\gcd(a, n) = 1$, then the smallest positive integer k such that $a \equiv g^k \pmod{n}$ is called the *index* of a to the base g modulo n and is denoted by $\text{ind}_{g,n}(a)$, or simply by $\text{ind}_g a$.

Clearly, by definition, we have

$$a \equiv g^{\text{ind}_g a} \pmod{n}. \tag{1.301}$$

The function $\text{ind}_g a$ is sometimes called the *discrete logarithm* and is denoted by $\log_g a$ so that

$$a \equiv g^{\log_g a} \pmod{n}. \tag{1.302}$$

Generally, the discrete logarithm is a computationally intractable problem; no efficient algorithm has been found for computing discrete logarithms and hence it has important applications in public key cryptography. We shall discuss some modern computer algorithms for computing general discrete logarithms (including elliptic curve analogues of discrete logarithms) in Chapter 2 and applications of the computational infeasibility of discrete logarithms in cryptography in Chapter 3.

Theorem 1.6.36 (Index theorem). If g is a primitive root modulo n, then $g^x \equiv g^y \pmod{n}$ if and only if $x \equiv y \pmod{\phi(n)}$.

Proof. Suppose first that $x \equiv y \pmod{\phi(n)}$. Then, $x = y + k\phi(n)$ for some integer k. Therefore,

$$\begin{aligned} g^x &\equiv g^{y+k\phi(n)} \pmod{n} \\ &\equiv g^y \cdot (g^{\phi(n)})^k \pmod{n} \\ &\equiv g^y \cdot 1^k \pmod{n} \\ &\equiv g^y \pmod{n}. \end{aligned}$$

The proof of the "only if" part of the theorem is left as an exercise. □

The properties of the function $\text{ind}_g a$ are very similar to those of the conventional real base logarithm function, as the following theorems indicate:

Theorem 1.6.37. Let g be a primitive root modulo the prime p, and $\gcd(a, p) = 1$. Then $g^k \equiv a \pmod{p}$ if and only if

$$k \equiv \text{ind}_g\, a \pmod{p-1}. \tag{1.303}$$

Theorem 1.6.38. Let n be a positive integer with primitive root g, and $\gcd(a, n) = \gcd(b, n) = 1$. Then

(1) $\text{ind}_g 1 \equiv 0 \pmod{\phi(n)}$.

(2) $\text{ind}_g(ab) \equiv \text{ind}_g a + \text{ind}_g b \pmod{\phi(n)}$.

(3) $\text{ind}_g a^k \equiv k \cdot \text{ind}_g a \pmod{\phi(n)}$, if k is a positive integer.

Example 1.6.41. Compute the index of 15 base 6 modulo 109, that is, $6^{\text{ind}_6 15} \bmod 109 = 15$. To find the index, we just successively perform the computation $6^k \pmod{109}$ for $k = 1, 2, 3, \cdots$ until we find a suitable k such that $6^k \pmod{109} = 15$:

$$\begin{array}{ll}
6^1 \equiv 6 \pmod{109} & 6^2 \equiv 36 \pmod{109} \\
6^3 \equiv 107 \pmod{109} & 6^4 \equiv 97 \pmod{109} \\
6^5 \equiv 37 \pmod{109} & 6^6 \equiv 4 \pmod{109} \\
6^7 \equiv 24 \pmod{109} & 6^8 \equiv 35 \pmod{109} \\
6^9 \equiv 101 \pmod{109} & 6^{10} \equiv 61 \pmod{109} \\
6^{11} \equiv 39 \pmod{109} & 6^{12} \equiv 16 \pmod{109} \\
6^{13} \equiv 96 \pmod{109} & 6^{14} \equiv 31 \pmod{109} \\
6^{15} \equiv 77 \pmod{109} & 6^{16} \equiv 26 \pmod{109} \\
6^{17} \equiv 47 \pmod{109} & 6^{18} \equiv 64 \pmod{109} \\
6^{19} \equiv 57 \pmod{109} & 6^{20} \equiv 15 \pmod{109}.
\end{array}$$

Since $k = 20$ is the smallest positive integer such that $6^{20} \equiv 15 \pmod{109}$, $\text{ind}_6 15 \bmod 109 = 20$.

In what follows, we shall study the congruences of the form $x^k \equiv a \pmod{n}$, where n is an integer with primitive roots and $\gcd(a, n) = 1$. First of all, we present a definition, which is the generalization of quadratic residues.

Definition 1.6.20. Let a, n and k be positive integers with $k \geq 2$. Suppose $\gcd(a, n) = 1$, then a is called a kth (higher) power residue of n if there is an x such that

$$x^k \equiv a \pmod{n}. \tag{1.304}$$

The set of all kth (higher) power residues is denoted by $K(k)_n$. If the congruence has no solution, then a is called a kth (higher) power nonresidue of n. The set of such a is denoted by $\overline{K(k)_n}$. For example, $K(9)_{126}$ would denote the set of the 9th power residues of 126, whereas $\overline{K(5)_{31}}$ the set of the 5th power nonresidue of 31.

Theorem 1.6.39 (kth power theorem). Let n be a positive integer having a primitive root, and suppose $\gcd(a, n) = 1$. Then the congruence (1.304) has a solution if and only if

$$a^{\phi(n)/\gcd(k,\phi(n))} \equiv 1 \pmod{n}. \tag{1.305}$$

If (1.304) is soluble, then it has exactly $\gcd(k, \phi(n))$ incongruent solutions.

Proof. Let x be a solution of $x^k \equiv a \pmod{n}$. Since $\gcd(a, n) = 1$, $\gcd(x, n) = 1$. Then

$$\begin{aligned}
a^{\phi(n)/\gcd(k,\phi(n))} &\equiv (x^k)^{\phi(n)/\gcd(k,\phi(n))} \\
&\equiv (x^{\phi(n)})^{k/\gcd(k,\phi(n))} \\
&\equiv 1^{k/\gcd(k,\phi(n))} \\
&\equiv 1 \pmod{n}.
\end{aligned}$$

Conversely, if $a^{\phi(n)/\gcd(k,\phi(n))} \equiv 1 \pmod{n}$, then $r^{(\text{ind}_r a)\phi(n)/\gcd(k,\phi(n))} \equiv 1 \pmod{n}$. Since $\text{ord}_n r = \phi(n)$, $\phi(n) \mid (\text{ind}_r a)\phi(n)/\gcd(k,\phi(n))$, and hence $d \mid \text{ind}_r a$ because $(\text{ind}_r a)/d$ must be an integer. Therefore, there are $\gcd(k,\phi(n))$ incongruent solutions to $k(\text{ind}_r x) \equiv (\text{ind}_r a) \pmod{n}$ and hence $\gcd(k,\phi(n))$ incongruent solutions to $x^k \equiv a \pmod{n}$. □

If n is a prime number, say, p, then we have:

Corollary 1.6.10. Suppose p is prime and $\gcd(a,p) = 1$. Then a is a kth power residue of p if and only if

$$a^{(p-1)/\gcd(k,(p-1))} \equiv 1 \pmod{p}. \tag{1.306}$$

Example 1.6.42. Determine whether or not 5 is a sixth power of 31, that is, decide whether or not the congruence

$$x^6 \equiv 5 \pmod{31}$$

has a solution. First of all, we compute

$$5^{(31-1)/\gcd(6,31-1)} \equiv 25 \not\equiv 1 \pmod{31}$$

since 31 is prime. By Corollary 1.6.10, 5 is not a sixth power of 31. That is, $5 \notin K(6)_{31}$. However,

$$5^{(31-1)/\gcd(7,31-1)} \equiv 1 \pmod{31}.$$

So, 5 is a seventh power of 31. That is, $5 \in K(7)_{31}$.

Exercise 1.6.5. Determine whether or not 5 is a seventh power of 359. That is, decide whether or not $5 \in K(7)_{359}$.

Exercise 1.6.6. Find the complete set of incongruent 16th power residues of 512. That is, find all the a's which satisfy $a \in K(16)_{512}$.

Now let us introduce a new symbol $\left(\frac{a}{p}\right)_k$, the kth power residue symbol, analogous to the Legendre symbol for quadratic residues (Ko and Sun, [125]).

Definition 1.6.21. Let p be a odd prime, $k > 1$, $k \mid p-1$ and $q = \frac{p-1}{k}$. Then the symbol

$$\left(\frac{\alpha}{p}\right)_k = \alpha^q \bmod p \tag{1.307}$$

is called the k power residue symbol modulo p, where $\alpha^q \bmod p$ represent the absolute smallest residue of α^q modulo p (the complete set of the absolute smallest residues modulo p are: $(p-1)/2, \cdots, -1, 0, 1, \cdots, (p-1/2))$.

Theorem 1.6.40. Let $\left(\frac{\alpha}{p}\right)_k$ be the kth power residue symbol. Then

(1) $p \mid a \Longrightarrow \left(\dfrac{a}{p}\right)_k = 0.$

(2) $a \equiv a_1 \pmod{p} \Longrightarrow \left(\dfrac{a}{p}\right)_k = \left(\dfrac{a_1}{p}\right)_k.$

(3) For $a_1, a_2 \in \mathbb{Z} \Longrightarrow \left(\dfrac{a_1 a_2}{p}\right)_k = \left(\dfrac{a_1}{p}\right)_k \left(\dfrac{a_2}{p}\right)_k.$

(4) $\text{ind}_g a \equiv b \pmod{k}, 0 \le b < k \Longrightarrow \left(\dfrac{a}{p}\right)_k \equiv g^{aq} \pmod{p}.$

(5) a is the kth power residue of $p \Longleftrightarrow \left(\dfrac{a}{p}\right)_k = 1.$

(6) $n = p_1^{\alpha_1} p_2^{\alpha_2} \cdots p_l^{\alpha_l} \Longrightarrow \left(\dfrac{n}{p}\right)_k = \left(\dfrac{p_1}{p}\right)_k^{\alpha_1} \left(\dfrac{p_2}{p}\right)_k^{\alpha_2} \cdots \left(\dfrac{p_l}{p}\right)_k^{\alpha_l}.$

Example 1.6.43. Let $p = 19$, $k = 3$ and $q = 6$. Then

$$\left(\frac{-1}{19}\right)_3 = \left(\frac{1}{19}\right)_3 = 1.$$

$$\left(\frac{2}{19}\right)_3 = 7.$$

$$\left(\frac{3}{19}\right)_3 = \left(\frac{-16}{19}\right)_3 = \left(\frac{-1}{19}\right)_3 \left(\frac{16}{19}\right)_3 = \left(\frac{-1}{19}\right)_3 \left(\frac{2}{19}\right)_3^4 = \left(\frac{2}{19}\right)_3 = 7.$$

$$\left(\frac{5}{19}\right)_3 = \left(\frac{24}{19}\right)_3 = \left(\frac{2}{19}\right)_3^3 \left(\frac{3}{19}\right)_3 = \left(\frac{3}{19}\right)_3 = 7.$$

$$\left(\frac{7}{19}\right)_3 = \left(\frac{45}{19}\right)_3 = \left(\frac{3}{19}\right)_3^2 \left(\frac{5}{19}\right)_3 = 7^3 \equiv 1.$$

$$\left(\frac{11}{19}\right)_3 = \left(\frac{30}{19}\right)_3 = \left(\frac{2}{19}\right)_3 \left(\frac{3}{19}\right)_3 \left(\frac{5}{19}\right)_3 = 7^3 \equiv 1.$$

$$\left(\frac{13}{19}\right)_3 = \left(\frac{32}{19}\right)_3 = \left(\frac{2}{19}\right)_3 = -8.$$

$$\left(\frac{17}{19}\right)_3 = \left(\frac{-2}{19}\right)_3 = \left(\frac{-1}{19}\right)_3 \left(\frac{2}{19}\right)_3 = 7.$$

All the above congruences are modular 19.

Exercise 1.6.7 (Research problem). Extend the Jacobi symbol for quadratic residues to the kth power residues.

1.7 Arithmetic of Elliptic Curves

As long as algebra and geometry have been separated, their progress has been slow and their uses limited; but when these two sciences have been united, they have lent each other mutual forces, and have marched together towards perfection.

AUGUSTUS DE MORGAN (1806–1871)

Elliptic curves have been studied by number theorists for about a century; not for applications in either mathematics or computing science but because of their intrinsic mathematical beauty and interest. In recent years, however, elliptic curves have found applications in many areas of mathematics and computer science. For example, by using the theory of elliptic curves, Lenstra [140] invented the powerful factoring method ECM, Atkin and Morain [12] designed the practical elliptic curve primality proving algorithm ECPP, Koblitz [126] and Miller [163] proposed the idea of elliptic public-key cryptosystems, and more interestingly, Wiles proved the famous Fermat's Last Theorem [254]. In this section, we shall provide some basic concepts and results on elliptic curves. In Chapter 2, we shall introduce some fast group operations on elliptic curves and algorithms for primality testing and factoring based on elliptic curves, and in Chapter 3, we shall introduce some applications of elliptic curves in cryptography.

1.7.1 Basic Concepts of Elliptic Curves

An elliptic curve is an algebraic curve given by a *cubic Diophantine equation*

$$y^2 = x^3 + ax + b. \tag{1.308}$$

More general cubics in x and y can be reduced to this form, known as Weierstrass normal form, by rational transformations. Two examples of elliptic curves are shown in Figure 1.11 (from left to right). The graph on the left is the graph of a *single* equation, namely $E_1 : y^2 = x^3 - 4x + 2$; even though it breaks apart into two pieces, we refer to it as a *single* curve. The graph on the right is given by the equation $E_2 : y^2 = x^3 - 3x + 3$. Note that an elliptic curve is not an *ellipse*, it is so named because it is related to the length of the perimeter of an ellipse; a more accurate name for an elliptic curve, in terms of *algebraic geometry*, is an *Abelian variety of dimension one*. It should be also noted that *quadratic* polynomial equations are fairly well understood by mathematicians today, but cubic equations still pose enough difficulties to be topics of current research. In what follows, we shall provide some more formal definitions of elliptic curves.

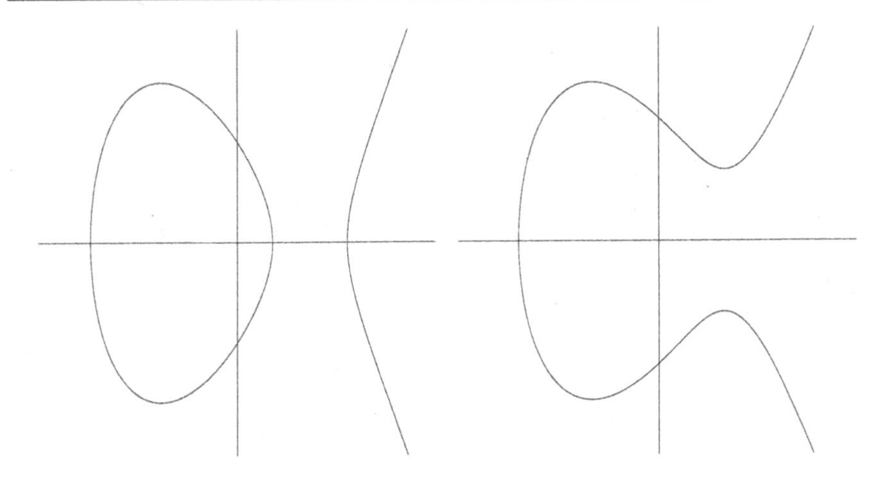

Figure 1.11. Two examples of elliptic curves

Definition 1.7.1. Let $\mathcal{K}$ be a field. Then the *characteristic* of the field $\mathcal{K}$ is 0 if

$$\underbrace{1 \oplus 1 \oplus \cdots \oplus 1}_{n \text{ summands}}$$

is never equal to 0 for any $n > 1$. Otherwise, the characteristic of the field $\mathcal{K}$ is the least positive integer n such that

$$\sum_{i=1}^{n} i = 0.$$

Example 1.7.1. The fields $\mathbb{Z}, \mathbb{Q}, \mathbb{R}$ and $\mathbb{C}$ all have characteristic 0, whereas the field $\mathbb{Z}/p\mathbb{Z}$ is of characteristic p, where p is prime.

Definition 1.7.2. Let $\mathcal{K}$ be a field (either the field $\mathbb{Q}$, $\mathbb{R}$, $\mathbb{C}$, or the finite field $\mathbb{F}_q$ with $q = p^\alpha$ elements), and $x^3 + ax + b$ with $a, b \in \mathcal{K}$ be a cubic polynomial. Then

(1) If $\mathcal{K}$ is a field of characteristic $\neq 2, 3$, then an *elliptic curve* over $\mathcal{K}$ is the set of points (x, y) with $x, y \in \mathcal{K}$ that satisfy the following cubic Diophantine equation:

$$E: \quad y^2 = x^3 + ax + b, \tag{1.309}$$

(where the cubic on the right-hand side has no multiple roots) together with a single element, denoted by $\mathcal{O}_E$, called the *point at infinity*.

(2) If $\mathcal{K}$ is a field of characteristic 2, then an *elliptic curve* over $\mathcal{K}$ is the set of points (x, y) with $x, y \in \mathcal{K}$ that satisfy one of the following cubic Diophantine equations:

$$\left.\begin{array}{ll} E: & y^2 + cy = x^3 + ax + b, \\ E: & y^2 + xy = x^3 + ax^2 + b, \end{array}\right\} \tag{1.310}$$

(here we do not care whether or not the cubic on the right-hand side has multiple roots) together with a *point at infinity* $\mathcal{O}_E$.

(3) If $\mathcal{K}$ is a field of characteristic 3, then an *elliptic curve* over $\mathcal{K}$ is the set of points (x, y) with $x, y \in \mathcal{K}$ that satisfy the cubic Diophantine equation:

$$E: \quad y^2 = x^3 + ax^2 + bx + c, \tag{1.311}$$

(where the cubic on the right-hand side has no multiple roots) together with a *point at infinity* $\mathcal{O}_E$.

In this book, we shall not consider the elliptic curves over a field of characteristic $= 2, 3$. We are now moving on to the definition of the notion of an elliptic curve over the ring $\mathbb{Z}/N\mathbb{Z}$, which are specifically useful in primality testing, integer factorization and public-key cryptography.

Definition 1.7.3. Let N be a positive integer with $\gcd(N, 6) = 1$. An *elliptic curve* over $\mathbb{Z}/N\mathbb{Z}$ is given by the following cubic Diophantine equation:

$$E: \quad y^2 = x^3 + ax + b, \tag{1.312}$$

where $a, b \in \mathbb{Z}$ and $\gcd(N, \ 4a^3 + 27b^2) = 1$. The set of points on E is the set of solutions in $(\mathbb{Z}/N\mathbb{Z})^2$ to the equation (1.312), together with a *point at infinity* $\mathcal{O}_E$.

Remark 1.7.1. The subject of elliptic curves is one of the jewels of 19th century mathematics, originated by Abel, Gauss, Jacobi and Legendre. Contrary to popular opinion, an elliptic curve (i.e., a nonsingular cubic curve) is not an ellipse; as Niven, Zuckerman and Montgomery [174] remarked, it is natural to express the arc length of an ellipse as an integral involving the square root of a quartic polynomial. By making a rational change of variables, this may be reduced to an integral involving the square root of a cubic polynomial. In general, an integral involving the square root of a quartic or cubic polynomial is called an elliptic integral. So, the word *elliptic* actually came from the theory of elliptic integrals of the form:

$$\int R(x, y)dx \tag{1.313}$$

where $R(x, y)$ is a rational function in x and y, and y^2 is a polynomial in x of degree 3 or 4 having no repeated roots. Such integrals were intensively studied in the 18th and 19th centuries. It is interesting to note that elliptic integrals serve as a motivation for the theory of elliptic functions, whilst elliptic functions parameterize elliptic curves. It is not our intention here to explain fully the theory of elliptic integrals and elliptic functions; interested readers are suggested to consult some more advanced texts, such as Cohen [50], Lang [137], and McKean and Moll [153] for more information.

1.7.2 Geometric Composition Laws of Elliptic Curves

The basic operation on an elliptic curve $E: y^2 = x^3 + ax + b$ is the addition of points on the curve. The geometric interpretation of addition of points on an elliptic curve is quite straightforward. Suppose E is an elliptic curve as shown in Figure 1.12. A straight line (non-vertical) L connecting points P and Q intersects the elliptic curve E at a third point R, and the point $P \oplus Q$ is the reflection of R in the X-axis. That is, if $R = (x_3, y_3)$, then $P \oplus Q = (x_3, -y_3)$ is the reflection of R in the X-axis. Note that a vertical line, such as L' or L'', meets the curve at two points (not necessarily distinct), and also at the point at infinity $\mathcal{O}_E$ (we may think of the point at infinity as lying far off in the direction of the Y-axis). The line at infinity meets the curve at the point $\mathcal{O}_E$ three times. Of course, the non-vertical line meets the curve in three points in the XY plane. Thus, every line meets the curve in three points.

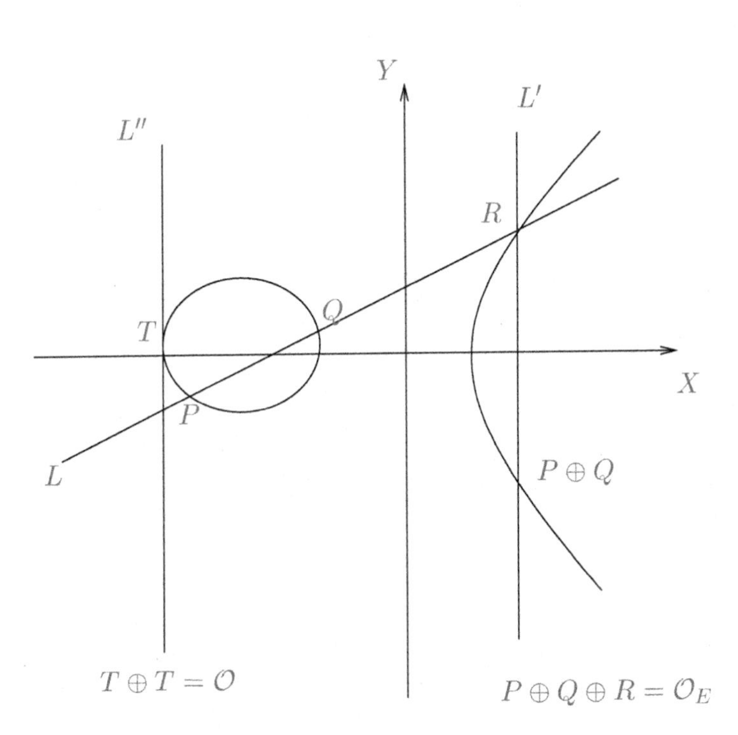

Figure 1.12. Geometric composition laws of an elliptic curve

As can be seen from Figure 1.12, an elliptic curve can have many rational points; any straight line connecting two of them intersects a third. The point at infinity $\mathcal{O}_E$ is the third point of intersection of any two points (not necessarily distinct) of a vertical line with the elliptic curve E. This makes it possible to generate all rational points out of just a few.

The above observations lead naturally to the following geometric composition law of elliptic curves [229].

Theorem 1.7.1 (Geometric composition law). Let $P, Q \in E$, L the line connecting P and Q (tangent line to E if $P = Q$), and R the third point of intersection of L with E. Let L' be the line connecting R and $\mathcal{O}_E$ (the point at infinity). Then the point $P \oplus Q$ is the third point on E such that L' intersects E at R, $\mathcal{O}_E$ and $P \oplus Q$.

1.7.3 Algebraic Computation Laws for Elliptic Curves

The geometric composition law gives us a clear idea how two points on an elliptic curve can be added together to find a third. However, to systematically perform the additions of points on elliptic curves on a computer, we will need an algebraic formula. The following result gives us a very convenient formula for computing points on an elliptic curve.

Theorem 1.7.2 (Algebraic computation law). Let $P_1 = (x_1, y_1)$, $P_2 = (x_2, y_2)$ be points on the elliptic curve:

$$E: \; y^2 = x^3 + ax + b, \tag{1.314}$$

then $P_3 = (x_3, y_3) = P_1 \oplus P_2$ on E may be computed by

$$P_1 \oplus P_2 = \begin{cases} \mathcal{O}_E, & \text{if } x_1 = x_2 \;\&\; y_1 = -y_2 \\ (x_3, y_3), & \text{otherwise.} \end{cases} \tag{1.315}$$

where

$$(x_3, y_3) = (\lambda^2 - x_1 - x_2, \;\; \lambda(x_1 - x_3) - y_1) \tag{1.316}$$

and

$$\lambda = \begin{cases} \dfrac{3x_1^2 + a}{2y_1}, & \text{if } P_1 = P_2, \\ \dfrac{y_2 - y_1}{x_2 - x_1}, & \text{otherwise.} \end{cases} \tag{1.317}$$

Example 1.7.2. Let E be the elliptic curve $y^2 = x^3 + 17$ over $\mathbb{Q}$, and let $P_1 = (x_1, y_1) = (-2, 3)$ and $P_2 = (x_2, y_2) = (1/4, \; 33/8)$ be two points on E. To find the third point P_3 on E, we perform the following computation:

$$\lambda = \frac{y_2 - y_1}{x_2 - x_1} = \frac{1}{2}$$
$$x_3 = \lambda^2 - x_1 - x_2 = 2$$
$$y_3 = \lambda(x_1 - x_3) - y_1 = -5.$$

So, $P_3 = P_1 \oplus P_2 = (x_3, y_3) = (2, -5)$.

Exercise 1.7.1. Find the points $P_1 \oplus P_2$ and $2P_1$ on the elliptic curve $E: y^2 = x^3 - 36x$, where $P_1 = (-3, 9)$ and $P_2 = (-2, 8)$.

Example 1.7.3. Let $P = (3, 2)$ be a point on the elliptic curve $E: y^2 = x^3 - 2x - 3$ over $\mathbb{Z}/7\mathbb{Z}$. Compute

$$10P = \underbrace{P \oplus P \oplus \cdots \oplus P}_{10 \text{ summands}} \pmod 7.$$

According to (1.316), we have:

$$\begin{aligned}
2P &= P \oplus P = (3,2) \oplus (3,2) = (2,6), \\
3P &= P \oplus 2P = (3,2) \oplus (2,6) = (4,2), \\
4P &= P \oplus 3P = (3,2) \oplus (4,2) = (0,5), \\
5P &= P \oplus 4P = (3,2) \oplus (0,5) = (5,0), \\
6P &= P \oplus 5P = (3,2) \oplus (5,0) = (0,2), \\
7P &= P \oplus 6P = (3,2) \oplus (0,2) = (4,5), \\
8P &= P \oplus 7P = (3,2) \oplus (4,5) = (2,1), \\
9P &= P \oplus 8P = (3,2) \oplus (2,1) = (3,5), \\
10P &= P \oplus 9P = (3,2) \oplus (3,5) = \mathcal{O}_E.
\end{aligned}$$

Example 1.7.4. Let $E: y^2 = x^3 + 17$ be the elliptic curve over $\mathbb{Q}$ and $P = (-2, 3)$ a point on E. Then

$2P = (8, -23)$

$3P = \left(\frac{19}{25}, \frac{522}{125}\right)$

$4P = \left(\frac{752}{529}, \frac{-54239}{12167}\right)$

$5P = \left(\frac{174598}{32761}, \frac{76943337}{5929741}\right)$

$6P = \left(\frac{-4471631}{3027600}, \frac{-19554357097}{5268024000}\right)$

$7P = \left(\frac{12870778678}{76545001}, \frac{1460185427995887}{669692213749}\right)$

$8P = \left(\frac{-3705032916448}{1556248765009}, \frac{3635193007425360001}{1941415665602432473}\right)$

$9P = \left(\frac{1508016107720305}{1146705139411225}, \frac{-185877155243117444053750 2}{3883091627056219156787 5}\right)$

$$10P = \left(\frac{2621479238320017368}{2155046648421950400 1}, \frac{4125080845025235054098132572 57}{10004260991388455752541474399 9}\right)$$

$$11P = \left(\frac{98386489129108787338 2478}{45577082245357611985 6081}, \frac{-16005818393035651701390378886102542 93}{30769453204705350935032590551794327 1}\right)$$

$$12P = \left(\frac{1727701779459733569579962592 1}{4630688543838991376029953600}, \frac{261632579225132155842970406236745469642671 9}{315114478121426726704392053642337633216000}\right)$$

Suppose now we are interested in measuring the *size* (or the height) of points on an elliptic curve E. One way to do this is to look at the numerator and denominator of the x-coordinates. If we write the coordinates of kP as

$$kP = \left(\frac{A_k}{B_k}, \frac{C_k}{D_k}\right), \tag{1.318}$$

we may define the height of these points as follows

$$H(kP) = \max(|A_k|, |B_k|). \tag{1.319}$$

For example, the values for various heights of points kP for $k = 1, 2, 3, \cdots, 38$ on the elliptic curve $E : y^2 = x^3 - 7x + 10$ for $P = (1, 2)$ are shown in Table 1.22. It is interesting to note that for large k, the height of kP looks like [230]:

$$D(H(kP)) \approx 0.1974k^2 \tag{1.320}$$

$$H(kP) \approx 10^{0.1974k^2} \approx (1.574)^{k^2} \tag{1.321}$$

where $D(H(kP))$ denotes the number of digits in $H(kP)$.

Remark 1.7.2. To provide greater flexibility, we may consider the following more general form of elliptic curves:

$$E : y^2 = x^3 + ax^2 + bx + c. \tag{1.322}$$

In order for E to be an elliptic curve, it is necessary and sufficient that

$$D = a^2b^2 - 4a^3c - 4b^3 + 18abc - 27c^2 \neq 0. \tag{1.323}$$

Thus,

$$P_3(x_3, y_3) = P_1(x_1, y_1) \oplus P_2(x_2, y_2),$$

on E may be computed by

$$(x_3, y_3) = (\lambda^2 - a - x_1 - x_2, \ \lambda(x_1 - x_3) - y_1) \tag{1.324}$$

where

$$\lambda = \begin{cases} (3x_1^2 + 2a + b)/2y_1, & \text{if } P_1 = P_2 \\ (y_2 - y_1)/(x_2 - x_1), & \text{otherwise.} \end{cases} \tag{1.325}$$

Exercise 1.7.2. Compute $10P$ on the elliptic curve $E : y^2 = x^3 - 7x + 10$ with $P = (1, 2)$.

Table 1.22. The height of points kP on $y^2 = x^3 - 7x + 10$ for $P = (1, 2)$

1

1

9

9

79

439

4861

8831

364121

13215591

147938569

1905671716

758845143289

31994400448399

3328831959482831

233184473054307329

10683181372399165448 1

1213657536294897179624 1

543600225189284171282984 9

19069091665162622624048485 21

9498372249961240678068207451 29

636933550541815376167294302395161

28030046471840093449814875979848644 41

1224829627942856195377997653151207742849

5889898459726150895441159594983293353052951

21656760900176595018121976228640953857941830039

10697411245907413316378690096308235371815981155860 9

169792488359083598660737845914423844266600668951942391

340971339018052719027316243362884546780938918466123647600 9

4810007152645419354924450068543685984502187600796776361837763 1

26538527030602388558747693790983815044266310156549166288079961552 1

2324890959139314651285076899470635081879395109755678116825582758717696

453049093921848404733903980671633092531297281136059705177335106029632866081

80091739901262220249200437828750466641383227470852465812903310892650574133131 19

433965859129085578432399684794789457422649568271710907205906642216055387948107116119

2203153760792594371848879623147671087103576337107738588940202283889658007520457153160320 9

469480703424951546533495086379703132404911393419039206374969237622743393556346405422120973428 9

3946874584403975972217072930685208913336466253842992742216668757421322010613389879614661600783821 21

1.7.4 Group Laws on Elliptic Curves

The points on an elliptic curve form an Abelian group with addition of points as the binary operation on the group. In this subsection, we shall study some group-theoretic properties of elliptic curves.

Theorem 1.7.3 (Group laws on elliptic curves). The geometric composition laws of elliptic curves have the following group-theoretic properties:

(1) If a line L intersects E at the (not necessary distinct) points P, Q, R, then
$$(P \oplus Q) \oplus R = \mathcal{O}_E.$$

(2) $P \oplus \mathcal{O}_E = P, \quad \forall P \in E.$

(3) $P \oplus Q = Q \oplus P, \quad \forall P, Q \in E.$

(4) Let $P \in E$, then there is a point of E, denoted $\ominus P$, such that
$$P \oplus (\ominus P) = \mathcal{O}_E.$$

(5) Let $P, Q, R \in E$, then
$$(P \oplus Q) \oplus R = P \oplus (Q \oplus R).$$

In other words, the geometric composition law makes E into an Abelian group with identity element $\mathcal{O}_E$. Moreover, if E is defined over a field $\mathcal{K}$, then
$$E(\mathcal{K}) = \{(x, y) \in \mathcal{K}^2 : y^2 = x^3 + ax + b\} \cup \{\mathcal{O}_E\}.$$
is a subgroup of E.

Example 1.7.5. Let $E(\mathbb{Q})$ be the set of rational points on E. Then $E(\mathbb{Q})$ with the addition operation defined on it forms an Abelian group.

We shall now introduce the important concept of the order of a point on E.

Definition 1.7.4. Let P be an element of the set $E(\mathbb{Q})$. Then P is said to have *order* k if
$$kP = \underbrace{P \oplus P \oplus \cdots \oplus P}_{k \text{ summands}}$$
with $k'P \neq \mathcal{O}_E$ for all $1 < k' < k$ (that is, k is the smallest integer such that $kP = \mathcal{O}_E$). If such a k exists, then P is said to have *finite order*, otherwise, it has *infinite order*.

Example 1.7.6. Let $P = (3, 2)$ be a point on the elliptic curve $E: y^2 = x^3 - 2x - 3$ over $\mathbb{Z}/7\mathbb{Z}$ (see Example 1.7.3). Since $10P = \mathcal{O}_E$ and $kP \neq \mathcal{O}_E$ for $k < 10$, P has order 10.

Example 1.7.7. Let $P = (-2, 3)$ be a point on the elliptic curve $E: y^2 = x^3 + 17$ over $\mathbb{Q}$ (see Example 1.7.4). Then P apparently has infinite order.

1.7.5 Number of Points on Elliptic Curves

As mentioned in the previous subsection, it is possible to generate all rational points of an elliptic curve out of just a few. In this subsection, we shall be concerned with the problem: *How many points (rational or integral) are there on an elliptic curve*? Let us first look at an example:

Example 1.7.8. Let E be the elliptic curve $y^2 = x^3 + 3x$ over $\mathbb{F}_5$, then

$$\mathcal{O}_E,\ (0,0),\ (1,2),\ (1,3),\ (2,2),\ (2,3),\ (3,1),\ (3,4),\ (4,1),\ (4,4)$$

are the 10 points on E. However, the elliptic curve $y^2 = 3x^3 + 2x$ over $\mathbb{F}_5$ has only two points:

$$\mathcal{O}_E,\ (0,0).$$

Exercise 1.7.3. Find the number of points on the following elliptic curves over $\mathbb{F}_{13}$:

$$(1) E_1 :\ y^2 = x^3 + 2x + 1, \qquad (2) E_2 :\ y^2 = x^3 + 4x.$$

How many points are there on an elliptic curve $E :\ y^2 = x^3 + ax + b$ over $\mathbb{F}_p$? The following theorem answers this question:

Theorem 1.7.4. There are

$$1 + p + \sum_{x \in \mathbb{F}_p} \left(\frac{x^3 + ax + b}{p} \right) = 1 + p + \epsilon \tag{1.326}$$

points on $E :\ y^2 = x^3 + ax + b$, including the point at infinity $\mathcal{O}_E$, where $\left(\dfrac{x^3 + ax + b}{p} \right)$ is the Legendre symbol.

The quantity ϵ in (1.326) is given in the following theorem, due to Hasse[35] in 1933:

Theorem 1.7.5 (Hasse).

$$|\epsilon| \le 2\sqrt{p}. \tag{1.327}$$

Example 1.7.9. Let $p = 5$, then $|\epsilon| \le 4$. Hence, we have between 2 and 10 points on an elliptic curve over $\mathbb{F}_5$. In fact, all the possibilities occur in the following elliptic curves given in Table 1.23.

35

Helmut Hasse (1898–1979) was born in Kassel, Germany. He was educated in Göttingen and Marburg, and subsequently worked in Kiel, Halle, Marburg, and Göttingen. In 1922 Hasse was appointed a lecturer at the University of Kiel, then three years later he was appointed professor at Halle, and in 1930 he was appointed a chair in Marburg. Hasse made significant contributions to the theory of elliptic curves; for example, he proved, among others, the analogue of the Riemann Hypothesis for zeta functions of elliptic curves. Note that Hasse also wrote a very influential book in number theory, *ZAHLENTHEORIE* in 1963, English translation in 1980.

Table 1.23. Number of points on elliptic curves over $\mathbb{F}_5$

Elliptic curve	Number of points
$y^2 = x^3 + 2x$	2
$y^2 = x^3 + 4x + 2$	3
$y^2 = x^3 + x$	4
$y^2 = x^3 + 3x + 2$	5
$y^2 = x^3 + 1$	6
$y^2 = x^3 + 2x + 1$	7
$y^2 = x^3 + 4x$	8
$y^2 = x^3 + x + 1$	9
$y^2 = x^3 + 3x$	10

A more general question is: how many rational points are there on an elliptic curve $E: y^2 = x^3 + ax + b$ over $\mathbb{Q}$? Mordell[36] solved this problem in 1922:

Theorem 1.7.6 (Mordell's finite basis theorem). Suppose that the cubic polynomial $f(x,y)$ has rational coefficients, and that the equation $f(x,y) = 0$ defines an elliptic curve E. Then the group $E(\mathbb{Q})$ of rational points on E is a finitely generated Abelian group.

In elementary language, this says that on any elliptic curve that contains a rational point, there exists a finite collection of rational points such that all other rational points can be generated by using the chord-and-tangent method. From a group-theoretic point of view, Mordell's theorem tells us that we can produce all of the rational points on E by starting from some finite set and using the group laws. It should be noted that for some cubic curves, we have tools to find this generating set, but unfortunately, there is no general method (i.e., algorithm) guaranteed to work for all cubic curves.

36

Louis Joel Mordell (1888–1972) was born in Philadelphia, Pennsylvania. He was educated at Cambridge and began research in number theory. He lectured at Manchester College of Technology from 1920 to 1922. During this time he discovered the famous *finite basis theorem*, which was suggested by Poincaré in 1901. In 1922 he moved to the University of Manchester where he remained until he succeeded Hardy at Cambridge in 1945. Together with Davenport, he initiated great advances of the geometry of numbers. Mordell was elected Fellow of the Royal Society and received the De Morgan Medal in 1941 and the Sylvester Medal in 1949. He was also the President of the London Mathematical Society from 1943 to 1945.

The fact that the Abelian group is finitely generated means that it consists of a finite "torsion subgroup" $\mathrm{E}_{\mathrm{tors}}$, consisting of the rational points of finite order, plus the subgroup generated by a finite number of points of infinite order:

$$E(\mathbb{Q}) \approx \mathrm{E}_{\mathrm{tors}} \oplus \mathbb{Z}^r.$$

The number r of generators needed for the infinite part is called the *rank* of $E(\mathbb{Q})$; it is zero if and only if the entire group of rational points is finite. The study of the rank r and other features of the group of points on an elliptic curve over $\mathbb{Q}$ is related to many interesting problems in number theory and arithmetic algebraic geometry, readers are suggested to consult, e.g., Silverman and Tate's book [228] for more information.

1.8 Bibliographic Notes and Further Reading

Elementary number theory is the oldest but it is still a lively subject in number theory; it is the basis for other branches of number theory, including algebraic number theory, geometric number theory, analytic number theory, logic number theory, probabilistic number theory, combinatorial number theory, algorithmic number theory, and applied number theory. In this chapter, we have provided a survey of basic concepts and results of elementary number theory. For those who desire a more detailed exposition in elementary number theory, the following classical texts are highly recommend (in order): Hardy and Wright [100], Niven et al. [174], Davenport [58], Baker [17], Hua [105]. and Dirichlet [68]. Other good references in elementary number theory include Anderson and Bell [8], Koblitz [128], Kumanduri and Romero [135], Mollin [164], Nathanson [172], Rose [210], Rosen [211], and Silverman [230]. The books by Ore [181] and Dickson [65] contain a wealthy source of the historical development of the subject, whilst Ribenboim [200] contains the new records (up to 1996) of research in number theory, particularly in the theory of prime numbers. Khinchin's book [119] gives an excellent introduction to continued fractions.

One of the important features of this chapter is that we have provided a rather lengthy section on the distribution of prime numbers. It includes approximations to $\pi(x)$ by $\frac{x}{\ln x}$, $\mathrm{Li}(x)$, and $R(x)$. It also contains a discussion of the Riemann ζ-function and relationships between the distribution of the complex zeros of $\zeta(s)$ and the distribution of prime numbers. The study of the real function $\pi(x)$ and its various approximations belongs to the field of *Analytic Number Theory*. This particular domain of number theory operates with very advanced methods of calculus and it is considered to be one of the most difficult fields of mathematics. Readers who are interested in Analytic Number Theory are referred to Apostol's book [11] or to the Open University text [180].

Another very important feature of this chapter is that we have provided a section on an introduction to elliptic curves. The study of elliptic curves belongs to the field of *algebraic geometry*, or more specifically *Diophantine geometry*, because we are essentially only interested in the integral or rational solutions of certain types of algebraic equations represented by elliptic curves. Elliptic curve theory is a rich and well studied area, with a wide range of results, including Wiles' proof of Fermat's Last Theorem. Remarkably enough, the theory of elliptic curves is not only applicable to mathematics, but also applicable to computing science, including primality testing, integer factorization and cryptography. For those who desire a more detailed exposition of elliptic curves, please refer to the following more comprehensive texts: Husemöeller [109], Koblitz [127], Silverman [229], and Silverman and Tate [228].

Number theory is intimately connected with abstract algebra, particularly with the theory of groups, rings and fields. In fact, number theory can be studied from an algebraic point of view. For this reason, much of the material in this chapter is presented in terms of algebraic language. Hence, readers may find it helpful to consult one of the following algebra books: Childs [49], Ellis [70], Fraleigh [76], Herstein [103], Hungerford [108], McEliece [152], or Rotman [212].

2. Computational/Algorithmic Number Theory

The problem of distinguishing prime numbers from composite, and of resolving composite numbers into their prime factors, is one of the most important and useful in all arithmetic. ... The dignity of science seems to demand that every aid to the solution of such an elegant and celebrated problem be zealously cultivated.

C. F. GAUSS (1777–1855)

Computational and algorithmic number theory are two very closely related subjects; they are both concerned with, among many others, computer algorithms, particularly efficient algorithms (including parallel and distributed algorithms, sometimes also including computer architectures), for solving different sorts of problems in number theory and in other areas, including computing and cryptography. Primality testing, integer factorization and discrete logarithms are, amongst many others, the most interesting, difficult and useful problems in number theory, computing and cryptography. In this chapter, we shall study both computational and algorithmic aspects of number theory. More specifically, we shall study various algorithms for primality testing, integer factorization and discrete logarithms that are particularly applicable and useful in computing and cryptography, as well as methods for many other problems in number theory, such as the Goldbach conjecture and the odd perfect number problem.

2.1 Introduction

In this section, we shall first present a brief introduction to algorithmic and computational number theory, and then provide a theoretical foundation of algorithms, including effective computability and computational complexity, which are useful in both algorithmic and computational number theory.

2.1.1 What is Computational/Algorithmic Number Theory?

Algorithmic number theory studies of algorithms (including parallel algorithms, sometimes also including computing architectures) for problems that arise in number theory. Primality testing, integer factorization, and discrete logarithms (including elliptic curve discrete logarithms) are, amongst many others, the most interesting, difficult and useful problems in number theory. Computational number theory, however, studies problems from elementary, algebraic geometric and analytic number theory which require the help of fast computers (particularly vector and parallel systems) and fast algorithms (particularly deterministic polynomial-time algorithms). It is clear that these two subjects are closely related each other; some people may well regard them as *one* single subject which belongs to both mathematics and computer science, whereas others may regard algorithmic number theory as a part of computer science and computational number theory a part of mathematics. In this chapter, we shall study both algorithmic and computational aspects of number theory.

Computational (or algorithmic) number theory is a relatively new branch of science, which has become a discipline in its own right during the past two decades. In computational (or algorithmic) number theory, all the problems studied are from number theory, but the methods for solving these problems can be either from mathematics, or computer science, or both. This makes computational number theory different from other branches of number theory such as algebraic number theory which uses algebraic methods to solve number-theoretic problems. Thus, computational (or algorithmic) number theory is an interdisciplinary subject of number theory and computer science, and the people working in this area often come from either mathematics or computer science. Its main purpose is to design efficient computer algorithms (and sometimes high-speed computer architectures) for large-scale numerical computations (including verifications) for number theory. Among its wide spectrum of activities, this new branch of number theory is concerned with problems such as the following:

(1) Primality testing: The fastest deterministic algorithm for primality testing is the APRCL algorithm (see Adleman, Pomerance and Rumely [3], and Cohen [50]), invented by Adleman, Pomerance, Rumely, Cohen and Lenstra, which runs in $\mathcal{O}(\log N)^{c \log\log\log N}$ and is possible to prove the primality of integers with 1000 digits in a not too unreasonable amount of time. At present, the most practical primality testing/proving algorithm is the elliptic curve primality proving algorithm ECPP, designed by Atkin and Morain [12], which can prove the primality of integers with several thousand digits in reasonable amount of time, for example, weeks of workstation time.

(2) Integer factorization: The fastest general algorithm for integer factorization is the Number Field Sieve (NFS), which under plausible assumptions

has the expected running time

$$\mathcal{O}\left(\exp\left(c\sqrt[3]{\log N}\sqrt[3]{(\log\log N)^2}\right)\right).$$

Clearly, NFS is still a subexponential-time algorithm, not a polynomial-time algorithm. The largest integer factored with NFS is the RSA-155 (August 1999), an integer with 155 digits.

(3) Discrete logarithms: over a finite field: This discrete logarithm problem (DLP) for the multiplicative group $\mathbb{F}_p^*$ is similar to that of integer factorization (although it is a little bit more difficult than integer factorization), and the methods for factoring (e.g., Number Field Sieve) are usually applicable to discrete logarithms. It should be noted, however, that there are quantum algorithms [227] that can be used to solve the integer factorization problem and the discrete logarithm problem in polynomial time on a quantum computer, although no one knows at present whether or not a practical quantum computer can be built.

(4) Elliptic curve discrete logarithms: Let $E/\mathbb{F}_p$ be an elliptic curve defined over a finite field, and let $P, Q \in E(\mathbb{F}_p)$ be two points on E. The elliptic curve discrete logarithm problem (ECDLP) asks to find an integer k such that $Q = kP$ in $E(\mathbb{F}_p)$. This problem is considered to be very difficult to solve if p is large, for which reason it has formed the basis for various cryptographic systems. Note that there are subexponential complexity *Index Calculus* algorithms such as the Number Field Sieve for discrete logarithms over a finite field, however, no practical Index Calculus method has been found for the Elliptic curve discrete logarithms, and more serious, it looks like that ECDLP does not admit an Index Calculus. Current research in ECDLP aims to develop new algorithms such as *Xedni Calculus* [231] that might be used to solve the ECDLP.

(5) Counting the numbers of primes, $\pi(x)$: The most recent record is $\pi(4 \cdot 10^{22}) = 783964159852157952242$, that is, there are exactly 783964159852157952242 prime numbers up to $4 \cdot 10^{22}$.

(6) Mersenne primes: There are now 39 known Mersenne primes. The largest is $2^{13466917} - 1$; it has 4053946 digits and was discovered by Cameron, Woltman and Kurowski, et al. in 2001. At present, we still do not know if there are infinitely many Mersenne primes.

(7) Odd perfect numbers: Even perfect numbers are in one-to-one correspondence with Mersenne primes. That is, once we find a Mersenne prime $2^p - 1$, we have an even perfect number $2^{p-1}(2^p - 1)$. All the known perfect numbers are even; we do not know if there exists an odd perfect number. Numerical results show that there are no odd perfect numbers up to 10^{300} (Brent, Cohen and Te Riele, [39]).

(8) Fermat numbers: Only the first five Fermat numbers (i.e., $F_n = 2^{2^n} + 1$ for $n = 0, 1, 2, 3, 4$) have been found prime, all the rest are either composite, or their primality is unknown. The complete prime factorizations

for F_n with $5 \leq n \leq 11$ have been obtained; the smallest not completely factored Fermat number, and indeed the most wanted number, is F_{12}.

(9) Amicable numbers: The first amicable pair $(220, 284)$ was known to the legendary Pythagoras 2500 years ago, but the second smallest amicable pair $(1184, 1210)$ was not found until 1866 by a 16-year old Italian school boy, Nicolo Paganini. Prior to Euler (1707–1783), only three amicable pairs were known. Although there are 2574378 known amicable pairs at present, we still do not know if there are infinitely many amicable pairs or not; we even do not have a general rule to generate all the amicable pairs.

(10) Riemann Hypothesis: The first $1,500,000,001$ nontrivial zeros of the Riemann ζ-function have been calculated, and they all lie on the critical line $\mathrm{Re}(s) = 1/2$, as conjectured by Riemann in 1859. However, we do not know if *all* the nontrivial zeros of the ζ-function lie on the critical line $\mathrm{Re}(s) = 1/2$, On 24 May 2000 the Clay Mathematics Institute of Cambridge, Massachusetts announced seven Millennium Prize Problems; The Riemann Hypothesis is one of them. It designated a one-million US dollar prize fund for the solution to each of these seven problems. (For an official description of the problem, see [29].)

(11) Goldbach's conjecture: It has been numerically verified that Goldbach's conjecture is true for even numbers $4 \leq n \leq 4 \cdot 10^{14}$ (see Deshouillers, Te Riele and Saouter [62], and Richstein [201]). The experimental results are in good agreement with the theoretical prediction made by Hardy and Littlewood. On 20 March 2000 the British publishing company Faber and Faber in London announced a one-million US dollar prize to any person who can prove Goldbach's Conjecture within the next two years (before midnight, 15 March 2002).

(12) Calculation of π: By using an analytic extension of a formula of Ramanujan, David and Gregory Chudnovsky in 1989 calculated π to one billion decimal digits. It is interesting to note that the string of digits 123456789 occurs shortly after the half-billionth digit.

(13) Waring's Problem: In 1770 the English mathematician Edward Waring conjectured that every integer can be written as the sum of $g(k)$ positive kth powers, where $g(k) = q + 2^k - 2$ with $3^k = q \cdot 2^k + r$. It is currently known that

$$g(2) = 4, \;\; g(3) = 9, \;\; g(4) = 19, g(5) = 37$$

$$g(k) = \left[\left(\frac{3}{2} \right)^k \right] + 2^k - 2, \qquad \text{for } 6 \leq k \leq 471600000.$$

(14) Primes in arithmetic progressions: An arithmetic progression of primes is a sequence of primes where each is the same amount more than the one before. For example, the sequence 5, 11, 17, 23 and 29 forms an

arithmetic progression of primes, since all the numbers in the sequence are prime, and the common difference is 6. It is conjectured that there should be arbitrarily long arithmetic progressions of primes, but no proof has been given so far. The longest known arithmetic progression contains 22 terms. The first term is 11410337850553 and the common difference is 4609098694200. This sequence of primes was discovered in March 1993 at Griffith University, Queensland, Australia.

As can be seen, the main theme in computational number theory is algorithms. In the next two subsections, we shall provide a theoretical foundation of algorithms, including effective computability and computational complexity.

2.1.2 Effective Computability

Algorithmic number theory emphasizes algorithmic aspects of number theory and aims at the design of *efficient algorithms* for solving various number-theoretic problems. But what is an algorithm? Remarkably enough, the word *algorithm* itself is interesting and has a very long history; it comes from the name of the Persian mathematician Abu Ja'far Muhammad ibn Musa al-Khwarizmi[1]. An algorithm may be defined as follows.

Definition 2.1.1. An *algorithm* is a finite sequence of well-described instructions with the following properties:

(1) There is no ambiguity in any instruction.
(2) After performing a particular instruction there is no ambiguity about which instruction is to be performed next.
(3) The instruction to stop is always reached after the execution of a finite number of instructions.

An algorithm is also called an *effective procedure*, since all of the operations to be performed in the algorithm must be sufficiently basic that they can in principle be done exactly and in a finite length of time by a man using pencil and paper (Knuth [122]). So, for us the two terms *algorithm* and *effective procedure* are synonymous and we shall use them interchangeably.

[1]

Abu Ja'far Muhammad ibn Musa al-Khwarizmi (about 780–850) was born in an area not far from Baghdad. He wrote his celebrated book *Hisab al-jabr w'al-muqabala* (from which our modern word *algebra* comes) while working as a scholar at the House of Wisdom (a center of study and research in the Islamic world of the ninth century) in Baghdad. In addition to this treatise, al-Khwarizmi wrote works on astronomy, on the Jewish calendar, and on the Hindu numeration system. The English word *algorithm* derives from *algorism*, which is the Latin form of al-Khwarizmi's name.

Definition 2.1.2. A function f is *computable* (or equivalently, a problem is *decidable/solvable*) if there exists an effective procedure (or algorithm), A_f, that produces the value of f correctly for each possible input; otherwise, the function is called *noncomputable* (or equivalently, the problem is *undecidable/unsolvable*).

Clearly, the notion here for computable functions is intuitive, but to show that a function is computable or noncomputable, we need a formalized notion for effective computability; otherwise, we cannot show that an effective procedure does not exist for a function under consideration. This can be achieved by an imaginary computing machine, named the *Turing machine (TM)* after its inventor Alan Turing[2], which can be defined as follows:

Definition 2.1.3. A (standard k-tape) *Turing machine (TM), M* (see Figure 2.1), is an algebraic system defined by

$$M = (Q, \Sigma, \Gamma, \delta, q_0, \Box, F) \tag{2.1}$$

where

(1) Q is a finite set of *internal states*.

[2]

Alan M. Turing (1912–1954) was born in London, England. He was educated in Sherborne, an English boarding school and King's College, Cambridge. In 1935, Turing became fascinated with the decision problem, a problem posed by the great German mathematician David Hilbert, which asked whether there is a general method that can be applied to any assertion to determine whether the assertion is true. The paper which made him famous "On Computable Numbers, with an Application to the Entscheidungsproblem (problem of decidability)" was published in the *Proceedings of the London Mathematics Society*, Vol 42, November 1936. It was in this paper that he proposed the very general computation model, now widely known as the *Turing machine*, which can compute any computable function. The paper attracted immediate attention and led to an invitation to Princeton (recommended by John von Neumann), where he worked with Alonzo Church. He took his PhD there in 1938; the subject of his thesis was "Systems of Logic based on Ordinals". During World War II Turing also led the successful effort in Bletchley Park (then the British Government's Cryptography School in Milton Keynes) to crack the German "Enigma" cipher, which Nazi Germany used to communicate with the U-boats in the North Atlantic. To commemorate Turing's original contribution, the *Association for Computing Machinery* in the U.S.A. created the Turing Award in 1966. The award is presented annually to an individual selected for contributions of a technical nature to the computing community that are judged to be of lasting and major importance to the field of computer science, and it is in fact regarded as the Nobel Prize of computer science. Turing committed suicide in 1954 after a conviction related to his homosexuality. Were it known that he had been a war hero (having deciphered Enigma), the prosecution would never have taken place, and this great man might still be alive today.

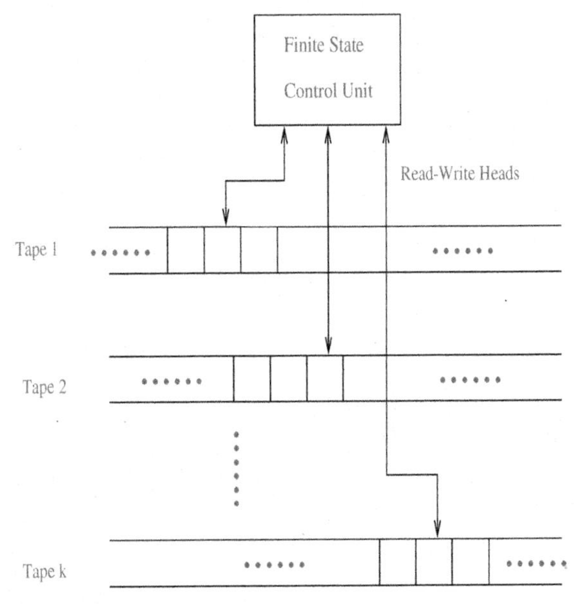

Figure 2.1. A standard Turing machine

(2) Σ is a finite set of symbols called the *input alphabet*. We assume that $\Sigma \subseteq \Gamma - \{\Box\}$.
(3) Γ is a finite set of symbols called the *tape alphabet*.
(4) δ is the transition function, which is defined by
   (i) if M is a deterministic Turing machine (DTM), then

$$\delta : \ Q \times \Gamma^k \to Q \times \Gamma^k \times \{L, R\}^k, \tag{2.2}$$

   (ii) if M is a nondeterministic Turing machine (NDTM), then

$$\delta : Q \times \Gamma^k \to 2^{Q \times \Gamma^k \times \{L,R\}^k}, \tag{2.3}$$

   where L and R specify the movement of the read-write head *left* or *right*. When $k = 1$, it is just a standard one-tape Turing machine.
(5) $\Box \in \Gamma$ is a special symbol called the *blank*.
(6) $q_0 \in Q$ is the *initial state*.
(7) $F \subseteq Q$ is the set of *final states*.

A *probabilistic Turing machine* is a type of nondeterministic Turing machine with distinguished states called *coin-tossing states*. For each coin-tossing state, the finite control unit specifies two possible legal next states. The computation of a probabilistic Turing machine is deterministic except that in coin-tossing states the machine tosses an unbiased coin to decide between the two *possible legal* next states.

The computation of a Turing machine is formalized by using the notion of an *instantaneous description*: Let M be a Turing machine, then any string $a_1...a_{k-1}q_1a_ka_{k+1}...a_n$, with $a_i \in \Gamma$ and $q_1 \in Q$, is an *instantaneous description* (ID) of M. A move

$$a_1...a_{k-1}q_1a_ka_{k+1}...a_n \vdash a_1...a_{k-1}bq_2a_{k+1}...a_n \tag{2.4}$$

is possible if

$$\delta(q_1, a_k) = (q_2, b, R). \tag{2.5}$$

A move

$$a_1...a_{k-1}q_1a_ka_{k+1}...a_n \vdash a_1...q_2a_{k-1}ba_{k+1}...a_n \tag{2.6}$$

is possible if

$$\delta(q_1, a_k) = (q_2, b, L). \tag{2.7}$$

M is said to halt, starting from some initial configuration $x_1q_ix_2$, if

$$x_1q_ix_2 \overset{*}{\vdash} y_1q_jay_2 \tag{2.8}$$

for any q_j and a, for which $\delta(q_j, a)$ is undefined. The sequence of configurations leading to a halt state is called a *computation*. If M never halts, then we represent it by

$$x_1q_ix_2 \overset{*}{\vdash} \infty, \tag{2.9}$$

indicating that, starting from the initial configuration $x_1q_ix_2$, the machine never halts. Thus, the Turing machine provides us with the simplest possible abstract model of computation in general. Moreover, any effectively computable function can be computed by a Turing machine, and there is no effective procedure that a Turing machine cannot perform. This leads to the following famous Church-Turing thesis, named after Church[3] and Turing:

> **The Church-Turing thesis.** A function is effectively *computable* if it can be computed by a Turing machine. That is, computable is Turing computable.

3

Alonzo Church (1903–1995) was born in Washington, D.C. Much of his professional life was centered around Princeton University. He received his first degree in 1924 and PhD in 1927, both from Princeton. He was a National Research Fellow in 1927–29, spending time at Harvard, Göttingen and Amsterdam. Church was a faculty member in Mathematics at Princeton University from 1929 until 1967 when he moved to the University of California at Los Angeles. He made substantial contributions to the theory of computability including his solution to the decision problem, his invention of the lambda-calculus, and his statement known as the Church-Turing thesis. He also supervised 31 doctoral students, including Alan Turing, Stephen Kleene, Martin Davis, Michael Rabin, Dana Scott and John Kemeny.

Remark 2.1.1. The Church-Turing thesis is a *thesis*, not a theorem, because it is not a mathematical result and cannot be proved mathematically; it just asserts that a certain intuitive notion (effective procedure) corresponds to a certain mathematical object (Turing machine). To prove it, we would have to compare effective procedures (an intuitive notion) and Turing machines (a formal notion). To do this, we would have to formalize the notion of an effective procedure. But then we would face the problem: is the introduced formalization equivalent to the intuitive notion? The solution of this problem would require a claim to the Church-Turing thesis, and so we would fall into an endless loop. Hence, the Church-Turing thesis has to remain as a *thesis*, not a *theorem*. Nevertheless, a tremendous amount of evidence has shown that the Church-Turing thesis is true, and researchers in computer science and also in mathematics generally believe the truth of the thesis. It is theoretically possible, however, that the Church-Turing thesis could be disproved at some future date, if someone were to propose an alternative model of computation that was provably capable of carrying out computations that cannot be carried out by any Turing machine; but this is not likely to happen.

The Church-Turing thesis thus provides us with a very powerful tool to distinguish which functions are computable and which are noncomputable: functions that can be computed by a Turing machine are computable, whereas functions that cannot be computed by a Turing machine are noncomputable. We can therefore classify all computational problems into two categories:

(1) Class of problems solvable by a Turing machine.

(2) Class of problems unsolvable by a Turing machine.

There are many unsolvable problems; the best known one is surprisingly concerned with the Turing machine itself: *given a Turing machine M and an input w, does M halt on w?* This is the so-called *halting problem* for Turing machines, and is unsolvable by a Turing machine. Of course, unsolvable problems do not only exist in the domain of Turing machines, but in virtually all fields of mathematics. It is not our purpose to discuss the uncomputability of Turing machines here; we shall restrict ourselves to Turing computability, particularly to *practical* Turing computability.

2.1.3 Computational Complexity

Effective computability studies theoretical computability, which does not imply any restrictions concerning the *efficiency* of computations; efficiency is often described in terms of *complexity*, which is essentially a measure of time and memory space needed to perform a computation (in this book we shall treat complexity primarily in terms of time). Effective computability does not mean *practical computability*. In fact, many problems, although solvable in theory, cannot be solved in any practical sense by a Turing machine due to excessive time requirements. For example, using the Sieve of Eratosthenes to find the nth prime, it is practical to compute the 10^{10}th prime, but it would never become practical to find the $10^{10^{10}}$-th prime. In this subsection, we shall give a brief introduction to the theory of *practically feasible computation* (practically feasible computation is also called *practically tractable computation*; we shall use the two terms interchangeably).

The time complexity (or the running time) of an algorithm is a function of the length of the input. An algorithm is of time complexity $t(n)$ if for all n and all inputs of length n, the execution of the algorithm takes at most $t(n)$ steps. More precisely, we have:

Definition 2.1.4. Let TM be a Turing machine which halts after m steps for an input of length n. Then the *time complexity function* or the *running time* associated with TM, denoted by $t_{\text{TM}}(n)$, is defined by

$$t_{\text{TM}}(n) = \max\{m: \text{ TM halts after } m \text{ steps for an input of length } n\}. \tag{2.10}$$

Let NDTM be a nondeterministic Turing machine. For an input w we denote by $s(w)$ the shortest halting computation starting from w. Then the *time complexity function* associated with NDTM, denoted by t_{NDTM}, is defined by

$$t_{\text{NDTM}}(n) = \max\{(1, m): w \text{ is of length } n \text{ and } s(w) \text{ has } m \text{ steps}\}. \tag{2.11}$$

Definition 2.1.5. A deterministic Turing machine (DTM) is called *polynomially bounded* if there exists a polynomial function $p(n) \in \mathcal{O}(n^k)$, for some positive integer k, such that

$$t_{\text{DTM}}(n) \leq p(n), \tag{2.12}$$

where n is the length of the input. A problem is called *polynomially solvable* if there is a polynomially bounded Turing machine that solves it. The class of all polynomially solvable problems is denoted by $\mathcal{P}$.

Definition 2.1.6. A deterministic Turing machine (DTM) is called *exponentially bounded* if there exists an exponential function $\exp(n) \in \mathcal{O}(a^n)$ for some constant $a > 1$ such that

$$t_{\text{DTM}}(n) \leq \exp(n), \qquad \text{for all } n, \tag{2.13}$$

where n is the length of the input. A problem is called *exponentially solvable* if there is an exponentially bounded Turing machine that solves it. The class of all exponentially solvable problems is denoted by $\mathcal{EXP}$.

Definition 2.1.7. A nondeterministic Turing machine (NDTM) is called *polynomially bounded* if there exists a polynomial function $p(n) \in \mathcal{O}(n^k)$, for some positive integer k, such that

$$t_{\text{NDTM}}(n) \leq p(n), \tag{2.14}$$

where n is the length of the input. The class of all problems solvable by a polynomially bounded nondeterministic Turing machine is denoted by $\mathcal{NP}$.

All different types of Turing machines, such as single-tape DTM, multitape DTM and NDTM are equivalent in computation power but may be different in efficiency. For example, let $t(n)$ be a function with $t(n) > n$. Then

(1) Every $t(n)$ time multitape deterministic Turing machine has an equivalent $\mathcal{O}(t^2(n))$ time single-tape deterministic Turing machine.

(2) Every $t(n)$ time single-tape nondeterministic Turing machine has an equivalent $2^{\mathcal{O}(t(n))}$ time single-tape deterministic Turing machine.

In complexity theory, it is common to concentrate on *decision problems*, i.e., those having a yes/no solution, since any decision problem can be treated as a language recognition problem.

Definition 2.1.8. An *alphabet* Σ is a finite set of symbols. A *language* L over Σ is any set of strings made up of symbols from Σ. We denote the empty string by ϵ, and the empty language by $\emptyset$. The language of all strings over Σ is denoted by Σ^*. We also define the *complement* of L by $\overline{L} = \Sigma^* - L$. We say a Turing machine M accepts a string $x \in \Sigma^*$ if, given input x, M outputs $M(x) = 1$, and otherwise $M(x) = 0$.

Within the framework of formal language theory, the complexity classes $\mathcal{P}$, $\mathcal{NP}$ and $\mathcal{EXP}$ defined above can then be re-defined as follows.

Definition 2.1.9. The class $\mathcal{P}$ consists of all languages L that have a *polynomially bounded* deterministic Turing machine (DTM), such that for any string $x \in \Sigma^*$,

$$\begin{cases} x \in L & \Longrightarrow & \text{DTM}(x) = 1, \\ x \notin L & \Longrightarrow & \text{DTM}(x) = 0. \end{cases}$$

The class $\mathcal{EXP}$ consists of all languages L that have an *exponentially bounded* deterministic Turing machine DTM, such that for any string $x \in \Sigma^*$,

$$\begin{cases} x \in L & \Longrightarrow & \text{DTM}(x) = 1, \\ x \notin L & \Longrightarrow & \text{DTM}(x) = 0. \end{cases}$$

The class $\mathcal{NP}$ consists of all languages L that have a *polynomially bounded* nondeterministic Turing machine (NDTM), such that for any string $x \in \Sigma^*$,

$$\begin{cases} x \in L & \Longrightarrow & \exists y \in \Sigma^*, \text{NDTM}(x, y) = 1, \text{ where } |y| \text{ is} \\ & & \quad \text{bounded by a polynomial in } |x|, \\ x \notin L & \Longrightarrow & \exists y \in \Sigma^*, \text{NDTM}(x, y) = 0. \end{cases}$$

For probabilistic Turing machines, we have the corresponding probabilistic complexity classes $\mathcal{RP}$, $\mathcal{BPP}$, and $\mathcal{ZPP}$.

Definition 2.1.10. The class $\mathcal{RP}$ (Randomized Polynomial) consists of all languages L that have a probabilistic Turing machine (PTM) running in expected polynomial time with one-sided error. That is, for any input $x \in \Sigma^*$,

$$\begin{cases} x \in L & \Longrightarrow & \text{Prob}[\text{PTM}(x) = 1] \geq 1/2, \\ x \notin L & \Longrightarrow & \text{Prob}[\text{PTM}(x) = 1] = 0. \end{cases}$$

Definition 2.1.11. The class $\mathcal{ZPP}$ (Zero-error Probabilistic Polynomial) is defined by $\mathcal{ZPP} = \mathcal{RP} \cap \text{co-}\mathcal{RP}$. That is, $\mathcal{ZPP}$ is the class of all languages L that have a probabilistic Turing machine (PTM) running in expected polynomial time with zero-sided error. That is, for any input $x \in \Sigma^*$,

$$\begin{cases} x \in L & \Longrightarrow & \text{Prob}[\text{PTM}(x) = 1] = 0, \\ x \notin L & \Longrightarrow & \text{Prob}[\text{PTM}(x) = 1] = 0. \end{cases}$$

Definition 2.1.12. The class $\mathcal{BPP}$ (Bounded-error Probabilistic Polynomial) consists of all languages L that have a probabilistic Turing machine (PTM) running in expected polynomial time with two-sided error. That is, for any input $x \in \Sigma^*$,

$$\begin{cases} x \in L & \Longrightarrow & \text{Prob}[\text{PTM}(x) = 1] \geq 3/4, \\ x \notin L & \Longrightarrow & \text{Prob}[\text{PTM}(x) = 1] \leq 1/4. \end{cases}$$

The space complexity classes $\mathcal{P}$-SPACE and $\mathcal{NP}$-SPACE can be defined analogously as $\mathcal{P}$ and $\mathcal{NP}$. It is clear that a time class is included in the corresponding space class since one unit is needed to the space by one square. Although it is not known whether or not $\mathcal{P} = \mathcal{NP}$, it is known that $\mathcal{P}$-SPACE $= \mathcal{NP}$-SPACE. It is generally believed that

$$\mathcal{P} \subseteq \mathcal{ZPP} \subseteq \mathcal{RP} \subseteq \begin{pmatrix} \mathcal{BPP} \\ \mathcal{NP} \end{pmatrix} \subseteq \mathcal{P}\text{-SPACE} \subseteq \mathcal{EXP}.$$

Besides the proper inclusion $\mathcal{P} \subset \mathcal{EXP}$, it is not known whether any of the other inclusions in the above hierarchy is proper. Note that the relationship of $\mathcal{BPP}$ and $\mathcal{NP}$ is not known, although it is believed that $\mathcal{NP} \not\subseteq \mathcal{BPP}$.

Remark 2.1.2. Although the complexity classes are defined in terms of decision problems, they can be used to classify the complexity of a broader class of problems, such as search or optimization problems. It should be also noted that complexity classes are not only referred to *problems*, but also to *algorithms*. For example, we can say that Euclid's algorithm is of polynomial complexity, since it can be performed in polynomial time, that is, Euclid's algorithm is in $\mathcal{P}$.

From a *practical computability* point of view, all algorithms can be classified into two categories:

(1) Efficient (good) algorithms: those algorithms that can be performed in polynomial time.

(2) Inefficient (bad) algorithms: those algorithms that can only be performed in exponential time.

The reason is fairly obvious: An exponential function grows much more quickly than a polynomial function does for large values of n. Algorithms of polynomial complexity are considerably more efficient than those of exponential complexity. More generally, there is a hierarchy of increasing orders:

$$\log n,\ n,\ n^2,\ n^3,\ \cdots,\ 2^n,\ 3^n,\ \cdots,\ n!,\ n^n.$$

Table 2.1 compares growth rates of complexity functions for different input values of n, whereas Table 2.2 compares execution times for algorithms of various complexities [79] (we assume that each step of the algorithm takes one microsecond of computer time to execute).

By examining these tables, one can see that exponential and factorial complexity functions grow faster than any polynomial functions when n is large. This gives us the idea that the running time of any practically feasible computation must be bounded by a *polynomial* in the length of the input, and leads to the Cook-Karp thesis, a quantitative refinement of the Church-Turing thesis. Similarly, all solvable problems can also be classified into two categories:

(1) Computationally tractable (or feasible).

(2) Computationally intractable (or infeasible).

It is widely believed, although no proof has been given, that problems in $\mathcal{P}$ are computationally tractable, whereas problems not in (beyond) $\mathcal{P}$ are

Table 2.1. Comparison of growth rates of complexity functions with input sizes

Input	**Complexity Function** f					
Size n	$\log n$	n	$n \log n$	n^2	2^n	$n!$
5	2	5	12	25	32	120
10	3	10	33	100	1024	3.6×10^6
5×10	6	50	282	2500	1.1×10^{15}	3×10^{64}
10^2	7	100	664	10^4	1.3×10^{30}	9.3×10^{157}
5×10^2	9	500	4483	25×10^4	3.3×10^{150}	1.2×10^{1134}
10^3	10	10^3	9966	10^6	1.1×10^{301}	4.0×10^{2567}
10^4	13	10^4	132877	10^8	1.9×10^{3010}	2.8×10^{35659}
10^5	17	10^5	1.6×10^6	10^8	(too large)	(too large)

Table 2.2. Comparison of several polynomial and exponential time complexity functions f

f	**Input Size** n					
	10	20	30	40	50	60
n	0.00001 second	0.00002 second	0.00003 second	0.00004 second	0.00005 second	0.00006 second
n^2	0.0001 second	0.0004 second	0.0009 second	0.0014 second	0.0025 second	0.0036 second
n^3	0.001 second	0.008 second	0.027 second	0.064 second	0.125 second	0.216 second
n^5	0.1 second	3.2 seconds	24.3 seconds	1.7 minutes	5.2 minutes	13.0 minutes
2^n	0.01 second	1.0 second	17.9 minutes	12.7 days	35.7 years	366 centuries
3^n	0.59 second	58 minutes	6.5 years	3855 centuries	2.3×10^8 centuries	1.3×10^{13} centuries
5^n	9.8 seconds	3 years	3×10^5 centuries	2.9×10^{12} centuries	2.8×10^{19} centuries	2.8×10^{26} centuries
$n!$	3.6 seconds	7.7×10^4 years	8.4×10^{14} centuries	2.6×10^{32} centuries	9.6×10^{48} centuries	2.6×10^{66} centuries

computationally intractable. This is the famous *Cook-Karp Thesis*, named after Stephen Cook[4] and Richard Karp[5]:

[4]

Stephen Cook (1939–) was born in Buffalo, New York, received his BSc degree from the University of Michigan in 1961, and his PhD from Harvard University in 1966. From 1966 to 1970 he was Assistant Professor at the University of California, Berkeley. He joined the faculty at the University of Toronto in 1970 as an Associate Professor, and was promoted to Professor in 1975 and University Professor in 1985. He is the author of over 50 research papers, including his famous 1971 paper "The Complexity of Theorem Proving Procedures" which introduced the theory of $\mathcal{NP}$-completeness. Cook was the 1982 recipient of the Turing award, is a Fellow of the Royal Society of Canada, and a member of the U.S. National Academy of Sciences and the American Academy of Arts and Sciences. (Photo by courtesy of Prof. Cook.)

[5]

Richard M. Karp (1935–) earned his PhD in applied mathematics from Harvard University in 1959. He has been a researcher at the IBM Thomas J. Watson Research Center in New York, and Professor of Computer Science in the University of California, Berkeley and University of Washington, Seattle. He is currently professor at UC Berkeley, returning from Washington in June 1999. Karp was the 1985 Turing award winner for his fundamental contributions to complexity theory, which extended the earlier work of Stephen Cook in $\mathcal{NP}$-completeness theory. He has been elected to membership of the U.S. National Academy of Sciences and National Academy of Engineering. (Photo by courtesy of Prof. Karp.)

> **The Cook-Karp thesis.** A problem is said to be *computationally tractable* (or *computationally feasible*) if it is in $\mathcal{P}$; a problem which is not in $\mathcal{P}$ is said to be *computationally intractable* (or *computationally infeasible*).

Whether or not $\mathcal{P} = \mathcal{NP}$ is one of the most important open problems in both computer science and mathematics, and in fact, it has been chosen to be one of the seven Millennium Prize Problems by the Clay Mathematics Institute, with one million US dollars prize for a proof or disproof of the problem (for the official description of the problem, see [52]).

Example 2.1.1. The following two problems are computationally intractable:

(1) The primality testing problem. The best deterministic algorithm to test n for primality runs in time $\mathcal{O}\left((\log n)^{c(\log\log\log n)}\right)$, which grows superpolynomially in input length $\log n$; we do not regard a superpolynomial as being a polynomial.

(2) The integer factorization problem. The best algorithm for factoring a general integer n runs in time $\mathcal{O}(\exp((\log n)^{1/3}(\log\log n)^{2/3})$, which grows subexponentially (but superpolynomially) in input length $\log n$.

How about problems in $\mathcal{NP}$? Are all the problems in $\mathcal{NP}$ tractable? Clearly, $\mathcal{P}$ is included in $\mathcal{NP}$, but it is a celebrated open problem as to whether or not $\mathcal{P} = \mathcal{NP}$. However there are many $\mathcal{NP}$-complete problems, which are significantly harder than other problems in $\mathcal{NP}$. A specific problem is *$\mathcal{NP}$-complete* if it is in $\mathcal{NP}$ and, moreover, it is $\mathcal{NP}$-hard[6]. It thus follows that $\mathcal{P} = \mathcal{NP}$ if an $\mathcal{NP}$-complete problem is in $\mathcal{P}$. It is generally conjectured that $\mathcal{P} \neq \mathcal{NP}$. Therefore, $\mathcal{NP}$-complete problems are considered to be intractable. Several hundred problems in mathematics, operations research and computer science have been proven to be $\mathcal{NP}$-complete. The following are just some of them:

(1) The *traveling salesman* problem (TSP): Given a complete graph $G = (V, E)$, with edge costs, and an integer k, is there a simple cycle that visits all vertices and has total cost $\leq k$?

(2) The Hamiltonian Cycle Problem: Given a network of cities and roads linking them, is there a route that starts and finishes at the same city and visits every other city exactly once?

---

[6] A problem is $\mathcal{NP}$-hard if all problems in $\mathcal{NP}$ are polynomial time *reducible* to it, even though it may not be in $\mathcal{NP}$ itself. A formal definition for this reduction is: for an arbitrary problem in $\mathcal{NP}$, there exists a polynomially bounded deterministic Turing machine that translates every instance of the arbitrary problem into an instance of the problem.

(3) The *clique* problem: A *clique*, in an undirected graph $G = (V, E)$ is a subset $V' \in V$ of vertices, each pair of which is connected by an edge in E. The size m of a clique is the number of vertices it contains. The clique problem is then: given a finite graph $G = (V, E)$ and positive integer $m \leq |v|$, does G have a clique of size m?

(4) The *binary partition* problem: Given $A = \{a_1, a_2, \cdots, a_n\}$ a set of integers written in binary notation, is there a subset A' such that

$$\sum_{a \in A'} a = \sum_{a \in A - A'} a?$$

(Note that if A is a set of integers written in unary notation, then it can be decided in polynomial time.)

(5) The *quadratic congruence* problem: Given positive integers a, b and c, is there a positive integer $x < c$ such that $x^2 \equiv a \pmod{b}$?

(6) The *quadratic Diophantine equation* problem: Given positive integers a, b and c, are there positive integers x and y such that $ax^2 + by = c$?

(7) The *subset-sum* problem: Given a finite set $S \subseteq \mathbb{N}$ and a target $t \in \mathbb{N}$, is there a subset $S' \subseteq S$ whose elements sum to t?

The integer factorization problem, however, is currently thought to be in $\mathcal{NP}$, not in $\mathcal{P}$, but no one has yet proven that it must be in $\mathcal{NP}$. The best reference for computational intractability is still the book by Garey and Johnson [79], although it is a little bit out of date.

2.1.4 Complexity of Number-Theoretic Algorithms

As mentioned previously, the time complexity of an algorithm is a function of the length of the input. If the input n is an integer, then its length is the number of bits in n:

$$\text{length}(n) = \text{number of bits in } n. \tag{2.15}$$

In computational number theory, the inputs are of course always integers, and hence our input lengths (or sizes) will be the total number of bits needed to represent the inputs of the algorithms, and our running times for these algorithms will count bit operations rather than arithmetic operations. Polynomial time algorithms counted by arithmetic operations are essentially useless in computational number theory, because they will be of exponential time if we count by bit operations. When we describe the number of bit operations needed to perform an algorithm, we are describing the *computational complexity* of this algorithm. In describing the number of bit operations needed to perform an algorithm, we will need some notations particularly the big-$\mathcal{O}$ notation.

Definition 2.1.13. Let f and g be positive real-valued functions. Then

(1) Big-$\mathcal{O}$ notation (denotes the upper bound of the complexity function f): $f(n) = \mathcal{O}(g(n))$ if there exists a real constant $c > 0$ such that $f(n) \le c \cdot g(n)$ for all sufficiently large n.
(2) Small-o notation (denotes the upper bound of the complexity function f, that is not asymptotically tight): $f(n) = \mathcal{O}(g(n)), \forall c > 0$ such that $f(n) < c \cdot g(n)$.
(3) Big-Ω notation (denotes the low bound of the complexity function f): $f(n) = \Omega(g(n))$ if there exists a real constant c such that $f(n) \ge \frac{1}{c} \cdot g(n)$.
(4) Big-Θ notation (denotes the tight bound of the complexity function f): $f(n) = \Theta(n)$ if $f(n) = \mathcal{O}(g(n))$ and $f(n) = \Omega(g(n))$.

In this book, we shall mainly use the big-$\mathcal{O}$ notation.

Example 2.1.2. Let $f(n) = n^3 + 8n^2 \log n + 14n - 1$, then with the big-$\mathcal{O}$ notation, we have $f(n) = \mathcal{O}(n^3)$.

Definition 2.1.14. Given integers p, q and b with $q = b^p$, then p is said to be the *logarithm* to the base b of the number q. We shorten this to

$$p = \log_b q. \tag{2.16}$$

Symbolically,

$$p = \log_b q \iff q = b^p. \tag{2.17}$$

If $b = 2$, then

$$p = \log_2 q \iff q = 2^p. \tag{2.18}$$

Note that while base 10 is common in high school algebra and base e is typically used in calculus; in computer science logs are always assumed to be base 2. In this book, we shall use the notation log to mean $\log_2$, and ln to mean $\log_e$.

Any integer $n \in \mathbb{N}$ to the base b can be written as follows:

$$\begin{aligned} n &= (d_{\beta-1}d_{\beta-2}\cdots d_1 d_0)_b \\ &= d_{\beta-1}b^{\beta-1} + d_{\beta-2}b^{\beta-2} + \cdots + d_1 b + d_0 \\ &= \sum_{i=\beta-1}^{0} d_i b^i, \end{aligned} \tag{2.19}$$

where d_i $(i = \beta-1, \beta-2, \cdots, 1, 0)$ are digits. If $d_{\beta-1} \neq 0$, we call n a β-digit base-b number. Clearly, any number $b^{\beta-1} \le n < b^\beta$ is a β-digit number to the base b. For example, $10^5 \le 780214 < 10^6$ is a 6-digit number to the base 10. By Definition 2.1.14, this gives the following formula for the number of base-b digits for n:

$$\text{number of digits of } n = \lfloor \log_b n \rfloor + 1 = \left\lfloor \frac{\ln n}{\ln b} \right\rfloor + 1. \tag{2.20}$$

(The notation $\lfloor x \rfloor$, where x is a real number, is defined to be the greatest integer less than or equal to x and called the *floor* of x, whereas $\lceil x \rceil$ is defined to be the least integer greater than or equal to x and called the *ceiling* of x. The notation $[x]$ is also used for $\lfloor x \rfloor$.) For example, let $n = 999$, then

$$\begin{aligned}\text{the number of digits of } 999 &= \lfloor \log_{10} 999 \rfloor + 1 \\ &= \left\lfloor \frac{\ln 999}{\ln 10} \right\rfloor + 1 \\ &= \lfloor 2.999565488 \rfloor + 1 \\ &= 2 + 1 = 3,\end{aligned}$$

$$\begin{aligned}\text{the number of bits of } 999 &= \lfloor \log_{2} 999 \rfloor + 1 \\ &= \left\lfloor \frac{\ln 999}{\ln 2} \right\rfloor + 1 \\ &= \lfloor 9.964340868 \rfloor + 1 \\ &= 9 + 1 = 10.\end{aligned}$$

It is easy to verify that 999 has 10 bits, since $999 = 1111100111$. Note that the word *bits* is short for *binary digits*, and usually refers to Shannon bits, in honour of the American scientist Claude Shannon[7].

Exercise 2.1.1. Find the number of digits and bits for the following numbers:

$$2^{67} - 1, \quad 3^{367} - 1, \quad 2^{511} - 1, \quad \frac{12^{157} - 1}{11}, \quad \frac{5^{128} + 1}{2 \cdot 257}.$$

In terms of the big-$\mathcal{O}$ notation, (2.15) can be rewritten as

$$\text{length}(n) = \lfloor \log_2 n \rfloor + 1 = \mathcal{O}(\log n). \tag{2.21}$$

[7]

Claude E. Shannon (1916–2001) was a graduate of Michigan and went to MIT to write his PhD in Boolean algebra, where he received his PhD in 1940. He joined Bell Telephones in 1941 remaining until 1972. He was also a Professor in Electrical Engineering at MIT from 1958 to 1980, and has been Professor Emeritus there since 1980. Shannon is the inventor of information theory, the first to apply Boolean algebra to the design of circuits, and the first to use "bits" to represent information. His paper "Communication Theory of Secrecy Systems", published in 1949, is regarded as one of the very first papers in modern secure communications.

Exercise 2.1.2. Estimate in terms of the big-$\mathcal{O}$ notation the number of bits in $n!$.

Now we are in a position to discuss the bit complexity of some basic arithmetic operations.

First let us look at the addition of two β-bit binary integers. (If one of the two integers has fewer bits than the other, we just fill in zeros on the left.) Consider the following example:

$$\begin{array}{rr} & 11101011000 \\ + & 01000110101 \\ \hline & 100110001101 \end{array}$$

Clearly, we must repeat the following steps β times:

(1) Starting on the right, look at the top and bottom bits, and also at whether there is a carry above the top bit.
(2) If both bits are 0 and there is no carry, then put down 0 and move on.
(3) If either one of the following occurs
  (i) both bits are 0 and there is a carry
  (ii) only one of the bits is 0 and there is no carry
  then put down 1 and move on.
(4) If either one of the following occurs
  (i) both bits are 1 and there is no carry
  (ii) only one of the bits is 0 and there is a carry
  then put down 0, put a carry on the next column, and move on.
(5) If both bits are 1 and there is a carry, then put down 1, put a carry on the next column, and move on.

Doing this procedure *once* is called a *bit operation*. So adding two β-bit numbers requires β bit operations. That is,

$$T(\beta\text{-bits} + \beta\text{-bits}) = \mathcal{O}(\beta) = \mathcal{O}(\log n).$$

Next let us observe the multiplication of two β-bit binary integers. Consider the following example:

$$\begin{array}{rr} & 11101011001 \\ \times & 01000110101 \\ \hline & 11101011001 \\ & 11101011001 \\ & 11101011001 \\ & 11101011001 \\ + & 11101011001 \\ \hline & 10000001101110110110 1 \end{array}$$

that is,

$$\begin{array}{l} 11101011001 + 1110101100100 = 10010010111101 \\ 10010010111101 + 111010110010000 = 100110101001001101 \\ 100110101001001101 + 1110101100100000 = 1001101001001101 \\ 11000010101101101 + 1110101100100000000 = 100000011011101101101. \end{array}$$

and hence,

$$11101011001 \cdot 1000110101 = 100000011011101101101.$$

The result can easily be verified to be correct, since

$$\begin{array}{l} 11101011001_2 = 1881, \\ 1000110101_2 = 565, \\ 1881 \cdot 565 = 1062765 = 100000011011101101101_2. \end{array}$$

The above example shows us that multiplying two β-bit integers requires at most β^2 bit operations. That is,

$$\begin{aligned} T(\beta\text{-bits} \times \beta\text{-bits}) &= \mathcal{O}(\beta^2) \\ &= \mathcal{O}(\log n)^2. \end{aligned}$$

How fast can we multiply two integers? Earlier attempts at improvements employed simple algebraic identities and resulted in a reduction to the following:

Theorem 2.1.1. There is an algorithm which can multiply two β-bit integers in

$$\begin{aligned} T(\beta\text{-bits} \times \beta\text{-bits}) &= \mathcal{O}(\beta^{\log_2 3}) \\ &= \mathcal{O}(\beta^{1.584962501}) \\ &= \mathcal{O}(\log n)^{1.584962501} \end{aligned}$$

bit operations.

However, Schönhage and Strassen in 1971 utilized some number-theoretic ideas and the Fast Fourier Transform (FFT) and obtained a much better result:

Theorem 2.1.2. There is an algorithm which can multiply two β-bit integers in

$$\begin{aligned} T(\beta\text{-bits} \times \beta\text{-bits}) &= \mathcal{O}(\beta \log \beta \log \log \beta) \\ &= \mathcal{O}(\log n \log \log n \log \log \log n) \end{aligned}$$

bit operations.

Definition 2.1.15. An algorithm is said to be of *polynomial complexity*[8], measured in terms of bit operations, if its required running time is

$$\mathcal{O}(\log N)^k, \tag{2.22}$$

for some constant k. An algorithm is said to be of *exponential complexity*, measured in terms of bit operations, if its required running time is

$$\mathcal{O}\left(N^{\epsilon}\right), \tag{2.23}$$

where $\epsilon < 1$ is a small positive real number.

Example 2.1.3. Let β be the number of bits needed to represent n. Then,

$$\beta = \lfloor \log n \rfloor + 1.$$

Suppose that the complexity of an algorithm, measured by arithmetic operations on an integer (input) n, is $\mathcal{O}(n)$. What is the complexity for this algorithm in terms of bit operations? Since for each arithmetic operation, $\mathcal{O}(\log n)^2$ bit operations will be needed,

$$\begin{aligned} \mathcal{O}(n) &= \mathcal{O}\left(n(\log n)^2\right) \\ &= \mathcal{O}\left(2^{\log n}(\log n)^2\right), \end{aligned}$$

Therefore, the algorithm is of polynomial complexity in arithmetic operations, but of exponential complexity in bit operations.

Remark 2.1.3. In some computational problems such as the Traveling Salesman Problem, and the problem of sorting a list, the complexities measured by arithmetic operations reflect the actual running times. However, in most of the computational problems in number theory, the complexities measured by bit operations reflect the actual running times. In this book, all the complexities will be measured in terms of bit operations, rather than arithmetic operations.

Let us finally observe the complexities of some other common operations in arithmetic and number theory.

(1) The computation of $q = \lfloor a/b \rfloor$, where a is a 2β-bit integer and b a β-bit integer, can be performed in $\mathcal{O}(\beta^2)$ bit operations. However, the number of bit operations needed for integer division can be related to the the

[8] More generally, an algorithm with an input containing integers $n_1, n_2, \cdots, n_r$ of lengths $\log n_1, \log n_2, \cdots, \log n_r$ bits, respectively, is said to be of polynomial complexity if there exist integers $k_1, k_2, \cdots, k_r$ such that the number of bit operations required to perform the algorithm is $\mathcal{O}\left((\log n_1)^{k_1}, (\log n_2)^{k_r}, \cdots, (\log n_r)^{k_r}\right)$. Thus, by a *large input*, we will always mean that an input contains *large integers*, rather than *many integers* as for sorting.

number of bit operations needed for integer multiplication. That is, the division of a 2β-bit integer by a β-bit integer can be done in $\mathcal{O}(M(n))$ bit operations, where $M(n)$ is the number of bit operations needed to multiply two β-bit integers.

(2) Euclid's algorithm for calculating $\gcd(M, N)$ where $M < N$ can be performed in $\mathcal{O}(\log M)^3$ bit operations. This follows from a theorem, due to the French mathematician Gabriel Lamé (1795–1870) in 1844 (see Cormen, Ceiserson and Rivest [54]), which states that the number of divisions necessary to compute the $\gcd(M, N)$ is at most five times the number of decimal digits of M. So it will perform $\mathcal{O}(\log M)$ arithmetic operations and $\mathcal{O}(\log M)^3$ bit operations (assuming that multiplication and division take $\mathcal{O}(\log n)^2$ bit operations).

(3) The computation of the Jacobi symbol $\left(\frac{M}{N}\right)$ with $1 \le M \le N$ can be performed in $\mathcal{O}(\log M)^3$ bit operations. This is derived from the reciprocity law for the Jacobi symbol. In fact, with a more effective method indicated by Lehmer, which avoids divisions, it is possible to compute both $\gcd(M, N)$ and $\left(\frac{M}{N}\right)$ in $\mathcal{O}(\log M)^2$ bit operations.

Exercise 2.1.3. Using the big-$\mathcal{O}$ notation, estimate the number of bit operations needed for the following operations.

(1) Let n be a β-bit integer written in binary. Estimate the time to convert n to decimal.

(2) Let $n!$ be the factorial $n \cdot (n-1) \cdots 2 \cdot 1$ and $\binom{n}{m}$ the binomial coefficient $\frac{n!}{(n-m)!m!}$. Estimate the time to compute $n!$ and $\binom{n}{m}$.

(3) Let A and B be $n \times n$ matrices, with entries a_{ij} and b_{ij} for $1 \le i, j \le n$, then AB is the $n \times n$ matrix with entries $c_{ij} = \sum_{k=1}^{n} a_{ik} b_{kj}$. Estimate the number of bit operations required to find AB directly from its definition.

(4) Suppose we want to test if a large odd number n is prime by trial division by all odd numbers up to $\sqrt{n}$. Estimate the number of bit operations this test will take. How about if we have a list of all primes up to $\sqrt{n}$? How many bit operations will be needed to test if n is prime by trial division by all the primes up to $\sqrt{n}$ (use the Prime Number Theorem)?

2.1.5 Fast Modular Exponentiations

A frequently occurring operation in elementary number-theoretic computation is that of raising one number to a power modulo another number,

$x^e \bmod n$, also known as *modular exponentiation*. The conventional method of *repeated multiplication* would take $\mathcal{O}(e \log n)^2$ bit operations, which is too slow when e is large. Fortunately, the method of *repeated squaring* will solve this problem efficiently using the binary representation of b. The idea of the repeated squaring method is as as follows:

Theorem 2.1.3. Suppose we want to compute $x^e \bmod n$ with $x, e, n \in \mathbb{N}$. Suppose moreover that the binary form of e is as follows:

$$e = \beta_k 2^k + \beta_{k-1} 2^{k-1} + \cdots + \beta_1 2^1 + \beta_0 2^0, \tag{2.24}$$

where each β_i $(i = 0, 1, 2, \cdots k)$ is either 0 or 1. Then we have

$$\begin{aligned} x^e &= x^{\beta_k 2^k + \beta_{k-1} 2^{k-1} + \cdots + \beta_1 2^1 + \beta_0 2^0} \\ &= \prod_{i=0}^{k} x^{\beta_i 2^i} \\ &= \prod_{i=0}^{k} \left(x^{2^i}\right)^{\beta_i}. \end{aligned} \tag{2.25}$$

Furthermore, by the exponentiation law

$$x^{2^{i+1}} = (x^{2^i})^2, \tag{2.26}$$

and so the final value of the exponentiation can be obtained by *repeated squaring and multiplication* operations.

Example 2.1.4. Suppose we wish to compute a^{100}; we first write $100_{10} = 1100100_2 := e_6 e_5 e_4 e_3 e_2 e_1 e_0$, and then compute

$$\begin{aligned} a^{100} &= ((((((a)^2 \cdot a)^2)^2)^2 \cdot a)^2)^2 \\ &\Rightarrow \; a,\ a^3,\ a^6,\ a^{12},\ a^{24},\ a^{25},\ a^{50},\ a^{100}. \end{aligned} \tag{2.27}$$

Note that for each e_i, if $e_i = 1$, we perform a *squaring* and a *multiplication* operation (except "$e_6 = 1$", for which we just write down a, as indicated in the first bracket), otherwise, we perform only a *squaring* operation. That is,

e_6	1	a	initialization
e_5	1	$(a)^2 \cdot a$	squaring and multiplication
e_4	0	$((a)^2 \cdot a)^2$	squaring
e_3	0	$(((a)^2 \cdot a)^2)^2$	squaring
e_2	1	$((((a)^2 \cdot a)^2)^2)^2 \cdot a$	squaring and multiplication
e_1	0	$(((((a)^2 \cdot a)^2)^2)^2 \cdot a)^2$	squaring
e_0	0	$((((((a)^2 \cdot a)^2)^2)^2 \cdot a)^2)^2$	squaring
		$\Vert$	
		a^{100}	

Exercise 2.1.4. Write down the similar expressions as in (2.27) for computing x^{931} and x^{6501}, and verify your results. (Hints: $931_{10} = 1110100011_2$ and $6501_{10} = 1100101100101$.)

We are now in a position to introduce a fast algorithm for modular exponentiations (note that we can simply remove the "mod n" operation if we only wish to compute the exponentiation $c = x^e$):

Algorithm 2.1.1 (Fast modular exponentiation $x^e \bmod n$). This algorithm will compute the modular exponentiation

$$c = x^e \bmod n,$$

where $x, e, n \in \mathbb{N}$ with $n > 1$. It requires at most $2 \log e$ and $2 \log e$ divisions (divisions are only needed for modular operations; they can be removed if only $c = x^e$ are required to be computed).

[1] [Precomputation] Let

$$e_{\beta-1}e_{\beta-2}\cdots e_1 e_0 \tag{2.28}$$

be the binary representation of e (i.e., e has β bits). For example, for $562 = 1000110010$, we have $\beta = 10$ and

$$\begin{array}{cccccccccc} 1 & 0 & 0 & 0 & 1 & 1 & 0 & 0 & 1 & 0 \\ \uparrow & \uparrow & \uparrow & \uparrow & \uparrow & \uparrow & \uparrow & \uparrow & \uparrow & \uparrow \\ e_9 & e_8 & e_7 & e_6 & e_5 & e_4 & e_3 & e_2 & e_1 & e_0 \end{array}$$

[2] [Initialization] Set $c \leftarrow 1$.

[3] [Modular Exponentiation] Compute $c = x^e \bmod n$ in the following way:

for i from $\beta - 1$ down to 0 do
  $c \leftarrow c^2 \bmod n$ (squaring)
  if $e_i = 1$ then
    $c \leftarrow c \cdot x \bmod n$ (multiplication)

[4] [Exit] Print c and terminate the algorithm.

Theorem 2.1.4. Let x, e and n be positive integers with $n > 1$. Then the modular exponentiation $x^e \bmod n$ can be computed in $\mathcal{O}(\log e)$ arithmetic operations and $\mathcal{O}\left((\log e)(\log n)^2\right)$ bit operations. That is,

$$\begin{aligned} \text{Time}(x^e \bmod n) &= \mathcal{O}_A(\log e), & (2.29) \\ &= \mathcal{O}_B\left((\log e)(\log n)^2\right). & (2.30) \end{aligned}$$

Proof. We first find the least positive residues of $x, x^2, x^4, \cdots, x^{2^k}$ modulo n, where $2^k \le e < 2^{k+1}$, by successively squaring and reducing modulo n. This requires a total of $\mathcal{O}\left((\log e)(\log n)^2\right)$ bit operations, since we perform $\mathcal{O}(\log e)$ squarings modulo n, each requiring $\mathcal{O}(\log n)^2$ bit operations. Next, we multiply together the least positive residues of the integers x^{2^i} corresponding to the binary bits of e which are equal to 1, and reduce modulo n. This also requires $\mathcal{O}\left((\log e)(\log n)^2\right)$ bit operations, since there are at most $\mathcal{O}(\log e)$ multiplications, each requiring $\mathcal{O}(\log n)^2$ bit operations. Therefore, a total of $\mathcal{O}\left((\log e)(\log n)^2\right)$ bit operations are needed to find the least positive residue of $x^e \bmod n$. □

Example 2.1.5. Use the above algorithm to compute $7^{9007} \bmod 561$ (here $x = 7$, $e = 9007$ and $m = 561$). By writing e in the binary form $e = e_{\beta-1}e_{\beta-2}\cdots e_1e_0$, we have

$$9007 = 10001100101111 = e_{13}e_{12}\cdots e_1e_0.$$

Now we just perform the following computations as described in Algorithm 2.1.1:

```
c ← 1
x ← 7
n ← 561
for i from β − 1 down to 0 do
    c ← c^2 mod n
    if e_i = 1 then c ← c · x mod n
print c; (now c = x^e mod n )
```

The values of (i, e_i, c) at each loop for i from 13 down to 0 are as follows:

13	12	11	10	9	8	7	6	5	4	3	2	1	0
1	0	0	0	1	1	0	0	1	0	1	1	1	1
7	49	157	526	160	241	298	166	469	49	538	337	46	226

So, at the end of the computation, the final result $c = 7^{9007} \bmod 561 = 226$ will be returned. It is clear that at most $2\log_2 9007$ multiplications and $2\log_2 9007$ divisions will be needed for the computation. In fact, only 22 multiplications and 22 divisions will be needed for this computation task.

Exercise 2.1.5. Use the fast exponentiation method to compute

$$c = 3^{4294967296} \bmod 4294967297$$

by completing the items marked with ? for F in the following table (note that $4294967296 = 1\underbrace{000\cdots00}_{32 \text{ zeros}}$ in binary):

i	32	31	30	29	28	27	······	2	1	0
e_i	1	0	0	0	0	0	······	0	0	0
c	3	9	81	6561	43046721	?	??????	?	?	?

Remark 2.1.4. The above fast exponentiation algorithm is about half as good as the best; more efficient algorithms are known. For example, Brickell,

et. al. [41] developed a more efficient algorithm, using precomputed values to reduce the number of multiplications needed. Their algorithm allows the computation of g^n for $n < N$ in time $\mathcal{O}(\log N/\log\log N)$. They also showed that their method can be parallelized, to compute powers in time $\mathcal{O}(\log\log N)$ with $\mathcal{O}(\log N/\log\log N)$ processors.

2.1.6 Fast Group Operations on Elliptic Curves

The most fundamental computations on elliptic curves are the group operations of the type

$$kP = \underbrace{P \oplus P \oplus \cdots \oplus P}_{k \text{ times}} \tag{2.31}$$

where $P = (x, y)$ is a point on an elliptic curve $E : y^2 = x^3 + ax + b$, and k a very large positive integer. Since the computation of kP is so fundamental in all elliptic curve related computations and applications, it is desirable that such computations be carried out as fast as possible. The basic idea of the fast computation of kP is as follows:

[1] Compute $2^i P$, for $i = 0, 1, 2, \cdots, \beta - 1$, with $\beta = \lfloor 1.442 \ln k + 1 \rfloor$.

[2] Add together suitable multiples of P, determined by the binary expansion of k, to get kP.

For example, to compute kP where $k = 232792560$, we first compute:

$$\beta = \lfloor 1.442 \ln k + 1 \rfloor = 28,$$

then compute $2^i P$, for $i = 0, 1, 2, \cdots, 27$ as follows:

$$\begin{array}{cccccccc} P \quad 2P & 2^2P & 2^3P & 2^4P & \cdots & 2^{25}P & 2^{26}P & 2^{27}P \\ & \| & \| & \| & & \| & \| & \| \\ & 2(2P) & 2(2^2P) & 2(2^3P) & \cdots & 2(2^{24}P) & 2(2^{25}P) & 2(2^{26}P) \end{array}$$

By the binary expansion of k,

$$k = 232792560_{10} = 1101111000000010000111110000_2 := e_{27}e_{26}e_{25}\cdots e_2e_1e_0,$$

we add only those multiples that correspond to 1:

$$\begin{array}{cccccccccccc} 1 & 1 & 1 & 1 & 1 & 1 & 1 & 1 & 1 & 1 & 1 & 1 \\ \updownarrow & \updownarrow & \updownarrow & \updownarrow & \updownarrow & \updownarrow & \updownarrow & \updownarrow & \updownarrow & \updownarrow & \updownarrow & \updownarrow \\ 2^{27} & 2^{26} & 2^{24} & 2^{23} & 2^{22} & 2^{21} & 2^{13} & 2^8 & 2^7 & 2^6 & 2^5 & 2^4 \end{array}$$

and ignore those multiples that correspond to 0:

$$2^{25}, 2^{20}, 2^{19}, 2^{18}, 2^{17}, 2^{16}, 2^{15}, 2^{14}, 2^{12}, 2^{11}, 2^{10}, 2^9, 2^3, 2^2, 2^1, 2^0.$$

Thus, we finally have:

$$kP = 2^{27}P \oplus 2^{26}P \oplus 2^{24}P \oplus 2^{23}P \oplus 2^{22}P \oplus 2^{21}P$$
$$\oplus 2^{13}P \oplus 2^{8}P \oplus 2^{7}P \oplus 2^{6}P \oplus 2^{5}P \oplus 2^{4}P$$
$$= 232792560P.$$

Remarkably enough, the idea of *repeated squaring* for fast exponentiations can be used almost directly for fast group operations (i.e., fast point additions) on elliptic curves. The idea of fast group additions is as follows: Let $e_{\beta-1}e_{\beta-2}\cdots e_1e_0$ be the binary representation of k. Then for i starting from $e_{\beta-1}$ down to e_0 ($e_{\beta-1}$ is always 1 and used for initialization), check whether or not $e_i = 1$. If $e_i = 1$, then perform a doubling and an addition group operation; otherwise, just perform a doubling operation. For example, to compute $89P$, since $89 = 1011001$, we have:

e_6	1	P	initialization
e_5	0	$2P$	doubling
e_4	1	$2(2P) + P$	doubling and addition
e_3	1	$2(2(2P) + P) + P$	doubling and addition
e_2	0	$2(2(2(2P) + P) + P)$	doubling
e_1	0	$2(2(2(2(2P) + P) + P))$	doubling
e_0	1	$2(2(2(2(2(2P) + P) + P))) + P$	doubling and addition
		$\Vert$	
		$89P$	

The following algorithm implements this idea of *repeated doubling and addition* for computing kP.

Algorithm 2.1.2 (Fast group operations kP on elliptic curves). This algorithm computes kP, where k is a large integer and P is assumed to be a point on an elliptic curve $E: \ y^2 = x^3 + ax + b$. (Note that we do not actually do the additions for the coordinates of P in this algorithm.)

[1] Write k in the binary expansion form $k = e_{\beta-1}e_{\beta-2}\cdots e_1e_0$, where each e_i is either 1 or 0. (Assume k has β bits.)

[2] Set $c \leftarrow 0$.

[3] Compute kP:

```
for i from β − 1 down to 0 do
    c ← 2c (doubling);
    if e_i = 1 then c ← c + P; (addition)
```

[4] Print c; (now $c = kP$)

Example 2.1.6. Use Algorithm 2.1.2 to compute $105P$. Let

$$k = 105 = 1101001 := e_6e_5e_4e_3e_2e_1e_0.$$

At the initial stage of the algorithm, we set $c = 0$. Now, we perform the following computation steps according to Algorithm 2.1.2:

$$\begin{array}{llll}
e_6 = 1: & c \leftarrow P + 2c & \Longrightarrow c \leftarrow P & \Longrightarrow c = P \\
e_5 = 1: & c \leftarrow P + 2c & \Longrightarrow c \leftarrow P + 2P & \Longrightarrow c = 3P \\
e_4 = 0: & c \leftarrow 2c & \Longrightarrow c \leftarrow 2(P + 2P) & \Longrightarrow c = 6P \\
e_3 = 1: & c \leftarrow P + 2c & \Longrightarrow c \leftarrow P + 2(2(P + 2P)) & \Longrightarrow c = 13P \\
e_2 = 0: & c \leftarrow 2c & \Longrightarrow c \leftarrow 2(P + 2(2(P + 2P))) & \Longrightarrow c = 26P \\
e_1 = 0: & c \leftarrow 2c & \Longrightarrow c \leftarrow 2(2(P + 2(2(P + 2P)))) & \Longrightarrow c = 52P \\
e_0 = 1: & c \leftarrow P + 2c & \Longrightarrow c \leftarrow P + 2(2(2(P + 2(2(P + 2P))))) & \Longrightarrow c = 105P.
\end{array}$$

That is, $P + 2(2(2(P + 2(2(P + 2P))))) = 105P$.

Example 2.1.7. Suppose we wish to compute $kP \bmod 1997$, where $k = 9007 = 10001100101111_2$. The computation can be summarized in the following table which shows the values of (i, e_i, c) for each execution of the "for" loop in Algorithm 2.1.2 (plus an additional modular operation "mod 1997" at the end of each loop):

13	12	11	10	9	8	7	6	...	2	1	0
1	0	0	0	1	1	0	0	...	1	1	1
P	2P	4P	8P	17P	35P	70P	140P	...	254P	509P	1019P

The final result of the computation is $c \equiv 1019P \pmod{1997}$. It is clear that the above computation will need at most $\log 9007$ arithmetic operations.

Note that Algorithm 2.1.2 does not actually calculate the coordinates (x, y) of kP on an elliptic curve over $\mathbb{Q}$ or over $\mathbb{Z}/N\mathbb{Z}$. To make Algorithm 2.1.2 a practically useful algorithm for point additions on an elliptic curve E, we must incorporate the actual coordinate addition $P_3(x_3, y_3) = P_1(x_1, y_1) + P_2(x_2, y_2)$ on E into the algorithm. To do this, we use the following formulas to compute x_3 and y_3 for P_3:

$$(x_3, y_3) = (\lambda^2 - x_1 - x_2,\ \lambda(x_1 - x_3) - y_1),$$

where

$$\lambda = \begin{cases} \dfrac{3x_1^2 + a}{2y_1} & \text{if } P_1 = P_2 \\[2ex] \dfrac{y_1 - y_2}{x_1 - x_2} & \text{otherwise.} \end{cases}$$

Algorithm 2.1.3 (Fast group operations kP on elliptic curves). This algorithm will compute the point $kP \bmod N$, where $k \in \mathbb{Z}^+$ and P is an initial point (x, y) on an elliptic curve $E : y^2 = x^3 + ax + b$ over $\mathbb{Z}/N\mathbb{Z}$; if we require E over $\mathbb{Q}$, just compute kP, rather than $kP \bmod N$. Let the initial point $P = (x_1, y_1)$, and the result point $P = (x_c, y_c)$.

[1] [Precomputation] Write k in the following binary expansion form $k = e_{\beta-1}e_{\beta-2}\cdots e_1e_0$. (Suppose k has β bits).

[2] [Initialization] Initialize the values for a, x_1 and y_1. Let $(x_c, y_c) = (x_1, y_1)$; this is exactly the computation task for e_1 (e_1 always equals 1).

[3] [Doublings and Additions] Computing $kP \bmod N$:

for i from $\beta - 2$ down to β do
  $m_1 \leftarrow 3x_c^2 + a \bmod N$
  $m_2 \leftarrow 2y_c \bmod N$
  $M \leftarrow m_1/m_2 \bmod N$
  $x_3 \leftarrow M^2 - 2x_c \bmod N$
  $y_3 \leftarrow M(x_c - x_3) - y_c \bmod N$
  $x_c \leftarrow x_3$
  $y_c \leftarrow y_3$
  if $e_i = 1$
   then $c \leftarrow 2c + P$
    $m_1 \leftarrow y_c - y_1 \bmod N$
    $m_2 \leftarrow x_c - x_1 \bmod N$
    $M \leftarrow m_1/m_2 \bmod N$
    $x_3 \leftarrow M^2 - x_1 - x_c \bmod N$
    $y_3 \leftarrow M(x_1 - x_3) - y_1 \bmod N$
    $x_c \leftarrow x_3$
    $y_c \leftarrow y_3$
   else $c \leftarrow 2c$

[4] [Exit] Print (c, x_c, y_c) and terminate the algorithm. (Note that this algorithm will stop whenever $m_1/m_2 \equiv \mathcal{O}_E \pmod{N}$, that is, it will stop whenever a modular inverse does not exit at any step of the computation.)

Exercise 2.1.6. Let

$$E : y^2 = x^3 - x - 1$$

be an elliptic curve over $\mathbb{Z}/1098413\mathbb{Z}$ and $P = (0, 1)$ a point on E. Use Algorithm 2.1.3 to compute the coordinates (x, y) of the points kP on E over $\mathbb{Z}/1098413\mathbb{Z}$ for $k = 8, 31, 92, 261, 513, 875, 7892, 10319P$. Find also the smallest integral values of k such that $kP = (467314, 689129)$ and $kP = (965302, 895958)$, respectively.

Theorem 2.1.5. Suppose that an elliptic curve E is defined by any one of the equations of (1.309), (1.310) and (1.311), over a finite field $\mathbb{F}_q$ with $q = p^r$ a prime power. Given $P \in E$, the coordinates of kP can be computed by Algorithm 2.1.3 in $\mathcal{O}(\log k)$ group operations and $\mathcal{O}\left((\log k)(\log p)^3\right)$ bit operations. That is,

$$\begin{aligned} \text{Time}(kP) &= \mathcal{O}_A(\log k), & (2.32) \\ &= \mathcal{O}_B\left((\log k)(\log q)^3\right). & (2.33) \end{aligned}$$

Note that both the fast modular exponentiation $a^k \bmod n$ and the elliptic curve group operation $kP \bmod n$ are very well suited for parallel computation. For example, a naive parallel algorithm to compute kP could be as follows:

```
begin parallel
    for i from i_0 to O(log k) do
        compute 2^i P
end parallel
compute Q = ∑ 2^i P
```

(It is assumed that we have sequentially tried all the small values up to i_0.) With this naive algorithm kP can be computed in $\mathcal{O}(\log\log k)$ group operations with $\mathcal{O}(\log k)$ processors. For example, at most 28 processors will be needed to compute $232792560P$ and at most 5 group operations will be needed for each of these processors. Brickell, Gordon and McCurley [41] developed a parallel algorithm for computing a^k in $\mathcal{O}(\log\log k)$ arithmetic operations and $\mathcal{O}(\log k/\log\log k)$ processors. It seems reasonable to conjecture that kP can also be computed in $\mathcal{O}(\log\log k)$ elliptic curve group operations with $\mathcal{O}(\log k/\log\log k)$ processors.

2.2 Algorithms for Primality Testing

It would be interesting to know, for example, what the situation is with the determination if a number is a prime, and in general how much we can reduce the number of steps from the method of simply trying for finite combinatorial problems.

KURT GÖDEL (1906–1978)

2.2.1 Deterministic and Rigorous Primality Tests

The *primality testing problem* (PTP) may be described as the following simple decision (i.e., yes/no) problem:

$$\left.\begin{array}{ll}\text{Input}: & n\in\mathbb{N}\ \text{ with }\ n>1.\\ \text{Output}: & \begin{cases}\text{Yes}, & \text{if } n\in\text{Primes},\\ \text{No}, & \text{otherwise}.\end{cases}\end{array}\right\}\tag{2.34}$$

In theory it is easy to determine if a given positive integer $n > 1$ is prime; simply verify that n is not divisible by any of the integers from 2 up to $n/2$ (the largest possible factor of n). Since any divisor of n is itself a product of primes, we need only check to see if n is divisible by the primes from 2 up to $n/2$. The following test however will reduce the amount of work considerably.

Theorem 2.2.1 (Primality test by trial divisions). Let $n > 1$. If n has no prime factor less than or equal to $\sqrt{n}$, then n is prime.

With this test we just try to divide n by each prime number from 2 up to $\lfloor\sqrt{n}\rfloor$ (this can be done by using the sieve of Eratosthenes, or by using a table containing prime numbers up to $\sqrt{n}$). It is easy to see that n is prime if and only if none of the trial divisors divides n. However, even this test is not practically useful for the test of primality for large numbers, since it is very inefficient needing $\mathcal{O}\left(2^{(\log n)/2}\right)$ bit operations.

In what follows, we shall introduce some other rigorous primality tests.

Theorem 2.2.2 (Lucas' converse of Fermat's little theorem, 1891). If there is an integer a such that

(1) $a^{n-1} \equiv 1 \pmod{n}$, and
(2) $a^{(n-1)/p} \not\equiv 1 \pmod{n}$, for each prime p of $n-1$.

Then n is prime.

Proof. Since $a^{n-1} \equiv 1 \pmod{n}$, Part (1) of Theorem 1.6.31 (see Chapter 1) tells us that $\text{ord}_n(a) \mid (n-1)$. We will show that $\text{ord}_n(a) = n-1$. Suppose that $\text{ord}_n(a) \neq n-1$. Since $\text{ord}_n(a) \mid (n-1)$, there is an integer k satisfying $n-1 = k \cdot \text{ord}_n(a)$. Since $\text{ord}_n(a) \neq n-1$, we know that $k > 1$. Let p be a prime factor of k. Then

$$x^{n-1}/q = x^{k/q \cdot \text{ord}_n(a)} = \left(x^{\text{ord}_n(a)}\right)^{k/q} \equiv 1 \pmod{n}.$$

However, this contradicts the hypothesis of the theorem, so we must have $\text{ord}_n(a) = n-1$. Now, since $\text{ord}_n(a) \le \phi(n)$ and $\phi(n) \le n-1$, it follows that $\phi(n) = n-1$. So, by Part (2) of Theorem 1.4.14, n must be prime. □

Example 2.2.1. Let $n = 2011$, then $2011 - 1 = 2 \cdot 3 \cdot 5 \cdot 67$. Note first 3 is a primitive root (in fact, the smallest) of 2011, since $\text{order}(3, 2011) = \phi(2011) = 2010$. So, we have

$$\begin{aligned} 3^{2011-1} &\equiv 1 \pmod{2011}, \\ 3^{(2010-1)/2} &\equiv -1 \not\equiv 1 \pmod{2011}, \\ 3^{(2010-1)/3} &\equiv 205 \not\equiv 1 \pmod{2011}, \\ 3^{(2010-1)/5} &\equiv 1328 \not\equiv 1 \pmod{2011}, \\ 3^{((2010-1)/67} &\equiv 1116 \not\equiv 1 \pmod{2011}. \end{aligned}$$

Thus, by Theorem 2.2.2, 2011 must be prime.

Remark 2.2.1. The defect of this test is that it requires the prime factorization of $n-1$, a problem of almost the same size as that of factoring n and a problem much harder than the primality testing of n.

Theorem 2.2.2 is equivalent to the following theorem:

Theorem 2.2.3. Let a and n be positive integers with $\gcd(a, n) = 1$. If

$$\text{ord}_n(a) = \phi(n) = n - 1, \tag{2.35}$$

then n is prime.

Example 2.2.2. Let $n = 3779$. We find for example that the integer $a = 19$ with $\gcd(19, 3779) = 1$ satisfies

(1) $\text{ord}_{3779}(19) = 3778$,

(2) $\phi(3779) = 3778$.

That is, $\text{ord}_{3779}(19) = \phi(3779) = 3778$. Thus by Theorem 2.2.3 3779 is prime.

Remark 2.2.2. If we know the value of $\phi(n)$, we can immediately determine whether or not n is prime, since by Part (2) of Theorem 1.4.14 we know that n is prime if and only if $\phi(n) = n-1$. Of course, this method is not practically useful, since to determine the primality of n, we need to find $\phi(n)$, but to find $\phi(n)$, we need to factor n, a problem even much harder than the primality testing of n.

It is possible to use different bases a_i (rather than a single base a) for different prime factors p_i of $n - 1$ in Theorem 2.2.2:

Theorem 2.2.4. If for each prime p_i of $n - 1$ there exists an integer a_i such that

(1) $a_i^{n-1} \equiv 1 \pmod{n}$, and

(2) $a_i^{(n-1)/p_i} \not\equiv 1 \pmod{n}$.

Then n is prime.

Proof. Suppose that $n - 1 = \prod_{i=1}^{k} p_i^{\alpha_i}$, with $\alpha_i > 0$, for $i = 1, 2, \cdots, k$. Let also $r_i = \text{ord}_n(a_i)$. Then $r_i \mid (n-1)$ and $r_i \nmid (n-1)/p_i$ implies that $p_i^{\alpha_i} \mid r_i$. But for each i, we have $r_i \mid \phi(n)$ and hence $p_i^{\alpha_i} \mid \phi(n)$. This gives us $(n-1) \mid \phi(n)$, so n must be prime. □

Example 2.2.3. Let again $n = 3779$, then $n - 1 = 2 \cdot 1889 = p_1 \cdot p_2$. For $p_1 = 2$ we choose $a_1 = 19$ and get

$$19^{3778} \equiv 1 \pmod{3779}, \quad 2^{3778/2} \equiv -1 \not\equiv 1 \pmod{3779}.$$

For $p_2 = 1889$ we choose $a_2 = 3$ and get

$$3^{3778} \equiv 1 \pmod{3779}, \quad 3^{3778/1889} \equiv 9 \not\equiv 1 \pmod{3779}.$$

So, by Theorem 2.2.4, 3779 is prime. Note that for $a = 3$, we have $2^{3778/2} \equiv 1 \pmod{3779}$ and $3^{3778/1889} \not\equiv 1 \pmod{3779}$, but it does not matter, since it is not necessary to have the same value of a for the prime factors 2 and 1889 of $n - 1$; a different value of a (e.g., $a = 2$) is allowed for the prime factor 1889 of $n - 1$.

It is interesting to note, although primality testing is difficult, the verification of primality is easy, since the primality (as well as the compositeness) of an integer n can be verified very quickly in polynomial time:

Theorem 2.2.5. If n is composite, it can be proved to be composite in $\mathcal{O}((\log n)^2)$ bit operations.

Proof. If n is composite, there are integers a and b with $1 < a < n$, $1 < b < n$ and $n = ab$. Hence, given the two integers a and b, we multiply a and b, and verify that $n = ab$. This takes $\mathcal{O}((\log n)^2)$ bit operations and proves that n is composite. □

Theorem 2.2.6. If n is prime, it can be proved to be prime in $\mathcal{O}((\log n)^4)$ bit operations.

Remark 2.2.3. It should be noted that Theorem 2.2.5 cannot be used for finding the short proof of primality, since the factorization of $n-1$ and the primitive root a of n are required.

Theorem 2.2.5 was discovered by Pratt [193] in 1975; he interpreted the result as showing that every prime has a *succinct primality certification*. For some primes, Pratt's certificate is considerably shorter. For example, if $p = 2^{2^k}+1$ is a Fermat number with $k \geq 1$, then p is prime if and only if

$$3^{(p-1)/2} \equiv -1 \pmod{p}. \tag{2.36}$$

This result, known as Papin's test, gives a Pratt certificate for Fermat primes. The work in verifying (2.36) is just $\mathcal{O}(p)$, since $2^k - 1 = \lfloor \log_2 p \rfloor - 1$. In fact, as Pomerance [189] showed, every prime p has an $\mathcal{O}(p)$ certificate. More precisely, he proved:

Theorem 2.2.7. For every prime p there is a proof that it is prime, which requires for its certification $(5/2 + o(1)) \log_2 p$ multiplications modulo p.

However, if we assume that the Riemann hypothesis is true, then there is a deterministic polynomial algorithm for primality testing (Miller, [162]). But as we do not know if the Riemann hypothesis is true, the complexity is uncertain.

The fastest unconditional, rigorous and deterministic algorithm is the ARRCL test, invented by Adleman, Pomerance, Rumely Cohen and Lenstra (see [3] and [50]); its running time is

$$\mathcal{O}\left((\log N)^{c(\log\log\log N)}\right),$$

where c is small positive real number. Although the exponent $c(\log\log\log N)$ is an extremely slowly growing function, it is not polynomial, but superpolynomial. Thus, the ARRCL test is of superpolynomial complexity.

Although no deterministic polynomial time algorithm has been found for primality testing, there do exist some efficient probabilistic algorithms for primality testing. In the next few subsections, we shall introduce some of these probabilistic algorithms.

2.2.2 Fermat's Pseudoprimality Test

This section will be concerned with the basic concepts of probable primes, pseudoprimes and pseudoprimality testing. Let $(\mathbb{Z}/n\mathbb{Z})^+$ denote the nonzero elements of $(\mathbb{Z}/n\mathbb{Z})$:

$$(\mathbb{Z}/n\mathbb{Z})^+ = \{1, 2, \cdots, n-1\}. \tag{2.37}$$

Clearly, if n is prime, then $(\mathbb{Z}/n\mathbb{Z})^+ = \mathbb{Z}/n\mathbb{Z}$.

Let us first re-examine Fermat's little theorem: if b is a positive integer, p a prime and $\gcd(b, p) = 1$, then

$$b^{p-1} \equiv 1 \pmod{p}. \tag{2.38}$$

The converse of Fermat's little theorem is: for some odd positive integer n, if $\gcd(b, n) = 1$ and

$$b^{n-1} \not\equiv 1 \pmod{n}, \tag{2.39}$$

then n is composite. So, if there exists an integer b with $1 < b < n$, $\gcd(b, n) = 1$ and $b^{n-1} \not\equiv 1 \pmod{n}$, then n must be composite. What happens if we find a number n such that $b^{n-1} \equiv 1 \pmod{n}$? Can we conclude that n is certainly a prime? The answer is unfortunately not, because n sometimes is indeed a prime, but sometimes is not! This leads to the following important concepts of probable primes and pseudoprimes.

Definition 2.2.1. We say that n is a base-b *probable prime* if

$$b^{n-1} \equiv 1 \pmod{n}. \tag{2.40}$$

A base-b probable prime n is called a base-b *pseudoprime* if n is composite. A base-b probable prime and a base-b pseudoprime are also called a base-b *Fermat probable prime* and a base-b *Fermat pseudoprime*, respectively.

Example 2.2.4. If $n = 1387$, we have $2^{341-1} \equiv 1 \pmod{341}$. Thus 341 is a base-2 probable prime. But since $341 = 11 \cdot 31$ is composite, it is a base-2 pseudoprime. The first few base-2 pseudoprimes are as follows: $341, 561, 645, 1105, 1387, 1729, 1905$.

Note that there are some composite numbers that satisfy (2.40) for every positive integer b, such that $\gcd(b, n) = 1$:

Definition 2.2.2. A composite number n that satisfies $b^{n-1} \equiv 1 \pmod{n}$ for every positive integer b such that $\gcd(b, n) = 1$, is called a *Carmichael number*, in honour of the American mathematician Carmichael[9].

Example 2.2.5. The first ten Carmichael numbers are as follows:

$$561, 1105, 1729, 2465, 2821, 6601, 8911, 10585, 15841, 29341.$$

It is usually much harder to show that a given integer (particularly when it is large) is a Carmichael number than to show that it is a base-b pseudoprime, as we can see from the following example.

Example 2.2.6. Show that 561 is a Carmichael number. Note that $561 = 3 \cdot 11 \cdot 17$. Thus $\gcd(b, 561) = 1$ implies that $\gcd(b, 3) = \gcd(b, 11) = \gcd(b, 17) = 1$. To show that $a^{560} \equiv 1 \pmod{561}$ for all b for which $\gcd(b, 3) = \gcd(b, 11) = \gcd(b, 17) = 1$, we use the Chinese Remainder Theorem and Fermat's little theorem, and get

$$\begin{aligned} b^2 \equiv 1 \pmod{3} &\Longrightarrow a^{560} = (a^2)^{280} \equiv 1 \pmod{3}, \\ b^{10} \equiv 1 \pmod{11} &\Longrightarrow a^{560} = (a^{10})^{56} \equiv 1 \pmod{11}, \\ b^{16} \equiv 1 \pmod{17} &\Longrightarrow a^{560} = (a^{16})^{35} \equiv 1 \pmod{17}. \end{aligned}$$

Hence $b^{560} \equiv 1 \pmod{561}$ for all b satisfying $\gcd(b, 561) = 1$. Therefore, 561 is a Carmichael number.

The largest known Carmichael number was found by H. Dubner in 1994; it has 8060 digits and is a product of three primes; it also has been known that there are 246683 Carmichael numbers up to 10^{16} (Pinch, [185]). Carmichael numbers are characterized by the following property.

Theorem 2.2.8. A composite integer $n > 2$ is a Carmichael number if and only if

$$n = \prod_{i=1}^{k} p_i, \quad k \geq 3$$

for all distinct odd primes p_i such that $\lambda(n) \mid n-1$, or equivalently $p_i - 1 \mid n-1$, for all nonnegative integers $i \leq k$.

Exercise 2.2.1. Use Theorem 2.2.8 to show that the integer 29341 is a Carmichael number, but 341 is not.

Fermat's little theorem implies that if n is prime, then n satisfies the congruence (2.40) for every a in $(\mathbb{Z}/n\mathbb{Z})^+$. Thus, if we can find an integer

---

[9] Robert Carmichael conjectured in 1912 that *there are infinitely many such numbers* that now bear his name. W. Alford, G. Granville and C. Pomerance [6] proved this conjecture in 1992.

$b \in (\mathbb{Z}/n\mathbb{Z})^+$ such that n does not satisfy the congruence (2.40), then n is certainly composite. Surprisingly, the converse almost holds, so that this criterion forms an *almost* perfect test for primality. The following is the algorithm for $b = 2$:

Algorithm 2.2.1 (Base-2 Fermat pseudoprimality test). This algorithm will test numbers from 3 up to j, say, $j = 10^{10}$ for primality. If it outputs n is composite, then n is certainly composite. Otherwise, n is almost surely prime.

[1] Initialize the values $i \geq 3$ and $j > i$. Set $n \leftarrow i$.

[2] If $2^n \pmod n = 2$, then n is a *base-2 probable prime*, else n is composite.

[3] $n \leftarrow n + 1$. If $n \leq j$ goto [2], else goto [4].

[4] Terminate the execution of the algorithm.

The above base-2 pseudoprimality test is also called Chinese test, since the Chinese mathematicians had this idea earlier than Fermat (Rosen [211]). Among the numbers below 2000 that can pass the Chinese test, only six are composites: 341, 561, 645, 1105, 1729 and 1905; all the rest are indeed primes. Further computation shows that such composite numbers seem to be rare. To exhibit quite how rare these are, note that up to 10^{10} there are around 450 million primes, but only about fifteen thousand base-2 pseudoprimes, while up to $2 \cdot 5 \times 10^{10}$ there are over a billion primes, and yet fewer than 22 thousand base-2 pseudoprimes. So, if we were to choose a random number $n < 2 \cdot 5 \times 10^{10}$ for which n divides $2^n - 2$, then there would be less than an one-in-fifty-thousand chance that our number would be composite. We quote the following comments on the usefulness of the Chinese test from Rosen [211]:

> *Because most composite integers are not pseudoprimes, it is possible to develop primality tests based on the original Chinese idea, together with extra observations.*

2.2.3 Strong Pseudoprimality Test

It this subsection we shall present an improved version of the pseudoprimality test discussed previously, called the strong pseudoprimality test, (or just strong test, for short).

Theorem 2.2.9. Let p be a prime. Then

$$x^2 \equiv 1 \pmod p \tag{2.41}$$

if and only if $x \equiv \pm 1 \pmod p$.

Proof. First notice that

$$\begin{aligned} x^2 \equiv \pm 1 \pmod{p} &\iff (x+1)(x-1) \equiv 0 \pmod{p} \\ &\iff p \mid (x+1)(x-1) \\ &\iff p \mid (x+1) \text{ or } p \mid (x-1) \\ &\iff x+1 \equiv 0 \pmod{p} \text{ or } x-1 \equiv 0 \pmod{p} \\ &\iff x \equiv -1 \pmod{p} \text{ or } x \equiv 1 \pmod{p}. \end{aligned}$$

Conversely, if either $x \equiv -1 \pmod{p}$ or $x \equiv 1 \pmod{p}$ holds, then $x^2 \equiv 1 \pmod{p}$. □

Definition 2.2.3. The number x is called a *nontrivial square root* of 1 modulo n if it satisfies (2.41) but $x \not\equiv \pm 1 \pmod{n}$.

Example 2.2.7. The number 6 is a nontrivial square root of 1 modulo 35. since $x^2 = 6^2 \equiv 1 \pmod{35}$, $x = 6 \not\equiv \pm 1 \pmod{35}$.

Corollary 2.2.1. If there exists a nontrivial square root of 1 modulo n, then n is composite.

Example 2.2.8. Show that 1387 is composite. Let $x = 2^{693}$. We have $x^2 = (2^{693})^2 \equiv 1 \pmod{1387}$, but $x = 2^{693} \equiv 512 \not\equiv \pm 1 \pmod{1387}$. So, 2^{693} is a nontrivial square root of 1 modulo 1387. Then by Corollary 2.2.1, 1387 is composite.

Now we are in a position to introduce the strong pseudoprimality test, an improved version of the (Fermat) pseudoprimality test.

Theorem 2.2.10 (Strong pseudoprimality test). Let $n = 1 + 2^j d$, with d odd, is prime. Then the so-called b-sequence

$$\{b^d,\ b^{2d},\ b^{4d},\ b^{8d},\ \cdots,\ b^{2^{j-1}d},\ b^{2^j d}\} \bmod n \tag{2.42}$$

has one of the following two forms:

$$(1,\ 1,\ \cdots,\ 1,\quad 1,\ 1,\ \cdots,\ 1), \tag{2.43}$$

$$(?,\ ?,\ \cdots,\ ?,\ -1,\ 1,\ \cdots,\ 1), \tag{2.44}$$

reduced to modulo n, for any $1 < b < n$. (The question mark "?" denotes a number different from ± 1.)

The correctness of the above theorem relies on Theorem 2.2.9: if n is prime, then the only solutions to $x^2 \equiv 1 \pmod{n}$ are $x \equiv \pm 1$. To use the strong pseudoprimality test on n, we first choose a base b, usually a small prime. Then we compute the b-sequence of n; write $n - 1$ as $2^j d$ where d is odd, compute $b^d \bmod n$, the first term of the b-sequence, and then square repeatedly to obtain the b-sequence of $j + 1$ numbers defined in (2.42), all reduced to modulo n. If n is prime, then the b-sequence of n will be of the form of either (2.43) or (2.44). If the b-sequence of n has any one of the following three forms

$$(?, \cdots, ?, 1, 1, \cdots, 1), \tag{2.45}$$

$$(?, \cdots, ?, ?, ?, \cdots, -1), \tag{2.46}$$

$$(?, \cdots, ?, ?, ?, \cdots, ?), \tag{2.47}$$

then n is *certainly* composite. However, a composite can masquerade as a prime for a few choices of base b, but not be "too many" (see Wagon [251]). The above idea leads naturally to a very efficient and also practically useful algorithm for (pseudo)primality testing:

Algorithm 2.2.2 (Strong pseudoprimality test). This algorithm will test n for primality with high probability:

[1] Let n be an odd number, and the base b a random number in the range $1 < b < n$. Find j and d with d odd, so that $n - 1 = 2^j d$.

[2] Set $i \leftarrow 0$ and $y \leftarrow b^d \pmod{n}$.

[3] If $i = 0$ and $y = 1$, or $y = n - 1$, then terminate the algorithm and output "n is probably prime". If $i > 0$ and $y = 1$ goto [5].

[4] $i \leftarrow i + 1$. If $i < j$, set $y \equiv y^2 \pmod{n}$ and return to [3].

[5] Terminate the algorithm and output "n is definitely not prime".

The strong pseudoprimality test is most often called the Miller-Rabin test, in honour of the computer scientists Miller[10] and Rabin[11]. It is also called the Miller-Selfridge-Rabin test, because Selfridge[12] used the test in 1974 before Miller first published the result (Mollin [164]).

Definition 2.2.4. A positive integer n with $n - 1 = d \cdot 2^j$ and d odd, is called a base-b ***strong probable prime*** if it passes the strong pseudoprimality test described above (i.e., the last term in sequence 2.42 is 1, and the first occurrence of 1 is either the first term or is preceded by -1). A base-b strong probable prime is called a base-b ***strong pseudoprime*** if it is a composite.

[10] Gary L. Miller is currently Professor in Computer Science at Carnegie-Mellon University working on computer algorithms. He received his PhD in Computer Science at the University of California, Berkeley in 1974.

[11] Michael O. Rabin received his MSc in 1953 from Hebrew University and his PhD in 1956 from Princeton University. He is currently Professor in Computer Science at Harvard University, working on the theory and application of computer algorithms. Prof. Rabin was a 1976 Turing Award co-recipient for his joint paper "Finite Automata and Their Decision Problem" with Dana S. Scott. Both were PhD students of Alonzo Church at Princeton.

[12] John L. Selfridge was born in Ketchican, Alaska in 1927. He obtained his PhD from the University of California at Los Angeles in 1958 and became a Professor at Pennsylvania State University six years later. Selfridge is one of the leading scientists in computational number theory and has made important contributions to the field.

If n is prime and $1 < b < n$, then n passes the test. The converse is usually true, as shown by the following theorem.

Theorem 2.2.11. Let $n > 1$ be an odd composite integer. Then n passes the strong test for at most $(n-1)/4$ bases b with $1 \leq b < n$.

Proof. The proof is rather lengthy, we thus only give a sketch of the proof. A more detailed proof can be found either in Section 8.4 of Rosen [211], or in Chapter V of Koblitz [128].

First note that if p is an odd prime, and α and q are positive integers then the number of incongruent solutions of the congruence

$$x^{q-1} \equiv 1 \pmod{p^\alpha}$$

is $\gcd(q,\ p^{\alpha-1}(p-1))$.

Let $n - 1 = d \cdot 2^j$, where d is an odd positive integer and j is a positive integer. For n to be a strong pseudoprime to the base b, either

$$b^d \equiv 1 \pmod{n}$$

or

$$b^{2^i d} \equiv -1 \pmod{n}$$

for some integer i with $0 < i < j-1$. In either case, we have

$$b^{n-1} \equiv 1 \pmod{n}.$$

Let the prime factorization of n be

$$n = p_1^{\alpha_1} p_2^{\alpha_2} \cdots p_k^{\alpha_k}.$$

By the assertion made at the beginning of the proof, we know that there are

$$\gcd(n-1,\ p_i^{\alpha_i}(p_i - 1)) = \gcd(n-1,\ p_i - 1)$$

incongruent solutions to the congruence

$$x^{n-1} \equiv 1 \pmod{p_i^{\alpha_i}}, \qquad i = 1, 2, \cdots, k.$$

Further, by the Chinese remainder theorem, we know that there are exactly

$$\prod_{i=1}^{k} \gcd(n-1,\ p_i - 1)$$

incongruent solutions to the congruence

$$x^{n-1} \equiv 1 \pmod{n}.$$

To prove the theorem, there are three cases to consider:

[1] the prime factorization of n contains a prime power $p_r^{\alpha_r}$ with exponent $\alpha_r \geq 2$;

[2] $n = pq$, with p and q distinct odd primes.

[3] $n = p_1 p_2 \cdots p_k$, with $p_1, p_2, \cdots, p_k$ distinct odd primes.

The second case can actually be included in the third case. We consider here only the first case. Since

$$\begin{aligned}\frac{p_r - 1}{p_r^{\alpha_r}} &= \frac{1}{p_r^{\alpha_r - 1}} - \frac{1}{p_r^{\alpha_r}} \\ &\leq \frac{2}{9}\end{aligned}$$

we have

$$\begin{aligned}\prod_{i=1}^{k} \gcd(n-1,\ p_i - 1) &\leq \prod_{i=1}^{k} (p_i - 1) \\ &\leq \left(\prod_{\substack{i=1 \\ i \neq r}}^{k} p_i\right)\left(\frac{2}{9} p_r^{\alpha_r}\right) \\ &\leq \frac{2}{9} n \\ &\leq \frac{n-1}{4} \qquad \text{for } n \geq 9.\end{aligned}$$

Thus, there are at most $(n-1)/4$ integers b, $1 < b < n-1$, for which n is a base-b strong pseudoprime and n can pass the strong test. □

A probabilistic interpretation of Theorem 2.2.11 is as follows:

Corollary 2.2.2. Let $n > 1$ be an odd composite integer and b be chosen randomly from $\{2, 3, \cdots, n-1\}$. Then the probability that n passes the strong test is less than $1/4$.

From Corollary 2.2.2, we can construct a simple, general purpose, polynomial time primality test which has a positive (but arbitrarily small) probability of giving the wrong answer. Suppose an error probability of ϵ is acceptable. Choose k such that $4^{-k} < \epsilon$, and select $b_1, b_2, \cdots, b_k$ randomly and independently from $\{2, 3, \cdots, n-1\}$. If n fails the strong test on b_i, $i = 1, 2, \cdots, k$, then n is a strong probable prime.

Theorem 2.2.12. The strong test (i.e., Algorithm 2.2.2) requires, for $n-1 = 2^j d$ with d odd and for k randomly selected bases, at most $k(2+j)\log n$ steps. If n is prime, then the result is always correct. If n is composite, then the probability that n passes all k tests is at most $1/4^k$.

Proof. The first two statements are obvious, only the last statement requires proof. An error will occur only when the n to be tested is composite and the bases $b_1, b_2, \cdots, b_k$ chosen in this particular run of the algorithm are all nonwitnesses. (An integer a is a *witness* to the compositeness of n if it is possible using a to prove that n is composite, otherwise it is a *nonwitness*). Since the probability of randomly selecting a nonwitness is smaller than $1/4$ (by Corollary 2.2.2), then the probability of independently selecting k nonwitnesses is smaller than $1/4^k$. Thus the probability that with any given number n, a particular run of the algorithm will produce an erroneous answer is smaller than $1/4^k$. □

In the following list, we give some values of k and $1/4^k$ for the purposes of comparison:

k	$1/4^k$
10	$< 10^{-6}$
25	$< 10^{-15}$
30	$< 10^{-18}$
50	$< 10^{-30}$
100	$< 10^{-60}$
168	$< 10^{-101}$
1000	$< 10^{-602}$

Let n be a composite positive integer. Using the strong test, if we pick 100 different integers between 1 and n at random and perform the strong test for each of these 100 bases, then the probability that n passes all the tests is less than $4^{-100} < 10^{-60}$, an extremely small number. In fact, it may be more likely that a computer error was made than that a composite integer passes all 100 tests. We conclude that for all *practical* purposes, we can test primality in polynomial time.

The Generalized Riemann Hypothesis (GRH) for the Dirichlet L-functions has the following important consequence (see also Niven, Zuckerman and Montgomery [174] or Rosen [211]):

Conjecture 2.2.1. For every composite positive integer n, there is a number (base) b with $1 < b \leq 2(\log n)^2$, such that n fails the strong test for the base b.

If this conjecture is true, then the following result provides a rapid primality test:

Theorem 2.2.13. If the Generalized Riemann Hypothesis (GRH) is true, then there is an algorithm to determine whether or not a positive integer n is prime using $\mathcal{O}((\log n)^5)$ bit operations.

Proof. Let b be a positive integer less than n. To perform the strong test for the base b on n takes $\mathcal{O}((\log n)^5)$ bit operations, because this test requires that we perform no more than $((\log n)^3)$ modular exponentiations, each using $\mathcal{O}((\log b)^2)$ bit operations. Assume that the GRH is true. If n is composite, then, by Conjecture 2.2.1, there is a base b with $1 < b \leq 2(\log n)^2$ such that n fails the strong test for b. To discover this b requires less than $\mathcal{O}((\log n)^3 > \cdot\mathcal{O}((\log n)^2) = \mathcal{O}((\log n)^5)$ bit operations. Hence, using $\mathcal{O}((\log n)^5)$ bit operations, we can determine whether n is composite or prime. □

Although very few composites can pass the strong pseudoprimality test, such numbers do exist. For example, the composite $n = 2047 = 23 \cdot 89$ can pass the base-2 strong pseudoprimality test, because $n - 1 = 2^1 \cdot 1023$, $d = 1023$ and the sequence (2.42) is $2^{1023} \equiv 1 \pmod{2047}$, $2^{2046} \equiv 1 \pmod{2047}$. So, $n = 2047$ is a *base-2 strong pseudoprime*. Thus, from a pure mathematics point of view, we cannot conclude that n is prime just by a strong primality test.

Another probabilistic test for primality similar to (although not as good as) the strong pseudoprimality test is Euler's pseudoprimality test; it uses the Jacobi symbol and relies on Euler's criterion (Theorem 1.6.26).

Theorem 2.2.14 (Euler's pseudoprimality test). Let n be a positive integer greater than 1 and choose at random k integers $b_1, b_2, \cdots, b_k$ with $0 < b_i < n$ and $\gcd(b_i, n) = 1$ and compute

$$b_i^{(n-1)/2} \equiv \left(\frac{b_i}{n}\right) \pmod{n}, \quad \text{for } i = 1, 2, \cdots, k. \tag{2.48}$$

If (2.48) fails to hold for any i, then n is composite. The probability that n is composite but (2.48) holds for every i is less than $1/2^k$.

Euler's pseudoprimality test is often called the Solovay-Strassen test, in honour of its inventors Solovay and Strassen [244]. If the positive integer $n > 1$ passes Euler's pseudoprimality test on base b, then n is called a base-b *Euler probable prime*. A Euler probable prime is called the base-b *Euler pseudoprime* if it is composite.

Example 2.2.9. Let $n = 1105 = 5 \cdot 13 \cdot 17$ and $b = 2$. Then we have $b^{(n-1)/2} \bmod n = 2^{(1105-1)/2} \bmod 1105 = 1$ and $\left(\frac{b}{n}\right) = \left(\frac{2}{1105}\right) = 1$. Thus, $b^{(n-1)/2} \equiv \left(\frac{2}{1105}\right) \pmod{n}$. Therefore, 1105 is a base-2 Euler pseudoprime. However, 1105 is not a base-2 strong pseudoprime.

Remark 2.2.4. Since every base-a strong pseudoprime is a base-a Euler pseudoprime, more composites pass Euler's pseudoprimality test than the strong pseudoprimality test, although both require $\mathcal{O}((\log n)^3)$ bit operations.

2.2.4 Lucas Pseudoprimality Test

In this subsection, we shall study Lucas sequences and their applications to primality testing.

Let a, b be non-zero integers and $D = a^2 - 4b$. Consider the equation $x^2 - ax + b = 0$; its discriminant is $D = a^2 - 4b$, and α and β are the two roots:

$$\left.\begin{aligned} \alpha &= \frac{a + \sqrt{D}}{2} \\ \beta &= \frac{a - \sqrt{D}}{2}. \end{aligned}\right\} \tag{2.49}$$

So

$$\left.\begin{aligned} \alpha + \beta &= a \\ \alpha - \beta &= \sqrt{D} \\ \alpha\beta &= b. \end{aligned}\right\} \tag{2.50}$$

We define the sequences (U_k) and (V_k) by

$$\left.\begin{aligned} U_k(a,b) &= \frac{\alpha^k - \beta^k}{\alpha - \beta} \\ V_k(a,b) &= \alpha^k + \beta^k. \end{aligned}\right\} \tag{2.51}$$

In particular, $U_0(a,b) = 0, U_1(a,b) = 1$, while $V_0(a,b) = 2, V_1(a,b) = a$. For $k \geq 2$, we also have

$$\left.\begin{aligned} U_k(a,b) &= aU_{k-1} - bU_{k-2} \\ V_k(a,b) &= aV_{k-1} - bV_{k-2}. \end{aligned}\right\} \tag{2.52}$$

The sequences

$$\left.\begin{aligned} U(a,b) &= (U_k(a,b))_{k \geq 0} \\ V(a,b) &= (V_k(a,b))_{k \geq 0} \end{aligned}\right\} \tag{2.53}$$

are called the *Lucas sequences* associated with the pair (a, b), in honour of the French mathematician Lucas[13]. Special cases of Lucas sequences were

---

13

François Edouard Lucas (1842–1891), was born in Amiens, France and was educated at the École Normale, one of the two most prestigious French institutions of the time. After finishing his studies, he worked as an assistant at the Paris Observatory, and later on, he became a mathematics teacher at three Paris secondary schools. He became the last *largest prime record* holder in the pre-computer age, by discovering the 12th Mersenne prime in 1876, though it was only confirmed in 1914. In the world mathematics community, Lucas is perhaps best known for his work on Lucas numbers, the Lucas test for Mersenne prime and the Tower of Hanoi problem.

considered by Fibonacci, Fermat, and Pell, among others. For example, the sequence $U_k(a, b)$, $k = 0, 1$, corresponding to $a = 1$, $b = -1$, was first considered by Fibonacci, and it begins as follows:

$$0, 1, 1, 2, 3, 5, 8, 13, 21, 34, 55, 89, 144, 233, 377, 610, 987, 1597, 2584, 4181, \cdots$$

These are called *Fibonacci numbers*, in honour of the Italian mathematician Fibonacci[14]. The companion sequence to the Fibonacci numbers, still with $a = 1$, $b = -1$, is the sequence of *Lucas numbers*: $V_0 = V_0(1, -1) = 2$, $V_1 = V_1(1, -1) = 1$, and it begins as follows:

$$2, 1, 3, 4, 7, 11, 18, 29, 47, 76, 123, 199, 322, 521, 843, 1364, 2207, 3571, \\ 5778, 9349, 15127, \cdots$$

If $a = 3$, $b = 2$, then the sequences obtained are

$$\begin{aligned} &U_k(3, 2) = 2^k - 1 : \\ &\qquad 0, 1, 3, 7, 15, 31, 63, 127, 255, 511, 1023, 2047, 4095, 8191, 16383, \cdots \\ &V_k(3, 2) = 2^k + 1 : \\ &\qquad 2, 3, 5, 9, 17, 33, 65, 129, 257, 513, 1025, 2049, 4097, 8193, 16385, \cdots \end{aligned}$$

for $k \geq 0$. The sequences associated with $a = 2$, $b = -1$ are called *Pell sequences*; they begin as follows:

$$U_k(2, -1) : 0, 1, 2, 5, 12, 29, 70, 169, 408, 985, 2378, 5741, 13860, 33461, 80782, \cdots$$

$$V_k(2, -1) : 2, 2, 6, 14, 34, 82, 198, 478, 1154, 2786, 6726, 16238, 39202, 94642, \cdots$$

Now we are in a position to study some analogues of pseudoprimes in which $a^{n-1} - 1$ is replaced by a Lucas sequence. Recall that odd composite numbers n for which

$$a^{n-1} \equiv 1 \pmod{n}$$

are called pseudoprimes to base a.

14

Leonardo Pisano Fibonacci (1170–1250), is better known by his nickname Fibonacci, but Fibonacci himself sometimes used the name Bigollo, which may mean good-for-nothing or a traveller. Fibonacci ended his travels around the year 1200 and at that time he returned to Pisa. There he wrote a number of important texts which played an important role in reviving ancient mathematical skills and he made significant contributions of his own. Fibonacci lived in the days before printing, so his books were hand written and the only way to have a copy of one of his books was to have another hand-written copy made. Of his books we still have copies of *Liber Abbaci* (1202), *Practica Geometriae* (1220), *Flos* (1225), and *Liber Quadratorum*. A problem in the third section of *Liber Abbaci* led to the introduction of Fibonacci numbers and the Fibonacci sequence for which Fibonacci is best remembered today.

Theorem 2.2.15 (Lucas theorem). Let a and b be integers and put $D = a^2 - 4b \neq 0$. Define the Lucas sequence $\{U_k\}$ with the parameters D, a, b by

$$U_k = \frac{\alpha^k - \beta^k}{\alpha - \beta}, \qquad k \geq 0 \tag{2.54}$$

where α and β are the two roots of $x^2 - ax + b = 0$. If p is an odd prime, $p \nmid b$ and $\left(\frac{D}{p}\right) = -1$, where $\left(\frac{D}{p}\right)$ is a Jacobi symbol, then $p \mid U_{p+1}$.

The above theorem can be used directly to construct a primality test, often called the *Lucas test*:

Corollary 2.2.3 (Converse of the Lucas theorem – Lucas test). Let n be an odd positive integer. If $n \nmid U_{n+1}$, then n is composite.

Just as there are Fermat probable primes and Fermat pseudoprimes, we also have the concepts of Lucas probable primes and Lucas pseudoprimes.

Definition 2.2.5. An odd positive integer n is called a *Lucas probable prime* with D, a and b, if $n \nmid b$, $\left(\frac{D}{n}\right) = -1$ and $n \mid U_{n+1}$. A Lucas probable prime n is called a *Lucas pseudoprime* if n is composite.

Another different but equivalent presentation of Theorem 2.2.15 is as follows:

Theorem 2.2.16. Let n be an odd positive integer, $\varepsilon(n)$ the Jacobi symbol $\left(\frac{D}{n}\right)$, and $\delta(n) = n - \varepsilon(n)$. If n is prime and $gcd(n, b) = 1$, then

$$U_{\delta(n)} \equiv 0 \pmod{n}. \tag{2.55}$$

If n is composite, but (2.55) still holds, then n is called a *Lucas pseudoprime with parameters a and b.*

Although Theorem 2.2.16 is true when $\left(\frac{D}{n}\right) = 1$, it is best to avoid this situation. A good way to avoid this situation is to select a suitable D such that $\left(\frac{D}{n}\right) = -1$. Two methods have been proposed (see Baillie and Wagstaff [18]):

(1) Let D be the first element of the sequence $5, -7, 9, -11, 13, \cdots$ for which $\left(\frac{D}{n}\right) = -1$. Let $a = 1$ and $b = (1 - D)/4$.

(2) Let D be the first element of the sequence $5, 9, 13, 17, 21, \cdots$ for which $\left(\frac{D}{n}\right) = -1$. Let a be the least odd number exceeding $\sqrt{D}$ and $b = (a^2 - D)/4$.

The first 10 Lucas pseudoprimes found by the first method are

$$323, 377, 1159, 1829, 3827, 5459, 5777, 9071, 10877,$$

and the first 10 Lucas pseudoprimes found by the second method are:

$$323, 377, 1349, 2033, 2651, 3569, 3599, 3653, 3827, 4991.$$

The most interesting thing about the Lucas test is that if we choose the parameters D, a and b as described in the second method, then the first 50 Carmichael numbers and several other base-2 Fermat pseudoprimes will never be Lucas pseudoprimes (Baillie and Wagstaff [18]). This leads to the general belief that a combination of a strong pseudoprimality test and a Lucas pseudoprimality test (or just a combined test, for short) might be an infallible test for primality. Since to date, no composites have been found to pass such a combined test, it is thus reasonable to conjecture that:

Conjecture 2.2.2. If n is a positive integer greater than 1 which can pass the combination of a strong pseudoprimality test and a Lucas test, then n is prime.

The advantage of the combination of a strong test and a Lucas test seems to be that the two probable prime tests are independent. That is, n being a probable prime of the first type does not affect the probability of n being a probable prime of the second type. In fact, if n is a strong pseudoprime (to a certain base), then n is less likely than a typical composite to be a Lucas pseudoprime (with the parameters a and b), provided a and b are chosen properly, and vice versa. If n passes both a strong test and a Lucas test, we can be more certain that it is prime than if it merely passes several strong tests, or several Lucas tests. Pomerance, Selfridge and Wagstaff [192] issued a challenge (with a total prize now \$620) for an example of a composite number which passes both a strong pseudoprimality test base 2 and a Lucas test, or a proof that no such number exists. At the moment, the prize is unclaimed; no counter-example has yet been found.

There is, however, a very efficient and deterministic Lucas test specifically for Mersenne primes, known as the *Lucas–Lehmer test*, after the French mathematician Lucas who discovered the basic idea in 1876 and the American mathematician Lehmer[15] who refined the method in 1930, based on the following theorem:

15

Derrick H. Lehmer (1905–1991), perhaps the father of computational number theory, was born in Berkeley, California. He received his bachelor's degree in physics from the University of California, Berkeley, whereupon he went to the University of Chicago for graduate studies in number theory with L. E. Dickson. But since he didn't like working under Dickson, he went to Brown University in Providence, Rhode Island to study for a PhD. He served as a faculty member in the California Institute of Technology, Lehigh

Theorem 2.2.17 (Lucas–Lehmer theorem for Mersenne primes M_n). Let $a = 2$ and $b = -2$. Consider the associated Lucas sequences $(U_k)_{k\geq 0}$, $(V_k)_{k\geq 0}$, with discriminant $D = 12$. Then $N = M_n$ is prime if and only if $N \mid V_{(N+1)/2}$.

Example 2.2.10. First we notice that the Lucas sequence $V_k(2,-2)$ begins as follows:

$$2, 2, 8, 20, 56, 152, 416, 1136, 3104, 8480, 23168, 63296, 172928, \\ 472448, 1290752, 3526400, 9634304, \cdots.$$

Now suppose we wish to test the primality of $N = 2^7 - 1$. Compute $V_{(N+1)/2}$ for $N = 2^7 - 1$:

$$\begin{aligned} V_{(N+1)/2} &= V_{2^7/2} \\ &= V_{64} \\ &= 8615517765800787268541087744 \\ &\equiv 0 \pmod{(2^7-1)}, \end{aligned}$$

so by Theorem 2.2.17, $N = 2^7 - 1$ is a prime.

For the purpose of computation, it is convenient to replace the Lucas sequence $(V_k)_{k\geq 0}$ by the following Lucas–Lehmer sequence $(L_k)_{k\geq 1}$, defined recursively as follows:

$$\left.\begin{aligned} L_0 &= 4 \\ L_{k+1} &= L_k^2 - 2. \end{aligned}\right\} \tag{2.56}$$

The Lucas–Lehmer sequence begins with

$$4, 14, 194, 37634, 1416317954, 2005956546822746114, \\ 4023861667741036022825635656102100994, \cdots.$$

The reason that we can replace the Lucas sequence $V_k(2,-2)$ by the Lucas–Lehmer sequence L_k is based on the following observations:

$$\left.\begin{aligned} L_0 &= V_2/2 \\ L_{k-1} &= V_{2^k}/2^{2^{k-1}}. \end{aligned}\right\} \tag{2.57}$$

University and the University of Cambridge before joining the Mathematics Department at Berkeley in 1940. He made many significant contributions to number theory, and also invented some special purpose devices for number-theoretic computations, some with his father who was also a mathematician at Berkeley. The breadth of Lehmer's mathematical work is best judged by the 17 subject headings he chose for the 1981 publication of his *Selected Papers*. He was interested in primality testing throughout his life. He is perhaps best known for his sharp and definitive form of the Lucas primality test for Mersenne primes. Lehmer was also involved throughout his life with the theory and practice of integer factorization. (Photo by courtesy of the American Mathematical Society.)

Example 2.2.11. The following example shows how to calculate the Lucas–Lehmer sequence (L_k):

$$
\begin{aligned}
L_0 &= V_2/4 = 8/2 = 4 \\
L_1 &= V_{2^2}/2^{2^{2-1}} \\
&= V_4/2^2 \\
&= 56/4 = 14 \\
L_2 &= V_{2^3}/2^{2^{3-1}} \\
&= V_8/2^4 \\
&= 3104/16 = 194 \\
L_3 &= V_{2^4}/2^{2^{4-1}} \\
&= V_{16}/2^8 \\
&= 9634304/256 = 37634 \\
L_4 &= V_{2^5}/2^{2^{5-1}} \\
&= V_{32}/2^{16} \\
&= 92819813433344/65536 = 1416317954 \\
&\cdots\cdots\cdots \\
&\cdots\cdots\cdots .
\end{aligned}
$$

So Theorem 2.2.17 can be rewritten as follows:

Theorem 2.2.18 (Lucas–Lehmer test for Mersenne primes M_n). Let n be an odd prime. Then $2^n - 1$ is prime if and only if M_n divides L_{n-2}. That is,

$$L_{n-2} \equiv 0 \pmod{(2^n - 1)}. \tag{2.58}$$

Proof. There are several ways to prove this theorem (see, for example, Knuth [123] and Ribenboim [198]). Here we follow Ribenboim [198]:

Let $L_0 = 4 = V_2/2$. Assume that $L_{k-1} = V_{2^k}/2^{2^{k-1}}$. Then

$$
\begin{aligned}
L_k &= L_{k-1}^2 - 2 \\
&= \frac{V_{2^k}^2}{2^{2^k}} - 2 \\
&= \frac{V_{2^{k+1}+2^{2^k+1}}}{2^{2^k}} - 2 \\
&= \frac{V_{2^{k+1}}}{2^{2^k}}.
\end{aligned}
$$

By Theorem 2.2.17, M_n is prime if and only if M_n divides

$$V_{(M_n+1)/2} = V_{2^{n-1}} = 2^{2^{n-2}} L_{n-2},$$

or equivalently, $L_{n-2} \equiv 0 \pmod{(2^n - 1)}$. □

Example 2.2.12. Suppose we wish to test the primality of $2^7 - 1$; we first compute the Lucas–Lehmer sequence $\{L_k\}$ for $2^7 - 1$ ($k = 0, 1, \cdots, p-2 = 5$):

$$\begin{aligned} L_0 &= 4 \\ L_1 &\equiv 14 \\ L_2 &\equiv 67 \\ L_3 &\equiv 42 \\ L_4 &\equiv 111 \\ L_5 &\equiv 0 \pmod{127}. \end{aligned}$$

Since $L_{7-2} \equiv 0 \pmod{(2^7 - 1)}$, $2^7 - 1$ is a prime.

Thus, a practical primality testing algorithm for Mersenne primes can then be derived as follows:

Algorithm 2.2.3. (Lucas-Lehmer Test for Mersenne Primes)

```
Initialize the value for p ∈ Primes
L ← 4
for i from 1 to p − 2 do
    L ← L^2 − 2 (mod (2^p − 1))
if L = 0 then 2^p − 1 is prime
         else 2^p − 1 is composite
```

Remark 2.2.5. The above Lucas–Lehmer test for Mersenne primes is very efficient, since the major step in the algorithm is to compute

$$L = L^2 - 2 \pmod{(2^p - 1)}$$

which can be performed in polynomial time. But still, the computation required to test a single Mersenne prime M_p increases with p to the order of $\mathcal{O}(p^3)$. Thus, to test M_{2r+1} would take approximately eight times as long as to test M_r with the same algorithm (Slowinski [241]). Historically, it has required about four times as much computation to discover the next Mersenne prime as to re-discover all previously known Mersenne primes. The search for Mersenne primes has been an accurate measure of computing power for the past two hundred years and, even in the modern era, it has been an accurate measure of computing power for new supercomputers.

2.2.5 Elliptic Curve Test

In this subsection, we introduce a novel application of elliptic curves to primality testing, called the elliptic curve test. Although the elliptic curve primality test is still probabilistic, its answer is always correct; only the running time is random. In practice, the expected running time is finite; it is possible that the algorithm does not terminate but the probability of that occurring is zero.

First let us introduce one of the very useful converses of Fermat's little theorem:

Theorem 2.2.19 (Pocklington's theorem). Let s be a divisor of $N-1$. Let a be an integer prime to N such that

$$\left.\begin{aligned} a^{N-1} &\equiv 1 \pmod N \\ \gcd(a^{(N-1)/q}, N) &= 1 \end{aligned}\right\} \tag{2.59}$$

for each prime divisor q of s. Then each prime divisor p of N satisfies

$$p \equiv 1 \pmod s. \tag{2.60}$$

Corollary 2.2.4. If $s > \sqrt{N} - 1$, then N is prime.

A similar theorem can be stated for elliptic curves as follows.

Theorem 2.2.20. Let N be an integer greater than 1 and relatively prime to 6, E an elliptic curve over $\mathbb{Z}/N\mathbb{Z}$, P a point on E, m and s two integers with $s \mid m$. Suppose we have found a point P on E that satisfies $mP = \mathcal{O}_E$, and that for each prime factor q of s, we have verified that $(m/q)P \neq \mathcal{O}_E$. Then, if p is a prime divisor of N, $|E(\mathbb{Z}/p\mathbb{Z})| \equiv 0 \pmod s$.

Corollary 2.2.5. If $s > (\sqrt[4]{N} + 1)^2$, then N is prime.

Combining the above theorem with Schoof's algorithm [221] which computes $|E(\mathbb{Z}/p\mathbb{Z})|$ in time $\mathcal{O}\left((\log p)^{8+\epsilon}\right)$, we obtain the following GK algorithm due to Goldwasser[16] and Kilian[17] (see Goldwasser and Kilian [85] and its new version [86]).

16

Shafi Goldwasser obtained her PhD in Computer Science from the University of California at Berkeley. She is currently the RSA Professor of Electrical Engineering and Computer Science at the Massachusetts Institute of Technology (MIT), a co-leader of the cryptography and information security group and a member of the complexity theory group within the Theory of Computation Group and the Laboratory for Computer Science. Goldwasser is also Professor of Computer Science at the Weizmann Institute of Science, Israel. (Photo by courtesy of Prof. Goldwasser.)

17 Joe Kilian is currently with the NEC research Institute in Princeton. He was a PhD student at the MIT's Laboratory for Computer Science, with Goldwasser as

Algorithm 2.2.4 (Goldwasser-Kilian Algorithm). For a given probable prime N, this algorithm will show whether or not N is indeed prime:

[1] choose a nonsingular elliptic curve E over $\mathbb{Z}/N\mathbb{Z}$, for which the number of points m satisfies $m = 2q$, with q a probable prime;

[2] if (E, m) satisfies the conditions of Theorem 2.2.20 with $s = m$, then N is prime, otherwise it is composite;

[3] perform the same primality proving procedure for q;

[4] Exit.

The running time of the GK algorithm is analyzed in the following two theorems (Atkin and Morain [12]):

Theorem 2.2.21. Suppose that there exist two positive constants c_1 and c_2 such that the number of primes in the interval $[x, x + \sqrt{2x}]$, where $x(\geq 2)$, is greater than $c_1\sqrt{x}(\log x)^{-c_2}$, then the GK algorithm proves the primality of N in expected time $\mathcal{O}\left((\log N)^{10+c_2}\right)$.

Theorem 2.2.22. There exist two positive constants c_3 and c_4 such that, for all $k \geq 2$, the proportion of prime numbers N of k bits for which the expected time of GK is bounded by $c_3(\log N)^{11}$ is at least

$$1 - c_4 2^{-k^{\frac{1}{\log \log k}}}.$$

A serious problem with the GK algorithm is that Schoof's algorithm seems almost impossible to implement. In order to avoid the use of Schoof's algorithm, Atkin[18] and Morain[19] in 1991 developed a new implementation method called ECPP (Elliptic Curve Primality Proving), which uses the properties of elliptic curves over finite fields related to complex multiplication

his thesis advisor. His thesis *Primality Testing and the Power of Noisy Communication Channels*, won the 1989 ACM Distinguished Dissertation Award and was published by the MIT Press under the title *Uses of Randomness in Algorithms and Protocols*, in 1990 (see Kilian [120]).

[18] A. O. L. Atkin is currently Professor Emeritus at the University of Illinois at Chicago. He received his PhD in Mathematics from the University of Cambridge in 1952. Together with Bryan Birch, he organized the very successful 1969 *Computers in Number Theory Conference* in Oxford, England.

[19]

François Morain is currently with LIX Laboratoire d'Informatique de l' École Polytechnique, France. He received his PhD in mathematics, more specifically in elliptic curve primality proving from the Université de Lyon I in 1990. The ECPP (Elliptic Curve Primality Proving) program, developed jointly with Atkin, is the most popular primality testing program for large numbers of several thousand digits in the public domain. (Photo by courtesy of Dr. Morain.)

(Atkin and Morain [12]). We summarize the principal properties of ECPP as follows.

Theorem 2.2.23. Let p be a rational prime number that splits as the product of two principal ideals in a field $\mathcal{K}$: $p = \pi\pi'$ with π, π' integers of $\mathcal{K}$. Then there exists an elliptic curve E defined over $\mathbb{Z}/p\mathbb{Z}$ having complex multiplication by the ring of integers of $\mathcal{K}$, whose cardinality is

$$m = N_{\mathcal{K}}(\pi - 1) = (\pi - 1)(\pi' - 1) = p + 1 - t$$

with $|t| \leq 2\sqrt{p}$ (Hasse's Theorem) and whose invariant is a root of a fixed polynomial $H_D(X)$ (depending only upon D) modulo p.

For more information on the computation of the polynomials H_D, readers are referred to Morain [168]. Note that there are also some other important improvements on the GK algorithm, notably the Adleman-Huang primality proving algorithm [4] using hyperelliptic curves.

In the GK algorithm, it begins by searching for a curve and then computes its number of points, but in the ECPP algorithm, it does exactly the opposite. The following is a brief description of the ECPP algorithm.

Algorithm 2.2.5 (ECPP Algorithm). Given a probable prime N, this algorithm will show whether or not N is indeed prime:

- [1] [Initialization] Set $i \leftarrow 0$ and $N_0 \leftarrow N$.
- [2] [Building the sequence] While $N_i > N_{small}$
  - [2.1] Find a D_i such that $N_i = \pi_i \pi_i'$ *in* $\mathcal{K} = \mathcal{Q}(\sqrt{-D_i})$;
  - [2.2] If one of the $w(-D_i)$ numbers $m_1, ..., m_w$ ($m_r = N_K(\omega_r - 1)$ where ω_r is a conjugate of π) is probably factored goto step [2.3] else goto [2.1];
  - [2.3] Store $\{i, N_i, D_i, \omega_r, m_r, F_i\}$ where $m_r = F_i N_{i+1}$. Here F_i is a completely factored integer and N_{i+1} a probable prime; set $i \leftarrow i + 1$ and goto step [2.1].
- [3] [Proving] For i from k down to 0
  - [3.1] Compute a root j of $H_{D_i}(X) \equiv 0 \pmod{N_i}$;
  - [3.2] Compute the equation of the curve E_i of the invariant j and whose cardinality modulo N_i is m_i;
  - [3.3] Find a point P_i on the curve E_i;
  - [3.4] Check the conditions of Theorem 2.2.23 with $s = N_{i+1}$ and $m = m_i$.
- [4] [Exit] Terminate the execution of the algorithm.

For the ECPP, only the following heuristic analysis is known (Morain [168]).

Theorem 2.2.24. The expected running time of the ECPP algorithm is roughly proportional to $\mathcal{O}\left((\log N)^{6+\epsilon}\right)$ for some $\epsilon > 0$.

One of the largest primes verified so far with the ECPP algorithm is

$$391587 \times 2^{216193} - 1$$

which has 65087 digits. However, in practice, we normally carry out a primality test in the following way.

Algorithm 2.2.6 (Practical Test). Given an odd integer n, this algorithm will make use of the probabilistic test and elliptic curve test to determine whether or not n is prime:

[1] [Primality Testing – Probabilistic Method] Use a combination of the strong pseudoprimality test and the Lucas pseudoprimality test to determine if n is a probable prime. If it is, go to [2], else report that n is composite and go to [3].

[2] [Primality Proving – Elliptic Curve Method] Use the elliptic curve method (e.g., ECPP) to test whether or not n is indeed a prime. If it is, then report that n is prime, otherwise report that n is composite.

[3] [Exit] Terminate the algorithm.

2.2.6 Historical Notes on Primality Testing

In this subsection, we summarize some computational complexity results of primality testing.

Determining if a given integer $N \in \mathbb{N}$ is composite is easily seen to be in $\mathcal{NP}$ – simply multiply a nontrivial pair of integers whose product is N. In 1975 Pratt showed that determining primality is also in $\mathcal{NP}$ by exhibiting polynomial-time verifiable certificates of primality. (The basic idea is to prove N prime by showing that the multiplicative group of integers modulo N is cyclic of order $N - 1$; this requires a generator for the group, as well as a primality proof for each prime factor of $N-1$, which can be found recursively.) Finding these certificates, however, requires the ability to factor, a problem even harder than primality testing.

Miller [162] in 1976 showed that primality could be tested in polynomial time if the following conjecture (a version of the Generalized Riemann Hypothesis GRH) holds:

Let $\gcd(a, N) = 1$, then when $x \to \infty$,

$$|\{p \text{ prime},\ p \le x,\ p \equiv a \pmod{N}\}| = \frac{\mathrm{Li}(x)}{\phi(N)} + \mathcal{O}\left(\sqrt{x}\ \log x\right).$$

Miller's result can be stated as follows (Bach, Giesbrecht and McInnes [14]). For $l \in \mathbb{N}$, let $v_2(l) = \max\{e : 2^e \mid l\}$. Also, for $a \in \mathbb{N}$, let $L(a)$ be the Boolean expression:

$$L(a) = [a^{N-1} \equiv 1 \pmod{N} \text{ and } \forall k < v_2(N-1),\\ a^{\frac{N-1}{2^k}} \equiv 1 \Longrightarrow a^{\frac{N-1}{2^{k+1}}} \equiv \pm 1 \pmod{N}]. \tag{2.61}$$

Now assume that N is odd, and restrict a to be nonzero modulo N. Miller showed that if the GRH holds, then there is some $c > 0$ such that N is prime if and only if for all such a with $1 \le a \le c(\log N)^2$, $L(a)$ holds. Because $L(a)$ can be checked in $\mathcal{O}((\log N)^3)$ steps, this shows that the set of primes is in $\mathcal{P}$ (assuming GRH).

Shortly after Miller published his results, Solovay and Strassen [244] noted that N is prime if and only if every a with $\gcd(a, N) = 1$ satisfies

$$M(a) := \left[a^{(N-1)/2} \equiv \left(\frac{a}{N}\right) \pmod{N}\right]. \tag{2.62}$$

Furthermore, they observed that if N is composite, then $M(a)$ holds for at most half of the residues modulo N. One could therefore obtain statistical evidence of primality by choosing a from $\{1, \cdots, N-1\}$ at random, and testing $M(a)$. Because $M(a)$ will always hold when N is prime, and can be checked in $\mathcal{O}\left((\log N)^3\right)$ steps, this shows that the set of composite numbers belongs to $\mathcal{RP}$. Rabin [195] showed that Miller's predicate $L(a)$ has the same property, and is somewhat more reliable. He proved that if a is chosen at random from $\{1, \cdots, N-1\}$ and N is composite, then the probability that $L(a)$ holds is at most $1/4$.

Recent work has the goal of proving primality quickly, unconditionally, and without error. The fastest known *deterministic* algorithm, abbreviated the APRCL test, originally invented in 1980 by Adleman, Pomerance and Rumely [3] (known as the APR test), but further simplified and improved in 1981 by Cohen and Lenstra [50] using the idea of Jacobi sums, can determine the primality of N in time $\mathcal{O}\left((\log N)^{c(\log\log\log N)}\right)$, for some suitable constant $c > 0$. The exponent $c(\log\log\log N)$ is an extremely slowly growing function; for example, for the APR test, if N has a million decimal digits, then $\log\log\log N$ is only about 2.68. Riesel in 1985 reported an algorithm based on this method, which, when implemented on the CDC Cyber 170-750 Computer, was able to deal with 100-digit numbers in about 30 seconds and 200-digit numbers in about 8 minutes. It is now possible to prove the primality of numbers with 1000 digits in a not too unreasonable amount of time. Of course, the APRCL test does not run in *polynomial* time, nor does it provide polynomial-length certificates of primality.

In 1986, another modern primality testing algorithm, based on the theory of elliptic curves was invented, first for theoretical purposes by Goldwasser and Kilian [85], and then considerably modified by Atkin and implemented by Atkin and Morain [12]; we normally call Atkin and Morain's version the ECPP (Elliptic Curves and Primality Proving) test. The ECPP test is also practical for numbers with 1000 digits, and possibly with several thousand digits. The *expected* running time for ECPP is $\mathcal{O}\left((\log N)^6\right)$, hence, is polynomial time, but this is only on average, since for some numbers the running time of ECPP could be much larger. A totally impractical version based on Abelian varieties (higher dimensional analogs of elliptic curves) was given by Adleman and Huang [4] in 1992; they proved that without hypothesis, primality testing can be done in random polynomial time. In other words, they proved that *there exists a random polynomial time algorithm that can prove whether or not a given number* $N > 1$ *is prime.* Note that both the ECPP test and the Adleman–Huang test belong to the probabilistic complexity class $\mathcal{ZPP}$, that is, they always give the correct answer; only running time depends on chance and is expected to be polynomial.

More recently, Konyagin and Pomerance proposed several algorithms that can find proofs of primality in deterministic polynomial time for *some* primes. Their results do not rely on any unproven assertions such as the Riemann Hypothesis, but their algorithms need the *complete* prime factorization of $p-1$ in order to determine the primality of p.

Finally, we summarize some of the main complexity results in primality testing as follows:

(1) Primes/Composites $\in \mathcal{EXP}$; just try all the possible divisors.

(2) Composites $\in \mathcal{NP}$; guess a divisor.

(3) Primes $\in \mathcal{NP}$; Pratt (1975).

(4) Primes $\in \mathcal{P}$; Miller (1976); assuming the Extended Riemann Hypothesis.

(5) Composites $\in \mathcal{RP}$; Rabin (1976); using Miller's randomized algorithm.

(6) Primes $\in$ super-$\mathcal{P}$; the APRCL test, due to Adleman, Pomerance and Rumely (1980), and Cohen and Lenstra (1981).

(7) Primes $\in \mathcal{ZPP}$; the Elliptic Curve Test, due to Goldwasser and Killian (1985), and to Atkin and Morain (1991); not yet proved to work on all primes.

(8) Primes $\in \mathcal{ZPP}$; Hyperelliptic Curve Test, due to Adleman and Huang (1992); does not rely on any hypothesis, but is totally non-practical.

(9) Primes $\in \mathcal{P}$; Konyagin and Pomerance (1997); only for some primes.

2.3 Algorithms for Integer Factorization

Of all the problems in the theory of numbers to which computers have been applied, probably none has been influenced more than of factoring.

Hugh C. Williams
The Influence of Computers in the Development of Number Theory [255]

2.3.1 Complexity of Integer Factorization

According to the Fundamental Theorem of Arithmetic (Theorem 1.2.8), any positive integer greater than one can be written uniquely in the following prime factorization (prime decomposition) form:

$$n = p_1^{\alpha_1} p_2^{\alpha_2} \cdots p_k^{\alpha_k}, \tag{2.63}$$

where $p_1 < p_2 < \cdots < p_k$ are primes and $\alpha_1, \alpha_2, \cdots, \alpha_k$ positive integers. The so-called *integer factorization problem* (IFP) is to find a nontrivial factor f (not necessarily prime) of a composite integer n. That is,

$$\left.\begin{array}{ll} \text{Input}: & n \in \mathbb{N}_{>1} \\ \text{Output}: & f \text{ such that } f \mid n. \end{array}\right\} \tag{2.64}$$

Clearly, if there is an algorithm to test whether or not an integer n is a prime, and an algorithm to find a nontrivial factor f of a composite integer n, then there is a simply recursive algorithm to compute the prime power decomposition of N expressed in (2.63), as follows:

(1) find a nontrivial factor f of N;
(2 apply the algorithm recursively to f and N/f;
(3) put the prime power decompositions of f and N/f together to get the prime power decomposition of N.

There are, in fact, many algorithms for primality testing and integer factorization; the only problem is that there is no known *efficient* (deterministic polynomial-time) algorithm for either primality testing or integer factorization.

Primality testing, and particularly integer factorization, are very important in mathematics. Gauss [82] wrote in 1801 the following famous statements in his most profound publication *Disquistiones Arithmeticae*:

The problem of distinguishing prime numbers from composite numbers and of resolving the latter into their prime factors is known to be one of the most important and useful in arithmetic. $\cdots$ *the dignity of science itself seems to require that every possible means be explored for the solution of a problem so elegant and so celebrated.*

But unfortunately, primality testing, and particularly integer factorization, are computationally intractable (Adleman [2]), as Knuth[20] explained in his encyclopaedic work [123]:

> *It is unfortunately not a simple matter to find this prime factorization of n, or to determine whether or not n is prime. $\cdots$ therefore we should avoid factoring large numbers whenever possible.*

In fact, no deterministic or randomized polynomial-time algorithm has been found for integer factorization, nor has anyone proved that there is not an efficient algorithm[21]. Despite this, remarkable progress has been made in recent years, and mathematicians (at least some) believe that efficient primality testing and/or integer factorization algorithms are somewhere around the corner waiting for discovery, although it is very hard to find such algorithms. Generally speaking, the most useful factoring algorithms fall into one of the following two main classes (Brent [37]):

(1) The running time depends mainly on the size of N, the number to be factored, and is not strongly dependent on the size of the factor p found. Examples are:

   (i) *Lehman's method* [139], which has a rigorous worst-case running time bound $\mathcal{O}\left(N^{1/3+\epsilon}\right)$.

   (ii) *Shanks' SQUare FOrm Factorization method* SQUFOF, which has expected running time $\mathcal{O}\left(N^{1/4}\right)$.

   (iii) *Shanks' class group method*, which has running time $\mathcal{O}\left(N^{1/5+\epsilon}\right)$.

---

[20] 

Donald E. Knuth (1938–), studied mathematics as an undergraduate at Case Institute of Technology, and received a PhD in Mathematics in 1963 from the California Institute of Technology. Knuth joined Stanford University as Professor of Computer Science in 1968, and is now Professor Emeritus there. Knuth received in 1974 the prestigious Turing Award from the Association for Computing Machinery (ACM) for his work in analysis of algorithms and particularly for his series of books, TAOCP. (Photo by courtesy of Prof. Knuth.)

[21] For primality testing, although we also do not have a *truly* deterministic polynomial-time algorithm, we do have randomized polynomial-time algorithms; this explains partly that integer factorization is much harder than primality testing, although both of them are computationally intractable (in the sense that no deterministic polynomial-time algorithm exists for both of them). The following fact about randomized computation is important in public-key cryptography, which will be studied in detail in the next chapter. A problem is said to be *easy* if there is a randomized polynomial-time algorithm to solve it, otherwise, it is *hard*. For example, there is a randomized polynomial-time algorithm for the test of primality of an integer, so the primality testing problem is regarded as *easy*. However, there is no randomized polynomial-time algorithm for factoring a large integer, so the integer factorization problem is *hard*.

(iv) *Continued FRACtion (CFRAC) method*, which under plausible assumptions has expected running time

$$\mathcal{O}\left(\exp\left(c\sqrt{\log N \log\log N}\right)\right) = \mathcal{O}\left(N^{c\sqrt{\log\log N/\log N}}\right),$$

where c is a constant (depending on the details of the algorithm); usually $c = \sqrt{2} \approx 1.414213562$.

(v) *Multiple Polynomial Quadratic Sieve (MPQS)*, which under plausible assumptions has expected running time

$$\mathcal{O}\left(\exp\left(c\sqrt{\log N \log\log N}\right)\right) = \mathcal{O}\left(N^{c\sqrt{\log\log N/\log N}}\right),$$

where c is a constant (depending on the details of the algorithm); usually $c = \dfrac{3}{2\sqrt{2}} \approx 1.060660172$.

(vi) *Number Field Sieve (NFS)*, which under plausible assumptions has the expected running time

$$\mathcal{O}\left(\exp\left(c\sqrt[3]{\log N}\sqrt[3]{(\log\log N)^2}\right)\right),$$

where $c = (64/9)^{1/3} \approx 1.922999427$ if GNFS (a general version of NFS) is used to factor an arbitrary integer N, whereas $c = (32/9)^{1/3} \approx 1.526285657$ if SNFS (a general version of NFS) is used to factor a special integer N such as $N = r^e \pm s$, where r and s are small, $r > 1$ and e is large. This is substantially and asymptotically faster than any other currently known factoring methods.

(2) The running time depends mainly on the size of p (the factor found) of N. (We can assume that $p \leq \sqrt{N}$.) Examples are:

(i) *Trial division*, which has running time $\mathcal{O}\left(p(\log N)^2\right)$.

(ii) *Pollard's ρ-method* (also known as Pollard's "*rho*" algorithm), which under plausible assumptions has expected running time $\mathcal{O}\left(p^{1/2}(\log N)^2\right)$.

(iii) *Lenstra's Elliptic Curve Method (ECM)*, which under plausible assumptions has expected running time

$$\mathcal{O}\left(\exp\left(c\sqrt{\log p \log\log p}\right) \cdot (\log N)^2\right),$$

where $c \approx 2$ is a constant (depending on the details of the algorithm).

The term $\mathcal{O}\left((\log N)^2\right)$ is a generous allowance for the cost of performing arithmetic operations on numbers which are $\mathcal{O}(\log N)$ or $\mathcal{O}\left((\log N)^2\right)$ bits long; these could theoretically be replaced by $\mathcal{O}\left((\log N)^{1+\epsilon}\right)$ for any $\epsilon > 0$.

In practice, algorithms in both categories are important. It is sometimes very difficult to say whether one method is better than another, but it is generally worth attempting to find small factors with algorithms in the second class before using the algorithms in the first class. That is, we could first try the *trial division algorithm*, then use some other methods such as NFS. This fact shows that the trial division method is still useful for integer factorization, even though it is simple. In the subsections that follow, we shall introduce some of the most useful and widely used factoring algorithms.

Remark 2.3.1. As mentioned previousely, an algorithm is of exponential complexity, if its required running time is

$$\mathcal{O}\left(N^{\epsilon}\right), \tag{2.65}$$

where a typical value for ϵ would be between 0.1 and 0.5. But note that we usually do not regard the type of complexity

$$\begin{aligned} \mathcal{O}\left(N^{\epsilon(N)}\right) &= \mathcal{O}\left(N^{c\sqrt{\log\log N/\log N}}\right) \\ &= \mathcal{O}\left(\exp\left(c\sqrt{\log N \log\log N}\right)\right) \end{aligned} \tag{2.66}$$

as a ***truly*** exponential complexity; we normally call it *subexponential complexity*. The relationship between the polynomial, superpolynomial, subexponential, and exponential complexities, together with some examples, can be shown as follows:

$$\begin{array}{ccccc} \mathcal{O}\left((\log N)^k\right) & \Longleftrightarrow & \text{Polynomial} & \Longleftarrow & \text{Euclid's algorithm} \\ \cap & & \cap & & \\ \mathcal{O}\left((\log N)^{c\log\log\log N}\right) & \Longleftrightarrow & \text{Superpolynomial} & \Longleftarrow & \text{APRCL test} \\ \cap & & \cap & & \\ \mathcal{O}\left(N^{c\sqrt{\log\log N/\log N}}\right) & \Longleftrightarrow & \text{Subexponential} & \Longleftarrow & \text{MPQS factoring} \\ \cap & & \cap & & \\ \mathcal{O}\left(N^{\epsilon}\right) & \Longleftrightarrow & \text{Exponential} & \Longleftarrow & \text{Trial divisions} \end{array}$$

Remark 2.3.2. It is sometimes convenient to use some short expressions to denote subexponential complexity; one such short expression is the following:

$$L_N(\gamma, c) \stackrel{\text{def}}{=} \exp\left(c(\log N)^{\gamma}(\log\log N)^{1-\gamma}\right). \tag{2.67}$$

So, in this notation, we could, for example, write

$$T(\text{CFRAC}) = \mathcal{O}\left(L_N\left(1/2,\ \sqrt{2}\right)\right), \tag{2.68}$$

$$T(\text{ECM}) = \mathcal{O}\left(L_p\left(1/2,\ \sqrt{2}\right)\cdot(\log N)^2\right), \tag{2.69}$$

$$T(\text{MPQS}) = \mathcal{O}\left(L_N\left(1/2,\ 3/(2\sqrt{2})\right)\right), \tag{2.70}$$

$$T(\text{GNFS}) = \mathcal{O}\left(L_N\left(1/3,\ \sqrt[3]{64/9}\right)\right), \tag{2.71}$$

$$T(\text{SNFS}) = \mathcal{O}\left(L_N\left(1/3,\ \sqrt[3]{32/9}\right)\right). \tag{2.72}$$

Note also that some authors prefer to use the following short notation

$$L(N) \stackrel{\text{def}}{=} e^{\sqrt{\log N \log\log N}}. \tag{2.73}$$

In this notation, we could, for example write

$$T(\text{CFRAC}) = \mathcal{O}\left(L(N)^{2+o(1)}\right), \tag{2.74}$$

$$T(\text{ECM}) = \mathcal{O}\left(L(p)^{2+o(1)} \cdot (\log N)^2\right), \tag{2.75}$$

$$T(\text{MPQS}) = \mathcal{O}\left(L(N)^{1+o(1)}\right). \tag{2.76}$$

In order to avoid any possible confusion, in this book we shall use the ordinary full notation.

Remark 2.3.3. For primality testing, although we still do not have a *truly* deterministic polynomial-time algorithm, we do have randomised polynomial-time algorithms. However, there is no known deterministic, or even randomised polynomial-time, algorithm for finding a factor of a given composite integer n. This empirical fact is of great interest in public-key cryptography.

2.3.2 Trial Division and Fermat Method

(I) Factoring by Trial Divisions. The simplest factoring algorithm is the trial division method, which tries all the possible divisors of n to obtain its complete prime factorization:

$$n = p_1 p_2 \cdots p_t, \qquad p_1 \le p_2 \le \cdots \le p_t. \tag{2.77}$$

The following is the algorithm:

Algorithm 2.3.1 (Factoring by trial divisions). This algorithm tries to factor an integer $n > 1$ using trial divisions by all the possible divisors of n.

[1] [Initialization] Input n and set $t \leftarrow 0,\ \ k \leftarrow 2$.

[2] [$n = 1$?] If $n = 1$, then goto [5].

[3] [Compute Remainder]

$q \leftarrow n/k$ and $r \leftarrow n \pmod{k}$.
If $r \neq 0$, goto [4].
$t \leftarrow t+1,\ \ p_t \leftarrow k,\ \ n \leftarrow q$, goto [2].

[4] [Factor Found?]

If $q > k$, then $k \leftarrow k+1$, goto [3].
$t \leftarrow t+1;\ \ p_t \leftarrow n$.

[5] [Exit] Terminate the algorithm.

Exercise 2.3.1. Use Algorithm 2.3.1 to factor $n = 2759$.

An immediate improvement of Algorithm 2.3.1 is to make use of an auxiliary sequence of *trial divisors*:

$$2 = d_0 < d_1 < d_2 < d_3 < \cdots \tag{2.78}$$

which includes all primes $\leq \sqrt{n}$ (possibly some composites as well if it is convenient to do so) and at least one value $d_k \geq \sqrt{n}$. The algorithm can be described as follows:

Algorithm 2.3.2 (Factoring by Trial Division). This algorithm tries to factor an integer $n > 1$ using trial divisions by an auxiliary sequence of trial divisors.

[1] [Initialization] Input n and set $t \leftarrow 0, \quad k \leftarrow 0$.

[2] [$n = 1$?] If $n = 1$, then goto [5].

[3] [Compute Remainder]

$q \leftarrow n/d_k$ and $r \leftarrow n \pmod{d_k}$.
If $r \neq 0$, goto [4].
$t \leftarrow t + 1, \quad p_t \leftarrow d_k, \quad n \leftarrow q$, goto [2].

[4] [Factor Found?]

If $q > d_k$, then $k \leftarrow k + 1$, and goto [3].
$t \leftarrow t + 1; \quad p_t \leftarrow n$.

[5] Exit: terminate the algorithm.

Exercise 2.3.2. Use Algorithm 2.3.2 to factor $n = 2759$; assume that we have the list L of all primes $\leq \lfloor\sqrt{2759}\rfloor = 52$ and at least one $\geq \sqrt{n}$, that is, $L = \{2, 3, 5, 7, 11, 13, 17, 19, 23, 29, 31, 37, 41, 43, 47, 53\}$.

Theorem 2.3.1. Algorithm 2.3.2 requires a running time in

$$\mathcal{O}\left(\max\left(p_{t-1}, \ \sqrt{p_t}\right)\right).$$

If a primality test between steps [2] and [3] were inserted, the running time would then be in $\mathcal{O}(p_{t-1})$, or $\mathcal{O}\left(\dfrac{p_{t-1}}{\ln p_{t-1}}\right)$ if one does trial division only by primes, where p_{t-1} is the second largest prime factor of n.

The trial division test is very useful for removing small factors, but it should not be used for factoring completely, except when n is very small, say, for example, $n < 10^8$.

(II) Fermat's Factoring Method. Now suppose n is any odd integer (if n were even we could repeatedly divide by 2 until an odd integer resulted). If $n = pq$, where $p \leq q$ are both odd, then by setting $x = \frac{1}{2}(p+q)$ and $y = \frac{1}{2}(q-p)$ we find that $n = x^2 - y^2 = (x+y)(x-y)$, or $y^2 = x^2 - n$. The following algorithm tries to find $n = pq$ using the above idea.

Algorithm 2.3.3 (Fermat's factoring algorithm). Given an odd integer $n > 1$, then this algorithm determines the largest factor $\leq \sqrt{n}$ of n.

[1] Input n and set $k \leftarrow \lfloor\sqrt{n}\rfloor + 1$, $y \leftarrow k \cdot k - n$, $d \leftarrow 1$

[2] If $\lfloor\sqrt{y}\rfloor = \sqrt{y}$ goto step [4] else $y \leftarrow y + 2 \cdot k + d$ and $d \leftarrow d + 2$

[3] If $\lfloor\sqrt{y}\rfloor < n/2$ goto step [2] else print "No Factor Found" and goto [5]

[4] $x \leftarrow \sqrt{n+y}$, $y \leftarrow \sqrt{y}$, print $x - y$ and $x + y$, the nontrivial factors of n

[5] Exit: terminate the algorithm.

Exercise 2.3.3. Use the Fermat method to factor $n = 278153$.

Theorem 2.3.2 (The Complexity of Fermat's Method). The Fermat method will try as many as $\dfrac{n+1}{2} - \sqrt{n}$ arithmetic steps to factor n, that is, it is of complexity $\mathcal{O}\left(\dfrac{n+1}{2} - \sqrt{n}\right)$.

2.3.3 Legendre's Congruence

In the next two subsections, we shall introduce three widely used *general* purpose integer factorization methods, namely, the continued fraction method (abbreviated CFRAC), the quadratic sieve (abbreviated QS) and the number field sieve (abbreviated NFS). By a general purpose factoring method, we mean one that will factor *any* integer of a given size in about the same time as any other of that size. The method will take as long, for example, to split a 100-digit number into the product of a 1-digit and a 99-digit prime, as it will to split a different number into the product of two 50-digit primes. These methods do not depend upon any special properties of the number or its factors.

The CFRAC method, as well as other powerful general purpose factoring methods such as the Quadratic Sieve (QS) and the Number Field Sieve (NFS), makes use of the simple but important observation that if we have two integers x and y such that

$$x^2 \equiv y^2 \pmod{N}, \; 0 < x < y < N, \; x \neq y, \; x + y \neq N, \tag{2.79}$$

then $\gcd(x-y, N)$ and $\gcd(x+y, N)$ are possibly nontrivial factors of N, because $N \mid (x+y)(x-y)$, but $N \nmid (x+y)$ and $N \nmid (x-y)$. The congruence (2.79) is often called Legendre's congruence. So, to use Legendre's congruence for factorization, we simply perform the following two steps:

[1] Find a nontrivial solution to the congruence $x^2 \equiv y^2 \pmod{N}$.

[2] Compute the factors d_1 and d_2 of N by using Euclid's algorithm:

$$(d_1, d_2) = (\gcd(x + y,\ N),\ \gcd(x - y,\ N)).$$

Example 2.3.1. Let $N = 119$. $12^2 \bmod 119 = 5^2 \bmod 119$. Then

$$(d_1, d_2) = (\gcd(12 + 5, 119),\ \ \gcd(12 - 5, 119)) = (17, 7).$$

In fact, $119 = 7 \cdot 17$.

The best method for constructing congruences of the form (2.79) starts by accumulating several congruences of the form

$$\left(A_i = \prod p_k^{e_k}\right) \equiv \left(B_i = \prod p_j^{e_j}\right) \pmod{N}. \tag{2.80}$$

Some of these congruences are then multiplied in order to generate squares on both sides (Montgomery [167]). We illustrate this idea in the following example.

Example 2.3.2. Let $N = 77$. Then, on the left hand side of the following table, we collect eight congruences of the form (2.80) over the prime factor base FB $= \{-1, 2, 3, 5\}$ (note that we include -1 as a "prime" factor); the right hand side of the table contains the exponent vector information of $v(A_i)$ and $v(B_i)$ modulo 2.

$45 = 3^2 \cdot 5$	$\equiv$	$-32 = -2^5$	$\Longleftrightarrow$	$(0\ 0\ 0\ 1) \equiv (1\ 1\ 0\ 0)$
$50 = 2 \cdot 5^2$	$\equiv$	$-27 = -3^3$	$\Longleftrightarrow$	$(0\ 1\ 0\ 0) \equiv (1\ 0\ 1\ 0)$
$72 = 2^3 \cdot 3^2$	$\equiv$	-5	$\Longleftrightarrow$	$(0\ 1\ 0\ 0) \equiv (1\ 0\ 0\ 1)$
$75 = 3 \cdot 5^2$	$\equiv$	-2	$\Longleftrightarrow$	$(0\ 0\ 1\ 0) \equiv (1\ 1\ 0\ 0)$
$80 = 2^4 \cdot 5$	$\equiv$	3	$\Longleftrightarrow$	$(0\ 0\ 0\ 1) \equiv (0\ 0\ 1\ 0)$
$125 = 5^3$	$\equiv$	$48 = 2^4 \cdot 3$	$\Longleftrightarrow$	$(0\ 0\ 0\ 1) \equiv (0\ 0\ 1\ 0)$
$320 = 2^6 \cdot 5$	$\equiv$	$243 = 3^5$	$\Longleftrightarrow$	$(0\ 0\ 0\ 1) \equiv (0\ 0\ 1\ 0)$
$384 = 2^7 \cdot 3$	$\equiv$	-1	$\Longleftrightarrow$	$(0\ 1\ 1\ 0) \equiv (1\ 0\ 0\ 0)$

Now we multiply some of these congruences in order to generate squares on both sides; both sides will be squares precisely when the sum of the exponent vectors is the zero vector modulo 2. We first multiply the sixth and seventh congruences and get:

$125 = 5^3$	$\equiv$	$48 = 2^4 \cdot 3$	$\Longleftrightarrow$	$(0\ 0\ 0\ 1)$	$\equiv$	$(0\ 0\ 1\ 0)$
$320 = 2^6 \cdot 5$	$\equiv$	$243 = 3^5$	$\Longleftrightarrow$	$(0\ 0\ 0\ 1)$	$\equiv$	$(0\ 0\ 1\ 0)$
				$\downarrow\downarrow\downarrow\downarrow$ $(0\ 0\ 0\ 0)$		$\downarrow\downarrow\downarrow\downarrow$ $(0\ 0\ 0\ 0)$

Since the sum of the exponent vectors is the zero vector modulo 2, we find squares on both sides:

$$5^3 \cdot 2^6 \cdot 5 \equiv 2^4 \cdot 3 \cdot 3^5 \iff (5^2 \cdot 2^3)^2 \equiv (2^2 \cdot 3^3)^2$$

and hence we have $\gcd(5^2 \cdot 2^3 \pm 2^2 \cdot 3^3,\ 77) = (77, 1)$, but this does not split 77, so we try to multiply some other congruences, for example, the fifth and the seventh, and get:

$80 = 2^4 \cdot 5$	$\equiv$	3	$\iff$	$(0\ 0\ 0\ 1)$	$\equiv$	$(0\ 0\ 1\ 0)$
$320 = 2^6 \cdot 5$	$\equiv$	$243 = 3^5$	$\iff$	$(0\ 0\ 0\ 1)$	$\equiv$	$(0\ 0\ 1\ 0)$
				$\downarrow\downarrow\downarrow\downarrow$ $(0\ 0\ 0\ 0)$		$\downarrow\downarrow\downarrow\downarrow$ $(0\ 0\ 0\ 0)$

The sum of the exponent vectors is the zero vector modulo 2, so we find

$$2^4 \cdot 5 \cdot 2^6 \cdot 5 \equiv 3 \cdot 3^5 \iff (2^5 \cdot 5)^2 \equiv (3^3)^2$$

and compute $\gcd(2^5 \cdot 5 \pm 3^3,\ 77) = (11, 7)$. This time, it splits 77. Once we split N, we stop the process. Just for the purpose of illustration, we try one more example, which will also split N.

$45 = 3^2 \cdot 5$	$\equiv$	$-32 = -2^5$	$\iff$	$(0\ 0\ 0\ 1)$	$\equiv$	$(1\ 1\ 0\ 0)$
$50 = 2 \cdot 5^2$	$\equiv$	$-27 = -3^3$	$\iff$	$(0\ 1\ 0\ 0)$	$\equiv$	$(1\ 0\ 1\ 0)$
$75 = 3 \cdot 5^2$	$\equiv$	-2	$\iff$	$(0\ 0\ 1\ 0)$	$\equiv$	$(1\ 1\ 0\ 0)$
$320 = 2^6 \cdot 5$	$\equiv$	$243 = 3^5$	$\iff$	$(0\ 0\ 0\ 1)$	$\equiv$	$(0\ 0\ 1\ 0)$
$384 = 2^7 \cdot 3$	$\equiv$	-1	$\iff$	$(0\ 1\ 1\ 0)$	$\equiv$	$(1\ 0\ 0\ 0)$
				$\downarrow\downarrow\downarrow\downarrow$ $(0\ 0\ 0\ 0)$		$\downarrow\downarrow\downarrow\downarrow$ $(0\ 0\ 0\ 0)$

So we have

$$3^2 \cdot 5 \cdot 2 \cdot 5^2 \cdot 3 \cdot 5^2 \cdot 2^6 \cdot 5 \cdot 2^7 \cdot 3 \equiv -2^5 \cdot -3^3 \cdot -2 \cdot 3^5 \cdot -1 \iff (2^7 \cdot 3^2 \cdot 5^3)^2 \equiv (2^3 \cdot 3^4)^2,$$

thus $\gcd(2^7 \cdot 3^2 \cdot 5^3 \pm 2^3 \cdot 3^4,\ 77) = (7, 11)$.

Based on the above idea, the trick, common to the CFRAC, QS and NFS, is to find a congruence (also called a *relation*) of the form

$$x_k^2 \equiv (-1)^{e_{0k}} p_1^{e_{1k}} p_2^{e_{2k}} \cdots p_m^{e_{mk}} \pmod{N}, \tag{2.81}$$

where each p_i is a "small" prime number (the set of all such p_i, for $1 \le i \le m$, forms a *factor base*, denoted by FB). If we find sufficiently many such congruences, by Gaussian elimination over $\mathbb{Z}/2\mathbb{Z}$ we may hope to find a relation of the form

$$\sum_{1 \le k \le n} \epsilon_k (e_{0k}, e_{1k}, e_{2k}, \cdots, e_{mk}) \equiv (0, 0, 0, \cdots, 0) \pmod{2}, \tag{2.82}$$

where ϵ is either 1 or 0, and then

$$x = \prod_{1 \le k \le n} x_k^{\epsilon_k}, \tag{2.83}$$

$$y = (-1)^{v_0} p_1^{v_1} p_2^{v_2} \cdots p_m^{v_m}, \tag{2.84}$$

where

$$\sum_k \epsilon_k(e_{0k}, e_{1k}, e_{2k}, \cdots, e_{mk}) = 2(v_0, v_1, v_2, \cdots, v_m). \tag{2.85}$$

It is clear that we now have $x^2 \equiv y^2 \pmod{N}$. This splits N if, in addition, $x \not\equiv \pm y \pmod{N}$.

Now we are in a position to introduce our first general purpose factoring method, the CFRAC method.

2.3.4 Continued FRACtion Method (CFRAC)

The continued fraction method is perhaps the first *modern, general* purpose integer factorization method, although its original idea may go back to M. Kraitchik in the 1920s, or even earlier to A. M. Legendre. It was used by D. H. Lehmer and R. E. Powers to devise a new technique in the 1930s, however the method was not very useful and applicable at the time because it was unsuitable for desk calculators. About 40 years later, it was first implemented on a computer by M. A. Morrison and J. Brillhart [169], who used it to successfully factor the seventh Fermat number

$$F_7 = 2^{2^7} + 1 = 59649589127497217 \cdot 5704689200685129054721$$

on the morning of 13 September 1970.

The Continued FRACtion (CFRAC) method looks for small values of $|W|$ such that $x^2 \equiv W \pmod{N}$ has a solution. Since W is small (specifically $W = \mathcal{O}(\sqrt{N}\,)$), it has a reasonably good chance of being a product of primes in our factor base FB. Now if W is small and $x^2 \equiv W \pmod{N}$, then we can write $x^2 = W + kNd^2$ for some k and d, hence $(x/d)^2 - kN = W/d^2$ will be small. In other words, the rational number x/d is an approximation of $\sqrt{kN}$. This suggests looking at the continued fraction expansion of $\sqrt{kN}$, since continued fraction expansions of real numbers give good rational approximations. This is exactly the idea behind the CFRAC method! We first obtain a sequence of approximations (i.e., convergents) P_i/Q_i to $\sqrt{kN}$ for a number of values of k, such that

$$\left|\sqrt{kN} - \frac{P_i}{Q_i}\right| \le \frac{1}{Q_i^2}. \tag{2.86}$$

Putting $W_i = P_i^2 - Q_i^2 kN$, then we have

$$W_i = (P_i + Q_i\sqrt{kN})(P_i - Q_i\sqrt{kN}) \sim 2Q_i\sqrt{kN}\frac{1}{Q_i} \sim 2\sqrt{kN}. \tag{2.87}$$

Hence, the $P_i^2 \bmod N$ are small and more likely to be smooth, as desired. Then, we try to factor the corresponding integers $W_i = P_i^2 - Q_i^2 kN$ over our factor base FB; at each success, we obtain a new congruence of the form

$$P_i^2 \equiv W_i \Longleftrightarrow x^2 \equiv (-1)^{e_0} p_1^{e_1} p_2^{e_2} \cdots p_m^{e_m} \pmod{N}. \tag{2.88}$$

Once we have obtained at least $m+2$ such congruences, by Gaussian elimination over $\mathbb{Z}/2\mathbb{Z}$ we have obtained a congruence $x^2 \equiv y^2 \pmod{N}$. That is, if $(x_1, e_{01}, e_{11}, \cdots, e_{m1}), \cdots, (x_r, e_{0r}, e_{1r}, \cdots, e_{mr})$ are solutions of (2.88) such that the vector sum

$$(e_{01}, e_{11}, \cdots, e_{m1}) + \cdots + (e_{0r}, e_{1r}, \cdots, e_{mr}) = (2e_0', 2e_1', \cdots, 2e_m') \tag{2.89}$$

is even in each component, then

$$x \equiv x_1 x_2 \cdots x_r \pmod{N} \tag{2.90}$$

$$y \equiv (-1)^{e_0'} p_1^{e_1'} \cdots p_m^{e_m'} \pmod{N} \tag{2.91}$$

is a solution to (2.79), except for the possibility that $x \equiv \pm y \pmod{N}$, and hence (usually) a nontrivial splitting of N.

Example 2.3.3. We now illustrate, by an example, the idea of CFRAC factoring. Let $N = 1037$. Then $\sqrt{1037} = [32, \overline{4, 1, 15, 3, 3, 15, 1, 4, 64}]$. The first ten continued fraction approximations of $\sqrt{1037}$ are:

Convergent P/Q	$P^2 - N \cdot Q^2 := W$
$32/1$	$-13 = -13$
$129/4$	$49 = 7^2$
$161/5$	$-4 = -2^2$
$2544/79 \equiv 470/79$	$19 = 19$
$7793/242 \equiv 534/242$	$-19 = -19$
$25923/805 \equiv 1035/805$	$4 = 2^2$
$396638/12317 \equiv 504/910$	$-49 = -7^2$
$422561/13122 \equiv 502/678$	$13 = 13$
$2086882/64805 \equiv 438/511$	$-1 = -1$
$133983009/4160642 \equiv 535/198$	$13 = 13$

Now we search for squares on both sides, either just by a single congruence, or by a combination (i.e., multiplying together) of several congruences and find that

$$\begin{aligned}
129^2 \equiv 7^2 &\Longleftrightarrow \gcd(1037,\ 129 \pm 7) = (17, 61) \\
1035^2 \equiv 2^2 &\Longleftrightarrow \gcd(1037,\ 1035 \pm 2) = (1037, 1) \\
129^2 \cdot 1035^2 \equiv 7^2 \cdot 2^2 &\Longleftrightarrow \gcd(1037,\ 129 \cdot 1035 \pm 7 \cdot 2) = (61, 17) \\
161^2 \cdot 504^2 \equiv (-1)^2 \cdot 2^2 \cdot 7^2 &\Longleftrightarrow \gcd(1037,\ 161 \cdot 504 \pm 2 \cdot 7) = (17, 61) \\
502^2 \cdot 535^2 \equiv 13^2 &\Longleftrightarrow \gcd(1037\ 502 \cdot 535 \pm 13) = (1037, 1).
\end{aligned}$$

Three of them yield a factorization of $1037 = 17 \cdot 61$.

Exercise 2.3.4. Use the continued fraction expansion

$$\sqrt{1711} = [41, \overline{2, 1, 2, 1, 13, 16, 2, 8, 1, 2, 2, 2, 2, 2, 1, 8, 2, 16, 13, 1, 2, 1, 2, 82}]$$

and the factor base $\mathrm{FB} = \{-1, 2, 3, 5\}$ to factor the integer 1711.

It is clear that the CFRAC factoring algorithm is essentially just a continued fraction algorithm for finding the continued fraction expansion $[q_0, q_1, \cdots, q_k, \cdots]$ of $\sqrt{kN}$, or the P_k and Q_k of such an expansion. In what follows, we shall briefly summarize the CFRAC method just discussed above in the following algorithmic form:

Algorithm 2.3.4 (CFRAC factoring). Given a positive integer N and a positive integer k such that kN is not a perfect square, this algorithm tries to find a factor of N by computing the continued fraction expansion of $\sqrt{kN}$.

[1] Let N be the integer to be factored and k any small integer (usually 1), and let the factor base, FB, be a set of small primes $\{p_1, p_2, \cdots, p_r\}$ chosen such that it is possible to find some integer x_i such that $x_i^2 \equiv kN \pmod{p_i}$. Usually, FB contains all such primes less than or equal to some limit. Note that the multiplier $k > 1$ is needed only when the period is short. For example, Morrison and Brillhart used $k = 257$ in factoring F_7.

[2] Compute the continued fraction expansion $[q_0, \overline{q_1, q_2, \cdots, q_r}]$ of $\sqrt{kN}$ for a number of values of k. This gives us good rational approximations P/Q. The recursion formulas to use for computing P/Q are as follows:

$$\begin{aligned} \frac{P_0}{Q_0} &= \frac{q_0}{1}, \\ \frac{P_1}{Q_1} &= \frac{q_0 q_1 + 1}{q_1}, \\ &\cdots\cdots \\ &\cdots\cdots \\ \frac{P_i}{Q_i} &= \frac{q_i P_{i-1} + P_{i-2}}{q_i Q_{i-1} + Q_{i-2}}, \quad i \geq 2. \end{aligned}$$

This can be done by a continued fraction algorithm such as Algorithm 1.2.23 introduced in subsection 1.2.5 of Chapter 1.

[3] Try to factor the corresponding integer $W = P^2 - Q^2 kN$ in our factor base FB. Since $W < 2\sqrt{kN}$, each of these W is only about half the length of kN. If we succeed, we get a new congruence. For each success, we obtain a congruence

$$x^2 \equiv (-1)^{e_0} p_1^{e_1} p_2^{e_2} \cdots p_m^{e_m} \pmod{N},$$

since, if P_i/Q_i is the ith continued fraction convergent to $\sqrt{kN}$ and $W_i = P_i^2 - N \cdot Q_i^2$, then

$$P_i^2 \equiv W_i \pmod{N}. \tag{2.92}$$

[4] Once we have obtained at least $m + 2$ such congruences, then by Gaussian elimination over $\mathbb{Z}/2\mathbb{Z}$ we obtain a congruence $x^2 \equiv y^2 \pmod{N}$. That is, if $(x_1, e_{01}, e_{11}, \cdots, e_{m1}), \cdots, (x_r, e_{0r}, e_{1r}, \cdots, e_{mr})$ are solutions of (2.88) such that the vector sum defined in (2.89) is even in each component, then

$$\begin{cases} x \equiv x_1 x_2 \cdots x_r \pmod{N} \\ y \equiv (-1)^{e'_0} p_1^{e'_1} \cdots p_m^{e'_m} \pmod{N} \end{cases}$$

is a solution to $x^2 \equiv y^2 \pmod{N}$, except for the possibility that $x \equiv \pm y \pmod{N}$, and hence we have

$$(d_1, d_2) = (\gcd(x+y,\ N),\ \ \gcd(x-y,\ N)),$$

which are then possibly nontrivial factors of N.

Conjecture 2.3.1 (The Complexity of the CFRAC Method). If N is the integer to be factored, then under certain reasonable heuristic assumptions, the CFRAC method will factor N in time

$$\mathcal{O}\left(\exp\left((\sqrt{2}+o(1))\sqrt{\log N \log\log N}\ \right)\right) = \mathcal{O}\left(N^{\sqrt{(2+o(1))\log\log N/\log N}}\ \right). \tag{2.93}$$

Remark 2.3.4. This is a conjecture, not a theorem, because it is supported by some heuristic assumptions which have not been proven (Cohen [50]).

2.3.5 Quadratic and Number Field Sieves (QS/NFS)

In this subsection, we shall briefly introduce two other powerful general purpose factoring methods: the Quadratic Sieve (QS) and the Number Field Sieve (NFS).

(I) The Quadratic Sieve (QS). The idea of the quadratic sieve (QS) was first introduced by Carl Pomerance[22] in 1982. QS is somewhat similar to CFRAC except that, instead of using continued fractions to produce the values for $W_k = P_k^2 - N \cdot Q_k^2$, it uses expressions of the form

$$W_k = (k + \lfloor\sqrt{N}\rfloor)^2 - N \equiv (k + \lfloor\sqrt{N}\rfloor)^2 \pmod{N}. \tag{2.94}$$

22

Carl Pomerance is currently Research Professor in the Department of Mathematics at the University of Georgia, U.S.A. He obtained his PhD at Harvard University in 1972 and has been a faculty member at Georgia since 1972. Pomerance has made several important contributions in computational number theory, particularly in primality testing and integer factorization; for example, the "P" in the APRCL primality test (the fastest known deterministic primality testing algorithm) stands for Pomerance. Pomerance co-authored 18 papers with the legendary Paul Erdös. (Photo by courtesy of Prof. Pomerance.)

Here, if $0 < k < L$, then

$$0 < W_k < (2L + 1)\sqrt{N} + L^2. \tag{2.95}$$

If we get

$$\prod_{i=1}^{t} W_{n_i} = y^2, \tag{2.96}$$

then we have $x^2 \equiv y^2 \pmod{N}$ with

$$x \equiv \prod_{i=1}^{t} (\lfloor \sqrt{N} \rfloor + n_i) \pmod{N}. \tag{2.97}$$

Once such x and y are found, there is a good chance that $\gcd(x - y, N)$ is a nontrivial factor of N. For the purpose of implementation, we can use the same set FB as that used in CFRAC and the same idea as that described above to arrange that (2.96) holds. The most widely used variation of the quadratic sieve is perhaps the multiple polynomial quadratic sieve (MPQS), proposed by Peter Montgomery [235] in 1986. MPQS has been used to obtain many spectacular factorizations. One such factorization is that of the 103-digit number

$$\frac{2^{361} + 1}{3 \cdot 174763} = 6874301617534827509350575768454356245025403 \cdot p_{61}.$$

The most recent record of the quadratic sieve is the factorization of the RSA-129, a 129 digit number. (It was estimated by Rivest [78] in 1977 that the running time required to factor numbers with about the same size as RSA-129 would be about 40 quadrillion years using the best algorithm and fastest computer at that time.)

Example 2.3.4. Use the quadratic sieve method (QS) to factor $N = 2041$. Let $W(x) = x^2 - N$, with $x = 43, 44, 45, 46$. Then we have:

$W(43)$	$=$	$-2^6 \cdot 3$
$W(44)$	$=$	$-3 \cdot 5 \cdot 7$
$W(45)$	$=$	-2^4
$W(46)$	$=$	$3 \cdot 5^2$

p	$W(43)$	$W(45)$	$W(46)$
-1	1	1	0
2	0	0	0
3	1	0	1
5	0	0	0

which leads to the following congruence:

$$(43 \cdot 45 \cdot 46)^2 \equiv (-1)^2 \cdot 2^{10} \cdot 3^2 \cdot 5^2 = (2^5 \cdot 3 \cdot 5)^2.$$

This congruence gives the factorization of $2041 = 13 \cdot 157$, since

$$\gcd(2041,\ 43 \cdot 45 \cdot 46 + 2^5 \cdot 3 \cdot 5) = 157, \quad \gcd(2041,\ 43 \cdot 45 \cdot 46 - 2^5 \cdot 3 \cdot 5) = 13.$$

Conjecture 2.3.2 (The complexity of the QS/MPQS Method). *If N is the integer to be factored, then under certain reasonable heuristic assumptions, the QS/MPQS method will factor N in time*

$$\mathcal{O}\left(\exp\left((1+o(1))\sqrt{\log N \log\log N}\,\right)\right)$$
$$= \mathcal{O}\left(N^{(1+o(1))\sqrt{\log\log N/\log N}}\,\right). \tag{2.98}$$

(II) The Number Field Sieve (NFS). Before introducing the number field sieve, NFS, it will be useful to briefly review some important milestones in the development of integer factorization methods. In 1970, it was barely possible to factor "hard" 20-digit numbers. In 1980, by using the CFRAC method, factoring of 50-digit numbers was becoming commonplace. In 1990, the QS method had doubled the length of the numbers that could be factored by CFRAC, with a record having 116 digits. In the spring of 1996, the NFS method had successfully split a 130-digit RSA challenge number in about 15% of the time the QS would have taken. At present, the number field sieve (NFS) is the champion of all known factoring methods. NFS was first proposed by John Pollard in a letter to A. M. Odlyzko, dated 31 August 1988, with copies to R. P. Brent, J. Brillhart, H. W. Lenstra, C. P. Schnorr and H. Suyama, outlining an idea of factoring certain big numbers via *algebraic number fields*. His original idea was not for any large composite, but for certain "pretty" composites that had the property that they were close to powers. He illustrated the idea with a factorization of the seventh Fermat number $F_7 = 2^{2^7} + 1$ which was first factored by CFRAC in 1970. He also speculated in the letter that "if F_9 is still unfactored, then it might be a candidate for this kind of method eventually?" The answer now is of course "yes", since F_9 was factored by NFS in 1990. It is worthwhile pointing out that NFS is not only a method suitable for factoring numbers in a special form like F_9, but also a general purpose factoring method for any integer of a given size. There are, in fact, two forms of NFS (see Huizing [107], and Lenstra and Lenstra [141]): the *special* NFS (SNFS), tailored specifically for integers of the form $N = c_1 r^t + c_2 s^u$, and the *general* NFS (GNFS), applicable to any arbitrary number.

The fundamental idea of the NFS is the same as that of the *quadratic sieve* (QS) introduced previously: by a sieving process we look for congruences modulo N by working over a factor base, and then we do a Gaussian elimination over $\mathbb{Z}/2\mathbb{Z}$ to obtain a congruence of squares $x^2 \equiv y^2 \pmod{N}$, and hence hopefully a factorization of N. Given an odd positive integer N, NFS has four main steps in factoring N:

[1] [Polynomial Selection] Select two irreducible polynomials $f(x)$ and $g(x)$ with small integer coefficients for which there exists an integer m, such that

$$f(m) \equiv g(m) \equiv 0 \pmod{N} \tag{2.99}$$

The polynomials should not have a common factor over $\mathbb{Q}$.

[2] [Sieving] Find pairs (a, b) such that $\gcd(a, b) = 1$ and both

$$b^{\deg(f)} f(a/b), \qquad b^{\deg(g)} g(a/b) \tag{2.100}$$

are smooth with respect to a chosen factor base. The expressions in (2.100) are the norms of the algebraic numbers $a - b\alpha$ and $a - b\beta$, multiplied by the leading coefficients of f and g, respectively. (α denotes a complex root of f and β a root of g). The principal ideals $a - b\alpha$ and $a - b\beta$ factor into products of prime ideals in the number field $\mathbb{Q}(\alpha)$ and $\mathbb{Q}(\beta)$, respectively.

[3] [Linear Algebra] Use techniques of linear algebra to find a set S of indices such that the two products

$$\prod_{i \in S} (a_i - b_i\alpha), \qquad \prod_{i \in S} (a_i - b_i\beta) \tag{2.101}$$

are both squares of products of prime ideals.

[4] [Square Root] Use the set S in (2.101) to find an algebraic numbers $\alpha' \in \mathbb{Q}(\alpha)$ and $\beta' \in \mathbb{Q}(\beta)$ such that

$$(\alpha')^2 = \prod_{i \in S} (a_i - b_i\alpha), \quad (\beta')^2 = \prod_{i \in S} (a_i - b_i\beta) \tag{2.102}$$

Prior to the NFS, all modern factoring methods had an expected running time of at best

$$\mathcal{O}\left(\exp\left((c + o(1))\sqrt{\log\log N / \log N}\right)\right). \tag{2.103}$$

For example, the multiple polynomial quadratic sieve (MPQS) takes time

$$\mathcal{O}\left(\exp\left((1 + o(1))\sqrt{\log\log N / \log N}\right)\right).$$

Because of the Canfield–Erdös–Pomerance theorem [50], some people even believed that this could not be improved, except maybe for the term $(c+o(1))$, but the invention of the NFS has changed this belief.

Conjecture 2.3.3 (Complexity of NFS). Under some reasonable heuristic assumptions, the NFS method can factor an integer N in time

$$\mathcal{O}\left(\exp\left((c + o(1))\sqrt[3]{\log N}\sqrt[3]{(\log\log N)^2}\right)\right), \tag{2.104}$$

where $c = (64/9)^{1/3} \approx 1.922999427$ if GNFS is used to factor an arbitrary integer N, whereas $c = (32/9)^{1/3} \approx 1.526285657$ if SNFS is used to factor a special integer N.

Example 2.3.5. The largest number ever factored by SNFS is

$$N = (12^{167} + 1)/13 = p_{75} \times p_{105}.$$

It was announced by P. Montgomery, S. Cavallar and H. te Riele at CWI in Amsterdam on 3 September 1997. They used the polynomials $f(x) = x^5 - 144$ and $g(x) = 12^{33}x + 1$ with common root $m \equiv 12^{134} \pmod{N}$. The factor base bound was 4.8 million for f and 12 million for g. Both large prime bounds were 150 million, with two large primes allowed on each side. They sieved over $|a| \leq 8.4$ million and $0 < b \leq 2.5$ million. The sieving lasted 10.3 calendar days; 85 SGI machines at CWI contributed a combined 13027719 relations in 560 machine-days. It took 1.6 more calendar days to process the data. This processing included 16 CPU-hours on a Cray C90 at SARA in Amsterdam to process a 1969262×1986500 matrix with 57942503 nonzero entries.

2.3.6 Polland's "*rho*" and "$p - 1$" Methods

In this and the next subsections, we shall introduce some special-purpose factoring methods. By *special-purpose* we mean that the methods depend, for their success, upon some special properties of the number being factored. The usual property is that the factors of the number are small. However, other properties might include the number or its factors having a special mathematical form. For example, if p is a prime number and if $2^p - 1$ is a composite number, then all of the factors of $2^p - 1$ must be congruent to 1 modulo $2p$. For example, $2^{11} - 1 = 23 \cdot 89$ and $23 \equiv 89 \equiv 1 \pmod{22}$. Certain factoring algorithms can take advantage of this special form of the factors. Special-purpose methods do not always succeed, but it is useful to try them first, before using the more powerful, general methods, such as CFRAC, MPQS or NFS.

(I) The "*rho*" Method. In 1975 John M. Pollard[23] [187] proposed a very efficient Monte Carlo method, now widely known as Pollard's "*rho*" or ρ-method, for finding a small nontrivial factor d of a large integer N. Trial division by all integers up to $\sqrt{N}$ is guaranteed to factor completely any number up to N. For the same amount of work, Pollard's "*rho*" method will factor any number up to N^2 (unless we are unlucky). The method uses an iteration of the form

[23] John M. Pollard is an English mathematician who has made several significant contributions to the field of computational number theory and is responsible for the factoring methods "*rho*", "$p - 1$" and the number field sieve (NFS). Pollard studied mathematics at the University of Cambridge, and later obtained a doctorate there (a rather unusual kind, based not on a thesis, but on published research work) for his contribution to computational number theory.

$$\left.\begin{array}{ll} x_0 = \text{random}(0,\ N-1), & \\ x_i \equiv f(x_{i-1}) \pmod{N}, & i = 1, 2, 3, \cdots \end{array}\right\} \quad (2.105)$$

where x_0 is a random starting value, N is the number to be factored, and $f \in \mathbb{Z}[x]$ is a polynomial with integer coefficients; usually, we just simply choose $f(x) = x^2 \pm a$ with $a \neq -2, 0$. Then starting from some initial value x_0, a "random" sequence $x_1, x_2, x_3, \cdots$ is computed modulo N in the following way:

$$\left.\begin{array}{l} x_1 = f(x_0), \\ x_2 = f(f(x_0)) = f(x_1), \\ x_3 = f(f(f(x_0))) = f(f(x_1)) = f(x_2), \\ \cdots\cdots \\ \cdots\cdots \\ x_i = f(f(x_{i-1})). \end{array}\right\} \quad (2.106)$$

Let d be a nontrivial divisor of N, where d is small compared with N. Since there are relatively few congruence classes modulo d (namely, d of them), there will probably exist integers x_i and x_j which lie in the same congruence class modulo d, but belong to different classes modulo N; in short, we will have

$$\left.\begin{array}{ll} x_i \equiv x_j & \pmod{d}, \\ x_i \not\equiv x_j & \pmod{N}. \end{array}\right\} \quad (2.107)$$

Since $d \mid (x_i - x_j)$ and $N \nmid (x_i - x_j)$, it follows that $\gcd(x_i - x_j,\ N)$ is a nontrivial factor of N. In practice, a divisor d of N is not known in advance, but it can most likely be detected by keeping track of the integers x_i, which we do know; we simply compare x_i with the earlier x_j, calculating $\gcd(x_i - x_j,\ N)$ until a nontrivial gcd occurs. The divisor obtained in this way is not necessarily the smallest factor of N and indeed it may not be prime. The possibility exists that when a gcd greater that 1 is found, it may also turn out to be equal to N itself, though this happens very rarely.

Example 2.3.6. For example, let $N = 1387 = 19 \cdot 73$, $f(x) = x^2 - 1$ and $x_1 = 2$. Then the "random" sequence $x_1, x_2, x_3, \cdots$ is as follows:

$$2, 3, 8, 63, 1194, \overline{1186, 177, 814, 996, 310, 396, 84, 120, 529, 1053, 595, 339}$$

where the repeated values are overlined. Now we find that

$$\begin{array}{l} x_3 \equiv 6 \pmod{19} \\ x_3 \equiv 63 \pmod{1387} \end{array}$$

$$\begin{array}{l} x_4 \equiv 16 \pmod{19} \\ x_4 \equiv 1194 \pmod{1387} \end{array}$$

$$\begin{array}{l} x_5 \equiv 8 \pmod{19} \\ x_5 \equiv 1186 \pmod{1387} \end{array}$$

$$\cdots\cdots$$

$$\cdots\cdots$$

So we have

$$\gcd(63-6, 1387) = \gcd(1194-16, 1387) = \gcd(1186-8, 1387) = \cdots = 19.$$

Of course, as mentioned earlier, d is not known in advance, but we can keep track of the integers x_i which we do know, and simply compare x_i with all the previous x_j with $j < i$, calculating $\gcd(x_i - x_j,\ N)$ until a nontrivial gcd occurs:

$$\begin{array}{l}
\gcd(x_1 - x_0,\ N) = \gcd(3-2,\ 1387) = 1 \\
\gcd(x_2 - x_1,\ N) = \gcd(8-3,\ 1387) = 1 \\
\gcd(x_2 - x_0,\ N) = \gcd(8-2,\ 1387) = 1 \\
\gcd(x_3 - x_2,\ N) = \gcd(63-8,\ 1387) = 1 \\
\gcd(x_3 - x_1,\ N) = \gcd(63-3,\ 1387) = 1 \\
\gcd(x_3 - x_0,\ N) = \gcd(63-2,\ 1387) = 1 \\
\gcd(x_4 - x_3,\ N) = \gcd(1194-63,\ 1387) = 1 \\
\gcd(x_4 - x_2,\ N) = \gcd(1194-8,\ 1387) = 1 \\
\gcd(x_4 - x_1,\ N) = \gcd(1194-3,\ 1387) = 1 \\
\gcd(x_4 - x_0,\ N) = \gcd(1194-2,\ 1387) = 1 \\
\gcd(x_5 - x_4,\ N) = \gcd(1186-1194,\ 1387) = 1 \\
\gcd(x_5 - x_3,\ N) = \gcd(1186-63,\ 1387) = 1 \\
\gcd(x_5 - x_2,\ N) = \gcd(1186-8,\ 1387) = 19.
\end{array}$$

So, after 13 comparisons and calculations, we eventually find the divisor 19.

As k increases, the task of computing $\gcd(x_i - x_j,\ N)$ for all $j < i$ becomes very time-consuming; for $n = 10^{50}$, the computation of $\gcd(x_i - x_j,\ N)$ would require about $1.5 \cdot 10^6$ bit operations, as the complexity for computing one gcd is $\mathcal{O}((\log n)^3)$. Pollard actually used Floyd's method to detect a cycle in a long sequence $\langle x_i \rangle$, which just looks at cases in which $x_i = x_{2i}$. To see how it works, suppose that $x_i \equiv x_j \pmod{n}$, then

$$\left.\begin{array}{l}
x_{i+1} \equiv f(x_i) \equiv f(x_j) \equiv x_{j+1} \pmod{d}, \\
x_{i+2} \equiv f(x_{i+1}) \equiv f(x_{j+1}) \equiv x_{j+2} \pmod{d}, \\
\quad \cdots\cdots \\
\quad \cdots\cdots \\
x_{i+k} \equiv f(x_{i+k-1}) \equiv f(x_{j+k-1}) \equiv x_{j+k} \pmod{d}.
\end{array}\right\} \tag{2.108}$$

If the period of the cycle is k with $k = j - i$, then $x_{2i} \equiv x_i \pmod{d}$. Hence, we only need look at $x_{2i} - x_i$ (or $x_i - x_{2i}$) for $i = 1, 2, \cdots$. That is, we only need to check one gcd for each i. Note that the sequence $x_0, x_1, x_2, \cdots$ modulo a prime number p, say, looks like a circle with a tail; it is from this behaviour that the method gets its name (see Figure 2.2 for a graphical sketch; it looks like the Greek letter ρ).

Example 2.3.7. Again, let $N = 1387 = 19 \cdot 73$, $f(x) = x^2 - 1$ and $x_1 = 2$. By comparing pairs x_i and x_{2i}, for $i = 1, 2, \cdots$, we have:

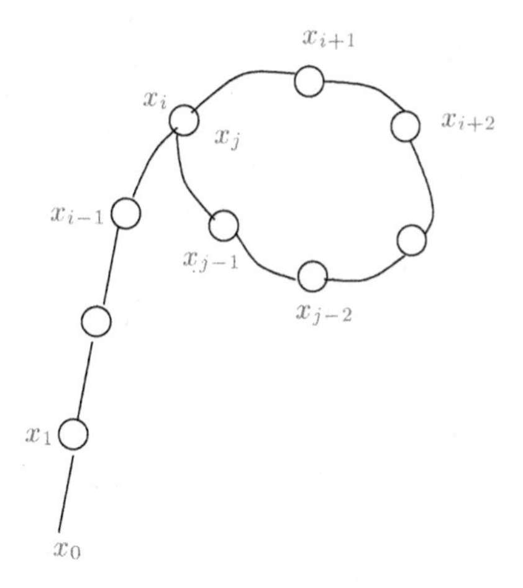

Figure 2.2. Illustration of the ρ-method

$$\begin{aligned}
\gcd(x_1 - x_2,\ N) &= \gcd(3 - 8,\ 1387) = 1 \\
\gcd(x_2 - x_4,\ N) &= \gcd(8 - 1194,\ 1387) = 1 \\
\gcd(x_3 - x_6,\ N) &= \gcd(63 - 177,\ 1387) = 19
\end{aligned}$$

So, only after 3 comparisons and gcd calculations, the divisor 19 of 1387 is found.

In what follows, we shall show that to compute $y_i = x_{2i}$, we do not need to compute $x_{i+1}, x_{i+2}, \cdots, x_{2i-1}$ until we get x_{2i}. Observe that

$$\left.\begin{array}{l}
y_1 = x_2 = f(x_1) = f(f(x_0)) = f(f(y_0)), \\
y_2 = x_4 = f(x_3) = f(f(x_2)) = f(f(y_1)), \\
y_3 = x_6 = f(x_6) = f(f(x_3)) = f(f(y_2)), \\
\qquad \cdots\cdots \\
\qquad \cdots\cdots \\
y_i = x_{2i} = f(f(y_{i-1})).
\end{array}\right\} \tag{2.109}$$

So at each step, we compute

$$\left.\begin{array}{l}
x_i = f(x_{i-1}) \pmod n, \\
y_i = f(f(y_{i-1}) \pmod n.
\end{array}\right\} \tag{2.110}$$

Therefore, only three evaluations of f will be required.

Example 2.3.8. Let once again $N = 1387 = 19 \cdot 73$, $f(x) = x^2 - 1$ and $x_0 = y_0 = 2$. By comparing pairs x_i and x_{2i}, for $i = 1, 2, \cdots$, we get:

$$\begin{aligned}
& f(y_0) = 2^2 - 1 = 3, \\
& f(f(y_0)) = 3^2 - 1 = 8 = y_1 \\
& \qquad \Longrightarrow \quad \gcd(y_1 - x_1,\ N) = \gcd(3 - 8,\ 1387) = 1
\end{aligned}$$

$$\begin{aligned}
&f(y_1) = 8^2 - 1 = 63,\\
&f(f(y_1)) = 63^2 - 1 = 1194 = y_2\\
&\qquad \Longrightarrow \gcd(y_2 - x_2,\ N) = \gcd(8 - 1194,\ 1387) = 1\\
&f(y_2) = 1194^2 - 1 \bmod 1387 = 1186,\\
&f(f(y_2)) = 1186^2 - 1 \bmod 1387 = 177 = y_3\\
&\qquad \Longrightarrow \gcd(y_3 - x_3,\ N) = \gcd(63 - 177,\ 1387) = 19.
\end{aligned}$$

The divisor 19 of 1387 is then found.

Remark 2.3.5. There is an even more efficient algorithm, due to Brent[24] [35], which looks only at the following differences and the corresponding gcd results:

$$\begin{array}{l}
x_1 - x_3 \Longrightarrow \gcd(x_1 - x_3,\ N)\\
x_3 - x_6 \Longrightarrow \gcd(x_3 - x_6,\ N)\\
x_3 - x_7 \Longrightarrow \gcd(x_3 - x_7,\ N)\\
x_7 - x_{12} \Longrightarrow \gcd(x_7 - x_{12},\ N)\\
x_7 - x_{13} \Longrightarrow \gcd(x_7 - x_{13},\ N)\\
x_7 - x_{14} \Longrightarrow \gcd(x_7 - x_{14},\ N)\\
x_7 - x_{15} \Longrightarrow \gcd(x_7 - x_{15},\ N)\\
\qquad \cdots\cdots\\
\qquad \cdots\cdots
\end{array}$$

and in general:

$$x_{2^n-1} - x_j, \quad 2^{n+1} - 2^{n-1} \le j \le 2^{n+1} - 1. \tag{2.111}$$

Brent's algorithm is about 24 percent faster than Pollard's original version.

Now we are in a position to present an algorithm for the ρ-method.

Algorithm 2.3.5 (Pollard's ρ-method). Let N be a composite integer greater than 1. This algorithm tries to find a nontrivial factor d of N, which is small compared with $\sqrt{N}$. Suppose the polynomial to use is $f(x) = x^2 + 1$.

24

Richard P. Brent was born in Melbourne, Australia in 1946. He obtained his BSc in Mathematics at Monash University, Melbourne, Australia in 1968, and his MSc and PhD at Stanford University, California in 1970 and 1971, respectively. Brent was Professor in Computer Science at the Australian National University for more than 25 years, and is currently Professor of Computing Science at Oxford University. Perhaps best known for his work in the improvement of Pollard's ρ-method and the factorization of the Fermat numbers F_8, F_{10} and F_{11}, he has also made significant contributions to the analysis of algorithms and parallel computations. It is interesting to note that he is descended from Hannah Ayscough, who was the mother of Sir Isaac Newton (1642-1727); Newton never married, and had no direct descendants. (Photo by courtesy of Prof. Brent.)

[1] [Initialization] Choose a seed, say $x_0 = 2$, a generating function, say $f(x) = x^2+1 \pmod N$. Choose also a value for t not much bigger than $\sqrt{d}$, perhaps $t < 100\sqrt{d}$.

[2] [Iteration and Computation] Compute x_i and y_i in the following way:

$$\begin{aligned}
x_1 &= f(x_0), \\
x_2 &= f(f(x_0)) = f(x_1), \\
x_3 &= f(f(f(x_0))) = f(f(x_1)) = f(x_2), \\
&\cdots\cdots \\
&\cdots\cdots \\
x_i &= f(x_{i-1}).
\end{aligned}$$

$$\begin{aligned}
y_1 &= x_2 = f(x_1) = f(f(x_0)) = f(f(y_0)), \\
y_1 &= x_4 = f(x_3) = f(f(x_2)) = f(f(y_1)), \\
y_3 &= x_6 = f(x_6) = f(f(x_3)) = f(f(y_2)), \\
&\cdots\cdots \\
&\cdots\cdots \\
y_i &= x_{2i} = f(f(y_{i-1})).
\end{aligned}$$

and simultaneously compare x_i and y_i by computing $d = \gcd(x_i - y_i,\ N)$.

[3] [Factor Found?] If $1 < d < N$, then d is a nontrivial factor of N, print d, and goto step [5].

[4] [Another Search?] If $x_i = y_i \pmod N$ for some i or $i \geq \sqrt{t}$, then goto step [2] to choose a new seed and a new generator and repeat.

[5] [Exit] Terminate the algorithm.

Now, let us move on to the complexity of the ρ-method. Let p be the smallest prime factor of N, and j the smallest positive index such that $x_{2j} \equiv x_j \pmod p$. Making some plausible assumptions, it is easy to show that the expected value of j is $\mathcal{O}(\sqrt{p})$. The argument is related to the well-known "birthday" paradox: suppose that $1 \leq k \leq n$ and that the numbers $x_1, x_2, \cdots, x_k$ are independently chosen from the set $\{1, 2, \cdots, n\}$. Then the probability that the numbers x_k are distinct is

$$\left(1 - \frac{1}{n}\right) \cdot \left(1 - \frac{2}{n}\right) \cdots \left(1 - \frac{k-1}{n}\right) \sim \exp\left(\frac{-k^2}{2n}\right). \tag{2.112}$$

Note that the x_i's are likely to be distinct if k is small compared with $\sqrt{n}$, but unlikely to be distinct if k is large compared with $\sqrt{n}$. Of course, we cannot work out x_i mod p, since we do not know p in advance, but we can detect x_j by taking greatest common divisors. We simply compute $d = \gcd(x_{2i} - x_i, N)$ for $i = 1, 2, \cdots$ and stop when a $d > 1$ is found.

Conjecture 2.3.4 (Complexity of the ρ-method). Let p be a prime dividing N and $p = \mathcal{O}(\sqrt{p})$, then the ρ-algorithm has has expected running time

$$\mathcal{O}(\sqrt{p}) = \mathcal{O}(\sqrt{p}\,(\log N)^2) = \mathcal{O}(N^{1/4}(\log N)^2) \tag{2.113}$$

to find the prime factor p of N.

Remark 2.3.6. The ρ-method is an improvement over trial division, because $\mathcal{O}(p) = \mathcal{O}(N^{1/4})$ divisions are needed for trial division to find a small factor p of N. But, of course, one disadvantage of the ρ-algorithm is that its running time is only a conjectured expected value, not a rigorous bound.

(II) The "$p-1$" Method. Pollard in 1974 invented also another simple but effective factoring algorithm, now widely known as Pollard's "$p-1$" method, which can be described as follows:

Algorithm 2.3.6. [Pollard's "$p-1$" Method] Let $N > 1$ be a composite number. This algorithm attempts to find a nontrivial factor of N.

[1] [Initialization) Pick out $a \in \mathbb{Z}/N\mathbb{Z}$ at random. Select a positive integer k that is divisible by many prime powers, for example, $k = \text{lcm}(1, 2, \cdots, B)$ for a suitable bound B (the larger B is the more likely the method will be to succeed in producing a factor, but the longer the method will take to work).

[2] [Exponentiation] Compute $a_k = a^k \bmod N$.

[3] [Compute GCD] Compute $d = \gcd(a_k - 1,\ N)$.

[4] [Factor Found?] If $1 < d < N$, then d is a nontrivial factor of N, output d and goto [6].

[5] [Start Over?] If d is not a nontrivial factor of N and if you still want to try more experiments, then goto [2] to start all over again with a new choice of a and/or a new choice of k, else goto [6].

[6] [Exit] Terminate the algorithm.

The "$p-1$" algorithm is usually successful in the fortunate case where N has a prime divisor p for which $p-1$ has no large prime factors. Suppose that $(p-1) \mid k$ and that $p \nmid a$. Since $|(\mathbb{Z}/p\mathbb{Z})^*| = p-1$, we have $a^k \equiv 1 \pmod{p}$, thus $p \mid \gcd(a_k - 1,\ N)$. In many cases, we have $p = \gcd(a_k - 1,\ N)$, so the method finds a nontrivial factor of N.

Example 2.3.9. Use the "$p-1$" method to factor the number $N = 540143$. Choose $B = 8$ and hence $k = 840$. Choose also $a = 2$. Then we have

$$\gcd(2^{840} - 1 \bmod 540143,\ 540143) = \gcd(53046,\ 540143) = 421.$$

Thus, 421 is a (prime) factor of 540143. In fact, $421 \cdot 1283$ is the complete prime factorization of 540143. It is interesting to note that, by using the "$p-1$" method, Baillie in 1980 found the prime factor

$$p_{25} = 1155685395246619182673033$$

of the Mersenne number $M_{257} = 2^{257} - 1$. In this case

$$p_{25} - 1 = 2^3 \cdot 3^2 \cdot 19^2 \cdot 47 \cdot 67 \cdot 257 \cdot 439 \cdot 119173 \cdot 1050151.$$

In the worst case, where $(p-1)/2$ is prime, the "$p-1$" algorithm is no better than trial division. Since the group has fixed order $p-1$ there is nothing to be done except try a different algorithm. Note that there is a similar method to "$p-1$", called "$p+1$", proposed by H. C. Williams in 1982. It is suitable for the case where N has a prime factor p for which $p+1$ has no large prime factors.

2.3.7 Lenstra's Elliptic Curve Method (ECM)

In Subsection 2.2.5, we discussed the application of elliptic curves to primality testing. In this subsection, we shall introduce a factoring method which uses of elliptic curves. The method is actually obtained from Pollard's "$p-1$" algorithm: if we can choose a *random* group G with order g close to p, we may be able to perform a computation similar to that involved in Pollard's "$p-1$" algorithm, working in G rather than in F_p. If all prime factors of g are less than the bound B then we find a factor of N. Otherwise, we repeat this procedure with a different group G (and hence, usually, a different g) until a factor is found. This is the motivation of the ECM method, invented by H. W. Lenstra[25] [140] in 1987.

Algorithm 2.3.7 (Lenstra's Elliptic Curve Method). Let $N > 1$ be a composite number, with $\gcd(N, 6) = 1$. This algorithm attempts to find a non-trivial factor of N. The method uses elliptic curves and is analogous to Pollard's "$p-1$" method.

[1] [Choose an Elliptic Curve] Choose a random pair (E, P), where E is an elliptic curve $y^2 = x^3+ax+b$ over $\mathbb{Z}/N\mathbb{Z}$, and $P(x, y) \in E(\mathbb{Z}/N\mathbb{Z})$ is a point on E. That is, choose $a, x, y \in \mathbb{Z}/N\mathbb{Z}$ at random, and set $b \leftarrow y^2 - x^3 - ax$. If $\gcd(4a^3 + 27b^2, N) \neq 1$, then E is not an elliptic curve, start all over and choose another pair (E, P).

[2] [Choose an Integer k] Just as in the "$p-1$" method, select a positive integer k that is divisible by many prime powers, for example, $k = \mathrm{lcm}(1, 2, \cdots, B)$ or $k = B!$ for a suitable bound B; the larger B is the more likely the method

---

25

Hendrik W. Lenstra Jr., perhaps best known for his invention of the ECM method for factoring, was born in 1949 in Zaandam, the Netherlands. He studied at the University of Amsterdam and obtained his PhD there in 1977. He was appointed Full Professor at the same university almost immediately after his PhD. In 1987 he left the Netherlands to become a Professor in the University of California at Berkeley while still keeping a strong mathematical connection with the Netherlands. Lenstra has received numerous awards, such as the Fulkerson Prize in 1985. (Photo by courtesy of Profs. Lenstra and W. A. Stein.)

will succeed in producing a factor, but the longer the method will take to work.

[3] [Calculate kP] Calculate the point $kP \in E(\mathbb{Z}/N\mathbb{Z})$. We use the following formula to compute $P_3(x_3, y_3) = P_1(x_1, y_1) + P_2(x_2, y_2) \bmod N$:

$$(x_3, y_3) = (\lambda^2 - x_1 - x_2 \mod N, \quad \lambda(x_1 - x_3) - y_1 \mod N),$$

where

$$\lambda = \begin{cases} \dfrac{m_1}{m_2} = \dfrac{3x_1^2 + a}{2y_1} \mod N & \text{if } P_1 = P_2 \\[2ex] \dfrac{m_1}{m_2} = \dfrac{y_1 - y_2}{x_1 - x_2} \mod N & \text{otherwise.} \end{cases}$$

The computation of $kP \bmod N$ can be done in $\mathcal{O}(\log k)$ doublings and additions.

[4] [Compute GCD] If $kP \equiv \mathcal{O}_E \pmod{N}$, then set $m_2 = z$ and compute $d = \gcd(z, N)$, else goto [1] to make a new choice for "a" or even for a new pair (E, P).

[5] [Factor Found?] If $1 < d < N$, then d is a nontrivial factor of N, output d and goto [7].

[6] [Start Over?] If d is not a nontrivial factor of N and if you still wish to try more elliptic curves, then goto [1] to start all over again, else goto [7].

[7] [Exit] Terminate the algorithm.

As for the "$p-1$" method, one can show that a given pair (E, P) is likely to be successful in the above algorithm if N has a prime factor p for which $\mathbb{Z}/p\mathbb{Z}$ is composed of small primes only. The probability for this to happen increases with the number of pairs (E, P) that one tries.

Example 2.3.10. Use the ECM method to factor the number $N = 187$.

[1] Choose $B = 3$, and hence $k = \operatorname{lcm}(1, 2, 3) = 6$. Let $P = (0, 5)$ be a point on the elliptic curve $E: \ y^2 = x^3 + x + 25$ which satisfies $\gcd(N, 4a^3 + 27b^2) = \gcd(187, 16879) = 1$ (note that here $a = 1$ and $b = 25$).

[2] Since $k = 6 = 110_2$, we compute $6P = 2(P + 2P)$ in the following way:

[2-1] Compute $2P = P + P = (0, 5) + (0, 5)$:

$$\begin{cases} \lambda = \dfrac{m_1}{m_2} = \dfrac{1}{10} \equiv 131 \pmod{187} \\ x_3 = 144 \pmod{187} \\ y_3 = 18 \pmod{187}. \end{cases}$$

So, $2P = (144, 18)$ with $m_2 = 10$ and $\lambda = 131$.

[2-2] Compute $3P = P + 2P = (0, 5) + (144, 18)$:

$$\begin{cases} \lambda = \dfrac{m_1}{m_2} = \dfrac{13}{144} \equiv 178 \pmod{187} \\ x_3 = 124 \pmod{187} \\ y_3 = 176 \pmod{187}. \end{cases}$$

So, $3P = (124, 176)$ with $m_2 = 144$ and $\lambda = 178$.

[2-3] Compute $6P = 2(3P) = 3P + 3P = (124, 176) + (124, 176)$:

$$\lambda = \frac{m_1}{m_2} = \frac{46129}{352} \equiv \frac{127}{165} \equiv \mathcal{O}_E \pmod{187}.$$

This time $m_1 = 127$ and $m_2 = 165$, so the modular inverse for $127/165$ modulo 187 does not exist; but this is exactly what we want! – this type of failure is called a "pretended failure". We now set $z = m_2 = 165$.

[3] Compute $d = \gcd(N, z) = \gcd(187, 165) = 11$. Since $1 < 11 < 187$, 11 is a (prime) factor of 187. In fact, $187 = 11 \cdot 17$.

In 1995 Richard Brent at the Australian National University completed the factorization of the tenth Fermat number using ECM:

$$2^{2^{10}} + 1 = 2^{1024} + 1 = 45592577 \cdot 6487031809 \cdot p_{40} \cdot p_{252}$$

where the 40-digit prime p_{40} was found using ECM, and p_{252} was proved to be a 252-digit prime. Other recent ECM-records include a 38-digit prime factor in the 112-digit composite $(11^{118} + 1)/(2 \cdot 61 \cdot 193121673)$, a 40-digit prime factor of $26^{126} + 1$, a 43-digit prime factor of the partition number $p(19997)$ and a 44-digit prime factor of the partition number $p(19069)$ in the RSA Factoring Challenge List, and a 47-digit prime in c_{135} of $5^{2^8} + 1 = 2 \cdot 1655809 \cdot p_{38} \cdot c_{135}$.

Conjecture 2.3.5 (Complexity of the ECM method). Let p be the smallest prime dividing N. Then the ECM method will find p of N, under some plausible assumptions, in expected running time

$$\mathcal{O}\left(\exp\left(\sqrt{(2 + o(1)) \log p \log\log p}\right) \cdot (\log N)^2\right) \tag{2.114}$$

In the worst case, when N is the product of two prime factors of the same order of magnitude, we have

$$\begin{aligned} &\mathcal{O}\left(\exp\left(\sqrt{(2 + o(1)) \log N \log\log N}\right)\right) \\ &= \mathcal{O}\left(N^{\sqrt{(2+o(1)) \log\log N / \log N}}\right). \end{aligned} \tag{2.115}$$

Remark 2.3.7. What is especially interesting about the ECM is that its running time depends very much on p (the factor found) of N, rather than N itself. So one advantage of the ECM is that one may use it, in a manner similar to trial division, to locate the smaller prime factors p of a number N which is much too large to factor completely.

2.4 Algorithms for Discrete Logarithms

The discrete logarithm problem has many different facets to it, and there are areas in which much remains to be discovered.

KEVIN S. MCCURLEY
Odds and Ends from Cryptology and Computational Number Theory [150]

The *discrete logarithm problem* (DLP) can be described as follows:

$$\left.\begin{array}{ll}\text{Input :} & a, b, n \in \mathbb{N} \\ \text{Output :} & x \in \mathbb{N} \text{ with } a^x \equiv b \pmod{n} \\ & \text{if such an } x \text{ exists,}\end{array}\right\} \qquad (2.116)$$

where the modulus n can either be a composite or a prime.

According to Adleman [1], the Russian mathematician Bouniakowsky developed a clever algorithm to solve the congruence $a^x \equiv b \pmod{n}$, with asymptotic complexity $\mathcal{O}(n)$ in 1870. Despite its long history, no efficient algorithm has ever emerged for the discrete logarithm problem. It is believed to be extremely hard, and harder than the integer factorization problem (IFP) even in the average case. The best known algorithm for DLP at present, using NFS and due to Gordon [90], requires an expected running time

$$\mathcal{O}\left(\exp\left(c(\log n)^{1/3}(\log\log n)^{2/3}\right)\right).$$

There are essentially three different categories of algorithms in use for computing discrete logarithms:

(1) Algorithms that work for arbitrary groups, that is, those that do not exploit any specific properties of groups; Shanks' baby-step giant-step method, Pollard's ρ-method (an analogue of Pollard's ρ-factoring method) and the λ-method (also known as wild and tame Kangaroos) are in this category.

(2) Algorithms that work well in finite groups for which the order of the groups has no large prime factors; more specifically, algorithms that work for groups with smooth orders. (A positive integer is called *smooth* if it has no large prime factors; it is called y-smooth if it has no large prime factors exceeding y.) The well-known Silver–Pohlig–Hellman algorithm based on the Chinese Remainder Theorem is in this category.

(3) Algorithms that exploit methods for representing group elements as products of elements from a relatively small set, called the factor base (it may also make use of the Chinese Remainder Theorem); the typical algorithms in this category are various forms of index calculus, including Number Field Sieve.

In the subsections that follow, we shall introduce the basic ideas of each of these three categories; more specifically, we shall introduce Shanks' baby-step giant-step algorithm, the Silver–Pohlig–Hellman algorithm, and the index calculus.

2.4.1 Shanks' Baby-Step Giant-Step Algorithm

Let G be a finite cyclic group of order n, a a generator of G and $b \in G$. The *obvious* algorithm for computing successive powers of a until b is found takes $\mathcal{O}(n)$ group operations. For example, to compute $x = \log_2 15 \pmod{19}$, we compute $2^x \bmod 19$ for $x = 0, 1, 2, \cdots, 19-1$ until $2^x \bmod 19 = 15$ for some x is found, that is:

x	0	1	2	3	4	5	6	7	8	9	10	11
a^x	1	2	4	8	16	13	7	14	9	18	17	15

So $\log_2 15 \pmod{19} = 11$. It is clear that when n is large, the algorithm is inefficient. In this section, we introduce a type of square root algorithm, called the baby-step giant-step algorithm, for taking discrete logarithms, which is better than the above mentioned *obvious* algorithm. The algorithm works on arbitrary groups, and according to Odlyzko [178], its original idea is due to Shanks[26].

26

Daniel Shanks (1917–1996), an American physicist and mathematician, responsible for the SQUFOF method for integer factorization and the baby-step giant-step method for taking discrete logarithms. He served as an editor of *Mathematics of Computation* from 1959 until his death. His book *Solved and Unsolved Problems in Number Theory* is one of the very popular books in number theory. (Photo by courtesy of the American Mathematical Society.)

Let $m = \lfloor \sqrt{n} \rfloor$. The baby-step giant-step algorithm is based on the observation that if $x = \log_a b$, then we can uniquely write $x = i + jm$, where $0 \le i, j < m$. For example, if $11 = \log_2 15 \bmod 19$, then $a = 2$, $b = 15$, $m = 5$, so we can write $11 = i + 5j$ for $0 \le i, j < m$. Clearly, here $i = 1$ and $j = 2$ so we have $11 = 1 + 5 \cdot 2$. Similarly, for $14 = \log_2 6 \bmod 19$ we can write $14 = 4 + 5 \cdot 2$, for $17 = \log_2 10 \bmod 19$ we can write $17 = 2 + 5 \cdot 3$, etc. The following is a description of the algorithm:

Algorithm 2.4.1 (Shanks' baby-step giant-step algorithm). This algorithm computes the discrete logarithm x of y to the base a, modulo n, such that $y = a^x \pmod{n}$:

[1] [Initialization] Computes $s = \lfloor \sqrt{n} \rfloor$.

[2] [Computing the baby step] Compute the first sequence (list), denoted by S, of pairs (ya^r, r), $r = 0, 1, 2, 3, \cdots, s-1$:

$$S = \{(y, 0), (ya, 1), (ya^2, 2), (ya^3, 3), \cdots, (ya^{s-1}, s-1) \bmod n\} \quad (2.117)$$

and sort S by ya^r, the first element of the pairs in S.

[3] [Computing the giant step] Compute the second sequence (list), denoted by T, of pairs (a^{ts}, ts), $t = 1, 2, 3, \cdots, s$:

$$T = \{(a^s, 1), (a^{2s}, 2), (a^{3s}, 3), \cdots, (a^{s^2}, s) \bmod n\} \quad (2.118)$$

and sort T by a^{ts}, the first element of the pairs in T.

[4] [Searching, comparing and computing] Search both lists S and T for a match $ya^r = a^{ts}$ with ya^r in S and a^{ts} in T, then compute $x = ts - r$. This x is the required value of $\log_a y \pmod{n}$.

This algorithm requires a table with $\mathcal{O}(m)$ entries ($m = \lfloor \sqrt{n} \rfloor$, where n is the modulus). Using a sorting algorithm, we can sort both the lists S and T in $\mathcal{O}(m \log m)$ operations. Thus this gives an algorithm for computing discrete logarithms that uses $\mathcal{O}(\sqrt{n} \log n)$ time and space for $\mathcal{O}(\sqrt{n})$ group elements. Note that Shanks' idea is originally for computing the order of a group element g in the group G, but here we use his idea to compute discrete logarithms. Note also that although this algorithm works on arbitrary groups, if the order of a group is larger than 10^{40}, it will be infeasible.

Example 2.4.1. Suppose we wish to compute the discrete logarithm $x = \log_2 6 \bmod 19$ such that $6 = 2^x \bmod 19$. According to Algorithm 2.4.1, we perform the following computations:

[1] $y = 6$, $a = 2$ and $n = 19$, $s = \lfloor \sqrt{19} \rfloor = 4$.

[2] Computing the baby step:

$$\begin{aligned} S &= \{(y,0), (ya,1), (ya^2,2), (ya^3,3) \bmod 19\} \\ &= \{(6,0), (6 \cdot 2, 1), (6 \cdot 2^2, 2), (6 \cdot 2^3, 3) \bmod 19\} \\ &= \{(6,0), (12,1), (5,2), (10,3)\} \\ &= \{(5,2), (6,0), (10,3), (12,1)\}. \end{aligned}$$

[3] Computing the giant step:

$$\begin{aligned} T &= \{(a^s, s), (a^{2s}, 2s), (a^{3s}, 3s), (a^{4s}, 4s) \bmod 19\} \\ &= \{(2^4, 4), (2^8, 8), (2^{12}, 12), (2^{16}, 16) \bmod 19\} \\ &= \{(16, 4), (9, 8), (11, 12), (5, 16)\} \\ &= \{(5, 16), (9, 8), (11, 12), (16, 4)\}. \end{aligned}$$

[4] Matching and computing: The number 5 is the common value of the first element in pairs of both lists S and T with $r = 2$ and $st = 16$, so $x = st - r = 16 - 2 = 14$. That is, $\log_2 6 \pmod{19} = 14$, or equivalently, $2^{14} \pmod{19} = 6$.

Example 2.4.2. Suppose now we wish to find the discrete logarithm $x \equiv \log_{59} 67 \pmod{113}$, such that $67 \equiv 59^x \pmod{113}$. Again by Algorithm 2.4.1, we have:

[1] $y = 67$, $a = 59$ and $n = 113$, $s = \lfloor \sqrt{113} \rfloor = 10$.

[2] Computing the baby step:

$$\begin{aligned} S &= \{(y, 0), (ya, 1), (ya^2, 2), (ya^3, 3), \cdots, (ya^9, 9) \bmod 113\} \\ &= \{(67, 0), (67 \cdot 59, 1), (67 \cdot 59^2, 2), (67 \cdot 59^3, 3), (67 \cdot 59^4, 4), \\ &\qquad (67 \cdot 59^5, 5), (67 \cdot 59^6, 6), (67 \cdot 59^7, 7), (67 \cdot 59^8, 8), \\ &\qquad (67 \cdot 59^9, 9) \bmod 113\} \\ &= \{(67, 0), (111, 1), (108, 2), (44, 3), (110, 4), (49, 5), (66, 6), \\ &\qquad (52, 7), (17, 8), (99, 9)\} \\ &= \{(17, 8), (44, 3), (49, 5), (52, 7), (66, 6), (67, 0), (99, 9), \\ &\qquad (108, 2), (110, 4), (111, 1)\}. \end{aligned}$$

[3] Computing the giant-step:

$$\begin{aligned} T &= \{(a^s, s), (a^{2s}, ss), (a^{3s}, 3s), \cdots (a^{10s}, 10s) \bmod 113\} \\ &= \{(59^{10}, 10), (59^{2\cdot 10}, 2 \cdot 10), (59^{3\cdot 10}, 3 \cdot 10), (59^{4\cdot 10}, 4 \cdot 10), \\ &\qquad (59^{5\cdot 10}, 5 \cdot 10), (59^{6\cdot 10}, 6 \cdot 10), (59^{7\cdot 10}, 7 \cdot 10), (59^{8\cdot 10}, 8 \cdot 10), \\ &\qquad (59^{9\cdot 10}, 9 \cdot 10) \bmod 113\} \\ &= \{(72, 10), (99, 20), (9, 30), (83, 40), (100, 50), (81, 60), \\ &\qquad (69, 70), (109, 80), (51, 90), (56, 100)\} \\ &= \{(9, 30), (51, 90), (56, 100), (69, 70), (72, 10), (81, 60), (83, 40), \\ &\qquad (99, 20), (100, 50), (109, 80)\}. \end{aligned}$$

[4] Matching and computing: The number 99 is the common value of the first element in pairs of both lists S and T with $r = 9$ and $st = 20$, so $x = st - r = 20 - 9 = 11$. That is, $\log_{59} 67 \pmod{113} = 11$, or equivalently, $59^{11} \pmod{113} = 67$.

Exercise 2.4.1. Use the baby-step giant-step algorithm to find the following discrete logarithms x:

(1) $x \equiv \log_3 5 \pmod{29}$.

(2) $x \equiv \log_5 96 \pmod{317}$.

(3) $x \equiv \log_{37} 15 \pmod{123}$.

Shanks' baby-step giant-step algorithm is a type of *square root method* for computing discrete logarithms. In 1978 Pollard [188] also gave two other types of square root methods, namely the ρ-method and the λ-method for taking discrete logarithms. Pollard's methods are probabilistic but remove the necessity of precomputing the lists S and T, as with Shanks' baby-step giant-step method. Again, Pollard's algorithm requires $\mathcal{O}(n)$ group operations and hence is infeasible if the order of the group G is larger than 10^{40}.

2.4.2 Silver–Pohlig–Hellman Algorithm

In 1978, Pohlig and Hellman [186] proposed an important special algorithm, now widely known as the Silver–Pohlig–Hellman algorithm for computing discrete logarithms over $\mathrm{GF}(q)$ with $\mathcal{O}(\sqrt{p})$ operations and a comparable amount of storage, where p is the largest prime factor of $q-1$. Pohlig and Hellman showed that if

$$q - 1 = \prod_{i=1}^{k} p_i^{\alpha_i}, \tag{2.119}$$

where the p_i are distinct primes and the α_i are natural numbers, and if $r_1, \cdots, r_k$ are any real numbers with $0 \le r_i \le 1$, then logarithms over $\mathrm{GF}(q)$ can be computed in

$$\mathcal{O}\left(\sum_{i=1}^{k}\left(\log q + p_i^{1-r_i}\left(1 + \log p_i^{r_i}\right)\right)\right) \tag{2.120}$$

field operations, using

$$\mathcal{O}\left(\log q \sum_{i=1}^{k}\left(1 + p_i^{r_i}\right)\right) \tag{2.121}$$

bits of memory, provided that a precomputation requiring

$$\mathcal{O}\left(\sum_{i=1}^{k} p_i^{r_i} \log p_i^{r_i} + \log q\right) \tag{2.122}$$

field operations is performed first. This algorithm is very efficient if q is "smooth", i.e., all the prime factors of $q-1$ are small. Here is a brief description of the algorithm:

Algorithm 2.4.2 (Silver–Pohlig–Hellman Algorithm). This algorithm computes the discrete logarithm $x = \log_a b \bmod q$:

[1] [Factor $q-1$ into its prime factorization form]

$$q - 1 = \prod_{i=1}^{k} p_1^{\alpha_1} p_2^{\alpha_2} \cdots p_k^{\alpha_k}.$$

[2] [Precompute the table $r_{p_i,j}$ for a given field]

$$r_{p_i,j} = a^{j(q-1)/p_i} \bmod q, \quad 0 \le j < p_i. \tag{2.123}$$

This only needs to be done once for any given field.

[3] [Compute the discrete logarithm of b to the base a modulo q, i.e., compute $x = \log_a b \bmod q$]

[3-1] Use an idea similar to that in the baby-step giant-step algorithm to find the individual discrete logarithms $x \bmod p_i^{\alpha_i}$: To compute $x \bmod p_i^{\alpha_i}$, we consider the representation of this number to the base p_i:

$$x \bmod p_i^{\alpha_i} = x_0 + x_1 p_i + \cdots + x_{\alpha_i - 1} p_i^{\alpha_i - 1}, \tag{2.124}$$

where $0 \le x_n < p_i - 1$.

[i] To find x_0, we compute $b^{(q-1)/p_i}$ which equals $r_{p_i,j}$ for some j, and set $x_0 = j$ for which

$$b^{(q-1)/p_i} \bmod q = r_{p_i,j}.$$

This is possible because

$$b^{(q-1)/p_i} \equiv a^{x(q-1)/p_i} \equiv a^{x_0(q-1)/p_i} \bmod q = r_{p_i,x_0}$$

[ii] To find x_1, compute $b_1 = ba^{-x_0}$. If

$$b_1^{(q-1)/p_i^2} \bmod q = r_{p_i,j}$$

then set $x_1 = j$. This is possible because

$$\begin{aligned} b_1^{(q-1)/p_i^2} &\equiv a^{(x-x_0)(q-1)/p_i^2} \equiv a^{(x_1+x_2p_i+\cdots)(q-1)/p_i} \\ &\equiv a^{x_1(q-1)/p_i} \bmod q = r_{p_i,x_1}. \end{aligned}$$

[iii] To obtain x_2, consider the number $b_2 = ba^{-x_0-x_1p_i}$ and compute

$$b_2^{(q-1)/p_i^3} \bmod q.$$

The procedure is carried on inductively to find all $x_0, x_1, \cdots, x_{\alpha_i-1}$.

[3-2] Use the Chinese Remainder Theorem to find the unique value of x from the congruences $x \bmod p_i^{\alpha_i}$.

We now give an example of how the above algorithm works:

Example 2.4.3. Suppose we wish to compute the discrete logarithm $x \equiv \log_2 62 \pmod{181}$. Now we have $a = 2$, $b = 62$ and $q = 181$ (2 is a generator of $\mathbb{F}_{181}^*$). We follow the computation steps described in the above algorithm:

[1] Factor $q - 1$ into its prime factorization form:

$$180 = 2^2 \cdot 3^2 \cdot 5.$$

[2] Use the following formula to precompute the table $r_{p_i,j}$ for the given field $\mathbb{F}_{181}^*$:

$$r_{p_i,j} = a^{j(q-1)/p_i} \bmod q, \quad 0 \le j < p_i.$$

This only needs to be done once for this field.

[2-1] Compute $r_{p_1,j} = a^{j(q-1)/p_1} \bmod q = 2^{90j} \bmod 181$ for $0 \le j < p_1 = 2$:

$$r_{2,0} = 2^{90 \cdot 0} \bmod 181 = 1,$$

$$r_{2,1} = 2^{90 \cdot 1} \bmod 181 = 180.$$

[2-2] Compute $r_{p_2,j} = a^{j(q-1)/p_2} \bmod q = 2^{60j} \bmod 181$ for $0 \le j < p_2 = 3$:

$$r_{3,0} = 2^{60 \cdot 0} \bmod 181 = 1,$$

$$r_{3,1} = 2^{60 \cdot 1} \bmod 181 = 48,$$

$$r_{3,2} = 2^{60 \cdot 2} \bmod 181 = 132.$$

[2-3] Compute $r_{p_3,j} = a^{j(q-1)/p_3} \bmod q = 2^{36j} \bmod 181$ for $0 \le j < p_3 = 5$:

$$r_{5,0} = 2^{36 \cdot 0} \bmod 181 = 1,$$

$$r_{5,1} = 2^{36 \cdot 1} \bmod 181 = 59,$$

$$r_{5,2} = 2^{36 \cdot 2} \bmod 181 = 42,$$

$$r_{5,3} = 2^{36 \cdot 3} \bmod 181 = 125,$$

$$r_{5,4} = 2^{36 \cdot 4} \bmod 181 = 135.$$

Construct the $r_{p_i,j}$ table as follows:

p_i	j				
	0	1	2	3	4
2	1	180			
3	1	48	132		
5	1	59	42	125	135

This table is manageable if all p_i are small.

[3] Compute the discrete logarithm of 62 to the base 2 modulo 181, that is, compute $x = \log_2 62 \bmod 181$. Here $a = 2$ and $b = 62$:

[3-1] Find the individual discrete logarithms $x \bmod p_i^{\alpha_i}$ using

$$x \bmod p_i^{\alpha_i} = x_0 + x_1 p_i + \cdots + x_{\alpha_i - 1} p_i^{\alpha_i - 1}, \quad 0 \le x_n < p_i - 1.$$

[i] Find the discrete logarithms $x \bmod p_1^{\alpha_1}$, i.e., $x \bmod 2^2$:

$$x \bmod 181 \Longleftrightarrow x \bmod 2^2 = x_0 + 2x_1.$$

[a] To find x_0, we compute

$$b^{(q-1)/p_1} \bmod q = 62^{180/2} \bmod 181 = 1 = r_{p_1,j} = r_{2,0}$$

hence $x_0 = 0$.
[b] To find x_1, compute first $b_1 = ba^{-x_0} = b = 62$, then compute

$$b_1^{(q-1)/p_1^2} \bmod q = 62^{180/4} \bmod 181 = 1 = r_{p_1,j} = r_{2,0}$$

hence $x_1 = 0$. So,

$$x \bmod 2^2 = x_0 + 2x_1 \Longrightarrow x \bmod 4 = 0.$$

[ii] Find the discrete logarithms $x \bmod p_2^{\alpha_2}$, that is, $x \bmod 3^2$:

$$x \bmod 181 \Longleftrightarrow x \bmod 3^2 = x_0 + 2x_1.$$

[a] To find x_0, we compute

$$b^{(q-1)/p_2} \bmod q = 62^{180/3} \bmod 181 = 48 = r_{p_2,j} = r_{3,1}$$

hence $x_0 = 1$.
[b] To find x_1, compute first $b_1 = ba^{-x_0} = 62 \cdot 2^{-1} = 31$, then compute

$$b_1^{(q-1)/p_2^2} \bmod q = 31^{180/3^2} \bmod 181 = 1 = r_{p_2,j} = r_{3,0}$$

hence $x_1 = 0$. So,

$$x \bmod 3^2 = x_0 + 2x_1 \Longrightarrow x \bmod 9 = 1.$$

[iii] Find the discrete logarithms $x \bmod p_3^{\alpha_3}$, that is, $x \bmod 5^1$:

$$x \bmod 181 \Longleftrightarrow x \bmod 5^1 = x_0.$$

To find x_0, we compute

$$b^{(q-1)/p_3} \bmod q = 62^{180/5} \bmod 181 = 1 = r_{p_3,j} = r_{5,0}$$

hence $x_0 = 0$. So we conclude that

$$x \bmod 5 = x_0 \Longrightarrow x \bmod 5 = 0.$$

[3-2] Find the x in

$$x \bmod 181$$

such that

$$\begin{cases} x \bmod 4 = 0, \\ x \bmod 9 = 1, \\ x \bmod 5 = 0. \end{cases}$$

To do this, we just use the Chinese Remainder Theorem to solve the following system of congruences:

$$\begin{cases} x \equiv 0 \pmod{4} \\ x \equiv 1 \pmod{9} \\ x \equiv 0 \pmod{5}. \end{cases}$$

The unique value of x for this system of congruences is $x = 100$. (This can be easily found by using, for example, the Maple function `chrem([0, 1, 0], [4,9, 5])`.) Thus, the value of x in the congruence $x \bmod 181$ is 100. Hence $x = \log_2 62 = 100$.

2.4.3 Index Calculus for Discrete Logarithms

In this subsection, we shall introduce several versions of index calculus for discrete logarithms for the multiplicative group $\mathbb{F}_p^*$ over the finite field $\mathbb{F}_p$, which are probabilistic, but have subexponential-time complexity.

In 1979, Adleman [1] proposed a general purpose, subexponential-time algorithm for taking discrete logarithms, called the *index calculus* for the multiplicative group $\mathbb{F}_p^*$, with the following expected running time:

$$\mathcal{O}\left(\exp\left(c\sqrt{\log p \log\log p}\right)\right).$$

His algorithm can be briefly described as follows:

Algorithm 2.4.3 (Adleman's index calculus). This algorithm tries to compute the discrete logarithm $x = \log_a b \bmod p$ with input a, b, p, where a and b are generators and p a prime:

[1] [Factoring] Factor $p-1$ into its prime factorization form:

$$p - 1 = p_1^{\alpha_1} p_2^{\alpha_2} \cdots p_k^{\alpha_k}$$

[2] [Computing] For each $p_k^{\alpha_k} \mid n$, carry out the following steps until m_l is obtained:

[2-1] [Guessing and checking] Find r_i, s_i such that $a^{r_i} \bmod p$ and $ba^{s_i} \bmod q$ are smooth with respect to the bound $2^{(\log p \log\log p)^{1/2}}$.

[2-2] [Gaussian elimination] Check if over the finite field $\mathbb{Z}_{p_l^{\alpha_l}}$, $ba^{s_i} \bmod q$ is dependent on

$$\{a^{r_1} \bmod p, \cdots, a^{r_i} \bmod p\}.$$

If yes, calculate β_j's such that

$$ba^{s_i} \bmod p \equiv \left(\sum_{j=1}^{i} \beta_j a^{r_i} \bmod p\right) \bmod p_l^{\alpha_l}$$

then

$$m_l = \left(\sum_{j=1}^{i} \beta_j r_j\right) \bmod p_l^{\alpha_l} - s_i$$

[3] [Chinese remaindering] Calculate and output x such that

$$x \equiv m_l \pmod{p_l^{\alpha_l}}, \quad l = 1, 2, \cdots k.$$

Note that the above algorithm can also be easily generalized to the case where p is not a prime, or a or b are not generators (Adleman [1]). The most widely used index calculus algorithm for discrete logarithms for $\mathbb{F}_p^*$ is, however, in the following form:

Algorithm 2.4.4 (Index calculus for discrete logarithms in $\mathbb{F}_p$). This algorithm tries to find an integer k such that

$$\alpha = \beta^k \quad \text{in } \mathbb{F}_p^*.$$

[1] [Factor base] Choose a set of multiplicatively independent integers (usually the first r primes):

$$\Gamma = \{\pi_1, \pi_2, \cdots, \pi_r\}. \tag{2.125}$$

[2] [Compute and factor $\beta^e \bmod p$] Randomly choose exponent $e < p$, compute $\beta^e \bmod p$, and attempt to factor it as a product:

$$\beta^e \bmod p = \pi_1^{m_1(e)} \pi_2^{m_2(e)} \cdots \pi_r^{m_r(e)}. \tag{2.126}$$

[3] Repeat Step [2] to find r independent factorization (2.126), then solve the system of congruences

$$e_i \equiv m_1(e_i) \log_\beta \pi_1 + \cdots + m_r(e_i) \log_\beta \pi_r \pmod{p-1} \tag{2.127}$$

for the quantities $\log_\beta \pi_j$.

[4] Randomly choose exponents $e < p$, compute $\alpha\beta^e \bmod p$, and attempt to factor it as a product:

$$\alpha\beta^e \bmod p = \pi_1^{k_1}\pi_2^{k_2}\cdots\pi_r^{k_r}. \tag{2.128}$$

When this is successful, the relation

$$\log_\beta \alpha + e \equiv k_1 \log_\beta \pi_1 + \cdots + k_r \log_\beta \pi_r \pmod{p-1} \tag{2.129}$$

gives the value of $k = \log_\beta \alpha$. If unsuccessful, choose a different e and go to step [2] and repeat.

Example 2.4.4. Use the index calculus described in Algorithm 2.4.4 to find the discrete logarithms $\log_{11} 7$ in $\mathbb{F}_{29}$. Here, $\alpha = 7, \beta = 11$ and $p = 29$, we wish to find $k = \log_\beta \alpha$ in $\mathbb{F}_{29}$. We follow exactly the computational procedures in Algorithm 2.4.4.

[1] [Factor base] Choose the factor base as follows:

$$\Gamma = \{2, 3, 5\}.$$

[2] [Compute and factor $\beta^e \bmod p$] Randomly choose exponent $e < p$, compute $\beta^e \bmod p$, and attempt to factor it as a product. This step needs to be repeated for several times in order to find r independent factorizations:

$$\begin{array}{lcl}
(1)\ \beta^2 = 11^2 & \equiv & 5 \pmod{29} \\
(2)\ \beta^3 = 11^3 & \equiv & 2 \cdot 23 \pmod{29} \ \ \text{(failure)} \\
(3)\ \beta^5 = 11^5 & \equiv & 2 \cdot 7 \pmod{29} \ \ \text{(failure)} \\
(4)\ \beta^6 = 11^6 & \equiv & 3^2 \pmod{29} \\
(5)\ \beta^7 = 11^7 & \equiv & 2^2 \cdot 3 \pmod{29} \\
(6)\ \beta^9 = 11^9 & \equiv & 2 \pmod{29}
\end{array}$$

[3] Solve the system of congruences for the quantities $\log_\beta \pi_j$:

(1) $\log_{11} 5 \equiv 2 \pmod{29} \Longrightarrow \log_{11} 5 \equiv 2 \pmod{28}$

(4) $2 \cdot \log_{11} 3 \equiv 6 \pmod{29} \Longrightarrow 2 \cdot \log_{11} 3 \equiv 6 \pmod{28}$

which does not uniquely determine $\log_{11} 3$ since $\gcd(2, 28) \neq 1$

(6) $\log_{11} 2 \equiv 9 \pmod{29} \Longrightarrow \log_{11} 2 \equiv 9 \pmod{28}$

(5) $2 \cdot \log_{11} 2 + \log_{11} 3 \equiv 7 \pmod{29} \Longrightarrow 2 \cdot \log_{11} 2 + \log_{11} 3 \equiv 7 \pmod{28}$

$\Longrightarrow \log_{11} 3 \equiv 17 \pmod{28}$

[4] Randomly choose exponent $e < p$, compute $\alpha\beta^e \bmod p$, and attempt to factor it as a product: $\alpha\beta^e \bmod p = \pi_1^{k_1}\pi_2^{k_2}\cdots\pi_r^{k_r}$.

$$\begin{aligned} \alpha\beta = 7 \cdot 11 &\equiv 19 \pmod{29} \ \text{(failure)} \\ \alpha\beta^2 = 7 \cdot 11^2 &\equiv 2 \cdot 3 \pmod{29} \end{aligned}$$

Thus, by the relation

$$\log_\beta \alpha + e \equiv k_1 \log_\beta \pi_1 + \cdots + k_r \log_\beta \pi_r \pmod{p-1}$$

we have

$$\log_{11} 7 \equiv \log_{11} 2 + \log_{11} 3 - 2 \equiv 9 + 17 - 2 \equiv 24 \pmod{28}.$$

The correctness of the above computation is, of course, ready to verify, since $11^{24} \equiv 7 \pmod{29}$.

Exercise 2.4.2. Use the index calculus described in Algorithm 2.4.4, to find the discrete logarithms $\log_{11} 15$ and $\log_{11} 27$ in $\mathbb{F}_{41}$.

Note that Gordon [90] in 1993 proposed an algorithm for computing discrete logarithms in $\mathbb{F}_p$. Gordon's algorithm is based on the Number Field Sieve (NFS) for integer factorization, with the heuristic expected running time

$$\mathcal{O}\left(\exp\left(c(\log p)^{1/3}(\log\log p)^{2/3}\right)\right),$$

the same as that used in factoring. The algorithm can be briefly described as follows:

Algorithm 2.4.5 (Gordon's NFS). This algorithm computes the discrete logarithm x such that $a^x \equiv b \pmod{p}$ with input a, b, p, where a and b are generators and p is prime:

[1] [Precomputation]: Find the discrete logarithms of a factor base of small rational primes, which must only be done once for a given p.

[2] [Compute individual logarithms]: Find the logarithm for each $b \in \mathbb{F}_p$ by finding the logarithms of a number of "medium-sized" primes.

[3] [Compute the final logarithm]: Combine all the individual logarithms (by using the Chinese Remainder Theorem) to find the logarithm of b.

Interested readers are referred to Gordon's paper [90] for more detailed information. Note also that Gordon, with co-author McCurley [89], discussed some implementation issues of massively parallel computations of discrete logarithms in $\mathbb{F}_q$, with $q = 2^n$.

2.4.4 Algorithms for Elliptic Curve Discrete Logarithms

Let $\mathbb{F}_p$ be a finite field with p elements (p prime), E an elliptic curve over $\mathbb{F}_p$, say, given by a Weierstrass equation

$$E: \ y^2 = x^3 + ax + b, \tag{2.130}$$

S and T the two pints in $E(\mathbb{F}_p)$. Then the elliptic curve discrete logarithm problem (ECDLP) is to find the integer k

$$k = \log_T S, \tag{2.131}$$

such that

$$S = kT. \tag{2.132}$$

In this subsection, we shall extend the (*index calculus* for the discrete logarithm problem (DLP) of multiplicative group $\mathbb{F}_p^*$ over finite field $\mathbb{F}_p$ to the ECDLP, more specifically and importantly, we shall study a new algorithm, called *xedni calculus* for the ECDLP.

To apply the index calculus for the DLP to the ECDLP, one would first lift the elliptic curve $E/\mathbb{F}_p$ to an elliptic curve $\mathcal{E}/\mathbb{Q}$, next attempt to lift various points from $E/\mathbb{F}_p$ to $\mathcal{E}/\mathbb{Q}$, and finally use relationships among these lifts to solve the ECDLP. The following is the algorithm:

Algorithm 2.4.6 (Index calculus for the ECDLP). This algorithm will try to find an integer k

$$k = \log_T S$$

such that

$$S = kT$$

where S and T are two pints on an elliptic curve

$$E: \ y^2 = x^3 + ax + b$$

over a finite field $\mathbb{F}_p$. (We denote E over $\mathbb{F}_p$ as $E/\mathbb{F}_p$.)

[1] Lift $E/\mathbb{F}_p$ to an elliptic curve $\mathcal{E}/\mathbb{Q}$ and fix a set of independent points:

$$\Gamma = \{P_1, P_2, \cdots, P_r\} \in \mathcal{E}(\mathbb{Q}). \tag{2.133}$$

[2] [Compute and lift $eT \in E(\mathbb{F}_p)$] Randomly choose integers $e < N_p$ (N_p denotes the number of points on E), computer $eT \in E(\mathbb{F}_p)$, lift it to a point in $\mathcal{E}(\mathbb{Q})$, and attempt to write the lift as a linear combination:

$$\text{Lift}(eT) = m_1(e)P_1 + m_2(e)P_2 + \cdots + m_r(e)P_r. \tag{2.134}$$

[3] Repeat Step [2] to find r independent expressions (2.134), then solve the system of congruences

$$e_i \equiv m_1(e_i) \log_T P_1 + \cdots + m_r(e_i) \log_T P_r \pmod{N_p} \tag{2.135}$$

for the quantities $\log_T P_j$.

[4] Randomly choose exponent $e < N_p$, compute $S + eT \in E(\mathbb{F}_p)$, lift it to point in $\mathcal{E}(\mathbb{Q})$, and attempt to write it as:

$$\text{Lift}(S + eT) = k_1 P_1 + k_2 P_2 + \cdots + k_r P_r. \tag{2.136}$$

When this is successful, the relation

$$\log_T S + e \equiv k_1 \log_T P_1 + \cdots + k_r \log_T P_r \pmod{N_p} \tag{2.137}$$

gives the value of $k = \log_T S$.

There are two difficulties in the above algorithm. First of all, one needs to lift $E(\mathbb{F}_p)$ to $\mathcal{E}(\mathbb{Q})$ having many independent rational points of small height. Secondly, one needs to lift points from $E(\mathbb{F}_p)$ to $\mathcal{E}(\mathbb{Q})$. Both of these two problems are very difficult, probably more difficult than the original ECDLP. Furthermore, even if one could find curves $\mathcal{E}(\mathbb{Q})$ of very high rank, there are good theoretical reasons for believing that the generators of $\mathcal{E}(\mathbb{Q})$ would never be small enough to allow the lifting problem to be solved in subexponential time. A conclusion made by Silverman and Suzuki [232] is that the index calculus will not work for solving the ECDLP because it is not possible to lift $E(\mathbb{F}_p)$ to a curve $\mathcal{E}(\mathbb{Q})$ having many independent points of sufficiently small height. For this reason, the ECDLP is believed to be much more harder than either the IFP (integer factorization problem) or the DLP in that no subexponential-time (general-purpose) algorithm is known.

In 1998, Joseph Silverman[27] proposed a new type of algorithm (although it has not yet been tested in practice) to attack the ECDLP [231]. He called it *xedni calculus* because it "stands index calculus on its head". The idea of the xedni calculus is as follows:

[1] Choose points in $E(\mathbb{F}_p)$ and lift them to pints in $\mathbb{Z}^2$.

[27]

Joseph H. Silverman is currently Professor of Mathematics at Brown University. He received his Ph.D. at Harvard University in Number theory in 1982. His research interests include number theory, elliptic curves, arithmetic and Diophantine geometry, number theoretic aspects of dynamical systems, and cryptography. Prof. Silverman is perhaps best known for his four books, all by Springer-Verlag: *The Arithmetic of Elliptic Curves*, 1986, *Arithmetic Geometry*, co-editored with Gary Cornell, 1986, *Rational Points on Elliptic Curves*, co-editored with John Tate, 1992, and *Advanced Topics in the Arithmetic of Elliptic Curves*, 1995.

[2] Choose a curve $\mathcal{E}(\mathbb{Q})$ containing the lift points; use Mestre's method [159] (in reverse) to make rank $\mathcal{E}(\mathbb{Q})$ small.

Whilst the index calculus works in reverse:

[1] Lift $E/\mathbb{F}_p$ to $\mathcal{E}(\mathbb{Q})$; use Mestre's method to make rank $\mathcal{E}(\mathbb{Q})$ large.

[2] Choose points in $E(\mathbb{F}_p)$ and try to lift them to pints in $\mathcal{E}(\mathbb{Q})$.

A brief description of the xedni algorithm is as follows (a completed and detailed description of the algorithm can be found in [231]).

Algorithm 2.4.7 (Xedni calculus for the ECDLP). Let $\mathbb{F}_p$ be a finite field with p elements (p prime), $E/\mathbb{F}_p$ an elliptic curve over $\mathbb{F}_p$, say, given by

$$E: \ y^2 + a_{p,1}xy + a_{p,3}y = x^3 + a_{p,2}x^2 + a_{p,4}x + a_{p,6}.$$

N_p the number of points in $E(\mathbb{F}_p)$, S and T the two points in $E(\mathbb{F}_p)$. This algorithm tries to find an integer k

$$k = \log_T S$$

such that

$$S = kT \quad \text{in } E(\mathbb{F}_p).$$

[1] Fix an integer $4 \le r \le 9$ and an integer M which is a product of small primes.

[2] Choose r points:

$$P_{M,i} = [x_{M,i}, y_{M,i}, z_{M,i}], \quad 1 \le i \le r \tag{2.138}$$

having integer coefficients and satisfying

- the first 4 points are $[1, 0, 0]$, $[0, 1, 0]$, $[0, 0, 1]$ and $[1, 1, 1]$.
- For every prime $l \mid M$, the matrix $\mathbf{B}(P_{M,1}, \cdots, P_{M,r})$ has maximal rank modulo l.

Further choose coefficients $u_{M,1}, \cdots, u_{M,10}$ such that the points $P_{M,1}, \cdots, P_{M,r}$ satisfy the congruence:

$$\begin{aligned} &u_{M,1}x^3 + u_{M,2}x^2y + u_{M,3}xy^2 + u_{M,4}y^3 + u_{M,5}x^2z + u_{M,6}xyz \\ &\quad + u_{M,7}y^2z + u_{M,8}xz^2 + u_{M,9}yz^2 + u_{M,10}z^3 \equiv 0 \pmod{M}. \end{aligned} \tag{2.139}$$

[3] Choose r random pairs of integers (s_i, t_i) satisfying $1 \le s_i, t_i < N_p$, and for each $1 \le i \le r$, compute the point $P_{p,i} = (x_{p,i}, y_{p,i})$ defined by

$$P_{p,i} = s_iS - t_iT \quad \text{in } E(\mathbb{F}_p). \tag{2.140}$$

[4] Make a change of variables in $\mathbb{P}^2$ of the form

$$\begin{pmatrix} X' \\ Y' \\ Z' \end{pmatrix} = \begin{pmatrix} a_{11} & a_{12} & a_{13} \\ a_{21} & a_{22} & a_{23} \\ a_{21} & a_{32} & a_{33} \end{pmatrix} \begin{pmatrix} X \\ Y \\ Z \end{pmatrix} \quad (2.141)$$

so that the first four points become

$$P_{p,1} = [1,0,0], \ P_{p,2} = [0,1,0], \ P_{p,3} = [0,0,1], \ P_{p,4} = [1,1,1].$$

The equation for E will then have the form:

$$u_{p,1}x^3 + u_{p,2}x^2y + u_{p,3}xy^2 + u_{p,4}y^3 + u_{p,5}x^2z + u_{p,6}xyz$$
$$+u_{p,7}y^2z + u_{p,8}xz^2 + u_{p,9}yz^2 + u_{p,10}z^3 = 0. \quad (2.142)$$

[5] Use the Chinese Remainder Theorem to find integers $u'_1, \cdots, u'_{10}$ satisfying

$$u'_i \equiv u_{p,i} \ (\text{mod } p) \text{ and } u'_i \equiv u_{M,i} \ (\text{mod } M) \text{ for all } 1 \leq i \leq 10. \quad (2.143)$$

[6] Lift the chosen points to $\mathbb{P}^2(\mathbb{Q})$. That is, choose points

$$P_i = [x_i, y_i, z_i], \quad 1 \leq i \leq r, \quad (2.144)$$

with integer coordinates satisfying

$$P_i \equiv P_{p,i} \ (\text{mod } p) \text{ and } P_i \equiv P_{M,i} \ (\text{mod } M) \text{ for all } 1 \leq i \leq r. \quad (2.145)$$

In particular, take $P_1 = [1,0,0], P_2 = [0,1,0], P_3 = [0,0,1], P_4 = [1,1,1]$.

[7] Let $\mathbf{B} = \mathbf{B}(P_1, \cdots, P_r)$ be the matrix of cubic monomials defined earlier. Consider the system of linear equations:

$$\mathbf{Bu} = 0. \quad (2.146)$$

Find a small integer solution $\mathbf{u} = [u_1, \cdots, u_{10}]$ to (2.146) which has the additional property

$$\mathbf{u} \equiv [u'_1, \cdots, u'_{10}] \ (\text{mod } M_p), \quad (2.147)$$

where $u'_1, \cdots, u'_{10}$ are the coefficients computed in Step [5]. Let $C_{\mathbf{u}}$ denote the associated cubic curve:

$$C_{\mathbf{u}}: \ u_1x^3 + u_2x^2y + u_3xy^2 + u_4y^3 + u_5x^2z + u_6xyz$$
$$+u_7y^2z + u_8xz^2 + u_9yz^2 + u_{10}z^3 = 0. \quad (2.148)$$

[8] Make a change of coordinates to put $C_{\mathbf{u}}$ into standard minimal Weierstrass form with the point $P_1 = [1,0,0]$ the point at infinity, $\mathcal{O}$. Write the resulting equation as

$$E_{\mathbf{u}}: \ y^2 + a_1xy + a_3y = x^3 + a_2x^2 + a_4x + a_6 \tag{2.149}$$

with $a_1, \cdots, a_6 \in \mathbb{Z}$, and let $Q_1, Q_2, \cdots, Q_r$ denote the images of $P_1, P_2, \cdots, P_r$ under this change of coordinates (so in particular, $Q_1 = \mathcal{O}$). Let $c_4(\mathbf{u})$, $c_6(\mathbf{u})$, and $\Delta(\mathbf{u})$ be the usual quantities in [229] associated to the equation (2.149).

[9] Check if the points $Q_1, Q_2, \cdots, Q_r \in E_{\mathbf{u}}(\mathbb{Q})$ are independent. If they are, return to Step [2] or [3]. Otherwise compute a relation of dependence

$$n_2Q_2 + n_3Q_3 + \cdots + n_rQ_r = \mathcal{O}, \tag{2.150}$$

set

$$n_1 = -n_2 - n_3 - \cdots - n_r, \tag{2.151}$$

and continue with the next step.

[10] Compute

$$s = \sum_{i=1}^{r} n_i s_i \quad \text{and} \quad t = \sum_{i=1}^{r} n_i t_i. \tag{2.152}$$

If $\gcd(s, N_p) > 1$, return to Step [2] or [3]. Otherwise compute an inverse $ss' \equiv 1 \pmod{N_p}$. Then

$$\log_T S \equiv s't \pmod{N_p}, \tag{2.153}$$

and the ECDLP is solved.

It is interesting to note that soon after Silverman proposed the xedni algorithm, Koblitz showed that a modified version of Silverman's xedni algorithm could be used to attack both the DLP (upon which the US government's Digital Signature Standard, DSS, is based) and IFP (upon which the security of RSA relies). This implied that if Silverman's algorithm turned out to be practical, it would break essentially all forms of public-key cryptography that are currently in practical use. Even if it is found not at all practical, it would be still interesting, because at least we know that IFP, DLP and ECDLP are not as different from each other as appears at first glance.

2.4.5 Algorithm for Root Finding Problem

There are three closely related problems in computational number theory:

(1) The *modular exponentiation problem* (see Subsection 2.1.5) – given the triple $\langle k, x, n \rangle$ to compute $y = x^k \pmod{n}$. This problem is relatively easy because it can be performed in polynomial time.

(2) The *discrete logarithm problem* (see the preceding four subsections) – given the triple $\langle x, y, n\rangle$ to find an exponent k such that $y = x^k \pmod{n}$. This problem is hard, since no polynomial time algorithm for it has been found yet.

(3) The *root finding problem* – given the triple $\langle k, y, n\rangle$ to find an x such that $y = x^k \pmod{n}$. This problem is slightly easier than the discrete logarithm problem, since there are efficient *randomized* algorithms for it, provided n is a prime power. However, for general n, even the problem for finding *square roots* modulo n is as difficult as the well-known integer factorization problem. It should be noted that if the value for $\phi(n)$ is known, then the kth root of y modulo n can be found fairly easily.

In what follows, we shall present an efficient and practical algorithm for computing k^{th} roots modulo n, provided $\phi(n)$ is known (Silverman [230]), and give an example to illustrate the use of the algorithm.

Algorithm 2.4.8 (Root finding algorithm). Given integers k, y and n, this algorithm tries to find an integer x such that $y \equiv x^k \pmod{n}$.

[1] Compute $\phi(n)$ (See Subsection 1.4.4).

[2] Find positive integers u and v such that $ku - \phi(n)v = 1$. (The linear Diophantine equation $ku - \phi(n)v = 1$ can be solved by using the continued fraction method – see Section 1.3.)

[3] Compute $x = y^u \pmod{n}$ (By using the fast exponentiation method – see Subsection 2.1.5).

How do we know $x = y^u$ is a solution to the congruence $x^k \equiv y \pmod{n}$? This can be verified by

$$\begin{aligned} x^k &= (y^u)^k \\ &= y^{ku} \\ &= y^{1+\phi(n)v} && \text{(by step [2])} \\ &= y \cdot (y^{\phi(n)})^v \\ &\equiv y && \text{(by Euler's Theorem).} \end{aligned}$$

Example 2.4.5. Find the 131th root of 758 modulo 1073. That is, find a solution to the congruence:

$$x^{131} \equiv 758 \pmod{1073}.$$

We follow the steps in Algorithm 2.4.8:

[1] Since $1073 = 29 \cdot 37$, $\phi(1073) = 28 \cdot 36 = 1008$.

[2] Solve the linear Diophantine equation:

$$131u - 1008v = 1.$$

Since $131/1008$ can be expanded as a finite continued fraction with convergents:

$$\left[0,\ \frac{1}{7}, \frac{1}{8}, \frac{3}{23}, \frac{10}{77}, \frac{13}{100}, \frac{23}{177}, \frac{36}{277}, \frac{131}{1008}\right],$$

we have

$$\begin{aligned} u &= (-1)^{n-1}q_{n-1} = (-1)^7 277 = -277, \\ v &= (-1)^{n-1}p_{n-1} = (-1)^7 36 = -36. \end{aligned}$$

Therefore,

$$131 \cdot (-277) - 1008 \cdot (-36) = -36287 + 35288 = 1.$$

Thus, $u = -277$ and $v = -36$. In order to get positive values for u and v, we modify this solution to:

$$u = -277 + 1008 = 731, \quad v = -36 + 131 = 95$$

with

$$131 \cdot 731 - 1008 \cdot 95 = 1.$$

[3] Finally compute

$$\begin{aligned} x &\equiv y^u \pmod{1073} \\ &\equiv (x^{131})^{731} \pmod{1073} \\ &\equiv 758^{731} \pmod{1073} \\ &\equiv 905 \pmod{1073}. \end{aligned}$$

Clearly, $x = 905$ is the required solution to

$$x^{131} \equiv 758 \pmod{1073},$$

since

$$905^{131} \equiv 758 \pmod{1073}.$$

Remark 2.4.1. The above method works only when $\phi(n)$ is known. It won't work if we cannot calculate $\phi(n)$, but it is exactly this *weakness* (unreasonable effectiveness, see Burr [44]) which is used by Rivest, Shamir, and Adleman [209] to construct their *unbreakable* cryptosystem (see Subsection 3.3.6 in Chapter 3 for a further discussion).

2.5 Quantum Number-Theoretic Algorithms

If we are to have any hope of sustaining the economic benefits to the national economy provided by sustaining Moore's law, we have no choice but to develop quantum switches and the means to interconnect them.

JOEL BIRNBAUM
Chief Scientist at Hewlett-Packard

In this section, we shall first introduce some basic concepts of quantum computation, including quantum computability and quantum complexity, then introduce some recently developed quantum algorithms for integer factorization and discrete algorithms.

2.5.1 Quantum Information and Computation

The idea that computers can be viewed as physical objects and computations as physical processes is revolutionary; it was first proposed by Benioff [23], Feynman[28] [74], Deutsch[29] [63] and others in the first half of the 1980s.

[28]

Richard Phillips Feynman (1918–1988) studied Physics at MIT and received his doctorate from Princeton in 1942. His doctoral work developed a new approach to quantum mechanics using the principle of least action. Feynman worked on the atomic bomb project at Princeton University (1941-42) and then at Los Alamos (1943-45). After World War II he was appointed to the chair of theoretical physics at Cornell University, then, in 1950, to the chair of theoretical physics at California Institute of Technology, where he remained for the rest of his career. Feynman was awarded the Nobel Prize in 1965 for introducing the so-called Feynman diagrams, the graphic analogues of the mathematical expressions needed to describe the behaviour of systems of interacting particles. Perhaps best known for his excellent text *The Feynman Lectures on Physics*, Feynman [75] also published posthumously a text on quantum computation, namely *Feynman Lectures on Computation*.

[29]

David Deutsch was born in Haifa, Israel, in 1953, and came to Britain in 1956. He obtained a BA in Natural Sciences from Cambridge University in 1974 and a doctorate in theoretical physics from Oxford University in 1978. Deutsch is currently with the Centre for Quantum Computation at Oxford University. He received the prestigious Paul Dirac Prize and Medal in 1998 for his pioneering work in quantum computation leading to the concept of quantum computers and for contributing to the understanding of how such devices might be constructed from quantum logic gates in quantum networks. He was also elected as a Distinguished Fellow of the British Computer Society in 1998. (Photo by courtesy of Dr. Deutsch.)

Quantum computers are machines that rely on characteristically quantum phenomena, such as quantum interference and quantum entanglement, in order to perform computation, whereas the classical theory of computation usually refers not to physics but to purely mathematical subjects. A conventional digital computer operates with bits (we may call them *Shannon bits*, since Shannon was the first to use bits to represent information) – the Boolean states 0 and 1 – and after each computation step the computer has a definite, exactly measurable state, that is, all bits are in the form 0 or 1 but not both. A quantum computer, a quantum analogue of a digital computer, operates with *quantum bits* (the quantum version of Shannon bit) involving quantum states. The state of a quantum computer is described as a *basis vector* in a *Hilbert space*[30], named after the German mathematician David Hilbert (1862–1943). More formally, we have:

Definition 2.5.1. A *qubit* is a quantum state $|\Psi\rangle$ of the form

$$|\Psi\rangle = \alpha\,|0\rangle + \beta\,|1\rangle, \tag{2.154}$$

where the amplitudes $\alpha, \beta \in \mathbb{C}$ such that $||\alpha|| + ||\beta|| = 1$, and $|0\rangle$ and $|1\rangle$ are *basis vectors* of the Hilbert space.

Note that state vectors are written in a special angular bracket notation called a "ket vector" $|\Psi\rangle$, an expression coined by Paul Dirac[31], who wanted a shorthand notation for writing formulae that arise in quantum mechanics. In a quantum computer, each qubit could be represented by the state of a simple 2-state quantum system such as the spin state of a spin-$\frac{1}{2}$ particle. The spin of such a particle, when measured, is always found to exist in one of two possible states $|+\frac{1}{2}\rangle$ (spin-up) and $|-\frac{1}{2}\rangle$ (spin-down). This *discreteness*

[30] Hilbert space is defined to be a complete inner-product space. The set of all sequences $x = (x_1, x_2, \cdots)$ of complex numbers (where $\sum_{i=1}^{\infty} |x_i|^2$ is finite) is a good example of a Hilbert space, where the sum $x + y$ is defined as $(x_1 + y_1, x_2 + y_2, \cdots)$, the product ax as $(ax_1, ax_2, \cdots)$, and the inner product as $(x, y) = \sum_{i=1}^{\infty} \overline{x}_i y_i$, where $\overline{x}_i$ is the complex conjugate of x_i, $x = (x_1, x_2, \cdots)$ and $y = (y_1, y_2, \cdots)$. In modern quantum mechanics all possible physical states of a system are considered to correspond to space vectors in a Hilbert space.

[31]

Paul Adrien Maurice Dirac (1902–1984), the creator of the complete theoretical formulation of quantum mechanics, was born in Bristol, England and studied electrical engineering at the University of Bristol before doing research in mathematics at St John's College at Cambridge. His first major contribution to quantum theory was a paper written in 1925. He published *The principles of Quantum Mechanics* in 1930 and for this work he was awarded the Nobel Prize for Physics in 1933. Dirac was appointed Lucasian Professor of Mathematics at the University of Cambridge in 1932, a post he held for 37 years. He was elected a fellow of the Royal Society in 1930 and was awarded the Society's Royal Medal in 1939.

is called *quantization*. Clearly, the two states can then be used to represent the binary value 1 and 0 (see Figure 2.3). The main difference between qubits

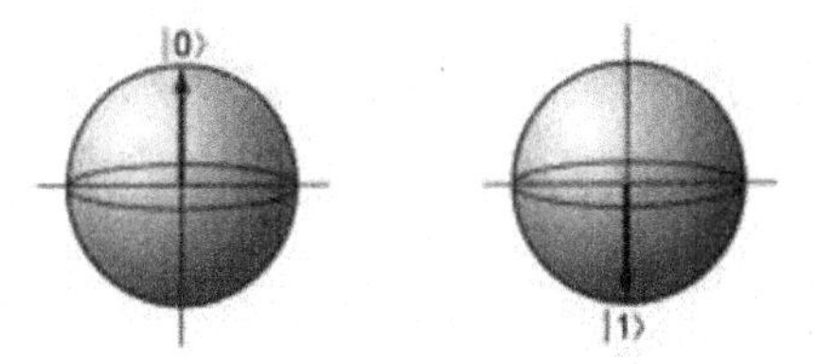

Figure 2.3. A qubit for the binary values 0 and 1 (picture by courtesy of Williams and Clearwater [258])

and classical bits is that a bit can only be set to either 0 and 1, while a qubit $|\Psi\rangle$ can take any (uncountable) quantum superposition of $|0\rangle$ and $|1\rangle$ (see Figure 2.4). That is, a qubit in a simple 2-state system can have two states

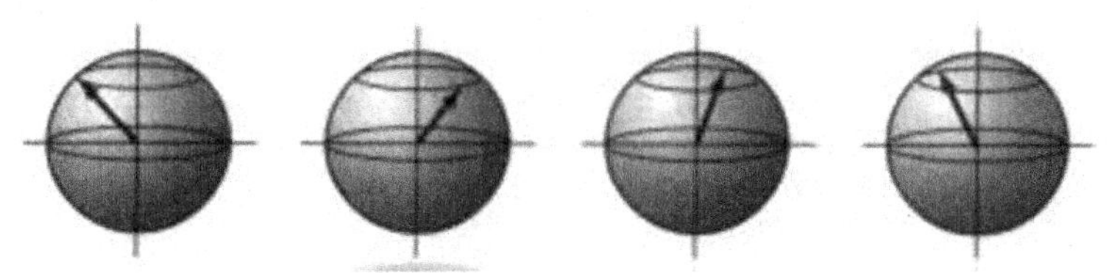

Figure 2.4. Each sphere represents a qubit with the same proportions of the $|0\rangle$ and $|1\rangle$ (picture by courtesy of Williams and Clearwater [258])

rather than just one allowed at a time as the classical Shannon bit. Moreover, if a 2-state quantum system can exist in any one of the states $|0\rangle$ and $|1\rangle$, it can also exist in the *superposed* state

$$|\Psi\rangle = \alpha_1 |0\rangle + \alpha_2 |1\rangle. \tag{2.155}$$

This is known as the *principle of superposition*. More generally, if a k-state quantum system can exist in any one of the following k eigenstates $|c_1\rangle, |c_1\rangle, \cdots, |c_k\rangle$, it can also exist in the *superposed* state

$$|\Psi\rangle = \sum_{i=0}^{2^k-1} \alpha_i |c_i\rangle, \tag{2.156}$$

where the amplitudes $\alpha_i \in \mathbb{C}$ are such that $\sum_i ||\alpha_i||^2 = 1$, and each $|c_i\rangle$ is a basis vector of the Hilbert space. Once we can encode the binary values 0 and 1 in the states of a physical system, we can make a complete memory of register out of a chain of such systems.

Definition 2.5.2. A *quantum register*, or more generally, a *quantum computer*, is an ordered set of a finite number of qubits.

In order to use a physical system to do computation, we must be able to change the state of the system; this is achieved by applying a sequence of unitary transformations to the state vector $|\Psi\rangle$ via a unitary matrix (a unitary matrix is one whose conjugate transpose is equal to its inverse). Suppose now a computation is performed on a one-bit quantum computer, then the superposition will be

$$|\Psi\rangle = \alpha\,|\,0\rangle + \beta\,|\,1\rangle\,, \tag{2.157}$$

where $\alpha, \beta \in \mathbb{C}$ are such that $||\alpha||^2 + ||\beta||^2 = 1$. The different possible states are $|\,0\rangle = \begin{pmatrix} 1 \\ 0 \end{pmatrix}$ and $|\,1\rangle = \begin{pmatrix} 0 \\ 1 \end{pmatrix}$. Let the unitary matrix M be

$$M = \frac{1}{\sqrt{2}} \begin{pmatrix} 1 & 1 \\ -1 & 1 \end{pmatrix}. \tag{2.158}$$

Then the quantum operations on a qubit can be written as follows:

$$M\,|\,0\rangle = \frac{1}{\sqrt{2}} \begin{pmatrix} 1 & 1 \\ -1 & 1 \end{pmatrix} \begin{pmatrix} 1 \\ 0 \end{pmatrix} = \frac{1}{\sqrt{2}}|\,0\rangle - \frac{1}{\sqrt{2}}|\,1\rangle = |\,0\rangle$$

$$M\,|\,1\rangle = \frac{1}{\sqrt{2}} \begin{pmatrix} 1 & 1 \\ -1 & 1 \end{pmatrix} \begin{pmatrix} 0 \\ 1 \end{pmatrix} = \frac{1}{\sqrt{2}}|\,0\rangle + \frac{1}{\sqrt{2}}|\,1\rangle = |\,1\rangle$$

which is actually the quantum gate (analogous to the classical logic gate):

$$|\,0\rangle \to \frac{1}{\sqrt{2}}|\,0\rangle - \frac{1}{\sqrt{2}}|\,1\rangle$$

$$|\,1\rangle \to \frac{1}{\sqrt{2}}|\,0\rangle + \frac{1}{\sqrt{2}}|\,1\rangle.$$

Logic gates can be regarded as logic operators. The NOT operator defined as

$$\text{NOT} = \begin{pmatrix} 0 & 1 \\ 1 & 0 \end{pmatrix}, \tag{2.159}$$

changes the state of its input as follows:

$$\text{NOT}\,|\,0\rangle = \begin{pmatrix} 0 & 1 \\ 1 & 0 \end{pmatrix} \begin{pmatrix} 1 \\ 0 \end{pmatrix} = \begin{pmatrix} 0 \\ 1 \end{pmatrix} = |\,1\rangle$$

$$\text{NOT}\,|\,1\rangle = \begin{pmatrix} 0 & 1 \\ 1 & 0 \end{pmatrix} \begin{pmatrix} 0 \\ 1 \end{pmatrix} = \begin{pmatrix} 1 \\ 0 \end{pmatrix} = |\,0\rangle.$$

Similarly, we can define the quantum gate of two bits as follows:

$$
\begin{aligned}
|\,00\rangle &\to |\,00\rangle \\
|\,01\rangle &\to |\,01\rangle \\
|\,10\rangle &\to \frac{1}{\sqrt{2}}|\,10\rangle + \frac{1}{\sqrt{2}}|\,11\rangle \\
|\,11\rangle &\to \frac{1}{\sqrt{2}}|\,10\rangle - \frac{1}{\sqrt{2}}|\,11\rangle
\end{aligned}
$$

or equivalently by giving the unitary matrix of the quantum operation:

$$
M = \begin{pmatrix} 1 & 0 & 0 & 0 \\ 0 & 1 & 0 & 0 \\ 0 & 0 & \frac{1}{\sqrt{2}} & \frac{1}{\sqrt{2}} \\ 0 & 0 & \frac{1}{\sqrt{2}} & -\frac{1}{\sqrt{2}} \end{pmatrix}. \tag{2.160}
$$

This matrix is actually the counterpart of the truth table of Boolean logic used for digital computers. Suppose now the computation is in the superposition of the states:

$$
\frac{1}{\sqrt{2}}|\,10\rangle - \frac{1}{\sqrt{2}}|\,11\rangle
$$

or

$$
\frac{1}{\sqrt{2}}|\,10\rangle + \frac{1}{\sqrt{2}}|\,11\rangle .
$$

Then using the unitary transformations defined in (2.160), we have

$$
\begin{aligned}
\frac{1}{\sqrt{2}}|\,10\rangle - \frac{1}{\sqrt{2}}|\,11\rangle &= \frac{1}{\sqrt{2}}\left(\frac{1}{\sqrt{2}}|\,10\rangle + \frac{1}{\sqrt{2}}|\,11\rangle\right) \\
&\quad -\frac{1}{\sqrt{2}}\left(\frac{1}{\sqrt{2}}|\,10\rangle - \frac{1}{\sqrt{2}}|\,11\rangle\right) \\
&= \frac{1}{2}(|\,10\rangle + |\,11\rangle) - \frac{1}{2}(|\,10\rangle + |\,11\rangle) \\
&= |\,11\rangle, \\
\frac{1}{\sqrt{2}}|\,10\rangle + \frac{1}{\sqrt{2}}|\,11\rangle &= \frac{1}{2}(|\,10\rangle + |\,11\rangle) + \frac{1}{2}(|\,10\rangle - |\,11\rangle) \\
&= |\,10\rangle.
\end{aligned}
$$

We have just introduced the very basic concepts of quantum computation, including quantum bits, quantum states, quantum registers, and quantum gates and quantum operations. Interested readers are advised to consult, for example, Williams and Clearwater's book [258] for more information.

2.5.2 Quantum Computability and Complexity

In this subsection, we shall give a brief introduction to some basic concepts of quantum computability and complexity within the theoretical framework of quantum Turing machines.

The first true quantum Turing machine (QTM) was proposed in 1985 by Deutsch [63]. A *quantum Turing machine* (QTM) is a quantum mechanical generalization of a probabilistic Turing machine (PTM), in which each cell on the tape can hold a *qubit* (quantum bit) whose state is represented as an arrow contained in a sphere (see Figure 2.5). Let $\overline{\mathbb{C}}$ be the set consisting of $\alpha \in \mathbb{C}$ such that there is a deterministic Turing machine that computes the real and imaginary parts of α to within 2^{-n} in time polynomial in n, then the quantum Turing machines can still be defined as an algebraic system

$$M = (Q, \Sigma, \Gamma, \delta, q_0, \Box, F) \tag{2.161}$$

where

$$\delta : Q \times \Gamma \to \overline{\mathbb{C}}^{Q \times \Gamma \times \{L,R\}}, \tag{2.162}$$

and the rest remains the same as a probabilistic Turing machine. Readers are suggested to consult Bernstein and Vazirani [27] for a more detailed discussion of quantum Turing machines. Quantum Turing machines open a new way to model our universe which is quantum physical, and offer new features of computation. However, quantum Turing machines do not offer more computation power than classical Turing machines. This leads to the following quantitative version of the Church-Turing thesis for quantum computation:

> **The Church-Turing thesis for quantum computation.** Any physical (quantum) computing device can be simulated by a Turing machine in a number of steps polynomial in the resources used by the computing device.

That is, from a computability point of view, a *quantum* Turing machine has no more computation power than a *classical* Turing machine. However, from a computational complexity point of view, a quantum Turing machine will be more efficient than a classical Turing machine. For example, the integer factorization and the discrete logarithm problems are intractable on classical Turing machines (as everybody knows at present), but they are tractable on quantum Turing machines.

Just as there are classical complexity classes, so are there quantum complexity classes. As quantum Turing machines are generalizations of probabilistic Turing machines, the quantum complexity classes resemble the probabilistic complexity classes. More specifically, we have:

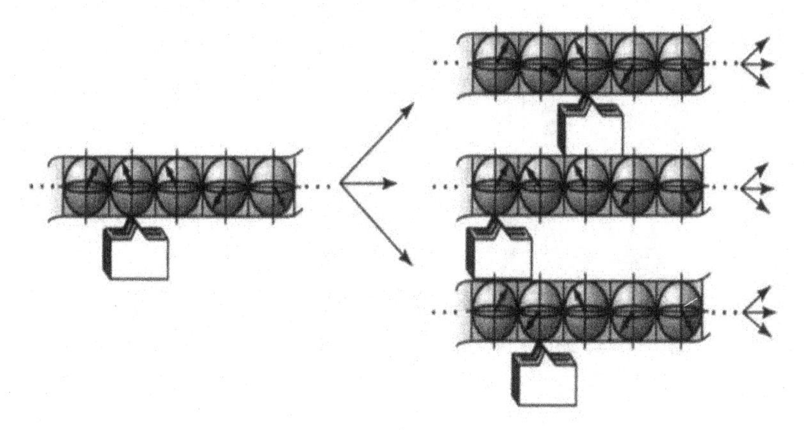

Figure 2.5. A quantum Turing machine (by courtesy of Williams and Clearwater [258])

(1) $\mathcal{QP}$ (quantum analogue of $\mathcal{P}$) the class of problems solvable, with certainty, in polynomial time on a quantum Turing machine. It can be shown that $\mathcal{P} \subset \mathcal{QP}$. That is, the quantum Turing machine can solve more problems *efficiently* than a classic Turing machine.
(2) $\mathcal{BQP}$ (quantum analogue of $\mathcal{BPP}$) is the class of problems solvable in polynomial time by a quantum Turing machine, possibly with a *bounded* probability $\epsilon < 1/3$ of error. It is known that $\mathcal{BPP} \subseteq \mathcal{BQP} \subseteq \mathcal{P}$-SPACE, and hence, it is not known whether quantum Turing machines are more powerful than probabilistic Turing machines.
(3) $\mathcal{ZQP}$ (quantum analogue of $\mathcal{ZPP}$) is the class of problems solvable in expected polynomial time with zero-error probability by a quantum Turing machine. It is clear that $\mathcal{ZPP} \subset \mathcal{ZQP}$.

2.5.3 Quantum Algorithm for Integer Factorization

In this and the next subsection, two quantum algorithms for integer factorization and discrete logarithms will be introduced.

In 1976, Miller [162] showed that, using randomization, one can factor an odd positive composite $n > 1$ if one can find the order of an element x modulo n (or more precisely, the order of an element x in the multiplicative group $\mathcal{G} = (\mathbb{Z}/n\mathbb{Z})^*$), denoted by $\text{ord}_n(x)$. The order r of x in the multiplicative group $\mathcal{G}$ (see Section 1.6.7 of Chapter 1), is the smallest positive integer r such that $x^r \equiv 1 \pmod{n}$. Finding the order of an element x in $\mathcal{G}$ is, in theory, not a problem: just keep multiplying until you get to "1", the identity element of the multiplicative group $\mathcal{G}$. For example, let $n = 179359$,

$x = 3 \in \mathcal{G}$, and $\mathcal{G} = (\mathbb{Z}/179359\mathbb{Z})^*$, such that $\gcd(3, 179359) = 1$. To find the order $r = \text{ord}_{179359}(3)$, we just keep multiplying until we get to "1":

$$\begin{array}{llll}
3^1 & \bmod & 179359 & = & 3 \\
3^2 & \bmod & 179359 & = & 9 \\
3^3 & \bmod & 179359 & = & 27 \\
 & \cdots\cdots\cdots & & & \\
 & \cdots\cdots\cdots & & & \\
3^{1000} & \bmod & 179359 & = & 31981 \\
3^{1001} & \bmod & 179359 & = & 95943 \\
3^{1002} & \bmod & 179359 & = & 108470 \\
 & \cdots\cdots\cdots & & & \\
 & \cdots\cdots\cdots & & & \\
3^{14716} & \bmod & 179359 & = & 99644 \\
3^{14717} & \bmod & 179359 & = & 119573 \\
3^{14718} & \bmod & 179359 & = & 1.
\end{array}$$

Thus, the order r of 3 in the multiplicative group $\mathcal{G} = (\mathbb{Z}/179359\mathbb{Z})^*$ is 14718, that is, $\text{ord}_{179359}(3) = 14718$. Once the order $\text{ord}_n(x)$ is found, it is then trivial to factor n by just calculating

$$\left\{\gcd(x^{r/2} + 1, n), \quad \gcd(x^{r/2} - 1, n)\right\}$$

which, as we have shown, can always be performed in polynomial time. For instance, for $x = 3$, $r = 14718$ and $n = 179359$, we have

$$\left\{\gcd(3^{14718/2} + 1, 179359), \quad \gcd(3^{14718/2} - 1, 179359)\right\} = (67, 2677),$$

and hence the factorization of

$$n = 179359 = 67 \cdot 2677.$$

If one of the factors is not prime, then we can invoke the above process recursively until a complete prime factorization of n is obtained. Of course, we can choose other elements x in $(\mathbb{Z}/179359\mathbb{Z})^*$, rather than 3. For example, we can choose $x = 5$. In this case, we have $\text{ord}_{179359}(5) = 29436$. Then we have

$$\left\{\gcd(5^{29436/2} + 1, 179359), \quad \gcd(5^{29436/2} - 1, 179359)\right\} = (2677, 67),$$

which also leads to the factorization of n: $179359 = 67 \cdot 2677$. However, in practice, the above computation for finding the order of $x \in (\mathbb{Z}/n\mathbb{Z})^*$ may not work, since for an element x in a large group $\mathcal{G}$ with n having more than 200 digits, the computation of r may require more than 10^{150} multiplications. Even if these multiplications could be carried out at the rate of 1000 billion per second, it would take approximately $3 \cdot 10^{80}$ years to arrive at the answer[32]. This explains partly why integer factorization is difficult. Fortunately, Shor[33] discovered in 1994 an efficient quantum algorithm to find the order of an element $x \in (\mathbb{Z}/n\mathbb{Z})^*$ and hence *possibly* the factorization of n. The main idea of Shor's method is as follows [258]. First of all, we create two quantum registers for our machine: Register-1 and Register-2. Of course, we can create just one single quantum memory register partitioned into two parts. Secondly, we create in Register-1, a superposition of the integers $a = 0, 1, 2, 3, \cdots$ which will be the arguments of $f(a) = x^a \pmod n$, and load Register-2 with all zeros. Thirdly, we compute in Register-2, $f(a) = x^a \pmod n$ for each input a. (Since the values of a are kept in Register-1, this can be done reversibly). Fourthly, we perform the discrete Fourier transform on Register-1. Finally we observe both registers of the machine and find the order r that satisfies $x^r \equiv 1 \pmod n$.

A few words at this point are needed about the relation between the Fourier transform and the order-finding (and eventually the factoring). As we know, any mathematical function can be described as a weighted *sum* of certain "basis" or "elementary building block" functions such as sines and cosines: $\sin x,\ \sin 2x, \cdots$ and $\cos x,\ \cos 2x, \cdots$. The Fourier transform of a function is the mathematical operation that translates the original function into this equivalent *sum* of sine and cosine functions. Simon [236] in 1994 showed that a quantum computer could obtain a sample from the Fourier transform of a function faster than any classical computer. Note that there exist a Fast (discrete) Fourier Transform (FFT) algorithm, developed by Coo-

[32] There is however a "quick" way to find the order of an element x in the multiplicative group $\mathcal{G}$ modulo n if the order $|\mathcal{G}|$ (where $|\mathcal{G}| = |(\mathbb{Z}/n\mathbb{Z})^*| = \phi(n)$) of $\mathcal{G}$ as well as the prime factorization of $|\mathcal{G}|$ are known, since, by Lagrange's theorem, $r = \mathrm{ord}_n(x)$ is a divisor of $|\mathcal{G}|$. Of course, as we know, the number $\lambda(n)$ is the largest possible order of an element x in the group $\mathcal{G}$. So, once we have the value of $\lambda(n)$, it is relatively easy to find $\mathrm{ord}_n(x)$, the order of the element $x \in \mathcal{G}$. For example, let $n = 179359$, then $\lambda(179359) = 29436$. Therefore, $\mathrm{ord}_{179359}(3) \le 29436$. In fact, $\mathrm{ord}_{179359}(3) = 14718$, which of course is a divisor of 29436. However, there are no efficient algorithms at present for calculating either $\phi(n)$ or $\lambda(n)$. Therefore, these two "quick" ways for computing $\mathrm{ord}_n(x)$ by either $\phi(n)$ or $\lambda(n)$ are essentially useless in practice.

[33]

Peter Shor, born in 1959, is a mathematician at the AT&T Research Laboratories in Florham Park, New Jersey. After studying at the California Institute of Technology he gained a PhD at the Massachusetts Institute of Technology. Before going to AT&T in 1986, he was a postdoctoral researcher for a year at the Mathematical Research Center in Berkeley, California. Perhaps best known for his 1994 work which shows that integer factorization can be performed in polynomial time on a quantum computer, Shor received the Nevanlinna Prize at the 1998 International Congress of Mathematicians, Berlin. (Photo by courtesy of Dr. Shor.)

ley and Tukey[34] in 1965 [53]; there exists also an efficient quantum algorithm for Fourier transform which is a quantum analog of the FFT. Shor first realized that if he could relate the problem of finding factors of a large number to that of finding the period of a function, then he could use Simon's idea for sampling from Fourier transform. Now we are in a position to give Shor's quantum algorithm for integer factorization.

Now we are in a position to give Shor's quantum algorithm for integer factorization.

Algorithm 2.5.1 (Quantum algorithm for integer factorization). Given integers x and n, the algorithm will find the order of x, i.e., the smallest positive integer r such that $x^r \equiv 1 \pmod{n}$. Assume our machine has two quantum registers: Register-1 and Register-2, which hold integers in binary form.

[1] [Initialize] Find a number q, a power of 2, with $n^2 < q < 2n^2$.

[2] [Prepare information for quantum registers] Put in Register-1 the uniform superposition of states representing numbers $a \pmod{q}$, and load Register-2 with all zeros. This leaves the machine in the state $|\Psi_1\rangle$:

$$|\Psi_1\rangle = \frac{1}{\sqrt{q}} \sum_{a=0}^{q-1} |a\rangle\,|0\rangle\,. \tag{2.163}$$

(Note that the joint state of both registers are represented by $|$ Register-1$\rangle$ and $|$ Register-2$\rangle$). What this step does is put each bit in Register-1 into the superposition

$$\frac{1}{\sqrt{2}}\left(|0\rangle + |1\rangle\right).$$

[3] [Create quantum-parallelly all powers] Compute $x^a \pmod{n}$ in Register-2. This leaves the machine in state $|\Psi_2\rangle$:

John Wilder Tukey (1915-2000) as educated at home by his parents who were both teachers; his formal education began only when he entered Brown University, where he earned his bachelor's and master's degrees in chemistry in 1936 and 1937, respectively. He then went to Princeton University in 1937 to study mathematics and obtained his doctorate in 1939. He was a faculty member at Princeton from 1939 to 1970, and in the same time he was a Member of Technical Staff at AT&T Bell Laboratories from 1945 to 1985. In 1965, in a paper with J. W. Cooley published in *Mathematics of Computation*, he introduced the important "Fast Fourier Transform" algorithm, a mathematical technique that greatly simplifies omputation for Fourier series and integrals. For many people this will be the work for which he is best known. However, it is only a small part of a large number of areas with which he made significant contributions.

$$|\Psi_2\rangle = \frac{1}{\sqrt{q}} \sum_{a=0}^{q-1} |a\rangle \, |x^a \pmod n\rangle . \tag{2.164}$$

This step can be done reversibly since all the a's were kept in Register-1.

[4] [Perform a quantum FFT] Apply FFT on Register-1. The FFT maps each state $|a\rangle$ to

$$\frac{1}{\sqrt{q}} \sum_{c=0}^{q-1} \exp(2\pi iac/q) \, |c\rangle . \tag{2.165}$$

That is, we apply the unitary matrix with the (a, c) entry equal to $\frac{1}{\sqrt{q}} \exp(2\pi iac/q)$. This leaves the machine in the state $|\Psi_3\rangle$:

$$|\Psi_3\rangle = \frac{1}{q} \sum_{a=0}^{q-1} \sum_{c=0}^{q-1} \exp(2\pi iac/q) \, |c\rangle \, |x^a \pmod n\rangle . \tag{2.166}$$

[5] [Detect periodicity in x^a] Observe the machine. For clarity, we observe both $|c\rangle$ in Register-1 and $|x^a \pmod n\rangle$ in Register-2, measure both arguments of this superposition, obtaining the values of $|c\rangle$ in the first argument and some $\left|x^k \pmod n\right\rangle$ as the answer for the second one ($0 < k < r$).

[6] [Extract r] Finally extract the required value of r. Given the pure state $|\Psi_3\rangle$, the probabilities of different results for this measurement will be given by the probability distribution:

$$\mathrm{Prob}(c, x^k) = \left| \frac{1}{q} \sum_{a=0}^{q-1} \exp(2\pi iac/q) \right|^2 \tag{2.167}$$

where the sum is over all values of a such that

$$x^a \equiv x^k \pmod n. \tag{2.168}$$

Independent of k, $\mathrm{Prob}(c, x^k)$ is periodic in c with period q/r; but since q is known, we can deduce r with just a few trial executions (this can be accomplished by using a continued fraction expansion).

[7] [Resolution] Once r is found, the factors of n can be *possibly* obtained from computing $\gcd(x^{r/2} - 1, n)$ and $\gcd(x^{r/2} + 1, n)$, that is, the pair of integers (a, b) satisfying

$$(a, b) = \{\gcd(x^{r/2} - 1, n), \ \gcd(x^{r/2} + 1, n)\}$$

could be the nontrivial factors of n. If it fails to produce a nontrivial factor of n, goto step [1] to choose a new base.

Steps [6] and [7] of the algorithm are purely classical computation and hence can be performed on a classical computer. Compared with the best known factoring algorithm NFS with asymptotic running time, as we already know,

$$\mathcal{O}\left(\exp\left(c(\log n)^{1/3}(\log\log n)^{2/3}\right)\right)$$

for some constant c depending on details of the implementation, the quantum factoring algorithm takes asymptotically

$$\mathcal{O}\left((\log n)^2(\log\log n)(\log\log\log n)\right)$$

steps on a quantum computer and $\mathcal{O}(\log n)$ amount of post-processing time on a classical computer that converts the output of the quantum computers. That is, Shor's quantum algorithm can factor integers in time $\mathcal{O}\left((\log n)^{2+\epsilon}\right)$.

It should be noted that Shor's factoring algorithm is probabilistic, not deterministic, that is, it can sometimes fail. In fact, it will fail if

(1) r is odd, in which case $r/2$ is not an integer.

(2) $x^{r/2} \equiv -1 \pmod{n}$, in which case the algorithm yields the trivial factors 1 and n.

For example, when $n = 21 = 3 \cdot 7$, we have the related values for the order of x modulo 21 for $x = 1, 2, \cdots, 20$ and $\gcd(x, 21) = 1$ (the order may not exist for some x when $\gcd(x, 21) \neq 1$) as shown in Table 2.3. Thus, of all the 20

Table 2.3. Various values about the order of x modulo n

x	r	r odd	$x^{r/2} \equiv -1 \pmod{n}$
1	1	Yes	
2	6		
4	3	Yes	
5	6		Yes
8	2		
10	6		
11	6		
13	2		
16	3	Yes	
17	6		Yes
19	6		
20	2		Yes

cases, Shor's algorithm only applies to 12 cases in which the order r exists[35] and of these 12 cases six (exactly half) will fail, since three have r odd and three $x^{r/2} \equiv -1 \pmod{n}$ (see Table 2.3). Thus, in this particular example,

[35] Note that the order r of x modulo 21 exists if and only if $\gcd(x, 21) = 1$ for $x = 1, 2, \cdots, 20$. Recall that (see Theorem 1.2.19 in Chapter 1) if two integers a and b are chosen at random, then $\text{Prob}[\gcd(a, b) = 1] = 0.6$. Thus, about 40% of the x will fail to produce an order r. For example, when $n = 21$, the order r will not exist for $x = 3, 6, 7, 9, 12, 14, 15, 18$.

about 70% of the values of x cannot lead to a successful factorization of n. Generally, when r exists (that is, $\mathbb{Z}/n\mathbb{Z}$ forms a multiplicative group), Shor's algorithm will produce a nontrivial factor of n with probability $\geq 1-1/2^{k-1}$, where k is the number of distinct odd prime factors of n. In the case $n = 21$, this probability is $1 - 1/2^{2-1} = 1/2$, which agrees with the calculation in Table 2.3. In public-key cryptography (see Chapter 3 of this book), however, the integers to be factored are specifically chosen with two prime factors, each having the same size, thus Shor's algorithm will fail for about 50% of the values of r, and hence is not very useful. The main problem here is that Shor's factoring algorithm is not really a factoring algorithm, but rather an algorithm for finding the order of element x modulo n, which will lead to a successful factorization of n for only about half of the values of r. In the author's opinion, a good quantum algorithm would be the quantum version of the best classical factoring algorithm such as Number Field Sieve (NFS) or Quadratic Sieve (QS).

2.5.4 Quantum Algorithms for Discrete Logarithms

It is clear that the finding of the order of x modulo n is related to the computation of discrete logarithms. Recall that the discrete logarithm problem may be described as: given a prime p, a generator g of the multiplicative group modulo p, and an x modulo p, find an integer r with $0 \leq r < p-1$ such that $g^r \equiv x \pmod{p}$. As a by-product, the quantum factoring algorithm can also be used, of course with some modifications, for the computation of discrete logarithms. The following is a sketch of Shor's algorithm for computing discrete logarithms.

Algorithm 2.5.2 (Quantum algorithm for discrete logarithms). Given $g, x \in \mathbb{N}$ and p prime. This algorithm will find the integer r such that $g^r \equiv x \pmod{p}$ if r exists. It uses three quantum registers.

[1] Find q a power of 2 such that q is close to p, that is, $p < q < 2p$.

[2] Put in the first two registers of the quantum computer the uniform superposition of all $|\, a\rangle$ and $|\, b\rangle \pmod{p-1}$, and compute $g^a x^{-b} \pmod{p}$ in the third register. This leaves the quantum computer in the state $|\, \Psi_1\rangle$:

$$|\, \Psi_1\rangle = \frac{1}{p-1}\sum_{a=0}^{p-2}\sum_{b=0}^{p-2} |\, a,\ b,\ g^a x^{-b} \pmod{p}\rangle . \tag{2.169}$$

[3] Use the Fourier transform A_q to map $|\, a\rangle \to |\, c\rangle$ and $|\, b\rangle \to |\, d\rangle$ with probability amplitude

$$\frac{1}{q}\exp\left(\frac{2\pi i}{q}(ac+bd)\right).$$

Thus, the state $|a, b\rangle$ will be changed to the state:

$$\frac{1}{q}\sum_{c=0}^{q-1}\sum_{d=0}^{q-1}\exp\left(\frac{2\pi i}{q}(ac+bd)\right)|c,\ d\rangle . \tag{2.170}$$

This leaves the machine in the state $|\Psi_2\rangle$:

$$|\Psi_2\rangle = \frac{1}{(p-1)q}\sum_{a,b=0}^{p-2}\sum_{c,d=0}^{q-1}\exp\left(\frac{2\pi i}{q}(ac+bd)\right)|c,\ d,\ g^a x^{-b} \pmod p\rangle . \tag{2.171}$$

[4] Observe the state of the quantum computer and extract the required information. The probability of observing a state $\left| c,\ d,\ g^k \pmod p \right\rangle$ is

$$\left|\frac{1}{(p-1)q}\sum_{a,b}\exp\left(\frac{2\pi i}{q}(ac+bd)\right)\right|^2 \tag{2.172}$$

where the sum is over all (a, b) such that

$$a - rb \equiv k \pmod{p-1}. \tag{2.173}$$

The better outputs (observed states) we get, the more chance of deducing r we will have; readers are referred to Shor's original paper for a justification.

The above quantum discrete logarithm algorithm uses only *two* modular exponentiations and *two* quantum Fourier transformations. It is significantly faster than any classical discrete logarithm algorithm.

As many important computational problems have been proven to be $\mathcal{NP}$-complete, quantum computers will not likely become widely useful unless they can solve $\mathcal{NP}$-complete problems. At present, we do not know whether or not a quantum computer can solve an $\mathcal{NP}$-complete problem, although there are some weak indications that quantum computers are not powerful enough to solve $\mathcal{NP}$-complete problems (Bennett et al., [26]). It is worthwhile pointing out that at present no-one knows how to build a quantum computer. Even if such a computer could *in principle* be constructed, there are still enormous technical issues to overcome before reaching this goal. Much work needs to be done! Despite the great difficulty of constructing a truly general-purpose quantum computer, it might be relatively easy to construct a special-purpose quantum factoring machine which could be used for code-breaking. History does have a tendency to repeat itself; were not the first digital computers used for code-breaking?

2.6 Miscellaneous Algorithms in Number Theory

We have, so far, introduced in this chapter three important types of algorithms for primality testing, integer factorization and discrete logarithms. There are, however, many other algorithms for solving different sorts of number-theoretic problems. This section aims to provide some algorithms for computing the exact value of $\pi(x)$, for verifying Goldbach's conjecture and for generating amicable numbers. Many important algorithms in computational number theory, such as those for computing the nontrivial zeros of the Riemann ζ-function and those for checking the odd perfect numbers, are omitted; it would be impossible for a single book to contain discussions on all sorts of algorithms in computational number theory.

2.6.1 Algorithms for Computing $\pi(x)$

In Section 1.5 of Chapter 1, we studied the asymptotic behaviour of the prime counting function $\pi(x)$ (recall that $\pi(x)$ is the number of primes up to x). In this subsection, we shall discuss some modern methods for calculating the exact values of $\pi(x)$.

The most straightforward method is, of course, to use the ancient sieve of Eratosthenes to find and count all the primes up to x. According to the Prime Number Theorem (PNT), it is not possible to have a method that computes $\pi(x)$ with less than about $x/\ln x$ arithmetic operations. Despite its time complexity, the sieve of Eratosthenes was for a very long time the practical way to compute $\pi(x)$. In the second half of the 19th century, the German astronomer Meissel[36] discovered a practical combinatorial method that does not need to find all primes $p \le x$. He used his method to compute by hand $\pi(10^8)$ and $\pi(10^9)$. In 1959, Lehmer extended and simplified Meissel's method (now widely known as the Meissel-Lehmer method, and he used the method on an IBM 701 computer to obtain the value of $\pi(10^{10})$. In 1985,

36

Daniel Friedrich Ernst Meissel (1826–1895) studied at the University of Berlin working under Jacobi. He also had contacts with Dirichlet. His doctorate was from Halle. He taught in a number of places, including in Kiel from 1871 until the end of his life. Meissel's mathematical work covers a number of areas. He worked on prime numbers giving the result that there are 50847478 primes less than 10^9. Lehmer showed, 70 years later, that this is 56 too few. In addition to other number theory work on Möbius inversion and the theory of partitions, Meissel wrote on Bessel functions, asymptotic analysis, refraction of light and the three body problem. His main skill was in numerical calculations and manipulation of complicated expressions.

Lagarias[37], Miller and Odlyzko[38] adapted the Meissel-Lehmer method and proved that it is possible to compute $\pi(x)$ with $\mathcal{O}\left(x^{2/3}/\ln x\right)$ operations and using $\mathcal{O}(x^{1/3}\ln^2\ln\ln x)$ space. They used their method to compute several values of $\pi(x)$ up to $x = 4\cdot 10^{16}$. More recently, Deleglise and Rivat [59] proposed a modified form of the Lagarias, Miller and Odlyzko method, which computes $\pi(x)$ with $\mathcal{O}\left(x^{2/3}/\ln^2 x\right)$ operations and using $\mathcal{O}(x^{1/3}\ln^3 x\ln\ln x)$ space. They used this method to compute several values of $\pi(x)$ for x up to 10^{19}. In what follows, we shall first introduce a simple form of Meissel's method:

Theorem 2.6.1. If $p_1, p_2, \cdots, p_k$ are the primes less than or equal to $\sqrt{n}$, then the formula for computing $\pi(x)$ is:

$$\begin{aligned}\pi(n) &= n - 1 + \pi(\sqrt{n}) - \left\{\left[\frac{n}{p_1}\right] + \left[\frac{n}{p_2}\right] + \cdots + \left[\frac{n}{p_k}\right]\right\} \\ &\quad + \left\{\left[\frac{n}{p_1p_2}\right] + \left[\frac{n}{p_1p_3}\right] + \cdots + \left[\frac{n}{p_1p_k}\right] + \left[\frac{n}{p_2p_3}\right] + \cdots + \left[\frac{n}{p_{k-1}p_k}\right]\right\} \\ &\quad - \left\{\left[\frac{n}{p_1p_2p_3}\right] + \cdots + \left[\frac{n}{p_{k-2}p_{k-1}p_k}\right]\right\} + \cdots + \\ &\quad (-1)^k\left[\frac{n}{p_1p_2\cdots p_k}\right]. \end{aligned} \tag{2.174}$$

37

Jeffrey C. Lagarias is a member of the Mathematics and Cryptography Research Department at AT&T Research Labs in Florham Park, New Jersey. He is a very active research scientist with more than 120 papers in number theory, Diophantine approximation, dynamical systems, harmonic analysis, discrete geometry, mathematical programming and optimization, computational complexity theory, cryptography and neural networks. (Photo by courtesy of Dr. Lagarias.)

38

A well-known scientist in computational number theory, computational complexity, coding and cryptography, Andrew M. Odlyzko studied Mathematics at the California Institute of Technology and obtained his PhD in Mathematics at the Massachusetts Institute of Technology in 1975. He is currently the head of the Mathematics and Cryptography Research Department at AT&T Research Labs in Florham Park, New Jersey. Odlyzko has made significant contributions to several central areas of number theory and cryptography. (Photo by courtesy of Schwarz and Wolfgang [223].)

Example 2.6.1. We shall show in this example how to use the Meissel's method to compute $\pi(129)$. First note that $\pi(\sqrt{129}) = 5$: the primes less than or equal to $\sqrt{129}$ are $2, 3, 5, 7$ and 11. By (2.174), we have:

$$\begin{aligned}
\pi(129) &= 129 - 1 + 5 \\
&\quad - \left\{ \left[\frac{129}{2}\right] + \left[\frac{129}{3}\right] + \left[\frac{129}{5}\right] + \left[\frac{129}{7}\right] + \left[\frac{129}{11}\right] \right\} \\
&\quad + \left[\frac{129}{2\cdot 3}\right] + \left[\frac{129}{2\cdot 5}\right] + \left[\frac{129}{2\cdot 7}\right] + \left[\frac{129}{2\cdot 11}\right] + \left[\frac{129}{3\cdot 5}\right] + \left[\frac{129}{3\cdot 7}\right] \\
&\quad + \left[\frac{129}{3\cdot 11}\right] + \left[\frac{129}{5\cdot 7}\right] + \left[\frac{129}{5\cdot 11}\right] - \left[\frac{129}{2\cdot 3\cdot 5}\right] \\
&\quad - \left[\frac{129}{2\cdot 3\cdot 7}\right] - \left[\frac{129}{2\cdot 3\cdot 11}\right] - \left[\frac{129}{2\cdot 5\cdot 7}\right] - \left[\frac{129}{2\cdot 5\cdot 11}\right] \\
&\quad - \left[\frac{129}{3\cdot 5\cdot 7}\right] - \left[\frac{129}{3\cdot 5\cdot 11}\right] - \left[\frac{129}{3\cdot 7\cdot 11}\right] - \left[\frac{129}{5\cdot 7\cdot 11}\right] \\
&\quad + \left[\frac{129}{2\cdot 3\cdot 5\cdot 7}\right] + \left[\frac{129}{2\cdot 3\cdot 5\cdot 11}\right] + \left[\frac{129}{2\cdot 3\cdot 7\cdot 11}\right] \\
&\quad + \left[\frac{129}{2\cdot 5\cdot 7\cdot 11}\right] + \left[\frac{129}{3\cdot 5\cdot 7\cdot 11}\right] - \left[\frac{129}{2\cdot 3\cdot 5\cdot 7\cdot 11}\right] \\
&= 129 - 1 + 5 - 64 - 43 - 25 - 18 - 11 + 21 + 12 \\
&\quad + 9 + 5 + 8 + 6 + 3 + 3 + 2 + 1 - 4 - 4 - 1 - 1 - 1 \\
&\quad - 0 - 1 - 0 - 0 - 0 + 0 + 0 + 0 + 0 + 0 - 0 \\
&= 31.
\end{aligned}$$

That is, there are exactly 31 primes up to 129. It is of course true, since the following are the only primes ≤ 129:

$$2,\ 3,\ 5,\ 7,\ 11,\ 13,\ 17,\ 19,\ 23,\ 29,\ 31,\ 37,\ 41,\ 43,\ 47,\ 53,\ 59,\ 61,\ 67,\ 71,$$
$$73,\ 79,\ 83,\ 89,\ 97,\ 101,\ 103,\ 107,\ 109,\ 113,\ 127.$$

We are now in a position to introduce a modern algorithm for computing $\pi(x)$, due to Meissel, Lehmer, Lagarias, Miller, Odlyzko, Deleglise and Rivat [59].

Theorem 2.6.2. Let $p_1, p_2, p_3, \cdots$ denote the primes $2, 3, 5, \cdots$ in increasing order. Let $\phi(x, a)$ denote the partial sieve function, which counts numbers $\leq x$ with all prime factors greater than p_a:

$$\phi(x, a) = |\{n \leq x; p \mid n \Longrightarrow p > p_a\}| \tag{2.175}$$

and let

$$P_k(x, a) = \#\{n \leq x;\ n = q_1 q_2 \cdots q_k \text{ and } q_1, q_2, \cdots q_k > p_a\} \tag{2.176}$$

which counts numbers $\leq x$ with exactly k prime factors, all larger than p_a. If we sort the numbers $\leq x$ by the number of their prime factors greater than p_a, we obtain

$$\phi(x, a) = P_0(x, a) + P_1(x, a) + \cdots + P_k(x, a) + \cdots \tag{2.177}$$

where the sum on the right-hand side has only finitely many non-zero terms, since $P_k(x, a) = 0$ for $P_a^k > x$. Finally let y be an integer such that $x^{1/3} \leq y \leq x^{1/2}$ and let $a = \pi(y)$. Then we have

$$\pi(x) = \phi(x, a) - P_2(x, a) + a - 1. \tag{2.178}$$

That is, $\pi(x)$ can be obtained by the calculation of $\phi(x, a)$ and $P_2(x, a)$.

Thus, the algorithm for computing $\pi(x)$ has just the following two main steps:

[1] Computing $P_2(x, a)$.

[2] Computing $\phi(x, a)$.

In what follows, we shall give a brief description of the two main computation steps in the algorithm.

[1] Computing $P_2(x, a)$. Note first that

[1-1] count all the prime pairs (p, q) such that $y < p \leq q$ and $pq \leq x$,

[1-2] $p \in [y + 1, \sqrt{x}]$,

[1-3] for each $p, q \in [p, x/p]$.

Then, we have

$$P_2(x, a) = \sum_{y < p \leq \sqrt{x}} \left(\pi\left(\frac{x}{p}\right) - \pi(p) + 1 \right). \tag{2.179}$$

The computation of $P_2(x, a)$ can be done in $\mathcal{O}(x/y \ln \ln x)$ time and $\mathcal{O}(y)$ space.

[2] Computing $\phi(x, a)$. The formula for computing $\phi(x, a)$ can be expressed as follows:

$$\phi(x, a) = S_0 + S, \tag{2.180}$$

where

$$S_0 = \sum_{n \le y} \mu(n) \left[\frac{x}{n}\right], \tag{2.181}$$

$$S = \sum_{n/\delta(n) \le y < n} \mu(n)\phi\left(\frac{x}{n}, \pi(\delta(n)) - 1\right), \tag{2.182}$$

and $\delta(n)$ denotes the smallest prime factor of n. The computation of S_0 can be done in $\mathcal{O}(y \ln \ln x)$ time which is negligible. Thus, the main computation of $\phi(x, a)$ is in computing S, which can be computed by

$$S = S_1 + S_2 + S_3, \tag{2.183}$$

where

$$S_1 = -\sum_{x^{1/3} < p \le y} \sum_{\substack{\delta(m) > p \\ m \le y < mp}} \mu(m)\phi\left(\frac{x}{mp}, \pi(p) - 1\right) \tag{2.184}$$

$$S_2 = -\sum_{x^{1/4} < p \le x^{1/3}} \sum_{\substack{\delta(m) > p \\ m \le y < mp}} \mu(m)\phi\left(\frac{x}{mp}, \pi(p) - 1\right) \tag{2.185}$$

$$S_3 = -\sum_{p \le x^{1/4}} \sum_{\substack{\delta(m) > p \\ m \le y < mp}} \mu(m)\phi\left(\frac{x}{mp}, \pi(p) - 1\right). \tag{2.186}$$

The computation of S_1 can be done in constant time which is negligible, since S_1 can be written as follows:

$$\begin{aligned} S_1 &= \sum_{x^{1/3} < p \le y} \sum_{p < q \le y} \phi\left(\frac{x}{pq}, \pi(p) - 1\right) \\ &= \frac{(\pi(y) - \pi(x^{1/3}))(\pi(y) - \pi(x^{1/3}) - 1)}{2}. \end{aligned} \tag{2.187}$$

However, the computations for S_2 and S_3 are more complicated and need to be broken down into several small steps in order to speed up the computation. For example, we can split S_2 in the following way:

$$\begin{aligned} S_2 &= \sum_{x^{1/4} < p \le x^{1/3}} \sum_{p < q \le y} \phi\left(\frac{x}{pq}, \pi(p) - 1\right) \\ &= U + V = U + V_1 + V_2 \\ &= U + V_1 + W_1 + W_2 + W_3 + W_4 + W_5, \end{aligned} \tag{2.188}$$

where

$$U = \sum_{\sqrt{x/y} < p < x^{1/3}} \left(\pi(y) - \pi\left(\frac{x}{p^2}\right)\right) \tag{2.189}$$

$$V_1 = \sum_{x^{1/4} < p \le x^{1/3}} \sum_{p < q \le \min(x/p^2, y)} (2 - \pi(p)) \tag{2.190}$$

$$W_1 = \sum_{x^{1/4}<p\leq x^x/p^2} \sum_{p<q\leq y} \pi\left(\frac{x}{pq}\right) \tag{2.191}$$

$$W_2 = \sum_{x/y^2<p\leq\sqrt{x/y}} \sum_{p<q\leq\sqrt{x/y}} \pi\left(\frac{x}{pq}\right) \tag{2.192}$$

$$W_3 = \sum_{x/y^2<p\leq\sqrt{x/y}} \sum_{\sqrt{x/p}<q\leq y} \pi\left(\frac{x}{pq}\right) \tag{2.193}$$

$$W_4 = \sum_{\sqrt{x/y}<p\leq x^{1/3}} \sum_{p<q\leq\sqrt{x/p}} \pi\left(\frac{x}{pq}\right) \tag{2.194}$$

$$W_5 = \sum_{\sqrt{x/y}<p\leq x^{1/3}} \sum_{\sqrt{x/p}<q\leq x/p^2} \pi\left(\frac{x}{pq}\right). \tag{2.195}$$

It is clear that the *significant* time and space are spent only on all computations of $P_2(x, a)$, W_1, W_2, W_3, W_4, W_5, S_3. As analyzed by Deleglise and Rivat [59], the space complexity $s(\cdot)$ and the time complexity $t(\cdot)$ of the algorithm are:

$$\left.\begin{aligned} \mathrm{s}(\pi(x)) &= \mathcal{O}(y) \\ \mathrm{t}(\pi(x)) &= \mathcal{O}\left(\frac{x}{y}\ln\ln x + \frac{x}{y}\ln x \ln\ln x + x^{1/4}y + \frac{x^{2/3}}{\ln^2 x}\right). \end{aligned}\right\} \tag{2.196}$$

If we choose $y = x^{1/3}\ln^3 x \ln\ln x$, then

$$\mathrm{t}(\pi(x)) = \frac{x^{2/3}}{\ln^2 x}. \tag{2.197}$$

2.6.2 Algorithms for Generating Amicable Pairs

Recall that a pair of positive integers (m, n) is an amicable pair if $\sigma(m) = \sigma(n) = m + n$. There are three essentially different methods for generating amicable pairs: exhaustive numerical method, algebraic assumption method and algebraic constructive method. In 1993, Herman J. J. te Riele [204] proposed a new method based on the following observation of Paul Erdös:

Proposition 2.6.1. Given a positive integer s, if $x_1, x_2, \cdots$ are integer solutions of the equation

$$\sigma(x) = s, \tag{2.198}$$

then any pair (x_i, x_j) for which $x_i + x_j = s$ is an amicable pair.

Proof. Since $\sigma(x_i) = \sigma(x_j) = s$ and $x_i + x_j = s$. □

Since this method has a strong connection with numerical methods, although it is not exhaustive, we call it a *seminumerical method.* From a computational point of view, it should not be too difficult to calculate all the integers within a certain range, which have the same σ-value. For example for $200 \leq x \leq 300$, all the following integers

$$204, \ 220, \ 224, \ 246, \ 284, \ 286$$

have the same σ-value 504, but of course, only $220 + 284 = 504$. Instead of randomly selecting the numbers $x_1, x_2, \cdots$ as candidates for the solution of (2.198), Te Riele used the so-called *smooth* numbers (i.e., numbers with only small prime factors) as candidates to the solution of (2.198). His idea is based on the number-theoretic fact that if $\sigma(p^\alpha) \mid s$ where p is a prime and α a positive integer, and if $\sigma(y) = s/\sigma(p^\alpha)$ has a solution y with $\gcd(y,p) = 1$, then $x = yp^\alpha$ is a solution to $\sigma(x) = s$. Based on this fact, a recursive algorithm can be designed to find a factor $\sigma(p^\alpha)$ of s and to solve $\sigma(y) = s/\sigma(p^\alpha)$ with $\gcd(p,y) = 1$. The following is the algorithm.

Algorithm 2.6.1 (Te Riele's algorithm for amicable numbers). For a given integer s, first find as many solutions x as possible to the equation $\sigma(x) = s$, where $2 \mid x$ or $3 \mid x$ but $6 \nmid x$, and then find amicable pairs among the pairs of solutions found. Two tables T_1 and T_2 are used to store the information concerning the triples (p, α, p^α); the ith triple from table T_1 (resp. T_2) is denoted by T_{1i} (resp. T_{2i}). Solutions are stored in $x_1, x_2, \cdots$. Choose upper bounds B_1 and B_2 for the σ-values admitted in T_1 and T_2, respectively.

[1] [Precomputation of Tables of $\sigma(p^\alpha)$-values]. Fill table T_1 with triples $(p, \alpha, \sigma(p^\alpha))$ for $p = 2, 3$ and for those integers $\alpha = 1, 2, \cdots$ for which $\sigma(p^\alpha) < B_1$. Similarly, fill table T_2 with all primes $p = 5, 7, \cdots$ and integers $\alpha = 1, 2, \cdots$ for which $\sigma(p^\alpha) < B_2$, such that the $\sigma(p^\alpha)$-values are in *increasing* order. Set i_{max} and j_{max} to the number of triples in Table T_1 and T_2, respectively.

[2] [Initialization]. Set $d \leftarrow 1$, $s_d \leftarrow s$, $i \leftarrow 0$, $n \leftarrow 0$. The current value of d indicates that the dth prime factor of x is being looked for (so $d-1$ prime factors have been found so far). The integer s_d is the current value of s for which $\sigma(x) = s$ is being solved; j_d $(d \geq 2)$ records the location in Table T_2 where a prime power factor of x has been found; p_d and e_d are the prime and the corresponding exponent in that prime power.

[3] [Select next triple from T_1]. Set $i \leftarrow i + 1$. If $i > i_{max}$ goto step [8], otherwise set $(p, \alpha, \sigma) \leftarrow T_{1i}$. If $\sigma \nmid s_1$, repeat step [3], otherwise set

$$p_1 \leftarrow p, \ \alpha_1 \leftarrow \alpha, \ d \leftarrow 2, \ s_2 \leftarrow s_1/\sigma, \ sq \leftarrow \sqrt{s_2}, \ j \leftarrow 0.$$

[4] [Select next triple from T_2]. Set $j \leftarrow j+1$. If $j > j_{max}$ goto step [5], otherwise set $(p, \alpha, \sigma) \leftarrow T_{2j}$. If $p = p_l$ for some $l \in \{2, 3, \cdots, d-1\}$, repeat step [4]. If $\sigma > sq$, goto step [6]. If $\sigma \mid s_d$, set

$$j_d \leftarrow j, \ p_d \leftarrow p, \ \alpha_d \leftarrow \alpha, \ s_{d+1} \leftarrow s_d/\sigma, \ sq \leftarrow \sqrt{s_{d+1}}, \ d \leftarrow d+1.$$

Repeat step [4].

[5] [Check if $s_d - 1$ is prime]. If $s_d - 1$ is prime, set

$$n \leftarrow n+1, \ x_n \leftarrow (s_d - 1) \prod_{k=1}^{d-1} p_k^{\alpha_k}.$$

Goto step [7].

[6] [Check if s_d occurs as σ-value in table T_2]. If there exists l and $T_{2l} = (p, \alpha, \sigma)$ with $\sigma = s_d$, set

$$n \leftarrow n+1, \ x_n \leftarrow p^{\alpha} \prod_{k=1}^{d-1} p_k^{\alpha_k}.$$

Goto step [7].

[7] [Decrease depth d]. Set $d = d - 1$. If $d = 1$, goto step [3], otherwise set $j = j_d$, $sq \leftarrow \sqrt{s_d}$; return to step [4].

[8] [Sort solutions and check pair sums]. Sort the solutions $x_1, x_2, \cdots, x_n$ in decreasing order and find amicable pairs, i.e., pairs (x_i, x_j) with $x_i + x_j = s$. Notice that the size of the members of amicable pairs found is close to $s/2$.

Example 2.6.2. Applying the above algorithm with $B_1 = 70$, $B_2 = 100$ to $s = 504$ yields the following five solutions

$$(x_1, x_2, x_3, x_4, x_5) = (286, 334, 220, 284, 224)$$

where $(x_3, x_4) = (220, 284)$ is an amicable pair since $x_3 + x_4 = s$. The equation $\sigma(x) = 504$ has five more solutions, viz., $x = \{204, 246, 415, 451, 503\}$, but the first two of them are divisible by 6, the next two of them have a smallest prime factor > 3 (namely, 5 and 11), whereas the last one (i.e., 503) is a prime. Such solutions were excluded from the algorithm.

Let N_s be the number of solutions x of $\sigma(x) = s$ found by the algorithm. What is the probability of finding an amicable pair (x_1, x_2) among the N_s solutions for which $x_1 + x_2 = s$? Te Riele's experiment shows that if $N_s \approx \sqrt{s}$, and if solutions are "randomly" distributed in $[1, s]$, then the values of $N_s/\sqrt{s}$ are in the range $(0.01, 0.43)$ (see Table 2.4 for more information).

Table 2.4. $N_s/\sqrt{s}$ for $s = i!$, $i = 8, 9, \cdots, 15$

s	N_s	$N_s/\sqrt{s}$
8!	30	0.1494
9!	187	0.3014
10!	593	0.3113
11!	1665	0.2635
12!	6999	0.3198
13!	25656	0.3251
14!	110137	0.3730
15!	442439	0.3869
$2 \cdot 14!$	163869	0.3924
$3 \cdot 14!$	200965	0.3930
$4 \cdot 14!$	236219	0.4000
$8 \cdot 14!$	331105	0.3965
$12 \cdot 14!$	449253	0.4392

2.6.3 Algorithms for Verifying Goldbach's Conjecture

As mentioned in Section 1 of Chapter 1, Goldbach in 1742 made a famous conjecture concerning the representation of an integer as a sum of prime numbers. Goldbach's conjecture, after some rephrasing, may be expressed as follows:

(1) Binary Goldbach Conjecture (BGC): Every even number ≥ 6 is the sum of two odd primes. For example, $6 = 3 + 3$, $8 = 3 + 5$, $10 = 3 + 7, 12 = 5 + 7, \cdots$.

(2) Ternary Goldbach Conjecture (TGC): Every odd number greater than 7 is the sum of three odd primes. For example, $9 = 3 + 3 + 3$, $11 = 3 + 3 + 5, 13 = 3 + 5 + 5, 15 = 3 + 5 + 7, \cdots$.

Clearly, the binary Goldbach conjecture (BGC) implies the ternary Goldbach conjecture (TGC). Much work has been done on this conjecture by many of famous mathematicians, including Hardy and Littlewood, though these conjectures still have not been completely solved yet. The best known results concerning Goldbach's Conjectures can be summarized as follows (here we let N_0 denote a sufficiently large even number, P_1, P_2, P_3 and P_4 be primes, E the even number ≥ 6, O the odd number ≥ 7, and GRH the Generalized Riemann Hypothesis):

(1) Binary Goldbach Conjecture:

(i) Theoretical Result:

(a) Unconditionally, every sufficiently large even number can be represented as a sum of one prime number and a product of at most two prime numbers. That is, $E = P_1 + P_2 \cdot P_3$ with $E \geq N_0$. This result was proved by J. R. Chen [46] in 1973.

(b) Assuming GRH, every even number can be represented as a sum of at most four prime numbers. That is, $E = P_1 + P_2 + P_3 + P_4$ under GRH. This result is a consequence of Kaniecki, and Deshouillers, Effinger, Te Riele and Zinoviev [62]. (Ramaré proved that unconditionally every even number can be represented as a sum of at most six prime numbers.)

(ii) Numerical Result: BGC is true up to $4 \cdot 10^{14}$ (Richstein [201]).

(2) Ternary Goldbach Conjecture:

(i) Theoretical Result:

(a) Unconditionally, TGC is true for all odd numbers $\geq 10^{43000}$; this is a refinement of Chen and Wang over Vinogradov's famous *three-prime theorem*.

(b) Assuming GRH, every odd number ≥ 7 can be represented as a sum of three prime numbers. That is, $O = P_1 + P_2 + P_3$ under GRH. This result is due to Deshouillers, Effinger, Te Riele and Zinoviev [61].

(ii) Numerical Result: The TGC is true up to 10^{20}. It was verified by Saouter [216] in 1995.

The above results are diagrammatically shown in Figure 2.6. Readers may also find the historic computation results (see Table 2.5) concerning the BGC interesting. In what follows, we shall introduce two algorithms for verifying Goldbach's conjecture.

Table 2.5. Historic computation results concerning BGC

Verified by	Date	Limit
A. Desboves	1855	10^4
N. Pipping	1940	10^5
M. K. Shen	1964	$3.3 \cdot 10^7$
M. L. Stein and P. R. Stein	1965	10^8
A. Granville, J. v. d. Lune and H. J. J. te Riele	1989	$2 \cdot 10^{10}$
M. Sinisalo	1993	$4 \cdot 10^{11}$
J. M. Deshouillers, H. J. J. te Riele and Y. Saouter	1989	10^{14}
J. Richstein	1998	$4 \cdot 10^{14}$

First, let us introduce an algorithm for verifying TGC, based on Saouter [216] who used it to verify TGC up to 10^{20}. Observe that, if n is an odd number, p a prime, and $n-p$ the sum of two primes, then n is the sum of three primes. It is already known that BGC is true up to $4 \cdot 10^{14}$ (Richstein [201]). Thus, if n is an odd number and there exists a prime p such that $n-p < 4 \cdot 10^{11}$, then n is the sum of three primes. So Saouter's algorithm just amounts to exhibiting an increasing sequence of primes

$$p_0, \ p_1, \ p_2, \ \cdots, \ p_l$$

such that

$$\left.\begin{array}{l} p_0 < 4 \cdot 10^{11}, \\ p_i - p_{i-1} < 4 \cdot 10^{11}, \quad i = 1, 2, \cdots, l \text{ and } p_l > 10^{20}. \end{array}\right\} \tag{2.199}$$

Saouter's algorithm has the following form:

Algorithm 2.6.2 (Saouter's algorithm for verifying the TGC). This algorithm verifies TGC in the interval $[4 \cdot 10^{11}, 10^{20}]$.

[1] (Initialization) Set $i = 0$ and let $p_i = 33 \cdot 2^{22} + 1 = 138412033$ (a prime number).

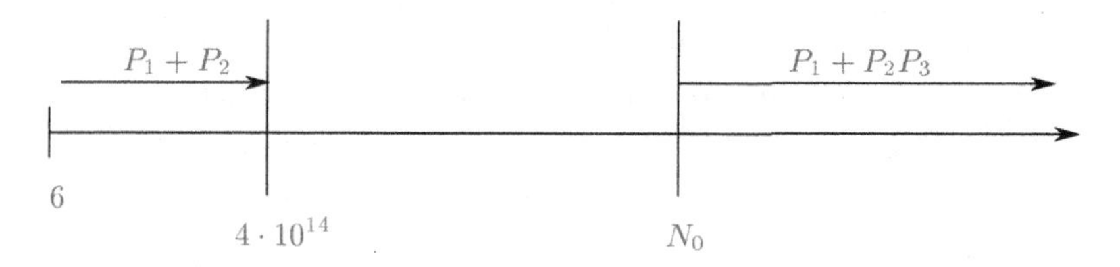

(i) Binary Golbach conjecture

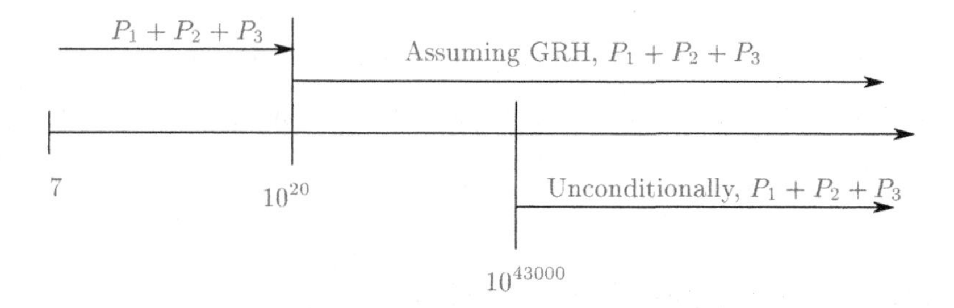

(ii) Tertnary Goldback conjecture

Figure 2.6. Best known results on Goldbach's conjecture

[2] (Is p_i Prime?) If p_i is a prime, then

$$\left.\begin{array}{l} i = i+1, \\ p_i = p_{i-1} + 95360 \cdot 2^{22} \end{array}\right\} \quad (2.200)$$

else

$$p_i = p_i - 10 \cdot 2^{22}. \quad (2.201)$$

[3] ($p_i < 10^{20}$?) If $p_i < 10^{20}$, then goto Step [2], else terminate the algorithm.

According to Saouter [216], if the algorithm terminates, then TGC is true up to 10^{20}. However the converse is not true.

Next comes the algorithm for verifying BGC, originally used by Shen but explicitly given by Deshouillers, Saouter and Te Riele in [62] (hence we called it the DSR algorithm), and also subsequently used by Richstein [201] for verifying BGC up to $4 \cdot 10^{14}$. The algorithm can be briefly described as follows:

Algorithm 2.6.3 (DSR algorithm for verifying the BGC). This algorithm tries to verify BGC on a given interval $[a, b]$ by finding two sets P and Q such that $P + Q$ covers all the even primes in the given interval $[a, b]$. Let P_i be the ith odd prime number.

[1] [Generating/Sieving the Primes in Sets P and Q] For every even number $e \in [a, b]$, find a prime q, close to a, for which $b - a$ is a prime. (This amounts to choosing for P the set of odd primes up to about $b - a$ and for Q the k largest primes $q_1 < q_2 < \cdots < q_k$ below a, for suitable k; both the generation of P and Q can be done by the sieve of Eratosthenes.)

[2] [Checking whether $P + Q$ Covers all the Even Numbers in $[a, b]$] Generate the sets

$$F_0 \subset F_1 \subseteq F_2 \subseteq \cdots$$

defined by

$$\left.\begin{array}{l} F_0 = \emptyset, \\ F_i = F_{i-1} \cup (P + q_{i+1}), \quad i = 1, 2, \cdots, \end{array}\right\} \quad (2.202)$$

until for some j the set F_j covers all the even numbers in $[a, b]$.

There are several different implementations of this classical algorithm on different types of machines, varying from SGI Workstations to the Cray C916 supercomputer. Interested readers are referred to Deshouillers, Te Riele and Saouter [62] and Richstein [201] for more information.

2.6.4 Algorithm for Finding Odd Perfect Numbers

Recall that a positive integer n is perfect if $\sigma(n) = 2n$. All known (in fact, thirty-seven at present) perfect numbers are even perfect, and there is a one-to-one correspondence between even perfect numbers and Mersenne primes, that is, $2^{p-1}(2^p - 1)$ is perfect whenever $2^p - 1$ is prime. However, no odd perfect number has ever been found, and the question of the existence of an odd perfect number has become one of the most notorious problems in number theory. In this subsection, we shall outline an algorithm, based on Brent, Cohen and Te Riele [39], for finding an odd perfect number less than a given bound B (if one exists) or proving that there is none, by checking each odd number $N < B$.

First note a simple fact, due to J. J. Sylvester, that any odd perfect number has at least *three* prime factors, since if $n = p^\alpha q^\beta$ is perfect, where p and q are distinct odd primes, then a contradiction is reached as follows:

$$\begin{aligned}
2 &= \frac{\sigma(n)}{n} \\
&= \frac{\sigma(p^\alpha)}{p^\alpha} \cdot \frac{\sigma(q^\beta)}{q^\beta} \\
&= \left(1 + \frac{1}{p} + \frac{1}{p^2} + \cdots + \frac{1}{p^\alpha}\right) \cdot \left(1 + \frac{1}{q} + \frac{1}{q^2} + \cdots + \frac{1}{q^\beta}\right) \\
&< \left(1 + \frac{1}{3} + \frac{1}{9} + \frac{1}{27} + \cdots\right) \cdot \left(1 + \frac{1}{5} + \frac{1}{25} + \frac{1}{125} + \cdots\right) \\
&= \frac{3}{2} \cdot \frac{5}{4} \\
&< 2.
\end{aligned}$$

Now let $n = \prod_{i=1}^{k} p_i^{\alpha_i}$, then $\sigma(n) = \prod_{i=1}^{k} \sigma(p_i^{\alpha_i})$. Hence,

$$2 = \frac{\sigma(n)}{n} < \prod_{i=1}^{k} \frac{1}{1 - 1/p_i}. \tag{2.203}$$

Using (2.203), it is easy to show that n must be divisible by a reasonably small prime $p = o(\log B)$, which gives a finite number of possible prime powers p^α. Many of the methods for odd perfect numbers are based on the simple fact that if n is an odd perfect number, and $p^\alpha \mid\mid n$, where p is prime and α is even, then $n \geq p^\alpha \sigma(p^\alpha) > p^{2\alpha}$. Methods based on this observation require the explicit factorization of $\sigma(p^\alpha)$ for large values of p^α. However, fewer factorizations would be required if it were known that $n > p^\gamma$ for $\gamma > p^{5\alpha/2}$, and in some cases, the exponent on p can be raised almost to 3α. By using these various techniques to restrict attention to prime powers $p^\alpha < B^{2/5}$, Brent, Cohen and Te Riele [39] were able to prove that *there are no odd perfect number less than* 10^{300}. Thus, an algorithm for odd perfect numbers can be outlined as follows:

Algorithm 2.6.4 (Algorithm for odd perfect numbers). Let n be an odd perfect number, and $p^\alpha \parallel n$, where p is prime and α is even. This algorithm tries to find an odd perfect number less than a given bound B (if one exists) or to prove that there is no odd perfect number less than B. The algorithm makes use of the simple fact that if $p^\alpha \parallel n$, then $\sigma(p^\alpha) \mid 2n$.

[1] Use various factoring algorithms such as ECM, MPQS and NFS, to factor the prime powers $p^\alpha < B^{2/5}$.

[2] If we deduce more and more primes which divide n, but eventually a contradiction to $n < B$ will occur, then there is no odd perfect number less than B.

[3] If we converge to a finite set of primes which do divide n, then n is an odd perfect number.

Clearly, the main task of this algorithm is the prime factorization of different, particularly large, values of p^α. In practice, step [3] in Algorithm 2.6.4 never seems to occur, and hence we obtain a tree of factorizations which prove that there is no odd perfect number less than B.

There are some other theoretical results about odd perfect numbers. For example, Hagis and Chein have independently showed that an odd perfect number is divisible by at least 8 distinct primes and Muskat that it is divisible by a prime power $> 10^{12}$. Hagis and McDaniel show that the largest prime divisor is greater than 1001100, and Pomerance that the next largest is greater than 138. Condict and Hagis have improved these bounds to $3 \cdot 10^5$ and 10^3, respectively. Pomerance has also shown that an odd perfect number with at most k factors is less than

$$(4k)^{(4k)^{2^{k^2}}}$$

but Heath-Brown [101] has improved this by showing that if n is an odd number with $\sigma(n) = an$, then $n < (4d)^{4^k}$, where d is the denominator of a, and k is the number of distinct prime factors of n. In particular, if n is an odd perfect number with k distinct prime factors, then $n < 4^{4^k}$.

2.7 Bibliographic Notes and Further Reading

Compared To elementary number theory, Computational/algorithmic number theory is a relatively new subject in number theory. In this chapter, we have introduced some fundamental issues of the computational/algorithmic aspects of number theory, more specifically, we have introduced various algorithms for primality testing, integer factorization, and discrete logarithms (including elliptic curve discrete logarithms), which will be useful in the next chapter on applications of number theory in computing and information technology, such as cryptography and network security. However, it should be noted that computational/algorithmic number theory covers a very wide range of topics, not just those of primality testing, integer factorization, and discrete logarithms.

Although computational/algorithmic number theory is new, it is very active; many new textbooks and monographs on the subject have already been published. For those who desire a more detailed exposition, we recommend the following references for further reading: Bach and Shallit [16], Brassard [33], Cohen [50], Giblin [83], Garrett [81], Knuth [123], Koblitz [128] and [129], Krana [134], Krishna, Krishna, Lin and Sun [133], Riesel [207], and Schroeder [222]. The book by Ribenboim [200] contains new records in number theory, including computational/algorithmic number theory. The paper by Silverman [234] provides a good survey as well as a nice introduction to the field. As computational/algorithmic number theory is an interdisciplinary subject of computer science (particularly *algorithms and complexity*) and number theory, readers are suggested to consult, for example, Bach [14], Garey and Johnson [79], Johnson [114], Lewis and Papadimitriou [142], Linz [143], Motwani and Raghavan [170], Rozenberg and Salomaa [214], and Yan [261] for more information on computability and complexity.

The proceedings of the international symposia on Algorithmic Number Theory (ANT), say, for example, the ANTS-III edited by Buhler [43]), often contain new developments in the subject; readers are strongly recommended to consult this series of proceedings regularly in order to update their knowledge of the subject.

Since Peter Shor [226] published his paper in 1994 on quantum factoring and discrete logarithms, quantum computing has become an increasingly important subject of research; there are at least 10 papers published every day in the world on this subject. Three classical references to quantum computation are Benioff [23], Deutsch [63] and Feynman [74]. Interested readers may also wish to consult Feynman's lectures on computation [75] and Williams and Clearwater's book [258] for more information. Serious readers in quantum computing are referred to a special section of *SIAM Journal on Computing* in **26**, (5)1997, which contains the following papers:

[1] U. Vazirani, "Introduction to Special Section on Quantum Computation", p 1409–1410.
[2] E. Bernstein and U. Vazirani, "Quantum Complexity Theory", pp 1411–1473.
[3] D. R. Simon, "On the Power of Quantum Computation", pp 1474–1483.
[4] P. Shor, "Polynomial-Time Algorithms for Prime Factorization and Discrete Logarithms on a Quantum Computer", pp 1484–1509.
[5] C. H. Bennett et al., "Strengths and Weakness of Quantum Computing", pp 1510–1523.

[6] L. M. Adleman et al., "Quantum Computability", pp 1524–1540.
[7] A. Barenco et al., "Stabilization of Quantum Computations by Symmetrization", pp 1541–1557.

We strongly recommend the interested reader to consult Shor's paper listed above for a full account of the quantum algorithms for integer factorization and discrete logarithms. However, beginners in quantum computing may find the papers by Bennett [24], Jozsa [115], Rieffel and Polak [202], and Scarani [217] useful.

3. Applied Number Theory in Computing/Cryptography

To the layman, a lot of math (like primality testing and factoring large numbers) may seem a frivolous waste of time. However, this research often pays off unexpectedly years later. Factoring and primality testing have become important because of the need to make electronic communications secure. ⋯ So, what used to be an esoteric playground for mathematicians has become applicable research.

DAVID GRIES AND FRED B. SCHNEIDER
A Logical Approach to Discrete Math [93]

The aim of this chapter is to introduce some novel applications of elementary and particularly algorithmic number theory to the design of computer (both hardware and software) systems, coding and cryptography, and information security, especially network/communication security.

3.1 Why Applied Number Theory?

The eminent American number theorist Leonard Dickson[1] once said:

Thank God that number theory is unsullied by any application.

1

Leonard Eugene Dickson (1874–1954), one of the key figures of 20th century mathematics, particularly number theory, was born in Independence, Iowa, a descendant of one William Dickson who had emigrated from Londonderry, Northern Ireland to Londonderry, New Hampshire in the 18th century. Dickson obtained his PhD in 1896 from the University of Chicago, the first PhD awarded in Mathematics by the institution. Following periods at the Universities of Leipzig, Paris, California and Texas, he returned to Chicago in 1900, becoming a full professor in 1910. One of the most productive of all mathematicians, Dickson wrote over 250 papers and 18 books, including the three volume 1600 page *History of the Theory of Numbers* [65]. It is amusing to note that he stopped to write papers and books in mathematics abruptly and completely on reaching the age of 65 in 1939 and devoted himself to his recreations, including bridge, tennis and billiards. (Photo by courtesy of the American Mathematical Society.)

The most famous English mathematician G. H. Hardy (1877–1947) also in his *Apology* ([98], page 120) expressed that

> If the theory of numbers could be employed for any practical and obviously honourable purpose, if it could be turned directly to the furtherance of human happiness or relief of human suffering, as physiology and even chemistry can, then surely neither Gauss or any other mathematician would have been so foolish as to decry or regret such applications.

He then further proudly stated on page 140 that

> Real mathematics has no effects on war. No one has yet discovered any warlike purpose to be served by the theory of numbers.

The above famous quotations made by the two greatest mathematicians of the 20th century may be true before 1950, but certainly are not true at the present time, since, e.g., number theory now can help the generals to plan their battles in a completely secret way. Remarkably enough, the great Russian mathematician Nikolay Lobachevsky (1792–1856) predicated nearly 200 years ago that

> There is no branch of mathematics, however abstract, which may not some day be applied to phenomena of the real world.

In fact, any branch of pure mathematics will eventually find real world applications. Number theory, for example, was considered the purest branch of pure mathematics, with no direct applications to the real world. The advent of digital computers and digital communications and particularly public-key cryptography revealed that number theory could provide unexpected answers to real-world problems. As showed in Schroeder [222] and Waldschmidt, Moussa, Luck and Itzykson [250], and Guterl [96], number theory has already been successfully applied to such diverse areas as physics, biology, chemistry, computing, engineering, coding and cryptography, random number generation, acoustics, communications, graphic design, and even music and business. It is also interesting to note that the eminent mathematician Shiing-Shen Chern (1911–), the 1980 Wolf Prize Winner, even considers number theory as a branch of applied mathematics [48] because of its strong applicability in other fields. Today, number theory is used widely in computing and information theory/technology, due in part to the invention of the high-speed computers based on e.g., the residue number systems and the cryptographic schemes based on e.g., large prime numbers. For example, the feasibility of several modern cryptographic schemes rests on our ability to find large primes easily, while their security rests on our inability to factor the product of large primes.

Number theory is generally considered to be laid in the discrete, finite side of mathematics, along with algebra and combinatorics, and is intimately connected to computing science and technology, since computers are basically finite machines; they have finite storage and can only deal with numbers of some finite length. Because of these features in computing, number theory is particularly useful and applicable to computing. For example, congruence theory has been used for devising systematic methods for storing computer files, generating random numbers, designing highly secure and reliable encryption schemes and even developing high-speed residue computers. Since most computer scientists are more interested in the applications of number theory in computing, rather than the number theory itself, in this chapter, we shall apply the number-theoretic results and algorithms from the previous two chapters to the design of fast computer architectures, and more secure, more reliable computer/network systems.

3.2 Computer Systems Design

> *··· virtually every theorem in elementary number theory arises in a natural, motivated way in connection with the problem of making computers do high-speed numerical calculations.*
>
> DONALD E. KNUTH
>
> Computer Science and its Relation to Mathematics [121]

3.2.1 Representing Numbers in Residue Number Systems

The way we do arithmetic on numbers is intimately related to the way we represent the numbers. There are essentially two different types of methods to represent numbers: *nonpositional* and *positional*. The Roman numerals i, ii, iii, iv, v, vi, vii, viii, ix, x, xi, xii, xiii, ··· are a classical example of a nonpositional number system; whereas the familiar *decimal* or *binary* number system are good examples of a positional number system. The positional number system using base b (or radix b) is defined by the rule

$$\begin{aligned}(\cdots a_3a_2a_1a_0a_{-1}a_{-2}a_{-3}\cdots)_b = \cdots + a_3b^3 + a_2b^2 + a_1b^1 + a_0b^0 \\ +a_{-1}b^{-1}a_{-2}b^{-2}a_{-3}b^{-3} + \cdots \qquad (3.1)\end{aligned}$$

It is clear that when $b = 10$, it is the decimal system, whereas when $b = 2$ we have the binary system. This type of positional number system is said to have a *fixed-base* or *fixed-radix*. A positional number system which is not fixed-base is said to be *mixed-base*. The number systems, residue number systems, we shall study in this section are a type of mixed-base system.

Let us first recall the Fundamental Theorem of Arithmetic: any positive integer $n \in \mathbb{N}_{>1}$ can be uniquely written as

$$n = p_1^{\alpha_1} p_2^{\alpha_2} \cdots p_k^{\alpha_k} = n_1 n_2 \cdots n_k \qquad (3.2)$$

where $p_1, p_2, \cdots, p_k$ are distinct primes, $\alpha_1, \alpha_2, \cdots, \alpha_k$ are natural numbers, $n_i = p_i^{\alpha_i}$, $i = 1, 2, \cdots, k$, and $\gcd(n_i, n_j) = 1$ for $i \neq j$. The prime decomposition of n can be used to represent any number in $\mathbb{Z}/n\mathbb{Z}$ in terms of the numbers in $\mathbb{Z}/n_i\mathbb{Z}$, for $i = 1, 2, \cdots, k$.

Definition 3.2.1. Let x be any number in $\mathbb{Z}/n\mathbb{Z}$ and

$$\left.\begin{array}{l} x \equiv a_1 \pmod{n_1} \\ x \equiv a_2 \pmod{n_2} \\ \cdots\cdots \\ \cdots\cdots \\ x \equiv a_k \pmod{n_k} \end{array}\right\} \tag{3.3}$$

then the k-tuple

$$\langle a_1, a_2, \cdots, a_k \rangle = \langle x \bmod n_1,\ x \bmod n_2,\ \cdots,\ x \bmod n_k \rangle \tag{3.4}$$

is called the residue (congruence, or modular) representation of x. For simplicity, we often write the residue representation of x as follows:

$$x \Longleftrightarrow \langle x \bmod n_1,\ x \bmod n_2,\ \cdots,\ x \bmod n_k \rangle \tag{3.5}$$

Example 3.2.1. Let $n_1 = 3, n_2 = 5, n_3 = 7$, then the residue representation of the integer 103 will be

$$\left\{\begin{array}{l} 103 \equiv 1 \pmod 3 \\ 103 \equiv 3 \pmod 5 \\ 103 \equiv 5 \pmod 7 \end{array}\right.$$

That is

$$103 \Longleftrightarrow (1, 3, 5).$$

Note that the residue representation of an integer x *wrt* moduli $n_1, n_2, \cdots, n_k$ is unique. However, the inverse is not true.

Example 3.2.2. Let again $n_1 = 3,\ n_2 = 5,\ n_3 = 7$, then all the numbers in the form

$$105t + 103, \quad \text{for } t \in \mathbb{N}$$

have the same residue representation $(1, 3, 5)$. That is,

$$105t + 103 \Longleftrightarrow (1, 3, 5).$$

Definition 3.2.2. Let $(\mathbb{Z}/n\mathbb{Z})^*$ be the "direct-product" decomposition of $\mathbb{Z}/n\mathbb{Z}$. That is,

$$(\mathbb{Z}/n\mathbb{Z})^* = \mathbb{Z}/n_1\mathbb{Z} \times \mathbb{Z}/n_2\mathbb{Z} \times \cdots \times \mathbb{Z}/n_k\mathbb{Z} \tag{3.6}$$

where $n_i = p_i^{\alpha_i}$ for $i = 1, 2, \cdots, k$ is the prime decomposition of n.

Theorem 3.2.1. Let $m_1 > 0, m_2 > 0, \cdots, m_k > 0$, and $\gcd(m_i, m_j) = 1$ with $0 < i < j \leq k$. Then two integers x and x' have the same residue representation if and only if

$$x \equiv x' \pmod{M} \tag{3.7}$$

where $M = m_1 m_2 \cdots m_k$.

So if we restrict $0 \leq x < M = m_1 m_2 \cdots m_k$, then different integers x will have different residue representation moduli $m_1, m_2, \cdots, m_k$.

Theorem 3.2.2. Let $f : \mathbb{Z}/n\mathbb{Z} \to (\mathbb{Z}/n\mathbb{Z})^*$ be such that for any $x \in \mathbb{Z}/n\mathbb{Z}$, we have

$$\begin{aligned} f(x) &= (a_1, a_2, \cdots, a_k) \\ &= (x \bmod n_1,\ x \bmod n_2,\ \cdots,\ x \bmod n_k) \end{aligned} \tag{3.8}$$

then f is a bijection (one-to-one and onto).

Remark 3.2.1. Theorem 3.2.1 is just another form of the Chinese Remainder Theorem.

Example 3.2.3. Let $m = 30$, so that $m_1 = 2$, $m_2 = 3$, $m_3 = 5$ with

$$(\mathbb{Z}/30\mathbb{Z})^* = \mathbb{Z}/2\mathbb{Z} \times \mathbb{Z}/3\mathbb{Z} \times \mathbb{Z}/5\mathbb{Z}.$$

Then the residue representations for integers in $\mathbb{Z}/30\mathbb{Z}$ will be:

$$\begin{array}{ll}
0 \Longleftrightarrow (0,0,0) & 1 \Longleftrightarrow (1,1,1) \\
2 \Longleftrightarrow (0,2,2) & 3 \Longleftrightarrow (1,0,3) \\
4 \Longleftrightarrow (0,1,4) & 5 \Longleftrightarrow (1.2,0) \\
6 \Longleftrightarrow (0,0,1) & 7 \Longleftrightarrow (1,1,2) \\
8 \Longleftrightarrow (0,2,3) & 9 \Longleftrightarrow (1,0,4) \\
10 \Longleftrightarrow (0,1,0) & 11 \Longleftrightarrow (1,2,1) \\
12 \Longleftrightarrow (0,0,2) & 13 \Longleftrightarrow (1,1,3) \\
14 \Longleftrightarrow (0,2,4) & 15 \Longleftrightarrow (1,0,0) \\
16 \Longleftrightarrow (0,1,1) & 17 \Longleftrightarrow (1,2,2) \\
18 \Longleftrightarrow (0,0,3) & 19 \Longleftrightarrow (1,1,4) \\
20 \Longleftrightarrow (0,2,0) & 21 \Longleftrightarrow (1,0,1) \\
22 \Longleftrightarrow (0,1,2) & 23 \Longleftrightarrow (1,2,3) \\
24 \Longleftrightarrow (0,0,4) & 25 \Longleftrightarrow (1,1,0) \\
26 \Longleftrightarrow (0,2,1) & 27 \Longleftrightarrow (1,0,2) \\
28 \Longleftrightarrow (0,1,3) & 29 \Longleftrightarrow (1,2,4).
\end{array}$$

Once the residue representation

$$(a_1, a_2, \cdots, a_k) = (x \bmod n_1,\ x \bmod n_2,\ \cdots,\ x \bmod n_k)$$

of an integer x is given, then we can uniquely solve x by using the Chinese Remainder Theorem (see the following example).

Example 3.2.4. Suppose we have the residue representations of x as follows:

$$x = (x \bmod 3,\ x \bmod 5,\ x \bmod 7) = (1, 3, 5).$$

Then we have

$$\begin{aligned} x &\equiv 1 \pmod 3, \\ x &\equiv 3 \pmod 5, \\ x &\equiv 5 \pmod 7). \end{aligned}$$

By using the Chinese Remainder Theorem, we get:

$$x = 103.$$

On most computers the word size is a large power of 2, with 2^{32} a common value. So to use residue arithmetic and the Chinese Remainder Theorem to do arithmetic, we need the moduli less than, say 2^{32}, pairwise relatively prime and multiplying together to give a large integer.

3.2.2 Fast Computations in Residue Number Systems

In this subsection, we shall discuss fast arithmetic operations in residue number systems. More specifically, we shall discuss the fast arithmetic operations of addition $+_n$, subtraction $-_n$ and multiplication $\cdot_n$ in $(\mathbb{Z}/n\mathbb{Z})^*$ in terms of the corresponding operations $+_{n_i}$ $-_{n_i}$ $\cdot_{n_i}$ in $\mathbb{Z}/n_i\mathbb{Z}$, for $i = 1, 2, \cdots, k$.

Definition 3.2.3. Let $x = (a_1, a_2, \cdots, a_k)$ and $y = (b_1, b_2, \cdots, b_k)$ in $\mathbb{Z}/n\mathbb{Z}$. Then

$$\begin{aligned} x + y &= (a_1, a_2, \cdots, a_k) \ +_n \ (b_1, b_2, \cdots, b_k) \\ &= f(x) \ +_n \ f(y) \\ &= ((x \bmod n_1) \ +_{n_1} \ (y \bmod n_1), \\ &\quad\ (x \bmod n_2) \ +_{n_2} \ (y \bmod n_2), \\ &\quad\ \cdots\cdots \\ &\quad\ \cdots\cdots \\ &\quad\ (x \bmod n_k) \ +_{n_k} \ (y \bmod n_k)). \end{aligned}$$

$$\begin{aligned} x - y &= (a_1, a_2, \cdots, a_k) \ -_n \ (b_1, b_2, \cdots, b_k) \\ &= f(x) \ -_n \ f(y) \\ &= ((x \bmod n_1) \ -_{n_1} \ (y \bmod n_1), \\ &\quad\ (x \bmod n_2) \ -_{n_2} \ (y \bmod n_2), \\ &\quad\ \cdots\cdots \\ &\quad\ \cdots\cdots \\ &\quad\ (x \bmod n_k) \ -_{n_k} \ (y \bmod n_k)). \end{aligned}$$

$$\begin{aligned}
x \cdot y &= (a_1, a_2, \cdots, a_k) \cdot_n (b_1, b_2, \cdots, b_k) \\
&= f(x) \odot_n f(y) \\
&= ((x \bmod n_1) \cdot_{n_1} (y \bmod n_1), \\
&\quad (x \bmod n_2) \cdot_{n_2} (y \bmod n_2), \\
&\quad \cdots\cdots \\
&\quad \cdots\cdots \\
&\quad (x \bmod n_k) \cdot_{n_k} (y \bmod n_k)).
\end{aligned}$$

Definition 3.2.4. Given groups $(\mathcal{G}, *)$ and $(\mathcal{H}, \star)$, a function $f : \mathcal{G} \to \mathcal{H}$ is called an *isomorphism* if the following conditions hold:

(1) f is one-to-one and onto.
(2) $f(a * b) = f(a) \star f(b)$, for all $a, b \in \mathcal{G}$.

We say that $(\mathcal{G}, *)$ is *isomorphic* to $(\mathcal{H}, \star)$ and write $(\mathcal{G}, *) \cong (\mathcal{H}, \star)$.

Example 3.2.5. Show that the function $f : (\mathbb{R}, +) \to (\mathbb{R}^+, \cdot)$ defined by $f(x) = 2^x$ is an isomorphism. First, we have:

(1) f is one-to-one, since $f(x) = f(y)$ implies $2^x = 2^y$, which implies $x = y$. Also f is onto, since for each $r \in \mathbb{R}^+$ there is $t \in \mathbb{R}$ such that $2^s = t$, namely $s = \log_2 t$.
(2) Let $a, b \in \mathbb{R}$, then $f(a + b) = 2^{a+b} = 2^a \cdot 2^b = f(a) \cdot f(b)$.

Therefore f is an isomorphism. That is

$$(\mathbb{R}, +) \cong (\mathbb{R}^+, \cdot), \quad f(x) = 2^x.$$

Theorem 3.2.3. Let $f : \mathbb{Z}/n\mathbb{Z} \to (\mathbb{Z}/n\mathbb{Z})^*$ defined by

$$f(x) = (x \bmod n_1,\ x \bmod n_2,\ \cdots,\ x \bmod n_k)$$

be one-to-one and onto. Then

(1) $(\mathbb{Z}/n\mathbb{Z},\ +_n) \cong ((\mathbb{Z}/n\mathbb{Z})^*,\ +_{n_i})$.
(2) $(\mathbb{Z}/n\mathbb{Z},\ -_n) \cong ((\mathbb{Z}/n\mathbb{Z})^*,\ -_{n_i})$.
(3) $(\mathbb{Z}/n\mathbb{Z},\ \cdot_n) \cong ((\mathbb{Z}/n\mathbb{Z})^*,\ \cdot_{n_i})$.

The above theorem tells us that the arithmetic operations $+_n$, $-_n$ and $\cdot_n$ in $\mathbb{Z}/n\mathbb{Z}$ can be done in $(\mathbb{Z}/n\mathbb{Z})^*$ by means of the corresponding operations $+_{n_i}$, $-_{n_i}$ and $\cdot_{n_i}$ in $(\mathbb{Z}/n_i\mathbb{Z})^*$, for $i = 1, 2, \cdots, k$. This is exactly what we need. In what follows, we shall give two examples of adding and multiplying two large integers in residue number systems. Later in the next subsection we shall also discuss its hardware implementation.

Example 3.2.6. Compute $z = x + y = 123684 + 413456$ on a computer of word size 100. Firstly we have

$$\begin{array}{ll} x \equiv 33 \pmod{99}, & y \equiv 32 \pmod{99}, \\ x \equiv \ 8 \pmod{98}, & y \equiv 92 \pmod{98}, \\ x \equiv \ 9 \pmod{97}, & y \equiv 42 \pmod{97}, \\ x \equiv 89 \pmod{95}, & y \equiv 16 \pmod{95}, \end{array}$$

so that

$$\left.\begin{array}{l} z = x + y \equiv 65 \pmod{99}, \\ z = x + y \equiv \ 2 \pmod{98}, \\ z = x + y \equiv 51 \pmod{97}, \\ z = x + y \equiv 10 \pmod{95}. \end{array}\right\} \tag{3.9}$$

Now, we use the Chinese Remainder Theorem to find

$$x + y \bmod (99 \times 98 \times 97 \times 95).$$

Note that the solution to (3.9) is

$$z \equiv \sum_{i=1}^{4} M_i M_i' z_i \pmod{m},$$

where

$$\begin{array}{l} m = m_1 m_2 m_3 m_4, \\ M_i = m/m_i, \\ M_i' M_i \equiv 1 \pmod{m_i}, \end{array}$$

for $i = 1, 2, 3, 4$. Now we have

$$M = 99 \times 98 \times 97 \times 95 = 89403930,$$

and

$$\begin{array}{l} M_1 = M/99 = 903070, \\ M_2 = M/98 = 912285, \\ M_3 = M/97 = 921690, \\ M_4 = M/95 = 941094. \end{array}$$

We need to find the inverse M_i', for $i = 1, 2, 3, 4$. To do this, we solve the following four congruences

$$\begin{array}{l} 903070 M_1' \equiv 91 M_1' \equiv 1 \pmod{99}, \\ 912285 M_2' \equiv \ 3 M_2' \equiv 1 \pmod{98}, \\ 921690 M_3' \equiv 93 M_3' \equiv 1 \pmod{97}, \\ 941094 M_4' \equiv 24 M_4' \equiv 1 \pmod{95}. \end{array}$$

We find that

$$\begin{array}{l} M_1' \equiv 37 \pmod{99}, \\ M_2' \equiv 38 \pmod{98}, \\ M_3' \equiv 24 \pmod{97}, \\ M_4' \equiv 4 \pmod{95}. \end{array}$$

Hence we get:

$$\begin{aligned}
x + y &\equiv \sum_{i=1}^{4} z_i M_i M_i' \pmod{m} \\
&\equiv 65 \times 903070 \times 37 + 2 \times 912285 \times 38 + 51 \times 921690 \times 24 \\
&\qquad +10 \times 941094 \times 4 \pmod{89403930} \\
&\equiv 3397886480 \pmod{89403930} \\
&\equiv 537140 \pmod{89403930}
\end{aligned}$$

Since $0 < x + y = 537140 < 89403930$, we conclude that $x + y = 537140$ is the correct answer.

Example 3.2.7. Suppose now we want to multiply $x = 123684$ and $y = 413456$ on a computer of word size 100. We then have

$$\begin{aligned}
x &\equiv 33 \pmod{99}, & y &\equiv 32 \pmod{99}, \\
x &\equiv 8 \pmod{98}, & y &\equiv 92 \pmod{98}, \\
x &\equiv 9 \pmod{97}, & y &\equiv 42 \pmod{97}, \\
x &\equiv 89 \pmod{95}, & y &\equiv 16 \pmod{95}, \\
x &\equiv 63 \pmod{89}, & y &\equiv 51 \pmod{89}, \\
x &\equiv 14 \pmod{83}, & y &\equiv 33 \pmod{83},
\end{aligned}$$

so that

$$\begin{aligned}
x \cdot y &\equiv 66 \pmod{99}, \\
x \cdot y &\equiv 50 \pmod{98}, \\
x \cdot y &\equiv 87 \pmod{97}, \\
x \cdot y &\equiv 94 \pmod{95}, \\
x \cdot y &\equiv 9 \pmod{89}, \\
x \cdot y &\equiv 47 \pmod{83}.
\end{aligned}$$

Now using the Chinese Remainder Theorem to solve the above system of congruences, we get

$$x \cdot y = 51137891904.$$

Since

$$0 < x \cdot y = 51137891904 < 803651926770 = n_1 n_2 n_3 n_4 n_5 n_6$$

we conclude that $x \cdot y = 51137891904$ is the correct answer.

In what follows, we shall present a general algorithm for residue arithmetic in $(\mathbb{Z}/n\mathbb{Z})^*$, where $n = n_1 n_2 \cdots n_k$, by means of the corresponding operations in $(\mathbb{Z}/n_i\mathbb{Z})^*$, for $i = 1, 2, \cdots, k$. Readers note that there are three different types of arithmetic:

(1) Integer arithmetic: arithmetic in $\mathbb{Z}$.

(2) Modular arithmetic: special integer arithmetic in $\mathbb{Z}/n\mathbb{Z}$.

(3) Residue arithmetic: special modular arithmetic in $(\mathbb{Z}/n\mathbb{Z})^*$.

In this book, we are actually more interested in the last two types of arithmetic.

Algorithm 3.2.1 (Residue arithmetic). This algorithm performs the residue arithmetic in $(\mathbb{Z}/n\mathbb{Z})^*$, where $n = n_1 n_2 \cdots n_k$:

[1] Convert integers to their residue representation: Represent integers, for example, x and y as elements of the group $(\mathbb{Z}/n\mathbb{Z})^*$, where

$$(\mathbb{Z}/n\mathbb{Z})^* = \mathbb{Z}/n_1\mathbb{Z} \times \mathbb{Z}/n_2\mathbb{Z} \times \cdots \times \mathbb{Z}/n_k\mathbb{Z}$$

by taking the congruence class of x or y modulo each n_i; for example, the following is the residue representation of x and y modulo each n_i:

$$(x \equiv x_1 \pmod{n_1},\ x \equiv x_2 \pmod{n_2},\ \cdots,\ x \equiv x_k \pmod{n_k}),$$

$$(y \equiv y_1 \pmod{n_1},\ y \equiv y_2 \pmod{n_2},\ \cdots,\ y \equiv y_k \pmod{n_k}).$$

[2] Perform the residue arithmetic for each $\mathbb{Z}/n_i\mathbb{Z}$: For example, if $\star$ denotes one of the three binary operations $+$, $-$ and $\cdot$, then we need to perform the following operations in $\mathbb{Z}/n_i\mathbb{Z}$:

$$(x_1 \star y_1 \pmod{n_1},\ x_2 \star y_2 \pmod{n_2},\ \cdots, x_k \star y_k \pmod{n_k}).$$

[3] Convert the residue representations back to integers: Use the Chinese Remainder Theorem to convert the computation results for each $\mathbb{Z}/n_i\mathbb{Z}$ into their integer form in $\mathbb{Z}/n\mathbb{Z}$

$$x \star y \equiv \sum_{i=1}^{k} M_i M_i' z_i \pmod{M},$$

where

$$\begin{aligned} M &= n_1 n_2 \cdots n_k, \\ M_i &= M/n_i, \\ M_i' &\equiv M_i^{-1} \pmod{n_i}, \\ z_i &\equiv x_i \star y_i \pmod{n_i}, \end{aligned}$$

for $i = 1, 2, \cdots, k$.

The above algorithm can be implemented entirely in special computer hardware, which is the subject matter of our next subsection.

3.2.3 Residue Computers

The conventional "binary computers" have a serious problem that restricts the speed of performing arithmetic operations, caused by, for example, the carry propagation and time delay. Fortunately, the residue number system (RNS) is not a fixed-base number system, and all arithmetic operations (except division) in RNS are inherently carry-free; that is, each digit in the computed result is a function of only the corresponding digits of the operands.

Consequently, addition, subtraction and multiplication can be performed in "residue computers" in less time than that needed in equivalent binary computers.

The construction of residue computers is much easier than that of binary computers; for example, to construct fast adders of a residue computer for

$$(\mathbb{Z}/n\mathbb{Z})^* = \mathbb{Z}/n_1\mathbb{Z} \times \mathbb{Z}/n_2\mathbb{Z} \times \cdots \times \mathbb{Z}/n_k\mathbb{Z}$$

it is sufficient to just construct some smaller adders for each $\mathbb{Z}/n_i\mathbb{Z}$, $(i = 1, 2, \cdots, k)$ (see Figure 3.1). More generally, we can construct residue com-

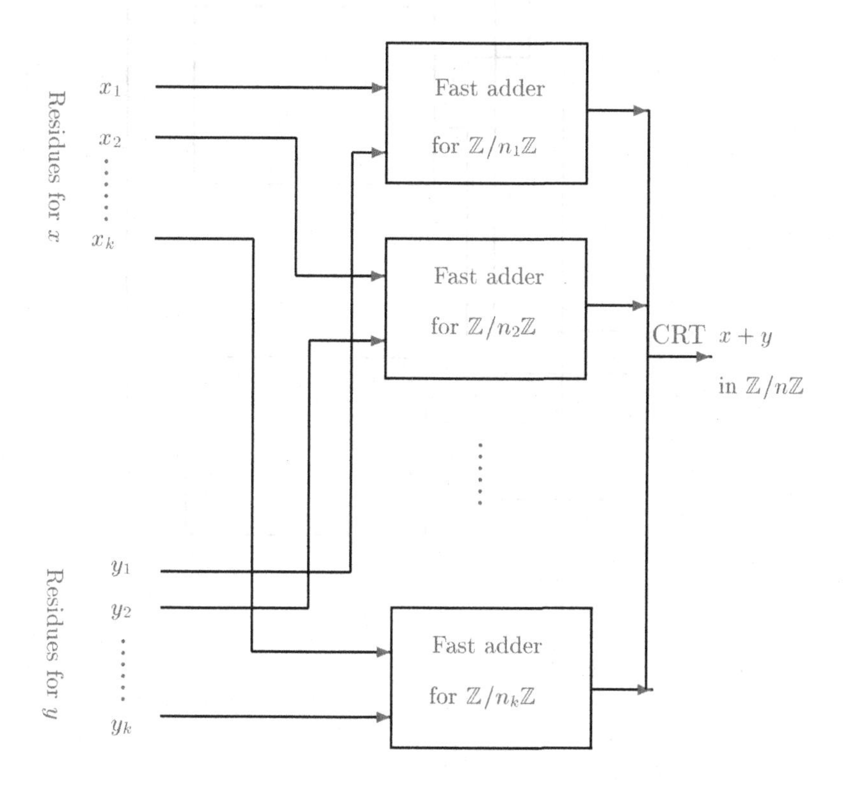

Figure 3.1. Fast adders for residue arithmetic

puters performing fast additions, subtractions and multiplications as in Figure 3.2. Since n_i is substantially less than n, computations in each $\mathbb{Z}/n_i\mathbb{Z}$ will certainly be much easier than those in $\mathbb{Z}/n\mathbb{Z}$. More importantly, additions, subtractions and multiplications in each $\mathbb{Z}/n_i\mathbb{Z}$ are carry-free, so residue computers will be substantially faster than conventional binary computers inherently with carry propagation. The idea of decomposing a large computation

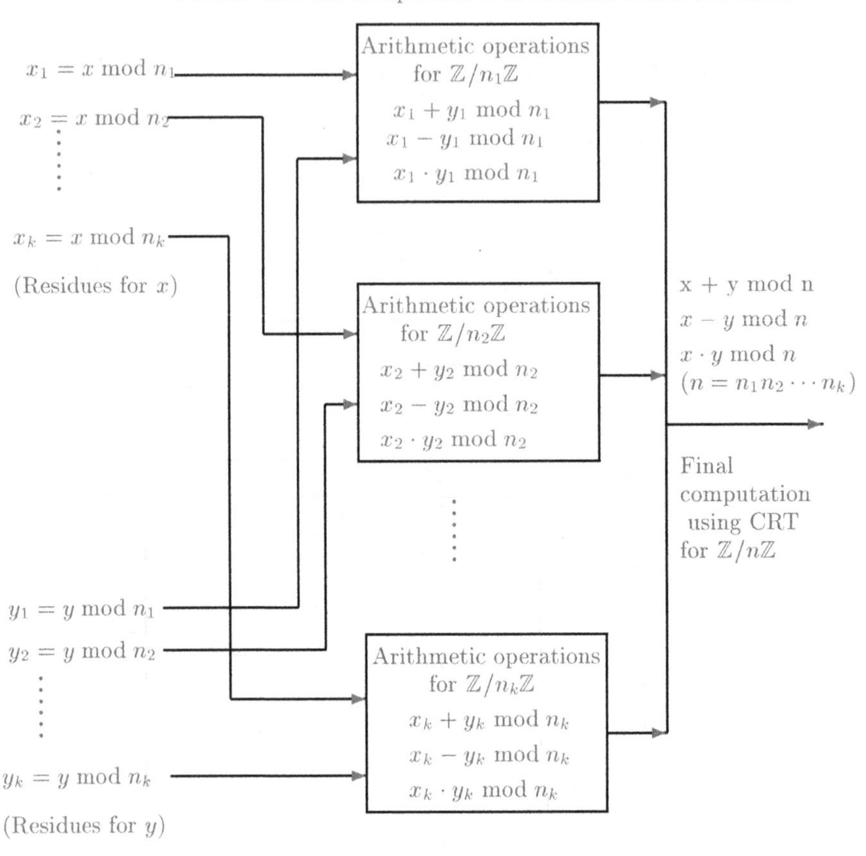

Figure 3.2. A model of residue computers

in $\mathbb{Z}/n\mathbb{Z}$ into several smaller computations in the $\mathbb{Z}/n_i\mathbb{Z}$ is exactly the idea of "divide-and-conquer" used in algorithm design. Of course, the central idea behind residue arithmetic and residue computers is the Chinese Remainder Theorem which enables us to combine separate results in each $\mathbb{Z}/n_i\mathbb{Z}$ to a final result in $\mathbb{Z}/n\mathbb{Z}$. So, if Euclid's algorithm is regarded as the first nontrivial algorithm, then the Chinese Remainder Theorem should be regarded as the first nontrivial divide-and-conquer algorithm.

Residue computers are a special type of high-speed computer, that has found many important applications in several central areas of computer science and electrical engineering, particularly in image and digital signal processing (Krishna, Krishna, Lin, and Sun [133]).

3.2.4 Complementary Arithmetic

The main memory of a computer is divided into a number of units of equal size, called *words*. Each word consists of $n = 2^m$ bits, where n is typically 16, 32, 64 or 128. Provided that a positive integer is not too large, it can then be represented simply by its binary form in a single word of the computer memory. For example, a 16-bit word could hold the positive values from 0 to 65535. The problem is then how to represent negative integers. There are a number of ways to do this; the most obvious way is the signed-magnitude representation.

Definition 3.2.5. In the signed-magnitude representation, the first bit of the m-bit word is used to denote the sign (0 for + and 1 for −), called the sign bit, and the remaining $m - 1$ bits are used to represent the size, or magnitude of the number in binary form.

Example 3.2.8. Let $m = 8$. Then to represent the integers +117 and −127 in signed-magnitude representation, we have:

$$+117 \quad \Longrightarrow \quad \underset{\text{sign bit}}{\overset{0}{\downarrow}} \qquad \underset{\text{number magnitude}}{\overset{1110101}{\downarrow}}$$

$$-127 \quad \Longrightarrow \quad \underset{\text{sign bit}}{\overset{1}{\downarrow}} \qquad \underset{\text{number magnitude.}}{\overset{1111111}{\downarrow}}$$

In a computer with 16-bit words, using the signed-magnitude representation, the largest integer that can be stored is

$$0\underbrace{111111111111111}_{15 \text{ ones}} = 2^{15} - 1 = 32767$$

and the smallest is

$$1\underbrace{111111111111111}_{15 \text{ ones}} = 1 - 2^{15} = -32767.$$

Example 3.2.9. Let $m = 8$. To compute $117 + (-127)$ and $117 + 127$ in the signed-magnitude representation, we have:

$$\begin{array}{rr} & 01110101 \\ + & 11111111 \\ \hline \end{array} \quad \Longrightarrow \quad \begin{array}{r} +1110101 \\ -1111111 \\ \hline -0001010 \end{array} \quad \Longrightarrow \quad 10001010 \quad \Longrightarrow \quad -10$$

$$\begin{array}{rr} & 01110101 \\ + & 01111111 \\ \hline \end{array} \Longrightarrow \begin{array}{r} +1110101 \\ +1111111 \\ \hline +1110100 \end{array} \Longrightarrow 01110100 \Longrightarrow 116.$$

Note that in the above computation the most significant bit is the sign bit that does not take part in the computation itself: we need first to convert the sign bit to either + or −, then perform the computation and convert the sign of the result into a sign bit. Note also that $117+127 = 116 \neq 244$; this is because the adder in a computer operates modulo n. Computers cannot deal with all integers but just a finite set of them, even using the multiple-precision representation. When two binary strings a and b are added together, the adder treats $n = 2^{m-1}$ as if it were 0! The computer sum $a \oplus b$ is not necessarily $a + b$, but $a + b$ modulo 2^{m-1}. If $a + b \geq 2^{m-1}$, then $a \oplus b = a + b - 2^{m-1}$. Since $2^{m-1} \equiv 0 \pmod{2^{m-1}}$, we have

$$a + b \equiv a \oplus b \pmod{2^{m-1}}. \tag{3.10}$$

Again in the above example, we have

$$117 + 127 = 224 \equiv 116 \pmod{2^{8-1}}.$$

While the signed-magnitude representation was used in several early computers, modern computers usually use either the one's or two's complement representation, rather than the signed-magnitude representation.

Definition 3.2.6. Let x be a binary number, then the complement of x, denoted by x', is obtained by replacing each 0 in x by 1 and each 1 in x by 0. In the one's complement representation, a positive integer is represented as in the signed-magnitude representation, whereas a negative integer is represented by the complement of the corresponding binary number. In the two's complement representation, a positive integer is represented as in the one's representation, but a negative integer is represented by adding one to its one's complement representation.

The range of a number (positive or negative) with m bits in one's complement representation is given by

$$1 - 2^{m-1}, 1 - 2^{m-1} + 1, \cdots, -1, -0, 0, 1, \cdots, 2^{m-1} - 1. \tag{3.11}$$

The range of a number with m bits in two's complement representation is given by

$$-2^{m-1}, -2^{m-1} + 1, \cdots, -2, -1, 0, 1, \cdots, 2^{m-1} - 1. \tag{3.12}$$

For example, let $m = 5$, then the range of a number in one's complement is

$$\begin{array}{ccccccccc} 1-2^{5-1}, & 1-2^{5-1}+1, & \cdots, & -1, & -0, & 0, & 1, & \cdots, & 2^{5-1}-1 \\ \downarrow & \downarrow & & \downarrow & \downarrow & \downarrow & \downarrow & & \downarrow \\ -15, & -14, & \cdots, & -1, & -0, & 0, & 1, & \cdots, & 15 \end{array}$$

and the range of a number in two's complement is

$$\begin{array}{cccccccccc} -2^{5-1}, & 1-2^{5-1}+1, & \cdots, & -2, & -1, & 0, & 1, & \cdots, & 2^{5-1}-1 \\ \downarrow & \downarrow & & \downarrow & \downarrow & \downarrow & \downarrow & & \downarrow \\ -16, & -15, & \cdots, & -2, & -1, & 0, & 1, & \cdots, & 15 \end{array}$$

One interesting observation about the two's complement representation is that it has only one representation for zero, whilst there are two zeros in, either the one's complement representation, or the signed-magnitude representation. For example, let $m = 5$, then in the one's complement representation, 00000 represents 0 but 11111 represents -0, whilst in the signed-magnitude representation, 00000 represents 0 but 10000 represents -0. Table 3.1 gives a comparison of different representations of numbers. By using either the one's complement or two's complement, rather than the signed-magnitude representation, the operation of subtraction is considerably simplified; this is the reason that modern computers use, either the one's complement, or two's complement, not the signed-magnitude representation.

Example 3.2.10. Let $a = 117$ and $b = -127$, compute $a + b$ in the one's complement representation. First note that in the signed-magnitude representation,

$$a = 01110101 \qquad\qquad b = 11111111$$

thus in the one's complement representation,

$$a' = 01110101 \qquad\qquad b' = 00000000$$

therefore, $a + b$ becomes

$$\begin{array}{rl} & 01110101 \\ + & 00000000 \\ \hline & 01110101 \quad \Longrightarrow \quad 10001010 \quad \Longrightarrow \quad -10. \end{array}$$

3.2.5 Hash Functions

Hashing is a very important technique in algorithm and database design, as well as in cryptography. In this subsection, we shall introduce an interesting application of number theory in hash function design.

Definition 3.2.7. Let k be the key of the file to be stored, and n be a positive integer. We define the hash function $h(k)$ by

$$h(k) \equiv k \pmod{n} \tag{3.13}$$

where $0 \le h(k) < n$, so that $h(k)$ is the least positive residue of k modulo n.

There are two fundamental problems here in the design a good hash function:

Table 3.1. Comparison of different representations of numbers

Pure Binary	Binary	Signed-Magnitude	1's Complement	2's Complement
0	00000	0	0	0
1	00001	1	1	1
2	00010	2	2	2
3	00011	3	3	3
4	00100	4	4	4
5	00101	5	5	5
6	00110	6	6	6
7	00111	7	7	7
8	01000	8	8	8
9	01001	9	9	9
10	01010	10	10	10
11	01011	11	11	11
12	01100	12	12	12
13	01101	13	13	13
14	01110	14	14	14
15	01111	15	15	15
16	10000	−0	−15	−16
17	10001	−1	−14	−15
18	10010	−2	−13	−14
19	10011	−3	−12	−13
20	10100	−4	−11	−12
21	10101	−5	−10	−11
22	10110	−6	−9	−10
23	10111	−7	−8	−9
24	11000	−8	−7	−8
25	11001	−9	−6	−7
26	11010	−10	−5	−6
27	11011	−11	−4	−5
28	11100	−12	−3	−4
29	11101	−13	−2	−3
30	11110	−14	−1	−2
31	11111	−15	−0	−1

(1) How to intelligently choose the value of n,

(2) How to avoid collisions.

The first problem can be solved (at least partially) by selecting n a prime close to the size of the memory. For example, if the memory size is 5000, we could pick n to be 4969, a prime close to 5000.

To solve the second problem, we could use the so-called *double hash* technique. The first hash function is the same as (3.13), defined previously, whilst the second hash function is taken as follows:

$$g(k) \equiv k + 1 \pmod{n - 2} \tag{3.14}$$

where $0 \le g(k) < n-1$, is such that $\gcd(h(k), n) = 1$. The probing sequence is defined as follows:

$$h_j(k) \equiv h(k) + j \cdot g(k) \pmod{n} \tag{3.15}$$

where $0 \le h_j(k) < n$. Since $\gcd(h(k), n) = 1$, as j runs through the integers $1, 2, 3, \cdots, n-1$, all memory locations will be traced out. Since n is prime, the ideal selection for the moduli $n-2$ would be also prime, that is, n and $n-2$ are twin primes.

Example 3.2.11. Suppose we wish to assign memory locations to files with the following index numbers:

$$\begin{array}{ll} k_1 = 197654291 & k_2 = 087365203 \\ k_3 = 528972276 & k_4 = 197354864 \\ k_5 = 873032731 & k_6 = 732975102 \\ k_7 = 216510386 & k_8 = 921001536 \\ k_9 = 933185952 & k_{10} = 109231931 \\ k_{11} = 132489973 & \end{array}$$

We first choose $n = 5881$, compute $h(k_i) = k_i \bmod n$, and get:

$$\begin{aligned} h(k_1) &= 197654291 \bmod 5881 = 5643 \\ h(k_2) &= 087365203 \bmod 5881 = 2948 \\ h(k_3) &= 528972276 \bmod 5881 = 5643 \\ h(k_4) &= 197354864 \bmod 5881 = 266 \\ h(k_5) &= 873032731 \bmod 5881 = 4162 \\ h(k_6) &= 732975102 \bmod 5881 = 2548 \\ h(k_7) &= 216510386 \bmod 5881 = 1371 \\ h(k_8) &= 921001536 \bmod 5881 = 1650 \\ h(k_9) &= 933185952 \bmod 5881 = 634 \\ h(k_{10}) &= 109231931 \bmod 5881 = 4162 \\ h(k_{11}) &= 132489973 \bmod 5881 = 2805 \end{aligned}$$

Since

$$\begin{aligned} h(k_1) &\equiv h(k_3) \equiv 5643 \pmod{5881}, \\ h(k_5) &\equiv h(k_{10}) \equiv 4162 \pmod{5881}. \end{aligned}$$

we then need to find new locations $h_1(k_3)$ and $h_1(k_{10})$ for the 3rd and the 10th record by the formula

$$h_1(k) \equiv h(k) + 1 \cdot g(k) \pmod{n}, \quad \text{with} \quad g(k) \equiv k + 1 \pmod{n-2}$$

as follows:

$$\begin{aligned} g(k_3) &= 1 + k_3 \bmod 5879 = 1 + 528972276 \bmod 5879 = 3373, \\ g(k_{10}) &= 1 + k_{10} \bmod 5879 = 1 + 109231931 \bmod 5879 = 112, \\ h_1(k_3) &= h(k_3) + 1 \cdot g(k_3) \bmod 5881 \\ &= 528972276 + 1 \cdot 3373 \bmod 5881 = 3222, \\ h_1(k_{10}) &= h(k_{10}) + 1 \cdot g(k_{10}) \bmod 5881 \\ &= 109231931 + 1 \cdot 112 \bmod 5881 = 4239. \end{aligned}$$

So finally we have:

Index Number	$h(k)$	$h_1(k)$
197654291	5643	
087365203	2948	
528972276	5643	3222
197354864	266	
873032731	4162	
732975102	2548	
216510386	1371	
921001536	1650	
933185952	634	
109231931	4162	4239
132489973	2805	

Since we can repeatedly compute $h(k), h_1(k), h_2(k), \cdots$, a suitable location for a record will be eventually found. However, by using the Chinese Remainder Theorem, it is possible to construct a collision-free hash function.

Definition 3.2.8. Let $W = \{w_0, w_1, \cdots, w_{m-1}\}$ and $I = \{0, 1, \cdots, (n-1)\}$ be sets with $n \geq m$. The hash function $h : W \to I$ is called a perfect hash function (PHF), if for all $x, y \in W$ and $x \neq y$, $h(x) \neq h(y)$. In particular, if $m = n$, h is called a minimal perfect hash function (MPHF). A minimal perfect hash function is also called a minimal collision-free hash function.

The MPHF technique is better than any existing information retrieval method, but the problem is that it is computationally intractable. Recent research shows, however, that we can use the Chinese Remainder Theorem to efficiently construct a MPHF. We describe in the following one such construction, due to Jaeschke [113].

Theorem 3.2.4. For a given finite set W (without loss of generality, we assume that W is a finite set of positive integers), there exist three constants C, D and E, such that the function h defined by

$$h(w) \equiv \lfloor C/(Dw + E) \rfloor \pmod{n-1}, \quad |W| = n - 1 \tag{3.16}$$

is a minimal perfect hash function.

The function is clearly a bijection from W onto the set I. The proof of this theorem can be done by using a generalization of the Chinese Remainder Theorem (CRT) for non-pairwise (i.e., not necessarily pairwise) relatively prime moduli. First note that for a given set $W = \{w_0, w_1, \cdots, w_{n-1}\}$ of positive integers there exist two integer constants D and E such that

$$Dw_0 + E, \; Dw_1 + E, \; \cdots, \; Dw_{n-1} + E$$

are pairwise relatively prime, so by the CRT there exists an integer C such that

$$\left.\begin{array}{l} C \equiv a_0 \pmod{(n-1)(Dw_0+E)} \\ C \equiv a_1 \pmod{(n-1)(Dw_1+E)} \\ \quad \cdots\cdots \\ \quad \cdots\cdots \\ C \equiv a_{n-1} \pmod{(n-1)(Dw_{n-1}+E)}. \end{array}\right\} \tag{3.17}$$

Finally, we introduce another type of hash function, called one-way hash function, also called message digest or fingerprint.

Definition 3.2.9. A one-way hash function maps a string (message) m of arbitrary length to an integer $d = H(m)$ with a fixed number of bits, called digest of m, that satisfies the following conditions:

[1] Given m, d is easy to compute.

[2] Given d, m is computationally infeasible to find.

A one-way hash function is said to be *collision resistant* if it is computationally infeasible to find two strings m_1 and m_2 that have the same digest d.

Several one-way hash functions believed to be collision resistant; the ones used most in practice are MD5, which produces a 128-bit digest, and SHA-1, which produces a 160-bit digest (MD stands for message digest and SHA stands for secure hash algorithm). The most important application of one-way collision resistant hash functions is to speed up the construction of digital signatures (we shall discuss digital signatures later), since we can sign the digest of the message, $d = H(m)$, rather than the message itself, m. That is,

$$S = D(H(m)), \tag{3.18}$$

where D is the digital signature algorithm.

3.2.6 Error Detection and Correction Methods

In this subsection, we shall discuss an interesting application of the theory of congruences in error detections and corrections.

It is evident that manipulating and transmitting bit strings can introduce errors. A simple error detection method, called *parity check* works in the following way (suppose the bit string to be sent is $x_1 x_2 \cdots x_n$):

[1] (Precomputation) Append to the bit string a *parity check bit* x_{n+1}

$$x_{n+1} \equiv x_1 + x_2 + \cdots + x_n \pmod{2}, \tag{3.19}$$

so that

$$x_{n+1} = \begin{cases} 0, & \text{if there is an even number of} \\ & \quad 1 \text{ in } x_1 x_2 \cdots x_n, \\ 1, & \text{otherwise.} \end{cases} \tag{3.20}$$

The appended string $x_1 x_2 \cdots x_n x_{n+1}$ should satisfy the following congruence

$$x_1 + x_2 + \cdots + x_n + x_{n+1} \equiv 0 \pmod{2}. \tag{3.21}$$

[2] (Error Detection) Suppose now we send the string $x = x_1 x_2 \cdots x_n x_{n+1}$ and the string $y = y_1 y_2 \cdots y_n y_{n+1}$ is received. If $x = y$, then there are no errors, but if $x \neq y$, there will be errors. We check whether or not

$$y_1 + y_2 + \cdots + y_n + y_{n+1} \equiv 0 \pmod{2} \tag{3.22}$$

holds. If this congruence fails, at least one error is present; but if it holds, errors may still exist. Clearly, we can detect an odd number of errors, but not an even number of errors.

The above method can be easily extended to checking for errors in strings of *digits*, rather than just bits. The use of check digits with identification numbers for error detection is now a standard practice. Notable examples include social security numbers, telephone numbers, serial numbers on currency predate computers, Universal Product Codes (UPC) on grocery items, and International Standard Book Numbers (ISBN) on published books. In what follows, we shall introduce a modulus 11 error correction and detection scheme for ISBN numbers.

Every recently published book has a 10-digit codeword called its International Standard Book Number (ISBN). This is a sequence of nine digits $x_1 x_2 \cdots x_9$, where each $x_i \in \{0, 1, 2, \cdots, 9\}$, together with a check digit $x_{10} \in \{0, 1, 2, \cdots, 9, X\}$ (we use the single letter X to represent the two digit number 10). This last digit x_{10} is included so as to give a check that the previous nine digits have been correctly transcribed; x_{10} can be obtained by

$$x_{10} = \sum_{i=1}^{9} i x_i \pmod{11}. \tag{3.23}$$

Note that if we arrange the ten digit ISBN number in the order of $x_{10} x_9 \cdots x_2 x_1$, then the check digit x_1 is determined by

$$x_1 = 11 - \sum_{i=10}^{2} i x_i \pmod{11}. \tag{3.24}$$

The whole 10-digit number satisfies the following so-called *check congruence*

$$\sum_{i=1}^{10} i x_i \equiv 0 \pmod{11}. \tag{3.25}$$

Example 3.2.12. The first nine digits of the ISBN number of the book by Ireland and Rosen [111] are as follows:

0-387-97329.

Find the check digit of this ISBN number. We first let

$$\begin{array}{ccccccccc} 0 & 3 & 8 & 7 & 9 & 7 & 3 & 2 & 9 \\ \updownarrow & \updownarrow & \updownarrow & \updownarrow & \updownarrow & \updownarrow & \updownarrow & \updownarrow & \updownarrow \\ x_1 & x_2 & x_3 & x_4 & x_5 & x_6 & x_7 & x_8 & x_9 \end{array}$$

Then

$$\begin{aligned} x_{10} &\equiv \sum_{i=1}^{9} ix_i \pmod{11} \\ &\equiv [1 \cdot 0 + 2 \cdot 3 + 3 \cdot 8 + 4 \cdot 7 + 5 \cdot 9 + \\ &\quad 6 \cdot 7 + 7 \cdot 3 + 8 \cdot 2 + 9 \cdot 9] \pmod{11} \\ &= 10 = X \end{aligned}$$

If we let

$$\begin{array}{ccccccccc} 0 & 3 & 8 & 7 & 9 & 7 & 3 & 2 & 9 \\ \updownarrow & \updownarrow & \updownarrow & \updownarrow & \updownarrow & \updownarrow & \updownarrow & \updownarrow & \updownarrow \\ x_{10} & x_9 & x_8 & x_7 & x_6 & x_5 & x_4 & x_3 & x_2 \end{array}$$

Then

$$\begin{aligned} x_1 &\equiv 11 - \sum_{i=10}^{2} ix_i \pmod{11} \\ &\equiv 11 - [10 \cdot 0 + 9 \cdot 3 + 8 \cdot 8 + 7 \cdot 7 + \\ &\quad 6 \cdot 9 + 5 \cdot 7 + 4 \cdot 3 + 3 \cdot 2 + 2 \cdot 9] \pmod{11} \\ &= 10 = X \end{aligned}$$

So the complete ISBN number of the book is

0-387-97329-X.

Generally speaking, the coefficients a_i, for $i = 1, 2, \cdots, n$ (or $i = n, n-1, \cdots, 1$) could be any numbers as long as the n digits satisfies the check congruence:

$$a_1x_1 + a_2x_2 + \cdots + a_nx_n \equiv 0 \pmod{m}.$$

Example 3.2.13. The ISBN number of the present book is

3-540-65472-0

and it satisfies its check congruence

$$\sum_{i=1}^{10} ix_i \equiv [1\cdot3+2\cdot5+3\cdot4+4\cdot0+5\cdot6+6\cdot5+7\cdot4+8\cdot7+9\cdot2+10\cdot0] \pmod{11}.$$

Suppose the first nine digits of the ISBN number are given and we are asked to find the check digit x_{10}, then we have

$$x_{10} = \sum_{i=1}^{9} ix_i = [1\cdot3+2\cdot5+3\cdot4+4\cdot0+5\cdot6+6\cdot5+7\cdot4+8\cdot7+9\cdot2] \pmod{11} = 0.$$

Example 3.2.14. Suppose a book whose ISBN number is as follows

$$9\text{-}810\text{-}x3422\text{-}8$$

where x is an unknown digit. What is x? To find the value for x, we perform the following computation:

$$[1\cdot9+2\cdot8+3\cdot1+4\cdot0+5x+6\cdot3+7\cdot4+8\cdot2+9\cdot2+10\cdot8] \pmod{11} = 1.$$

So, we have

$$1+5x \equiv 0 \pmod{11}.$$

To solve this linear congruence, we get

$$\begin{aligned} x &\equiv -\frac{1}{5} \pmod{11} \\ &\equiv -9 \pmod{11} \qquad \left(\text{since } \frac{1}{5} \equiv 9 \pmod{11}\right) \\ &\equiv 2 \pmod{11}. \end{aligned}$$

Thus, $x = 2$.

Exercise 3.2.1. Find the value of x in each of the following ISBN numbers:

0-201-07981-x,
0-8053-x340-2,
0-19-8x3171-0.

The ISBN code can detect

(1) 100% of all single digit errors,

(2) 100% of double errors created by the transposition of two digits.

The detection process is as follows. Let $\mathbf{x} = x_1x_2\cdots x_{10}$ be the original codeword sent, $\mathbf{y} = y_1y_2\cdots y_{10}$ the received string, and $S = \sum_{i=1}^{10} iy_i$. If $S \equiv 0 \pmod{11}$, then $\mathbf{y}$ is the legitimate codeword and we assume it is correct, whereas if $S \not\equiv 0 \pmod{11}$, then we have detected error(s):

(1) Suppose the received string $\mathbf{y} = y_1 y_2 \cdots y_{10}$ is the same as $\mathbf{x} = x_1 x_2 \cdots x_{10}$ except that the $y_k = x_k + a$ with $1 \le k \le 10$ and $a \ne 0$. Then

$$S = \sum_{i=1}^{10} i y_i = \sum_{i=1}^{10} i x_i + ka \not\equiv 0 \pmod{11},$$

since k and a are all non-zero elements in $\mathbb{Z}/11\mathbb{Z}$.

(2) Suppose the received string $\mathbf{y} = y_1 y_2 \cdots y_{10}$ is the same as $\mathbf{x} = x_1 x_2 \cdot x_{10}$ except that y_j and x_k have been transposed. Then

$$\begin{aligned} S &= \sum_{i=1}^{10} i y_i = \sum_{i=1}^{10} i x_i + (k-j)x_j + (j-k)x_k \\ &= (k-j)(x_j - x_k) \not\equiv 0 \pmod{11}, \quad \text{if } k \ne j \text{ and } x_j \ne x_k. \end{aligned}$$

Note that since $\mathbb{Z}/11\mathbb{Z}$ is a field, the product of two non-zero elements is also non-zero but this does not hold in $\mathbb{Z}/10\mathbb{Z}$, which is only a ring (say, for example, $2 \cdot 5 \equiv 0 \pmod{10}$); this is why we work with modulo 11 rather than modulo 10. Note also that the ISBN code cannot be used to correct errors unless we know that just one digit is in error. Interested readers are suggested to consult Gallian [77] and Hill [104] for more information about error detection and correction codes.

We now move to the introduction of another interesting error detection technique for programs (Brent [38]). The Galileo spacecraft is somewhere near Jupiter, but its main radio antenna is not working, so communication with it is slow. Suppose we want to check that a critical program in Galileo's memory is correct. How can we do this without transmitting the whole program from/to Galileo? The following is a method (possibly the simplest method) for checking out Galileo's program based on some simple number-theoretic ideas; the method was first proposed by Michael Rabin:

Let P_g be the program in Galileo and P_e the program on Earth, each represented as an integer. Assuming P_e is correct, this algorithm will try to determine whether or not P_g is correct:

[1] Choose a prime number $10^9 < p < 2 \cdot 10^9$ and transmit p (p has no more than 32 bits) to Galileo and ask it to compute $r_g \leftarrow P_g \bmod p$ and send the remainder r_g back to Earth (r_g has no more than 32 bits).

[2] On Earth, we compute $r_e \leftarrow P_e \bmod p$, and check if $r_g = r_e$.

[3] If $r_g \ne r_e$, we conclude that $P_g \ne P_e$. That is, Galileo's program has been corrupted!

[4] If $r_g = r_e$, we conclude that P_g is *probably* correct. That is, if P_g is not correct, there is only a small probability of $< 10^{-9}$ that $r_g = r_e$. If this error probability is too large to accept for the quality-assurance team, just goto step [1] to start the process all over again, else terminate the algorithm by saying that P_g is "almost surely" correct! It is clear that if we repeat the

process, for example, ten times on ten different random primes, then the error probability will be less than 10^{-90}, an extremely small number.

Clearly the idea underlying the method for program testing is exactly the same as that of the probabilistic method for primality testing.

3.2.7 Random Number Generation

Anyone who considers arithmetic methods of producing random digits is, of course, in a state of sin.

JOHN VON NEUMANN (1903–1957)

"Random" numbers have a great many uses in, e.g., numerical simulations, sampling, numerical analysis, testing computer chips for defects, decision making, coding and cryptography, and programming slot machines, etc. They are a valuable resource: in some cases, they can speed up computations, they can improve the rate of communication of partial information between two users, and they can also be used to solve problems in asynchronous distributed computation that is impossible to solve by deterministic means. A *sequence of numbers* is random if each number in the sequence is independent of the preceding numbers; there are no patterns to help us to predict any number of the sequence. Of course, truly *random* numbers are hard to come by, or even impossible to get. Thus, the so-called random numbers are actually *pseudorandom numbers*. Since the invention of the first electronic computer, researchers have been trying to find *efficient* ways to generate random numbers on a computer. We have, in fact, already seen some applications of random numbers in this book; for example, Pollard's ρ-method, introduced in Chapter 2, uses random numbers in finding prime factorization of large integers. In this subsection, we shall briefly introduce some methods for generating random numbers based on linear congruences.

Firstly, let us introduce an arithmetic method, called the *middle-square method*, suggested by John von Neumann[2] in 1946. The algorithmic description of the method is as follows:

[2]

John von Neumann (1903–1957) was born in Budapest, Hungary, but lived in the U.S.A. from 1930 onwards. He is one of the legendary figures of 20th century mathematics. He made important contributions to logic, quantum physics, optimization theory and game theory. His lifelong interest in mechanical devices led to his being involved crucially in the initial development of the modern electronic computer and the important concept of the *stored program*. He was also involved in the development of the first atomic bomb.

Algorithm 3.2.2 (Von Neumann's middle-square method). This algorithm uses the so-called middle-square method to generate random numbers:

[1] Let m be the number of random numbers we wish to generate (all with, for example, 10 digits), and set $i \leftarrow 0$.

[2] Randomly choose a starting 10-digit number n_0.

[3] Square n_i to get an intermediate number M, with 20 or less digits.

[4] Set $i = i+1$ and take the middle ten digits of M as the new random number n_i.

[5] If $i < m$ then goto step [3] to generate a new random number, else stop the generating process.

Example 3.2.15. Let $n_0 = 9524101765$, and $m = 10$. Then by Algorithm 3.2.2 we have

$$\begin{array}{lll}
9524101765^2 = 90708514430076115225 & \Longrightarrow & n_1 = 5144300761 \\
5144300761^2 = 26463830319625179121 & \Longrightarrow & n_2 = 8303196251 \\
8303196251^2 = 68943067982620455001 & \Longrightarrow & n_3 = 0679826204 \\
0679826204^2 = 462163667645049616 & \Longrightarrow & n_4 = 6366764504 \\
6366764504^2 = 40535690249394366016 & \Longrightarrow & n_5 = 6902493943 \\
6902493943^2 = 47644422633151687249 & \Longrightarrow & n_6 = 4226331516 \\
4226331516^2 = 17861878083134858256 & \Longrightarrow & n_7 = 8780831348 \\
8780831348^2 = 77102999162019497104 & \Longrightarrow & n_8 = 9991620194 \\
9991620194^2 = 99832474101148597636 & \Longrightarrow & n_9 = 4741011485 \\
4741011485^2 = 22477189900901905225 & \Longrightarrow & n_{10} = 1899009019.
\end{array}$$

A serious problem with the middle-square method is that for many choices of the initial integer, the method produces the same small set of numbers over and over. For example, working with numbers that have four digits, staring from 4100, we obtain the sequence

$$8100, 6100, 2100, 4100, 8100, 6100, 2100, \cdots.$$

In what follows, we shall introduce some methods based on congruence theory, which can generate a sequence of numbers that appear to be *essentially* random.

Congruence theory is useful in generating a list of random numbers. At present, the most popular random number generators in use are special cases of the so-called *linear congruential generator* (LCG for short), introduced first by D. H. Lehmer in 1949. In the linear congruential method, we first choose four "magic" numbers as follows:

n:	the modulus;	$n > 0$
x_0:	the seed;	$0 \le x_0 \le n$
a:	the multiplier;	$0 \le a \le n$
b:	the increment;	$0 \le b \le n$

then the sequence of random numbers is defined recursively by:

$$x_j \equiv ax_{j-1} + b \pmod{n}, \quad j > 0, \tag{3.26}$$

for $1 \leq j \leq l$, where $l \in \mathbb{N}$ is the least value such that $x_{l+1} \equiv x_j \pmod{n}$ for some $j \leq l$. We call l the period length of the LCG generator. Clearly, the maximum length of distinct random numbers generated by the LCG is the modulus n. The best random number generator is, of course, the one that has the maximum length of distinct random numbers. Knuth gives a necessary and sufficient condition for a LCG to have maximum length:

Theorem 3.2.5 (Knuth [123]). A LCG has period length $l = n$ if and only if $\gcd(b, n) = 1$, $a \equiv 1 \pmod{p}$ for all primes $p \mid n$ and $a \equiv 1 \pmod{4}$ if $4 \mid n$.

Note that the parameter a is sometimes set to be 1; in that case, the LCG is just a "plain" linear congruential generator. When a is set to be greater than 1, it is sometimes called a multiplicative linear congruential generator. Now we are in a position to give an algorithm for a LCG.

Algorithm 3.2.3 (Linear Congruential Generator). This algorithm will generate a sequence of random numbers $\{x_1, x_2, \cdots, x_n\}$.

[1] [Initialization] Input x_0, a, b, n and k (here k is just the number of random numbers the user wishes to generate; we can simply set $k = n$). Set $j \leftarrow 1$.

[2] [Random Number Generation] Compute $x_j \leftarrow (ax_{j-1} + b) \pmod{n}$, and print x_j.

[3] [Increase j] $j \leftarrow j + 1$. If $j \geq k$, then goto Step [4], else goto Step [2].

[4] [Exit] Terminate the algorithm.

Example 3.2.16. Let $x_0 = 5$, $a = 11$, $b = 73$, $n = 1399$ and $k = 10$. Then by Algorithm 3.2.3 we have:

$$\begin{array}{lcllcl}
x_0 & = & 5 & & & \\
x_1 & \equiv & ax_0 + b \pmod{n} & \Longrightarrow & & x_1 = 128 \\
x_2 & \equiv & ax_1 + b \pmod{n} & \Longrightarrow & & x_2 = 82 \\
x_3 & \equiv & ax_2 + b \pmod{n} & \Longrightarrow & & x_3 = 975 \\
x_4 & \equiv & ax_3 + b \pmod{n} & \Longrightarrow & & x_4 = 1005 \\
x_5 & \equiv & ax_4 + b \pmod{n} & \Longrightarrow & & x_5 = 1335 \\
x_6 & \equiv & ax_5 + b \pmod{n} & \Longrightarrow & & x_6 = 768 \\
x_7 & \equiv & ax_6 + b \pmod{n} & \Longrightarrow & & x_7 = 127 \\
x_8 & \equiv & ax_7 + b \pmod{n} & \Longrightarrow & & x_8 = 71 \\
x_9 & \equiv & ax_8 + b \pmod{n} & \Longrightarrow & & x_9 = 854 \\
x_{10} & \equiv & ax_9 + b \pmod{n} & \Longrightarrow & & x_{10} = 1073 \\
 & & \cdots\cdots & & \cdots\cdots & \\
 & & \cdots\cdots & & \cdots\cdots & \\
x_{231} & \equiv & ax_{230} + b \pmod{n} & \Longrightarrow & & x_{231} = 1149
\end{array}$$

$$
\begin{array}{lllll}
x_{232} & \equiv & ax_{231} + b \pmod{n} & \Longrightarrow & x_{232} = 121 \\
x_{233} & \equiv & ax_{232} + b \pmod{n} & \Longrightarrow & x_{233} = 5 \\
x_{234} & \equiv & ax_{233} + b \pmod{n} & \Longrightarrow & x_{234} = 128.
\end{array}
$$

So the length of this random number sequence

$$
\begin{aligned}
&(x_1, x_2, x_3, x_4, x_5, x_6, x_7, x_8, x_9, x_{10}, \cdots, x_{231}, x_{232}, x_{233}) \\
&= (128, 82, 975, 1005, 1335, 768, 127, 71, 854, 1073, \cdots, 1149, 121, 5)
\end{aligned}
$$

generated by the LCG

$$
\begin{aligned}
&x_0 \equiv 5 \pmod{1399}, \\
&x_j \equiv 11 \cdot x_{j-1} + 73 \pmod{1399}, \quad j = 1, 2, \cdots,
\end{aligned}
$$

is 233, i.e., $l = 233$.

Normally, we could set $n = 2^r$, $a = 2^i + 1$ with $i < r$, and $b = 1$. Thus, Equation (3.26) becomes

$$
x_j \equiv (2^k + 1)x_{j-1} + 1 \pmod{2^r}, \quad j = 1, 2, \cdots. \tag{3.27}
$$

To make a LCG a good random number generator, it is necessary to find good values for all the four magic numbers (not just the modulus n) that define the linear congruential sequence. Interested readers are invited to consult [123] for a thorough discussion about the choice of the parameters. There are many congruential generators based on the linear congruential generator:

(1) Power generator:

$$
x_j \equiv (x_{j-1})^d \pmod{n}, \quad j = 1, 2, \cdots \tag{3.28}
$$

where (d, n) are parameters describing the generator and x_0 is the seed. There are two important special cases of the power generator, both occurring when $n = pq$ is a product of two distinct odd primes.

(i) The RSA[3] Generator: This case occurs when $\gcd(d, \phi(n)) = 1$, where $\phi(n)$ is Euler's ϕ-function. The map $x \mapsto x^d \pmod{n}$ is one-to-one on $(\mathbb{Z}/n\mathbb{Z})^*$, and this operation is the encryption operation of the RSA public-key cryptosystem, where the pair (d, n) is publicly known. This special case of the power generator is called the *RSA generator*. For example, let $p = 13$, $q = 23$ and $d = 17$, so that $n = 299$, $\phi(299) = 264$ and $\gcd(299, 17) = 1$. Let also $x_0 = 6$. Then by the RSA generator

$$
\begin{aligned}
&x_0 = 6, \\
&x_j = x_{j-1}^{17} \pmod{299}, \quad j = 1, 2, \cdots,
\end{aligned}
$$

[3] RSA stands for three computer scientists Rivest, Shamir and Adleman [209], who invented the widely used RSA public-key cryptosystem in the 1970s, which will be studied in the next section. The RSA generator has essentially the same idea as the RSA cryptosystem.

we have the following random sequence:

$$
\begin{array}{lllll}
x_1 & \equiv & x_0^{17} \pmod{299} & \Longrightarrow & x_1 \equiv 6^{17} \equiv 288 \pmod{299} \\
x_2 & \equiv & x_1^{17} \pmod{299} & \Longrightarrow & x_2 \equiv 288^{17} \equiv 32 \pmod{299} \\
x_3 & \equiv & x_2^{17} \pmod{299} & \Longrightarrow & x_3 \equiv 32^{17} \equiv 210 \pmod{299} \\
x_4 & \equiv & x_3^{17} \pmod{299} & \Longrightarrow & x_4 \equiv 210^{17} \equiv 292 \pmod{299} \\
x_5 & \equiv & x_4^{17} \pmod{299} & \Longrightarrow & x_6 \equiv 292^{17} \equiv 119 \pmod{299} \\
x_6 & \equiv & x_5^{17} \pmod{299} & \Longrightarrow & x_6 \equiv 119^{17} \equiv 71 \pmod{299} \\
x_7 & \equiv & x_6^{17} \pmod{299} & \Longrightarrow & x_7 \equiv 71^{17} \equiv 41 \pmod{299} \\
x_8 & \equiv & x_7^{17} \pmod{299} & \Longrightarrow & x_8 \equiv 41^{17} \equiv 123 \pmod{299} \\
x_9 & \equiv & x_7^{17} \pmod{299} & \Longrightarrow & x_9 \equiv 123^{17} \equiv 197 \pmod{299} \\
x_{10} & \equiv & x_7^{17} \pmod{299} & \Longrightarrow & x_{10} \equiv 197^{17} \equiv 6 \pmod{299} \\
x_{11} & \equiv & x_{10}^{17} \pmod{299} & \Longrightarrow & x_{11} \equiv 6^{17} \equiv 288 \pmod{299}.
\end{array}
$$

Thus, the length of this random number sequence generated by the RSA generator is 10. That is $l = 10$.

(ii) The square generator: This case occurs when $d = 2$ and $n = pq$ with $p \equiv q \equiv 3 \pmod 4$; we call this the *square generator*. In this case, the mapping $x_i \mapsto (x_{j-1})^2 \pmod n$ is four-to-one on $(\mathbb{Z}/n\mathbb{Z})^*$. An even more special case of the square generator is the *quadratic residues generator*:

$$y \equiv x^2 \pmod n \tag{3.29}$$

for some x.

(2) Discrete exponential generator:

$$x_j \equiv g^{x_{j-1}} \pmod n, \quad j = 1, 2, \cdots \tag{3.30}$$

where (g, n) are parameters describing the generator and x_0 the seed. A special case of the discrete exponential generator is that when n is an odd prime p, and g is a primitive root modulo p; then the problem of recovering x_{j-1} given by (x_j, g, n) is the well-known hard *discrete logarithm problem*.

Note that simpler sequences of random numbers can be combined to produce complicated ones by using hashg and composition functions. For more information on this topic, see Lagarias [136] and the references therein.

In some cases, for example, in stream-cipher cryptography (Zeng [265]), a stream of random bits rather than a sequence of random digits (numbers) will be needed. We list in the following some of the widely used random bit generators (more random bit generators can be found, for example, in Lagarias [136]):

(1) RSA bit generator: Given $k \geq 2$ and $m \geq 1$, select odd primes p and q uniformly from the range $2^k \leq p,\ q < 2^{k+1}$ and form $n = pq$. Select e uniformly from $[1, n]$ subject to $\gcd(e, \phi(n)) = 1$. Set

$$x_j \equiv (x_{j-1})^e \pmod{n}, \quad j = 1, 2, \cdots \tag{3.31}$$

and let the bit z_j be given by

$$z_j \equiv x_j \pmod{2}, \quad j = 1, 2, \cdots . \tag{3.32}$$

Then $\{z_j : 1 \leq j \leq k^m + m\}$ are the random bits generated by the seed x_0 of the length $2k$ bits.

(2) Rabin's modified bit generator: Let $k \geq 2$, and select odd primes p and q uniformly from primes in the range $2^k \leq p,\ q < 2^{k+1}$ and form $n = pq$, such that $p \equiv q \equiv 3 \pmod{4}$ (this assumption is used to guarantee that -1 is a quadratic nonresidue for both p and q). Let

$$x_j = \begin{cases} (x_{j-1})^2 \pmod{n}, & \text{if it lies in } [0, n/2), \\ n - (x_{j-1})^2 \pmod{n}, & \text{otherwise,} \end{cases} \tag{3.33}$$

so that $0 \leq x_j < n/2$, and the bit z_j be given by

$$z_j \equiv x_j \pmod{2}, \quad j = 1, 2, \cdots . \tag{3.34}$$

Then $\{z_j : 1 \leq j \leq k^m + m\}$ are the random bits generated by the seed x_0 of the length $2k$ bits.

(3) Discrete exponential bit generator Let $k \geq 2$ and $m \geq 1$, and select an odd prime p uniformly from primes in the range $[2^k,\ 2^{k+1}]$, provided with a complete factorization of $p - 1$ and a primitive root g. Set

$$x_j \equiv g^{x_{j-1}} \pmod{p}, \quad j = 1, 2, \cdots \tag{3.35}$$

and let the bit z_j be the most significant bit

$$z_j \equiv \left\lceil \frac{x_j}{2^k} \right\rceil \pmod{2}. \tag{3.36}$$

Then $\{z_j : 1 \leq j \leq k^m + m\}$ are the random bits generated by the seed x_0.

(4) Elliptic curve bit generator: Elliptic curves, as we have already seen, have applications in primality testing and integer factorization. It is interesting to note that elliptic curves can also be used to generate random bits; interested readers are referred to Kaliski [116] for more information.

3.3 Cryptography and Information Security

Modern cryptography depends heavily on number theory, with primality testing, factoring, discrete logarithms (indices), and elliptic curves being perhaps the most prominent subject areas.

MARTIN HELLMAN
Foreword to the present book

Cryptography was concerned initially with providing secrecy for written messages. Its principles apply equally well to securing data flow between computers, to digitized speech, and to encrypting facsimile and television signals. For example, most satellites routinely encrypt the data flow to and from ground stations to provide both privacy and security for their subscribers. In this section, we shall introduce some basic concepts and techniques of cryptography and discuss their applications to computer-based information security.

3.3.1 Introduction

Cryptography (from the Greek *Kryptós*, "hidden", and *gráphein*, "to write") is the study of the principles and techniques by which information can be concealed in ciphertexts and later revealed by legitimate users employing the secret key, but in which it is either impossible or computationally infeasible for an unauthorized person to do so. Cryptanalysis (from the Greek *Kryptós* and *analýein*, "to loosen") is the science (and art) of recovering information from ciphertexts without knowledge of the key. Both terms are subordinate to the more general term *cryptology* (from the Greek *Kryptós* and *lógos*, "word"). That is,

$$\text{Cryptology} \stackrel{\text{def}}{=} \text{Cryptography} + \text{Cryptanalysis},$$

and

$$\text{Cryptography} \stackrel{\text{def}}{=} \text{Encryption} + \text{Decryption}.$$

Modern cryptography, however, is the study of "mathematical" systems for solving the following two main types of security problems:

(1) privacy,

(2) authentication.

A privacy system prevents the extraction of information by unauthorized parties from messages transmitted over a public and often insecure channel, thus assuring the sender of a message that it will only be read by the intended receiver. An authentication system prevents the unauthorized injection of messages into a public channel, assuring the receiver of a message of the legitimacy of its sender. It is interesting to note that the computational engine, designed and built by a British group led by Alan Turing at Bletchley Park, Milton Keynes to crack the German ENIGMA code is considered to be among the very first real electronic computers; thus one could argue that modern cryptography is the mother (or at least the midwife) of modern computer science.

There are essentially two different types of cryptographic systems (cryptosystems):

(1) *Secret-key* cryptographic systems (also called symmetric cryptosystems),
(2) *Public-key* cryptographic systems (also called asymmetric cryptosystems).

Before discussing these two types of different cryptosystems, we present some notation:

(1) *Message space* $\mathcal{M}$: a set of strings (plaintext messages) over some alphabet, that needs to be encrypted.
(2) *Ciphertext space* $\mathcal{C}$: a set of strings (ciphertexts) over some alphabet, that has been encrypted.
(3) *Key space* $\mathcal{K}$: a set of strings (keys) over some alphabet, which includes
    (i) The *encryption key* e_k.
    (ii) The *decryption key* d_k.
(4) The *encryption process (algorithm)* E: $E_{e_k}(M) = C$.
(5) The *decryption process (algorithm)* D: $D_{d_k}(C) = M$.
    The algorithms E and D must have the property that

$$D_{d_k}(C) = D_{d_k}(E_{e_k}(M)) = M.$$

3.3.2 Secret-Key Cryptography

The legend that every cipher is breakable is of course absurd, though still widespread among people who should know better.

J. E. Littlewood
Mathematics with Minimum 'Raw Material' [144]

In a conventional secret-key cryptosystem (see Figure 3.3), the same key (i.e., $e_k = d_k = k \in \mathcal{K}$), called the *secret key*, is used in both encryption and decryption. By same key we mean that someone who has enough information to encrypt messages automatically has enough information to decrypt messages as well. This is why we call it secret-key cryptosystem, or symmetric

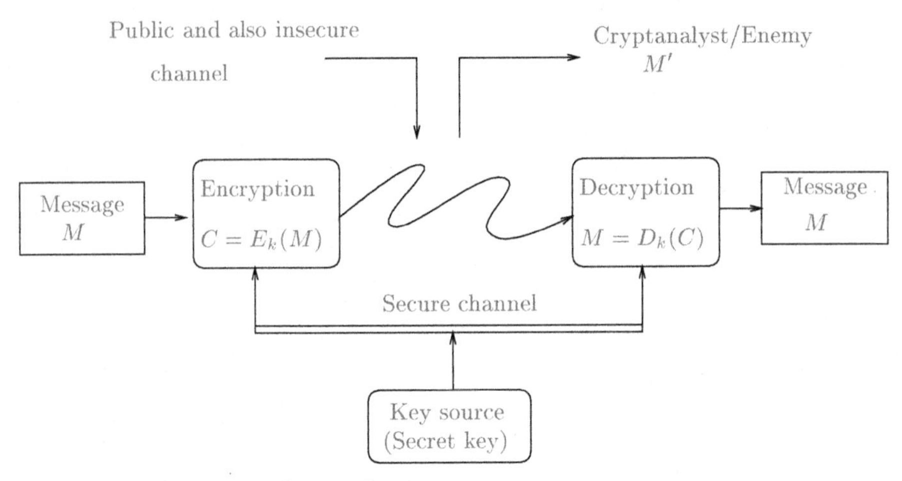

Figure 3.3. Conventional secret-key cryptosystems

cryptosystem. The sender uses an invertible transformation f defined by

$$f : \mathcal{M} \xrightarrow{k} \mathcal{C}, \tag{3.37}$$

to produce the cipher text

$$C = E_k(M), \quad M \in \mathcal{M} \text{ and } C \in \mathcal{C}, \tag{3.38}$$

and transmits it over the public *insecure* channel to the receiver. The key k should also be transmitted to the legitimate receiver for decryption but via a *secure* channel. Since the legitimate receiver knows the key k, he can decrypt C by a transformation f^{-1} defined by

$$f^{-1} : \mathcal{C} \xrightarrow{k} \mathcal{M}, \tag{3.39}$$

and obtain

$$D_k(C) = D_k(E_k(M)) = M, \quad C \in \mathcal{C} \text{ and } M \in \mathcal{M}, \tag{3.40}$$

the original plain-text message. There are many different types of secret-key cryptographic systems. In what follows, we shall introduce some of these systems. (Note that the terms cryptographic systems, cryptographic schemes, or ciphers are essentially the same concepts, and we shall use them interchangeably in this chapter.)

(I) Stream (Bit) Ciphers. In stream ciphers, the message units are bits, and the key is usually produced by a random bit generator (see Figure 3.4). The plaintext is encrypted on a bit-by-bit basis:

$$\begin{array}{c|cccccccccccccccccc}
M & 0 & 1 & 1 & 0 & 0 & 0 & 1 & 1 & 1 & 1 & 1 & 1 & 1 & 0 & 1 & 0 & 1 & 0\cdots \\
K & 1 & 0 & 0 & 1 & 1 & 0 & 0 & 1 & 0 & 0 & 0 & 1 & 0 & 1 & 1 & 1 & 0 & 1\cdots \\
\hline
C & 1 & 1 & 1 & 1 & 1 & 0 & 1 & 0 & 1 & 1 & 1 & 0 & 1 & 1 & 0 & 1 & 1 & 1\cdots
\end{array}$$

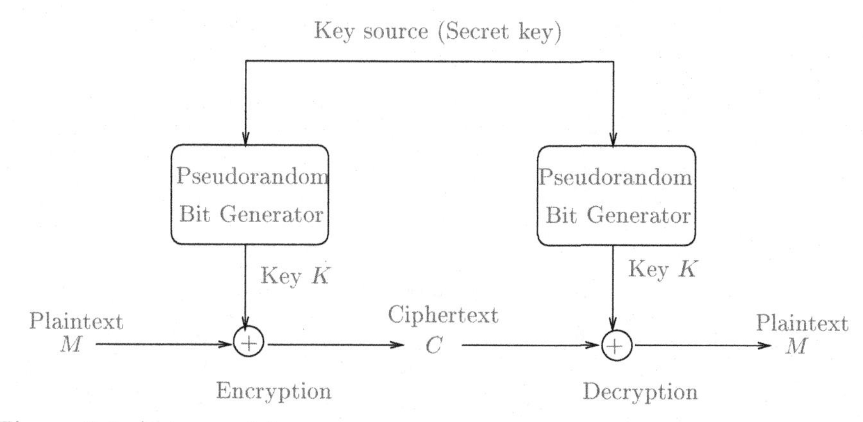

Figure 3.4. A stream cipher

The key is fed into the random bit generator to create a long sequence of binary signals. This "key-stream" K is then mixed with the plaintext stream M, usually by a bit-wise XOR (Exclusive-OR, or modulo-2 addition) to produce the ciphertext stream C. The decryption is done by XORing with the same key stream, using the same random bit generator and seed:

C	1	1	1	1	1	0	1	0	1	1	1	0	1	1	0	1	1	1 ···
K	1	0	0	1	1	0	0	1	0	0	0	1	0	1	1	1	0	1 ···
M	0	1	1	0	0	0	1	1	1	1	1	1	1	0	1	0	1	0 ···

(II) Monographic (Character) Ciphers. Earlier ciphers (cryptosystems) were based on transforming each letter of the plaintext into a different letter to produce the ciphertext. Such ciphers are called *character*, *substitution* or *monographic ciphers*, since each letter is shifted individually to another letter by a substitution. First of all, let us define the numerical equivalents, as in Table 3.2, of the 26 English capital letters, since our operations will be on

Table 3.2. Numerical equivalents of English capital letters

A	B	C	D	E	F	G	H	I	J	K	L	M
↕	↕	↕	↕	↕	↕	↕	↕	↕	↕	↕	↕	↕
0	1	2	3	4	5	6	7	8	9	10	11	12
N	O	P	Q	R	S	T	U	V	W	X	Y	Z
↕	↕	↕	↕	↕	↕	↕	↕	↕	↕	↕	↕	↕
13	14	15	16	17	18	19	20	21	22	23	24	25

the numerical equivalents of letters, rather than the letters themselves. The following are some typical character ciphers.

(1) Caesar[4] cipher: A simple *Caesar cipher* uses the following substitution transformation:

$$f_3 = E_3(m) \equiv m + 3 \pmod{26}, \quad 0 \le m \in \mathcal{M} \le 25, \tag{3.41}$$

and

$$f_3^{-1} = D_3(c) \equiv c - 3 \pmod{26}, \quad 0 \le c \in \mathcal{C} \le 25, \tag{3.42}$$

where 3 is the key for both encryption and decryption. Clearly, the corresponding letters of the Caesar cipher will be obtained from those in Table 3.2 by moving three letters forward, as described in Table 3.3. Mathematically, in encryption we just perform a mapping $m \mapsto m + 3 \bmod 26$ on the plaintext, whereas in decryption a mapping $c \mapsto c - 3 \bmod 26$ on the ciphertext.

Table 3.3. The corresponding letters of the Caesar cipher

$\mathcal{M}$	A	B	C	D	E	F	G	H	I	J	K	L	M
	↕	↕	↕	↕	↕	↕	↕	↕	↕	↕	↕	↕	↕
Shift	3	4	5	6	7	8	9	10	11	12	13	14	15
	↕	↕	↕	↕	↕	↕	↕	↕	↕	↕	↕	↕	↕
$\mathcal{C}$	D	E	F	G	H	I	J	K	L	M	N	O	P
$\mathcal{M}$	N	O	P	Q	R	S	T	U	V	W	X	Y	Z
	↕	↕	↕	↕	↕	↕	↕	↕	↕	↕	↕	↕	↕
Shift	16	17	18	19	20	21	22	23	24	25	0	1	2
	↕	↕	↕	↕	↕	↕	↕	↕	↕	↕	↕	↕	↕
$\mathcal{C}$	Q	R	S	T	U	V	W	X	Y	Z	A	B	C

[4]

Julius Caesar (100–44 BC) was a celebrated Roman general, statesman, orator and reformer. The Caesar cipher, involving in replacing each letter of the alphabet with the letter standing three places further down the alphabet, was apparently used by Caesar (he used the cipher in both his domestic and military efforts), but he was also supposed to have invented the cipher himself. Although the Caesar cipher was a simple cipher and particularly simple to crack, it is a useful vehicle for explaining cryptographic principles. (Photo by courtesy of Dr. Singh.)

(2) Shift transformations: Slightly more general transformations are the following so-called *shift transformations*:

$$f_k = E_k(m) \equiv m + k \pmod{26}, \quad 0 \le k, m \le 25, \tag{3.43}$$

and

$$f_k^{-1} = D_k(c) \equiv c - k \pmod{26}, \quad 0 \le k, c \le 25, \tag{3.44}$$

(3) Affine transformations: More general transformations are the following so-called *affine transformations*:

$$f_{(a,b)} = E_{(a,b)}(m) \equiv am + b \pmod{26}, \tag{3.45}$$

with $a, b \in \mathbb{Z}$ the key, $0 \le a, b, m \le 26$ and $\gcd(a, 26) = 1$, together with

$$f_{(a,b)}^{-1} = D_{(a,b)}(c) \equiv a^{-1}(c - b) \pmod{26}, \tag{3.46}$$

where a^{-1} is the multiplicative inverse of a modulo 26 (even more generally, the modulus 26 could be any number greater than 26, but normally chosen to be a prime number).

Example 3.3.1. In character ciphers, we have

E_3(IBM) = LEP,
E_4(NIST) = QLVW,
E_7(ENCRYPTION) = LUJXFWAPYU.

D_4(YWHEBKNJEW) = CALIFORNIA,
D_5(ZIBGVIY) = ENGLAND,
D_6(XYWLSJNCIH) = DECRYPTION.

Exercise 3.3.1. Decrypt the following character ciphertexts:

D_7(VHFFNGBVTMBHG),
D_9(JVTLIZKP).

Example 3.3.2. Use the following affine transformations

$$f_{(7,21)} \equiv 7m + 21 \pmod{26}$$

and

$$f_{(7,21)}^{-1} \equiv 7^{-1}(c - 21) \pmod{26}$$

to encrypt the message SECURITY and decrypt the message VLXIJH. To encrypt the message, we have

$$\begin{array}{lll}
S = 18, & 7 \cdot 18 + 21 \bmod 26 = 17, & S \Rightarrow R, \\
E = 4, & 7 \cdot 4 + 21 \bmod 26 = 23, & E \Rightarrow X, \\
C = 2, & 7 \cdot 2 + 21 \bmod 26 = 9, & C \Rightarrow J, \\
U = 20, & 7 \cdot 20 + 21 \bmod 26 = 5, & U \Rightarrow F, \\
R = 17, & 7 \cdot 17 + 21 \bmod 26 = 10, & R \Rightarrow K, \\
I = 8, & 7 \cdot 8 + 21 \bmod 26 = 25, & I \Rightarrow Z, \\
T = 19, & 7 \cdot 19 + 21 \bmod 26 = 24, & T \Rightarrow Y, \\
Y = 24, & 7 \cdot 24 + 21 \bmod 26 = 7, & Y \Rightarrow H.
\end{array}$$

Thus, $E_{(7,21)}(\text{SECURITY}) = \text{RXJFKZYH}$. Similarly, to decrypt the message VLXIJH, we have

$$\begin{array}{lll} V = 21, & 7^{-1} \cdot (21 - 21) \bmod 26 = 0, & V \Rightarrow A, \\ L = 11, & 7^{-1} \cdot (11 - 21) \bmod 26 = 6, & L \Rightarrow G, \\ X = 23, & 7^{-1} \cdot (13 - 21) \bmod 26 = 4, & X \Rightarrow E, \\ I = 8, & 7^{-1} \cdot (8 - 21) \bmod 26 = 13, & I \Rightarrow N, \\ J = 9, & 7^{-1} \cdot (9 - 21) \bmod 26 = 2, & J \Rightarrow C, \\ H = 7, & 7^{-1} \cdot (7 - 21) \bmod 26 = 24, & H \Rightarrow Y. \end{array}$$

Thus, $D_{(7,21)}(\text{VLXIJH}) = \text{AGENCY}$.

Exercise 3.3.2. Use the affine transformation

$$f_{(11,23)} = 11m + 23 \pmod{26}$$

to encrypt the message THE NATIONAL SECURITY AGENCY. Use also the inverse transformation

$$f^{-1}_{(11,23)} = 11^{-1}(c - 23) \pmod{26}$$

to verify your result.

(III) Polygraphic (Block) Ciphers. Monographic ciphers can be made more secure by splitting the plaintext into groups of letters (rather than a single letter) and then performing the encryption and decryption on these groups of letters. This block technique is called *block ciphering*. Block cipher is also called a *polygraphic cipher*. Block ciphers may be described as follows:

(1) Split the message M into blocks of n-letters (when $n = 2$ it is called a *digraphic cipher*) $M_1, M_2, \cdots, M_j$; each block M_i for $1 \le i \le j$ is a block consisting of n letters.

(2) Translate the letters into their numerical equivalents and form the ciphertext:

$$\mathbf{C}_i \equiv \mathbf{A}\mathbf{M}_i + \mathbf{B} \pmod{N}, \quad i = 1, 2, \cdots, j \tag{3.47}$$

where $(\mathbf{A}, \mathbf{B})$ is the key, $\mathbf{A}$ is an invertible $n \times n$ matrix with $\gcd(\det(\mathbf{A}), N) = 1$, $\mathbf{B} = (B_1, B_2, \cdots, B_n)^T$, $\mathbf{C}_i = (c_1, c_2, \cdots, c_n)^T$ and $\mathbf{M}_i = (m_1, m_2, \cdots, m_n)^T$. For simplicity, we just consider

$$\mathbf{C}_i \equiv \mathbf{A}\mathbf{M}_i \pmod{26}. \tag{3.48}$$

(3) For decryption, we perform

$$\mathbf{M}_i \equiv \mathbf{A}^{-1}(\mathbf{C}_i - \mathbf{B}) \pmod{N}, \tag{3.49}$$

where $\mathbf{A}^{-1}$ is the inverse matrix of $\mathbf{A}$. Again, for simplicity, we just consider

$$\mathbf{M}_i \equiv \mathbf{A}^{-1}\mathbf{C}_i \pmod{26}. \tag{3.50}$$

Example 3.3.3. Let

$$M = \text{YOUR PIN NO IS FOUR ONE TWO SIX}$$

be the plaintext and $n = 3$. Let also the encryption matrix be

$$\mathbf{A} = \begin{pmatrix} 11 & 2 & 19 \\ 5 & 23 & 25 \\ 20 & 7 & 17 \end{pmatrix}.$$

Then the encryption and decryption of the message can be described as follows:

(1) Split the message M into blocks of 3-letters and translate these letters into their numerical equivalents:

$$\begin{array}{ccccccccccccccc} \text{Y} & \text{O} & \text{U} & & \text{R} & \text{P} & \text{I} & & \text{N} & \text{N} & \text{O} & & \text{I} & \text{S} & \text{F} \\ \updownarrow & \updownarrow & \updownarrow & & \updownarrow & \updownarrow & \updownarrow & & \updownarrow & \updownarrow & \updownarrow & & \updownarrow & \updownarrow & \updownarrow \\ 24 & 14 & 20 & & 17 & 15 & 8 & & 13 & 13 & 14 & & 8 & 18 & 5 \end{array}$$

$$\begin{array}{ccccccccccccccc} \text{O} & \text{U} & \text{R} & & \text{O} & \text{N} & \text{E} & & \text{T} & \text{W} & \text{O} & & \text{S} & \text{I} & \text{X} \\ \updownarrow & \updownarrow & \updownarrow & & \updownarrow & \updownarrow & \updownarrow & & \updownarrow & \updownarrow & \updownarrow & & \updownarrow & \updownarrow & \updownarrow \\ 14 & 20 & 17 & & 14 & 13 & 4 & & 19 & 22 & 14 & & 18 & 8 & 23 \end{array}$$

(2) Encrypt these nine blocks in the following way:

$$\mathbf{C}_1 = \mathbf{A}\begin{pmatrix} 24 \\ 14 \\ 20 \end{pmatrix} = \begin{pmatrix} 22 \\ 6 \\ 8 \end{pmatrix}, \quad \mathbf{C}_2 = \mathbf{A}\begin{pmatrix} 17 \\ 15 \\ 8 \end{pmatrix} = \begin{pmatrix} 5 \\ 6 \\ 9 \end{pmatrix},$$

$$\mathbf{C}_3 = \mathbf{A}\begin{pmatrix} 13 \\ 13 \\ 14 \end{pmatrix} = \begin{pmatrix} 19 \\ 12 \\ 17 \end{pmatrix}, \quad \mathbf{C}_4 = \mathbf{A}\begin{pmatrix} 8 \\ 18 \\ 5 \end{pmatrix} = \begin{pmatrix} 11 \\ 7 \\ 7 \end{pmatrix},$$

$$\mathbf{C}_5 = \mathbf{A}\begin{pmatrix} 14 \\ 20 \\ 17 \end{pmatrix} = \begin{pmatrix} 23 \\ 19 \\ 7 \end{pmatrix}, \quad \mathbf{C}_6 = \mathbf{A}\begin{pmatrix} 14 \\ 13 \\ 4 \end{pmatrix} = \begin{pmatrix} 22 \\ 1 \\ 23 \end{pmatrix},$$

$$\mathbf{C}_7 = \mathbf{A}\begin{pmatrix} 19 \\ 22 \\ 14 \end{pmatrix} = \begin{pmatrix} 25 \\ 15 \\ 18 \end{pmatrix}, \quad \mathbf{C}_8 = \mathbf{A}\begin{pmatrix} 18 \\ 8 \\ 23 \end{pmatrix} = \begin{pmatrix} 1 \\ 17 \\ 1 \end{pmatrix}.$$

(3) Translating these into letters, we get the ciphertext C:

$$\begin{array}{ccccccccccccccc} 22 & 6 & 8 & & 5 & 6 & 9 & & 19 & 12 & 17 & & 11 & 7 & 7 \\ \updownarrow & \updownarrow & \updownarrow & & \updownarrow & \updownarrow & \updownarrow & & \updownarrow & \updownarrow & \updownarrow & & \updownarrow & \updownarrow & \updownarrow \\ \mathrm{W} & \mathrm{G} & \mathrm{I} & & \mathrm{F} & \mathrm{G} & \mathrm{J} & & \mathrm{T} & \mathrm{M} & \mathrm{R} & & \mathrm{L} & \mathrm{H} & \mathrm{H} \end{array}$$

$$\begin{array}{ccccccccccccccc} 23 & 19 & 7 & & 22 & 1 & 23 & & 25 & 15 & 18 & & 1 & 17 & 1 \\ \updownarrow & \updownarrow & \updownarrow & & \updownarrow & \updownarrow & \updownarrow & & \updownarrow & \updownarrow & \updownarrow & & \updownarrow & \updownarrow & \updownarrow \\ \mathrm{X} & \mathrm{T} & \mathrm{H} & & \mathrm{W} & \mathrm{B} & \mathrm{X} & & \mathrm{Z} & \mathrm{P} & \mathrm{S} & & \mathrm{B} & \mathrm{R} & \mathrm{B} \end{array}$$

(4) To recover the message M from C, we first compute A^{-1} modulo 26:

$$\mathbf{A}^{-1} = \begin{pmatrix} 11 & 2 & 19 \\ 5 & 23 & 25 \\ 20 & 7 & 17 \end{pmatrix}^{-1} = \begin{pmatrix} 10 & 23 & 7 \\ 15 & 9 & 22 \\ 5 & 9 & 21 \end{pmatrix}.$$

and then perform $\mathbf{C}_i = \mathbf{A}^{-1}\mathbf{C}_i$ as follows:

$$\mathbf{M}_1 = \mathbf{A}^{-1}\begin{pmatrix} 22 \\ 6 \\ 8 \end{pmatrix} = \begin{pmatrix} 24 \\ 14 \\ 20 \end{pmatrix}, \quad \mathbf{M}_2 = \mathbf{A}^{-1}\begin{pmatrix} 5 \\ 6 \\ 9 \end{pmatrix} = \begin{pmatrix} 17 \\ 15 \\ 8 \end{pmatrix},$$

$$\mathbf{M}_3 = \mathbf{A}^{-1}\begin{pmatrix} 19 \\ 12 \\ 17 \end{pmatrix} = \begin{pmatrix} 13 \\ 13 \\ 14 \end{pmatrix}, \quad \mathbf{M}_4 = \mathbf{A}^{-1}\begin{pmatrix} 11 \\ 7 \\ 7 \end{pmatrix} = \begin{pmatrix} 8 \\ 18 \\ 5 \end{pmatrix},$$

$$\mathbf{M}_5 = \mathbf{A}^{-1}\begin{pmatrix} 23 \\ 19 \\ 7 \end{pmatrix} = \begin{pmatrix} 14 \\ 20 \\ 17 \end{pmatrix}, \quad \mathbf{M}_6 = \mathbf{A}^{-1}\begin{pmatrix} 22 \\ 1 \\ 23 \end{pmatrix} = \begin{pmatrix} 14 \\ 13 \\ 4 \end{pmatrix},$$

$$\mathbf{M}_7 = \mathbf{A}^{-1}\begin{pmatrix} 25 \\ 15 \\ 18 \end{pmatrix} = \begin{pmatrix} 19 \\ 22 \\ 14 \end{pmatrix}, \quad \mathbf{M}_8 = \mathbf{A}^{-1}\begin{pmatrix} 1 \\ 17 \\ 1 \end{pmatrix} = \begin{pmatrix} 18 \\ 8 \\ 23 \end{pmatrix}.$$

So, we have:

$$\begin{array}{ccccccccccccccc} 24 & 14 & 20 & & 17 & 15 & 8 & & 13 & 13 & 14 & & 8 & 18 & 5 \\ \updownarrow & \updownarrow & \updownarrow & & \updownarrow & \updownarrow & \updownarrow & & \updownarrow & \updownarrow & \updownarrow & & \updownarrow & \updownarrow & \updownarrow \\ \mathrm{Y} & \mathrm{O} & \mathrm{U} & & \mathrm{R} & \mathrm{P} & \mathrm{I} & & \mathrm{N} & \mathrm{N} & \mathrm{O} & & \mathrm{I} & \mathrm{S} & \mathrm{F} \end{array}$$

$$\begin{array}{ccccccccccccccc} 14 & 20 & 17 & & 14 & 13 & 4 & & 19 & 22 & 14 & & 18 & 8 & 23 \\ \updownarrow & \updownarrow & \updownarrow & & \updownarrow & \updownarrow & \updownarrow & & \updownarrow & \updownarrow & \updownarrow & & \updownarrow & \updownarrow & \updownarrow \\ \mathrm{O} & \mathrm{U} & \mathrm{R} & & \mathrm{O} & \mathrm{N} & \mathrm{E} & & \mathrm{T} & \mathrm{W} & \mathrm{O} & & \mathrm{S} & \mathrm{I} & \mathrm{X} \end{array}$$

which is the original message.

Exercise 3.3.3. Let

$$\mathbf{A} = \begin{pmatrix} 3 & 13 & 21 & 9 \\ 15 & 10 & 6 & 25 \\ 10 & 17 & 4 & 8 \\ 1 & 23 & 7 & 2 \end{pmatrix} \quad \text{and} \quad \mathbf{B} = \begin{pmatrix} 1 \\ 21 \\ 8 \\ 17 \end{pmatrix}.$$

Use the block transformation

$$C_i \equiv \mathbf{A}M_i + \mathbf{B} \pmod{26}$$

to encrypt the following message

PLEASE SEND ME THE BOOK, MY CREDIT CARD NO IS
SIX ONE TWO ONE THREE EIGHT SIX ZERO
ONE SIX EIGHT FOUR NINE SEVEN ZERO TWO.

Use

$$M_i \equiv \mathbf{A}^{-1}(C_i - \mathbf{B}) \pmod{26}$$

to verify your result, where

$$\mathbf{A}^{-1} = \begin{pmatrix} 26 & 13 & 20 & 5 \\ 0 & 10 & 11 & 0 \\ 9 & 11 & 15 & 22 \\ 9 & 22 & 6 & 25 \end{pmatrix}.$$

(IV) Exponentiation Ciphers. The exponentiation cipher, invented by Pohlig and Hellman in 1976, may be described as follows. Let p be a prime number, M the numerical equivalent of the plaintext, where each letter of the plaintext is replaced by its two digit equivalent, as defined in Table 3.4. Subdivide M into blocks M_i such that $0 < M_i < p$. Let k be an integer with $0 < k < p$ and $\gcd(k, p-1) = 1$. Then the encryption transformation for M_i is defined by

$$C_i = E_k(M_i) \equiv M_i^k \pmod{p}, \tag{3.51}$$

Table 3.4. Two digit equivalents of letters

⊔	A	B	C	D	E	F	G	H	I	J	K	L	M
↕	↕	↕	↕	↕	↕	↕	↕	↕	↕	↕	↕	↕	↕
00	01	02	03	04	05	06	07	08	09	10	11	12	13
N	O	P	Q	R	S	T	U	V	W	X	Y	Z	
↕	↕	↕	↕	↕	↕	↕	↕	↕	↕	↕	↕	↕	
14	15	16	17	18	19	20	21	22	23	24	25	26	

and the decryption transformation by

$$M_i = D_{k^{-1}}(C_i) \equiv C_i^{k^{-1}} \equiv (M_i^k)^{k^{-1}} \equiv M_i \pmod{p}, \tag{3.52}$$

where $k \cdot k^{-1} \equiv 1 \pmod{p-1}$.

Example 3.3.4. Let $p = 7951$ and $k = 91$ such that $\gcd(7951-1, 91) = 1$. Suppose we wish to encrypt the message

$M =$ ENCRYPTION REGULATION MOVES TO A STEP CLOSER

using the exponentiation cipher. Firstly, we convert all the letters in the message to their numerical equivalents via Table 3.4

05 14 03 18 25 16 20 09 15 14 00 18 05 07 21 12 01 20 09 15 14 00
13 15 22 05 19 00 20 15 00 01 00 19 20 05 16 00 03 12 15 19 05 18

and group them into blocks with four digits

0514 0318 2516 2009 1514 0018 0507 2112 0120 0915 1400
1315 2205 1900 2015 0001 0019 2005 1600 0312 1519 0518

Then we perform the following computation

$$
\begin{array}{ll}
C_1 = 0514^{91} \bmod 7951 = 2174 & C_2 = 0318^{91} \bmod 7951 = 4468 \\
C_3 = 2516^{91} \bmod 7951 = 7889 & C_4 = 2009^{91} \bmod 7951 = 6582 \\
C_5 = 1514^{91} \bmod 7951 = 924 & C_6 = 0018^{91} \bmod 7951 = 5460 \\
C_7 = 0507^{91} \bmod 7951 = 7868 & C_8 = 2112^{91} \bmod 7951 = 7319 \\
C_9 = 0120^{91} \bmod 7951 = 726 & C_{10} = 915^{91} \bmod 7951 = 2890 \\
C_{11} = 1400^{91} \bmod 7951 = 7114 & C_{12} = 1315^{91} \bmod 7951 = 5463 \\
C_{13} = 2205^{91} \bmod 7951 = 5000 & C_{14} = 1900^{91} \bmod 7951 = 438 \\
C_{15} = 2015^{91} \bmod 7951 = 2300 & C_{16} = 0001^{91} \bmod 7951 = 1 \\
C_{17} = 0019^{91} \bmod 7951 = 1607 & C_{18} = 2005^{91} \bmod 7951 = 3509 \\
C_{19} = 1600^{91} \bmod 7951 = 7143 & C_{20} = 0312^{91} \bmod 7951 = 5648 \\
C_{21} = 1519^{91} \bmod 7951 = 3937 & C_{22} = 0518^{91} \bmod 7951 = 4736.
\end{array}
$$

So, the ciphertext of M is

2174 4468 7889 6582 0924 5460 7868 7319 0726 2890 7114
5463 5000 0438 2300 0001 1607 3509 7143 5648 3937 5064.

To decrypt the ciphertext C back to the plaintext M, since the secret key $k = 91$ and the prime modulus $p = 7951$ are known, we compute the multiplicative inverse k^{-1} of k modulo $p - 1$ as follows:

$$k^{-1} \equiv \frac{1}{k} \pmod{p-1} \equiv \frac{1}{91} \pmod{7950} \equiv 961 \pmod{7950}.$$

Thus, we have

$$
\begin{array}{ll}
M_1 = 2174^{961} \bmod 7951 = 514 & M_2 = 4468^{961} \bmod 7951 = 318 \\
M_3 = 7889^{961} \bmod 7951 = 2516 & M_4 = 6582^{961} \bmod 7951 = 2009 \\
M_5 = 924^{961} \bmod 7951 = 1514 & M_6 = 5460^{961} \bmod 7951 = 18 \\
M_7 = 7868^{961} \bmod 7951 = 507 & M_8 = 7319^{961} \bmod 7951 = 2112 \\
M_9 = 726^{961} \bmod 7951 = 120 & M_{10} = 2890^{961} \bmod 7951 = 915 \\
M_{11} = 7114^{961} \bmod 7951 = 1400 & M_{12} = 5463^{961} \bmod 7951 = 1315 \\
M_{13} = 5000^{961} \bmod 7951 = 2205 & M_{14} = 438^{961} \bmod 7951 = 1900 \\
M_{15} = 2300^{961} \bmod 7951 = 2015 & M_{16} = 1^{961} \bmod 7951 = 1 \\
M_{17} = 1607^{961} \bmod 7951 = 19 & M_{18} = 3509^{961} \bmod 7951 = 2005 \\
M_{19} = 7143^{961} \bmod 7951 = 1600 & M_{20} = 5648^{961} \bmod 7951 = 312 \\
M_{21} = 3937^{961} \bmod 7951 = 1519 & M_{22} = 4736^{961} \bmod 7951 = 518.
\end{array}
$$

Therefore, we have recovered the original message.

Exercise 3.3.4. Let $p = 9137$ and $k = 73$ so that $\gcd(p-1, k) = 1$ and $k^{-1} \bmod (p-1) = 750$. Use the exponentiation transformation $C = M^k \bmod p$ to encrypt the following message:

THE CESG IS THE UK NATIONAL TECHNICAL AUTHORITY

ON INFORMATION SECURITY.

THE NSA IS THE OFFICIAL INTELLIGENCE-GATHERING

ORGANIZATION OF THE UNITED STATES.

Use also $M = C^{k^{-1}} \bmod p$ to verify your result.

Exercise 3.3.5 (A challenge problem). The following cryptogram was presented by Édouard Lucas at the 1891 meeting of the French Association for Advancement of Science (see Williams, [257]); it has never been decrypted, and hence is suitable as a challenge to the interested reader.

XSJOD	PEFOC	XCXFM	RDZME
JZCOA	YUMTZ	LTDNJ	HBUSQ
XTFLK	XCBDY	GYJKK	QBSAH
QHXPE	DBMLI	ZOYVQ	PRETL
TPMUK	XGHIV	ARLAH	SPGGP

VBQYH	TVJYJ	NXFFX	BVLCZ
LEFXF	VDMUB	QBIJV	ZGGAI
TRYQB	AIDEZ	EZEDX	KS

3.3.3 Data/Advanced Encryption Standard (DES/AES)

The most popular secret-key cryptographic scheme in use (by both governments and private companies) is the Data Encryption Standard (DES) — DES was designed at IBM and approved in 1977 as a standard by the U.S. National Bureau of Standards (NBS), now called the National Institute of Standards and Technology (NIST). This standard, first issued in 1977 (FIPS 46 – Federal Information Processing Standard 46), is reviewed every five years. It is currently specified in FIPS 46-2. NIST is proposing to replace FIPS 46-2 with FIPS 46-3 to provide for the use of Triple DES (TDES) as specified in the American National Standards Institute (ANSI) X9.52 standard. Comments were sought from industry, government agencies, and the public on the draft of FIPS 46-3 before 15 April 15, 1999.

The standard (algorithm) uses a product transformation of transpositions, substitutions, and non-linear operations. They are applied for 16 iterations to each block of a message; the message is split into 64-bit message blocks. The key used is composed of 56 bits taken from a 64-bit key which includes 8 parity bits. The algorithm is used in reverse to decrypt each ciphertext block and the same key is used for both encryption and decryption. The algorithm itself is shown schematically in Figure 3.5, where the $\oplus$ is the "exclusive or" (XOR) operator. The DES algorithm takes as input a 64-bit message (plaintext) M and a 56-bit key K, and produces a 64-bit ciphertext C. DES first applies an initial fixed bit-permutation (IP) to M to obtain M'. This permutation has no apparent cryptographic significance. Second, DES divides M' into a 32-bit left half L_0 and 32-bit right half R_0. Third, DES executes the following operations for $i = 1, 2, \cdots, 16$ (there are 16 "rounds"):

$$\left.\begin{aligned} L_i &= R_{i-1}, \\ R_i &= L_{i-1} \oplus f(R_{i-1},\ K_i), \end{aligned}\right\} \tag{3.53}$$

where f is a function that takes a 32-bit right half and a 48-bit "round key" and produces a 32-bit output. Each round key K_i contains a different subset of the 56-bit key bits. Finally, the pre-ciphertext $C' = (R_{16}, L_{16})$ is permuted according to IP^{-1} to obtain the final ciphertext C. To decrypt, the algorithm is run in reverse: a permutation, 16 XOR rounds using the round key in reverse order, and a final permutation that recovers the plaintext. All of this extensive bit manipulations can be incorporated into the logic of a single

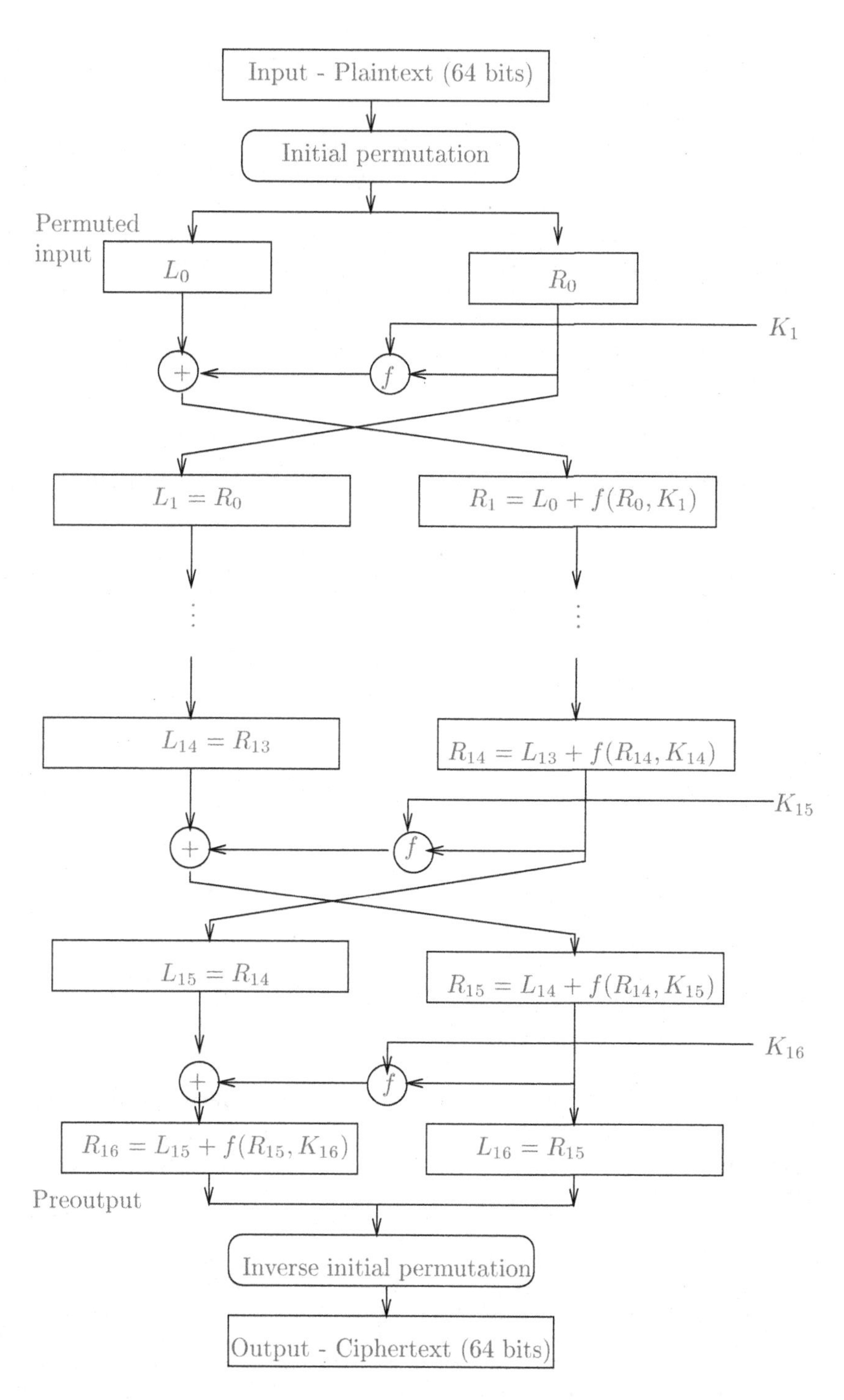

Figure 3.5. The Data Encryption Standard (DES) algorithm

special-purpose microchip, so DES can be implemented very efficiently. However, the DES cracking project being undertaken by the Electronic Frontier Foundation is able to break the encryption for 56 bit DES in about 22 hours. As a result, NIST has recommended that businesses use Triple DES[5] (TDES), which involves three different DES encryption and decryption operations. Let $E_K(M)$ and $D_K(C)$ represent the DES encryption and decryption of M and C using DES key K, respectively. Each TDES encryption/decryption operation (as specified in ANSI X9.52) is a compound operation of DES encryption and decryption operations. The following operations are used in TDES:

(1) **TDES encryption operation**: the transformation of a 64-bit block M into a 64-bit block C is defined as follows:

$$C = E_{K_3}(D_{K_2}(E_{K_1}(M))). \tag{3.54}$$

(2) **TDES decryption operation**: the transformation of a 64-bit block C into a 64-bit block M is defined as follows:

$$M = D_{K_1}(E_{K_2}(D_{K_3}(C))). \tag{3.55}$$

There are three options for the TDES *key bundle* (K_1, K_2, K_3):

(1) K_1, K_2, and K_3 are independent keys.

(2) K_1, K_2 are independent keys and $K_3 = K_1$.

(3) $K_1 = K_2 = K_3$.

For example, if option 2 is chosen, then the TDES encryption and decryption are as follows:

$$C = E_{K_1}(D_{K_2}(E_{K_1}(M))), \tag{3.56}$$

$$M = D_{K_1}(E_{K_2}(D_{K_1}(C))). \tag{3.57}$$

Interested readers are suggested to consult the current NIST report FIPS 46-3 [173] for the new standard of the TDES.

It is interesting to note that some experts say DES is still secure when used properly. However, Edward Roback at the NIST said that the DES, which uses 56-bit encryption keys, is no longer sufficiently difficult to decrypt. For example, in February 1998, a team of engineers used a distributed "brute force" decryption program to break a 56-bit DES key in 39 days, about three

[5] Triple DES is a type of *multiple encryption*. Multiple encryption is a combination technique aimed to improve the security of a block algorithm. It uses an algorithm to encrypt the same plaintext block multiple times with multiple keys. The simplest multiple encryption is the so-called *double encryption* in which an algorithm is used to encrypt a block twice with two different keys – first encrypt a block with the first key, and then encrypt the resulting ciphertext with the second key: $C = E_{k_2}(E_{k_1}(M))$. The decryption is just the reverse process of the encryption: $M = D_{k_1}(D_{k_2}(C))$.

times faster than it took another team just the year before, and more recently, the team cracked DES in just over 22 hours earlier this year.

The U.S. Department of Commerce's NIST had issued a formal call on 12 September 1997 for companies, universities, and other organizations to submit algorithm proposals for a new generation encryption standard for protecting sensitive data well into the 21st century. This new Advanced Encryption Standard (AES) will replace the DES and support encryption key size up to 256 bits and must be available royalty-free throughout the world. On 20 August 1998 at the First AES Candidate Conference (AES1), NIST announced fifteen (15) official AES candidate algorithms submitted by researchers from twelve (12) different countries, including the United States, Australia, France, Germany, Japan, Norway and the United Kingdom. Since then, cryptographers have tried to find ways to *attack* the different algorithms, looking for weaknesses that would compromise the encrypted information. Shortly after the Second AES Candidate Conference (AES2) on 22–23 March 1999 in Rome, Italy, NIST announced on 9 August 1999 that the following five (5) contenders had been chosen as finalist for the AES, all are block ciphers:

(1) **MARS**: Developed by International Business Machines (IBM) Corporation of Armonk, New York, USA.
(2) **RC6**: Developed by RSA Laboratories of Bedford, Massachusetts, USA.
(3) **Rijndael**: Developed by Joan Daemen and Vincent Rijmen of Belgium.
(4) **Serpent**: Developed by Ross Anderson, Eli Biham and Lars Knudsen of the United Kingdom, Israel and Norway, respectively.
(5) **Twofish**: Developed by Bruce Schneier, John Kelsey, Doug Whiting, David Wagner Chris Hall and Niels Ferguson, of Counterpane Systems, Minneapolis, USA.

These five finalist algorithms had received further analysis during a second, more in-depth review period (August 1999–May 2000) in the selection of the final algorithm for the FIPS (Federal Information Processing Standard) AES. On 2 October 2000, the algorithm Rijndael, developed by Joan Daemen (Proton World International, Belgium) and Vincent Rijmen (Katholieke Universiteit Leuven, Belgium) was finally chosen to be the AES. The strong points of Rijndael are a simple and elegant design, efficient and fast on modern processors, but also compact in hardware and on smartcards. These features make Rijndael suitable for a wide range of applications. It will be used to protect sensitive but 'unclassified' electronic information of the US government. During the last year, a large number of products and applications has been AES-enabled. Therefore, it is very likely to become a worldwide de facto standard in numerous other applications such as Internet security, bank cards and ATMs.

3.3.4 Public-Key Cryptography

An obvious requirement of a good cryptographic system is that secret messages should be easy to encrypt and decrypt for legitimate users, and these processes (or, at least, decryption) should be hard for everyone else. Number Theory has turned out to be an excellent source of computational problems that have both easy and (apparently) hard aspects and that can be used as the backbone of several cryptographic systems.

CARL POMERANCE
Cryptology and Computational Number Theory [191]

In their seminal paper "New Directions in Cryptography" [66], Diffie[6] and Hellman[7], both in the Department of Electrical Engineering at Stanford University at the time, first proposed the idea and the concept of public-key

[6]

Whitfield Diffie (1944–), a Distinguished Engineer at Sun Microsystems in Palo Alto, California, is perhaps best known for his 1975 discovery of the concept of public-key cryptography, for which he was awarded a Doctorate in Technical Sciences (Honoris Causa) by the Swiss Federal Institute of Technology in 1992. He received a BSc degree in mathematics from the Massachusetts Institute of Technology in 1965. Prior to becoming interested in cryptography, he worked on the development of the Mathlab symbolic manipulation system — sponsored jointly at Mitre and the MIT Artificial Intelligence Laboratory — and later on proof of correctness of computer programs at Stanford University. Diffie was the recipient of the IEEE Information Theory Society Best Paper Award 1979 for the paper *New Directions in Cryptography* [66], the IEEE Donald E. Fink award 1981 for expository writing for the paper *Privacy and Authentication* [67] (both papers co-authored with Martin Hellman), and the National Computer Systems Security Award for 1996. (Photo by courtesy of Dr. Simon Singh.)

[7]

Martin E. Hellman (1945–), the father of modern (public key) cryptography, received his BEng from New York University in 1966, and his MSc and PhD from Stanford University in 1967 and 1969, respectively, all in Electrical Engineering. Hellman was on the research staff at IBM's Watson Research Center from 1968-69 and on the faculty of Electrical Engineering at MIT from 1969-71. He returned to Stanford as a faculty member in 1971, where he served on the regular faculty until becoming Professor Emeritus in 1996. He has authored over 60 technical papers, five U.S. and a number of foreign patents. His work, particularly the invention of public key cryptography, has been covered in the popular media including Scientific American and Time magazine. He was the recipient of an IEEE Centennial Medal (1984). Notice that Diffie, Hellman and Merkle are the three joint inventors of public-key cryptography, with Diffie and Merkle as Hellman's research assistant and PhD student. (Photo by courtesy of Prof. Hellman.)

cryptography as well as digital signatures; they also proposed in the same time a key-exchange protocol, based on the hard *discrete logarithm problem*, for two parties to form a common private key over the insecure channel (see Subsection 3.3.2).

Figure 3.6. The DHM crypto years: (Left to right) Merkle, Hellman and Diffie (Photo by courtesy of Dr. Simon Singh)

It should be noted that Ralph Merkle[8], deserves equal credit with Diffie and Hellman for the invention of public key cryptography. Although his paper *Secure Communication Over Insecure Channels* [158] was published in 1978,

8

Ralph C. Merkle (1952–) studied Computer Science at the University of California at Berkeley with a B.A. in 1974 and a M.S. in 1977, and obtained his PhD in Electrical Engineering at Stanford University in 1979 with the thesis entitled *Secrecy, Authentication, and Public Key Systems*, with Prof. Martin Hellman as his thesis advisor. Merkle co-invented public-key cryptography, received the 1997 ACM Kanellakis Award (along with Leonard Adleman, Whitfield Diffie, Martin Hellman, Ronald Rivest and Adi Shamir), the 1998 Feynman Prize in Nanotechnology for theory, the 1999 IEEE Kobayashi Award, and the 2000 RSA Award in Mathematics. He is currently a Principal Fellow at Zyvex, working on molecular manufacturing (also known as nanotechnology). (Photo by courtesy of Dr. Merkle.)

two years later than Diffie and Hellman's paper *New Directions in Cryptography*, it was submitted in August 1975. Also, his conception of *public key distribution* occurred in the Fall of 1974, again before Diffie and Hellman conceived of *public key cryptosystems*.

Remarkably enough, just about one or two years later, three MIT computer scientists, Rivest, Shamir, and Adleman, proposed in 1978 a practical public-key cryptosystem based on primality testing and integer factorization, now widely known as RSA cryptosystem (see Subsection 3.3.6). More specifically, they based on their encryption and decryption on mod-n arithmetic, where n is the product of two large prime numbers p and q. A special case based on mod-p arithmetic with p prime, now known as exponential cipher, had already been studied by Pohlig and Hellman in 1978 [176].

It is interesting to note that in December 1997 the Communication-Electronics Security Group (CESG) of the British Government Communications Headquarters (GCHQ), claimed that public-key cryptography was conceived by Ellis[9] in 1970 and implemented by two of his colleagues Cocks[10]

9

James H. Ellis (1924–1997) was conceived in Britain but was born in Australia. While still a baby, he returned to and grew up in London. He studied Physics at Imperial College, London and worked in the Post Office Research Station at Dollis Hill. In 1965, Ellis, together with the cryptographic division at Dollis Hill, moved to Cheltenham to join the newly formed Communication-Electronics Security Group (CESG), a special section of the GCHQ, devoted to ensuring the security of British communications. Ellis was unpredictable, introverted and a rather quirky worker, he was never put in charge of any of the important CESG research groups, and he even didn't really fit into the day-to-day business of CESG. Nevertheless, he was a foremost British government cryptographer. Ellis had a good reputation as a cryptoguru, and if other researchers found themselves with impossible problems, they would knock his door in the hope that his vast knowledge and originality would provide a solution. It was probably because of this reputation that the British military asked him in the beginning of 1969 to investigate the key distribution problem, that led him to have the idea of the non-secret encryption.

10

Clifford C. Cocks studied mathematics, specialized in number theory, at the University of Cambridge and joined the CESG in September 1973. While as a school student in Manchester Grammar School, he represented Britain at the International Mathematical Olympiad in Moscow in 1968 and won a Silver prize. Before joining CESG he knew very little about encryption and its intimate connection with military and diplomatic communications, so his mentor, Nick Patterson at CESG told him Ellis's idea for public-key cryptography. "Because I had been working in number theory, it was natural to think about one-way functions, something you could do but not undo. Prime numbers and factoring was a natural candidate," explained by Cocks. It did not take him too long to formulate a special case of the RSA public key cryptography.

and Williamson[11] between 1973 and 1976 in CESG, by releasing the following five papers:

[1] James H. Ellis, *The Possibility of Non-Secret Encryption*, January 1970, 9 pages.

[2] Clifford C. Cocks, *A Note on Non-Secret Encryption*, 20 November 1973, 2 pages.

[3] Malcolm J. Williamson, *Non-Secret Encryption Using a Finite Field*, 21 January 1974, 2 pages.

[4] Malcolm Williamson, *Thoughts on Cheaper Non-Secret Encryption*, 10 August 1976, 3 pages.

[5] James Ellis, *The Story of Non-Secret Encryption*, 1987, 9 pages.

The US Government's National Security Agency (NSA) also made a similar claim that they had public-key cryptography a decade earlier. It must be pointed out that there are apparently two parallel universes in cryptography, the public and the secret worlds. The CESG and even the NSA people certainly deserve some kind of credit, but according to the "first to publish, not first to keep secret" rule, the *full* credit of the invention of public-key cryptography goes to Diffie, Hellman and Merkle (along with Rivest, Shamir and Adleman for their first practical implementation). It must also be pointed out that Diffie and Hellman [66] in the same time also proposed the marvelous idea of digital signatures, and in implementing their RSA cryptosystem, Rivest, Shmire and Adleman also implemented the idea of digital signatures, whereas none of the CESG released papers showed any evidence that they had any thought of digital signatures, which is half of the Diffie-Hellman-Merkle public-key cryptography invention!

In a public-key (non-secret key) cryptosystem (see Figure 3.7), the encryption key e_k and decryption key d_k are different, that is, $e_k \neq d_k$ (this is why we call public-key cryptosystems *asymmetric key cryptosystems*). Since e_k is

---

11

Malcolm J. Williamson also attended the Manchester Grammar School and studied mathematics at the University of Cambridge, but joined the CESG in September 1974. Same as Clifford Cocks, Malcolm Williamson also represented Britain at the International Mathematical Olympiad in Moscow in 1968 but won a Gold prize. When Cocks first explained his work on public-key cryptography to Williamson, Williamson really didn't believe it and tried to prove that Cocks had made a mistake and that public-key cryptography did not really exist. Remarkably enough, Williamson failed to find a mistake, instead he found another solution to the problem of key distribution, at roughly the same time that Prof. Martin Hellman discovered it. (Photos of Ellis, Cocks and Williamson by courtesy of Dr. Simon Singh.)

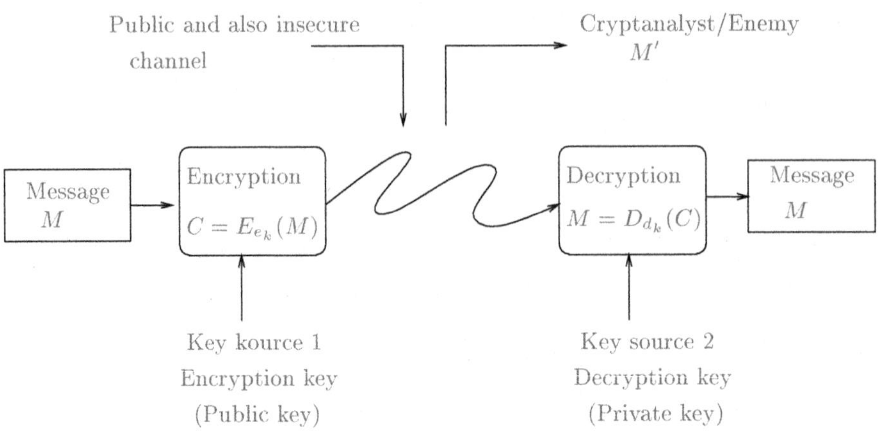

Figure 3.7. Modern public-key cryptosystems ($e_k \neq d_k$)

only used for encryption, it can be made public; only d_k must be kept a secret for decryption. To distinguish public-key cryptosystems from secret-key cryptosystems, e_k is called the *public key*, and d_k the *private key*; only the key used in secret-key cryptosystems is called the *secret key*. The implementation of public-key cryptosystems is based on *trapdoor one-way functions*.

Definition 3.3.1. Let S and T be finite sets. A one-way function

$$f: S \to T \tag{3.58}$$

is an invertible function satisfying

(1) f is easy to compute, that is, given $x \in S$, $y = f(x)$ is easy to compute.

(2) f^{-1}, the inverse function of f, is difficult to compute, that is, given $y \in T$, $x = f^{-1}(y)$ is difficult to compute.

(3) f^{-1} is easy to compute when a trapdoor (i.e., a secret string of information associated with the function) becomes available.

A function f satisfying only the first two conditions is also called a one-to-one one-way function. If f satisfies further the third condition, it is called a *trapdoor one-way function*.

Example 3.3.5. The following functions are one-way functions:

(1) $f: pq \mapsto n$ is a one-way function, where p and q are prime numbers. The function f is easy to compute since the multiplication of p and q can be done in polynomial time. However, the computation of f^{-1}, the inverse of f is an extremely difficult problem (this is the well-known difficult *integer factorization problem*); there is no efficient algorithm to determine p and q from their product pq, in fact, the fastest factoring algorithm NFS runs in subexponential time.

(2) $f_{g,N} : x \mapsto g^x \bmod N$ is a one-way function. The function f is easy to compute since the modular exponentiation $g^x \bmod N$ can be performed in polynomial time. But the computation of f^{-1}, the inverse of f is an extremely difficult problem (this is the well-known difficult *discrete logarithm problem*); there is no efficient method to determine x from the knowledge of $g^x \bmod N$ and g and N.

(3) $f_{k,N} : x \mapsto x^k \bmod N$ is a trapdoor one-way function, where $N = pq$ with p and q primes, and $kk' \equiv 1 \pmod{\phi(N)}$. It is obvious that f is easy to compute since the modular exponentiation $x^k \bmod N$ can be done in polynomial time, but f^{-1}, the inverse of f (i.e., the kth root of x modulo N) is difficult to compute. However, if k', the trapdoor is given, f can be easily inverted, since $(x^k)^{k'} = x$.

Remark 3.3.1. The discrete logarithm problem and the integer factorization problem are the most important difficult number-theoretic problems on which to build one-way functions in practice. Of course, there might exist some other problems which can be used to build one-way functions. One such problem is the so-called Quadratic Residuosity Problem (QRP), that can be simply stated as follows (recall that an integer a is a quadratic residue modulo n if $\gcd(a, n) = 1$ and if there exists a solution x to the congruence $x^2 \equiv a \pmod{n}$):

Given integers a and n, decide if a is a quadratic residue modulo n.

If $n = p$ is an odd prime, then by Euler's criterion (Theorem 1.6.26), a is a quadratic residue of p if and only if $a^{(p-1)/2} \equiv 1 \pmod{p}$. What about if n is an odd composite? In this case, we know that a is a quadratic residue of n if and only if it is quadratic residue modulo every prime dividing n. It is evident that if $\left(\frac{a}{n}\right) = -1$, then $\left(\frac{a}{p_i}\right) = -1$ for some i, and a is a quadratic nonresidue modulo n. On the other hand, even if $\left(\frac{a}{n}\right) = 1$, it may be possible for a to be a quadratic nonresidue modulo n. This is precisely the case that is regarded by some researchers as an intractable problem, since the only method we know for determining quadratic residuosity in this case requires that we first factor n. Because of our inability to solve the quadratic residuosity problem without factoring, several researchers have proposed cryptosystems whose security is based on the difficulty of determining quadratic residuosity. Whether it is in fact intractable (or at least equivalent to factoring in some sense) remains a very interesting question (McCurley [151]). We shall introduce an encryption scheme based the QRP in Section 3.3.7. There are also some analogues such as elliptic curve analogues of discrete logarithms, which can be used to build one-way functions in public-key cryptosystems; we shall introduce these analogues and their cryptosystems in later sections of this chapter.

Remark 3.3.2. Public-key cryptosystems have some important advantages over secret-key cryptosystems in the distribution of the keys. However, when a large amount of information has to be communicated, it may be that the use of public-key cryptography would be too slow, whereas the use of secret-key cryptography could be impossible for the lack of a shared secret key. In practice, it is better to combine the secret-key and public-key cryptography into a single cryptosystem for secure communications. Such a combined system is often called a *hybrid cryptosystem*. A hybrid cryptosystem uses a public-key cryptosystem once at the beginning of the communication to share a short piece of information that is then used as the key for encryption and decryption by means of a "conventional" secret-key cryptosystem in later stages. Such a cryptosystem is essentially a secret-key cryptosystem but still enjoys the advantages of the public-key cryptosystems.

3.3.5 Discrete Logarithm Based Cryptosystems

The Diffie-Hellman-Merkle scheme, the first public-key cryptographic scheme, is based on the intractable discrete logarithm problem, which can be described as follows:

$$\begin{array}{ll} \text{Input}: & a, b, n \in \mathbb{N} \\ \text{Output}: & x \in \mathbb{N} \text{ with } a^x \equiv b \pmod{n} \\ & \text{if such a } x \text{ exists} \end{array}$$

The Diffie-Hellman-Merkle scheme has found widespread use in practical cryptosystems, as for example in the optional security features of the NFS file system of SunOS operating system. In this subsection, we shall introduce some discrete logarithm based cryptosystems.

(I) The Diffie-Hellman-Merkle Key-Exchange Protocol. Diffie and Hellman [66] in 1976 proposed for the first time a public-key cryptographic scheme based on the difficult discrete logarithm problem. Their scheme was not a public key cryptographic system (first proposed in [66]), but rather a public key distribution system as proposed by Merkle [158]. Such a public key distribution scheme does not send secret messages directly, but rather allows the two parties to agree on a common private key over public networks to be used later in exchanging messages through conventional cryptography. Thus, the Diffie-Hellman-Merkle scheme has the nice property that a very fast scheme such as DES or AES can be used for actual encryption, yet it still enjoys one of the main advantages of public-key cryptography. The Diffie-Hellman-Merkle key-exchange protocol works in the following way (see also Figure 3.8):

(1) A prime q and a generator g are made public (assume all users have agreed upon a finite group over a fixed finite field $\mathbb{F}_q$),

(2) Alice chooses a random number $a \in \{1, 2, \cdots, q-1\}$ and sends $g^a \bmod q$ to Bob,

(3) Bob chooses a random number $b \in \{1, 2, \cdots, q-1\}$ and sends $g^b \bmod q$ to Alice,

(4) Alice and Bob both compute $g^{ab} \bmod q$ and use this as a private key for future communications.

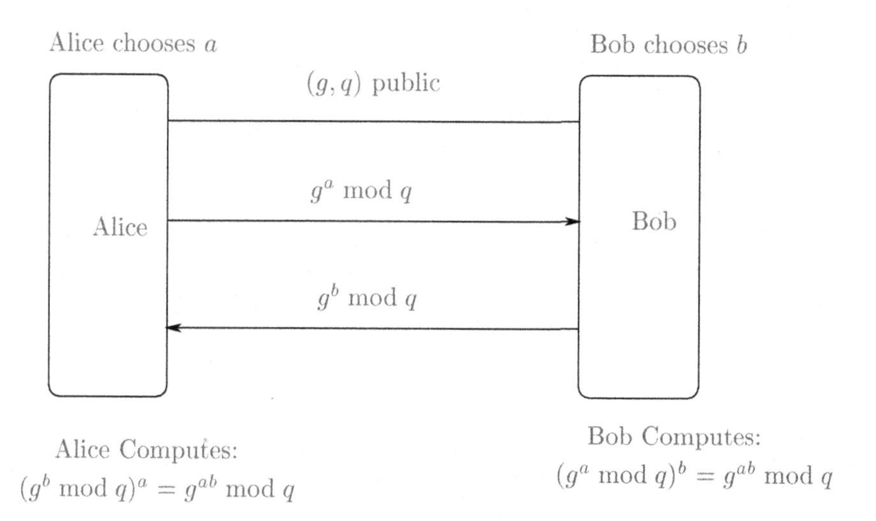

Figure 3.8. The Diffie-Hellman-Merkel key-exchange scheme

Clearly, an eavesdropper has g, q, $g^a \bmod q$ and $g^b \bmod q$, so if he can take discrete logarithms, he can calculate $g^{ab} \bmod q$ and understand communications. That is, if the eavesdropper can use his knowledge of g, q, $g^a \bmod q$ and $g^b \bmod q$ to recover the integer a, then he can easily break the Diffie-Hellman-Merkle system. So, the security of the Diffie-Hellman-Merkle system is based on the following assumption:

> **Diffie-Hellman-Merkle Assumption**: It is computationally infeasible to compute g^{ab} from g^a and g^b.

In theory, there could be a way to use knowledge of g^a and g^b to find g^{ab}. But at present we simply cannot imagine a way to go from g^a and g^b to g^{ab} without essentially solving the discrete logarithm problem.

Example 3.3.6. The following example, taken from McCurley [150], shows how the Diffie-Hellman-Merkle scheme works in a real situation:

(1) Let $q = (7^{149} - 1)/6$ and $p = 2 \cdot 739 \cdot q + 1$. (It can be shown that both p and q are primes.)

(2) Alice chooses a random number residue x modulo p, computes $7^x \pmod p$, and sends the result to Bob, keeping x secret.

(3) B receives

$$\begin{aligned} 7^x = \ & 127402180119973946824269244334322849749382042586931621654 \\ & 557735290322914679095998681860978813046595166455458144280 \\ & 588076766033781 \end{aligned}$$

(4) Bob chooses a random number residue y modulo p, computes $7^y \pmod p$, and sends the result to Alice, keeping y secret.

(5) Alice receives

$$\begin{aligned} 7^y = \ & 180162285287453102444782834836799895015967046695346697313 \\ & 025121734059953772058475958176910625380692101651848662362 \\ & 137934026803049 \end{aligned}$$

(6) Now both Alice and Bob can compute the private key $7^{xy} \pmod p$.

McCurley offered a prize of $100 in 1989 to the first person to find the private key constructed from the above communication.

Remark 3.3.3. McCurley's 129-digit discrete logarithm challenge was actually solved on 25 January 1998 using the NFS method, by two German computer scientists, Damian Weber at the Institut für Techno -und Wirtschaftsmathematik in Kaiserslautern and Thomas F. Denny at the Debis IT Security Services in Bonn.

As we have already mentioned earlier the Diffie-Hellman-Merkle scheme is not intended to be used for actual secure communications, but for key-exchanges. There are, however, several other cryptosystems based on discrete logarithms, that can be used for secure message transmissions.

(II) The ElGamal Cryptosystem for Secure Communications. In 1985, ElGamal proposed a public-key cryptosystem based on discrete logarithms:

(1) A prime q and a generator $g \in \mathbb{F}_q^*$ are made public.

(2) Alice chooses a private integer $a = a_A \in \{1, 2, \cdots, q-1\}$. This a is the private decryption key. The public encryption key is $g^a \in \mathbb{F}_q$.

(3) Suppose now Bob wishes to send a message to Alice. He chooses a random number $b \in \{1, 2, \cdots, q-1\}$ and sends Alice the following pair of elements of $\mathbb{F}_q$:

$$(g^b, \ Mg^{ab})$$

where M is the message.

(4) Since Alice knows the private decryption key a, she can recover M from this pair by computing $g^{ab} \pmod{q}$ and dividing this result into the second element, i.e., Mg^{ab}.

Remark 3.3.4. Someone who can solve the discrete logarithm problem in $\mathbb{F}_q$ breaks the cryptosystem by finding the secret decryption key a from the public encryption key g^a. In theory, there could be a way to use knowledge of g^a and g^b to find g^{ab} and hence break the cipher without solving the discrete logarithm problem. But as we have already seen in the Diffie-Hellman scheme, there is no known way to go from g^a and g^b to g^{ab} without essentially solving the discrete logarithm problem. So, the ElGamal cryptosystem is equivalent to the Diffie–Hellman key-exchange system.

(III) The Massey–Omura Cryptosystem for Message Transmissions. This is another popular cryptosystem based on discrete logarithms; it works in the following way:

(1) All the users have agreed upon a finite group over a fixed finite field $\mathbb{F}_q$ with q a prime power.

(2) Each user secretly selects a random integer e between 0 and $q-1$ such that $\gcd(e, q-1) = 1$, and computes $d = e^{-1} \bmod (q-1)$ by using the extended Euclidean algorithm.

(3) Now suppose that user Alice wishes to send a secure message M to user Bob, then they follow the following procedure:
   (i) Alice first sends M^{e_A} to Bob,
   (ii) On receiving Alice's message, Bob sends $M^{e_A e_B}$ back to Alice (note that at this point, Bob cannot read Alice's message M),
   (iii) Alice sends $M^{e_A e_B d_A} = M^{e_B}$ to Bob,
   (iv) Bob then computes $M^{d_B e_B} = M$, and hence recovers Alice's original message M.

3.3.6 RSA Public-Key Cryptosystem

In 1978, just shortly after Diffie and Hellman proposed the first public-key exchange protocol at Stanford, three MIT researchers Rivest[12], Shamir[13] and Adleman[14] proposed the first practical public-key cryptosystem, now widely known as the RSA public-key cryptosystem. The RSA cryptosystem is based on the following assumption:

> **RSA Assumption**: It is not so difficult to find two large prime numbers, but it is very difficult to factor a large composite into its prime factorization form.

12

Ronald L. Rivest (1948–) is currently the Webster Professor of Electrical Engineering and Computer Science in the Department of Electrical Engineering and Computer Science (EECS) at the Massachusetts Institute of Technology (MIT), an Associate Director of the MIT's Laboratory for Computer Science, and a leader of the lab's Cryptography and Information Security Group. He obtained a B.A. in Mathematics from Yale University in 1969, and a Ph.D. in Computer Science from Stanford University in 1974. Professor Rivest is an inventor of the RSA public-key cryptosystem, and a founder of RSA Data Security (now a subsidiary of Security Dynamics). He has worked extensively in the areas of cryptography, computer algorithms, machine learning and VLSI design. (Photo by courtesy of Prof. Rivest.)

13

Adi Shamir (Born 1952) is currently Professor in the Department of Applied Mathematics and Computer Science at the Weizmann Institute of Science, Israel. He obtained his PhD in Computer Science from the Weizmann Institute of Science in 1977, with Prof. Zohar Manna on "Fixedpoints of Recursive Programs", and did his postdoc with Prof. Mike Paterson for a year in Computer Science at Warwick University in England. He participated in developing the RSA public-key cryptosystem, the Fiat-Shamir identification scheme, polynomial secret sharing schemes, visual cryptosystems, lattice attacks on knapsack cryptosystems, differential cryptanalysis, fault attacks on smart cards, algebraic attacks on multivariate cryptosystems and numerous other cryptographic schemes and techniques. (Photo by courtesy of Prof. Shamir.)

14

Leonard Adleman (Born 1945) received his BSc in mathematics and PhD in computer science both from the University of California at Berkeley in 1972 and 1976, respectively. He is currently Professor in the Department of Computer Science at the University of Southern California. His main research activities are in theoretical computer science with particular emphasis on the complexity of number theoretic problems. Recently he has also been involved in the development of DNA biological computers. (Photo by courtesy of Prof. Adleman.)

The system works as follows:

$$\left.\begin{array}{l} C \equiv M^e \pmod N \\ M \equiv C^d \pmod N \end{array}\right\} \tag{3.59}$$

where

(1) M is the plaintext.
(2) C is the ciphertext.
(3) $N = pq$ is the modulus, with p and q large and distinct primes.
(4) e is the *public* encryption exponent (key) and d the *private* decryption exponent (key) , with $ed \equiv 1 \pmod{\phi(N)}$. $\langle N, e \rangle$ should be made public, but d (as well as $\phi(N)$) should be kept secret.

Figure 3.9. The RSA crypto years: (Left to right) Shamir, Rivest and Adleman (Photo by courtesy of Prof. Adleman)

Clearly, the function $f : M \to C$ is a one-way trap-door function, since it is easy to compute by the fast exponentiation method, but its inverse $f^{-1} : C \to M$ is difficult to compute, because for those who do not know the private decryption key (the trap-door information) d, they will have to factor n and to compute $\phi(n)$ in order to find d. However, for those who know d, then the computation of f^{-1} is as easy as of f. This exactly the idea of RSA cryptography.

Suppose now the sender, say, for example, Alice wants to send a message M to the receiver, say, for example, Bob. Bob will have already chosen a one-way trapdoor function f described above, and published his *public-key* (e, N), so we can assume that both Alice and any potential adversary know (e, N). Alice splits the message M into blocks of $\lfloor \log N \rfloor$ bits or less (padded on the right with zeros for the last block), and treats each block as an integer $x \in \{0, 1, 2, \cdots, N-1\}$. Alice computes

$$y \equiv x^e \pmod{N} \tag{3.60}$$

and transmits y to Bob. Bob, who knows the private key d, computes

$$x \equiv y^d \pmod{N} \tag{3.61}$$

where $ed \equiv 1 \pmod{\phi(N)}$. An adversary who intercepts the encrypted message should be unable to decrypt it without knowledge of d. There is no known way of cracking the RSA system without essentially factoring N, so it is clear that the security of the RSA system depends on the difficulty of factoring N. Some authors, for example, Woll [259] observed that finding the RSA decryption key d is random polynomial-time equivalent to factorization. More recently, Pinch [184] showed that an algorithm $A(N, e)$ for obtaining d given N and e can be turned into an algorithm which obtains p and q with positive probability.

Example 3.3.7. Suppose the message to be encrypted is "Please wait for me". Let $N = 5515596313 = 71593 \cdot 77041$. Let also $e = 1757316971$ with $\gcd(e, N) = 1$. Then $d \equiv 1/1757316971 \equiv 2674607171 \pmod{(71593 - 1)(77041 - 1)}$. To encrypt the message, we first translate the message into its numerical equivalent by the letter-digit encoding scheme described in Table 3.4 as follows:

$$M = 1612050119050023010920061518001305.$$

Then we split it into 4 blocks, each with 10 digits, padded on the right with zeros for the last block:

$$M = (M_1, M_2, M_3, M_4) = (1612050119\ 0500230109\ 2000061518\ 0013050000).$$

Now, we have

$$\begin{aligned}
C_1 &\equiv 1612050119^{1757316971} \equiv 763222127 \pmod{5515596313} \\
C_2 &\equiv 0500230109^{1757316971} \equiv 1991534528 \pmod{5515596313} \\
C_3 &\equiv 2000061518^{1757316971} \equiv 74882553 \pmod{5515596313} \\
C_4 &\equiv 0013050000^{1757316971} \equiv 3895624854 \pmod{5515596313}
\end{aligned}$$

That is,

$$C = (C_1, C_2, C_3, C_4) = (763222127, 1991534528, 74882553, 3895624854).$$

To decrypt the cipher text, we perform:

$$\begin{aligned} M_1 &\equiv 763222127^{2674607171} \equiv 1612050119 \pmod{5515596313} \\ M_2 &\equiv 1991534528^{2674607171} \equiv 500230109 \pmod{5515596313} \\ M_3 &\equiv 74882553^{2674607171} \equiv 2000061518 \pmod{5515596313} \\ M_4 &\equiv 3895624854^{2674607171} \equiv 13050000 \pmod{5515596313} \end{aligned}$$

By padding the necessary zeros on the left of some blocks, we get

$$M = (M_1, M_2, M_3, M_4) = (1612050119\ 0500230109\ 2000061518\ 0013050000)$$

which is ""Please wait for me", the original plaintext message.

Example 3.3.8. We now give a reasonably large RSA example. In one of his series of Mathematical Games, Martin Gardner [78] reported an RSA challenge with US$100 to decrypt the following message C:

96869613754622061477140922254355882905759991124574_
431987469512093081629822514570835693147662288398 9_
62801339199055182994515781515 4.

The public key consists of a pair of integers (e, N), where $e = 9007$ and N is a "random" 129-digit number (called RSA-129):

114381625757888867669235779976146612010218296721 2_
423625625618429357069352457338978305971235639587 0_
50589890751475992900268795435 41.

The RSA-129 was factored by Derek Atkins, Michael Graff, Arjen K. Lenstra, Paul Leyland et al. on 2 April 1994 to win the $100 prize offered by RSA in 1977. Its two prime factors are as follows:

3490529510847650949147849619903898133417764638493_
387843990820577,

3276913299326670954996198819083446141317764296799_
2942539798288533.

They used the double large prime variation of the Multiple Polynomial Quadratic Sieve (MPQS) factoring method. The sieving step took approximately 5000 mips years, and was carried out in 8 months by about 600 volunteers from more than 20 countries, on all continents except Antarctica. As we have explained in the previous example, to encrypt an RSA-encrypted message, we only need to use the public key (N, e) to compute

$$x^e \equiv y \pmod{N}.$$

But decrypting an RSA-message requires factorization of N if one does not know the secret decryption key. This means that if we can factor N, then we can compute the secret key d, and get back the original message by calculating

$$y^d \equiv x \pmod{N}.$$

Since now we know the prime factorization of N, it is trivial to compute the secret key $d = 1/e \bmod \phi(N)$, which in fact is

1066986143685780244428687713289201547807099066339_
3786280122622449663106312591177447087334016859746_
230655396854451327710905360 6095.

So we shall be able to compute

$$C^d \equiv M \pmod{N}$$

without any problem. To use the fast exponential method to compute $C^d \bmod N$, we first write d in its binary form $d_1 d_2 \cdots d_{\text{size}}$ (where size is the number of the bits of d) as follows:

$d = d_1 d_2 \cdots d_{426} =$
1001110110011111100101001100100010000001000000111010011110010011 0_
0100111101001110000000000000011111110100001101010110001011101111_
0101000011111011000000100000111011010101011110101010011111101 10_
1101000011111101000000111101001100010110010110011010010100011 00_
1001110101100001011101001010110100000111000000011100011101010 10_
0110111010001111010011100011010110101010100100111010100010011 11_
00000100111010011000110111110101100100011001111

and perform the following computation:

```
M ← 1
for i from 1 to 426 do
    M ← M^2 mod N
    if d_i = 1 then M ← M · C mod N
print M
```

which gives the plaintext M:

200805001301070903002315180419000118050019172105 0_
1130919080015191909061801070 5

and hence the original message:

THE MAGIC WORDS ARE SQUEAMISH OSSIFRAGE

via the encoding alphabet $\sqcup = 00, A = 01, B = 02, \cdots, Z = 26$. Of course, by the public encryption key $e = 9007$, we can compute $M^e \equiv C \pmod{N}$; first write e in the binary form $e = e_1 e_2 \cdots e_{14} = 10001100101111$, then perform the following procedure:

```
C ← 1
for i from 1 to 14 do
    C ← C^2 mod N
    if e_i = 1 then C ← C · M mod N
print C
```

which gives the encrypted text C at the beginning of this example:

9686961375462206147714092225435588290575999112457_
4319874695120930816298225145708356931476622883989_
628013391990551829945157815154.

Remark 3.3.5. In fact, anyone who can factor the integer RSA-129 can decrypt the message. Thus, decrypting the message is essentially factoring the 129-digit integer. The factorization of RSA-129 implies that it is possible to factor a random 129-digit integer. It should be also noted that on 10 April 1996, Arjen Lenstra et al. also factored the following RSA-130:

1807082088687404805951656164405905566278102516769_
4013491701270214500566625402440483873411275908123_
03371781887966563182013214880557

which has the following two prime factors:

3968599945959745429016112616288378606757644911281_
0064832555157243,

4553449864673597218840368689727440886435630126320_
5069600999044599.

This factorization was found using the Number Field Sieve (NFS) factoring algorithm, and beats the above mentioned 129-digit record by the Quadratic Sieve (QS) factoring algorithm. The amount of computer time spent on this 130-digit NFS-record is only a fraction of what was spent on the old 129-digit QS-record. More recently a group led by Peter Montgomery and Herman te Riele found in February 1999 that the RSA-140:

2129024631825875754749788201627151749780670396327_
7216278233383215381949984056495911366573853021918_
316783107387995317230889569230873441936471

can be written as the product of two 70-digit primes:

3398717423028438554530123627613875835633986495969_
597423490929302771479,

6264200187401285096151654948264442219302037178623_
509019111660653946049.

This factorization was found using the Number Field Sieve (NFS) factoring algorithm, and beats the 130-digit record that was set in April 1996, also with the help of NFS. The amount of computer time spent on this new 140-digit NFS-record is prudently estimated to be equivalent to 2000 mips years. For the old 130-digit NFS-record, this effort is estimated to be 1000 mips years (Te Riele [205]). Even more recently (August 26, 1999), Herman te Riele and Stefania Cavallar et al. successfully factored (again using NFS) the RSA-155, a number with 155 digits and 512 bits, which can be written as the product of two 78-digit primes:

$$1026395928297411057720541965739916759007165678080\_$$
$$38066803341933521790711307779,$$

$$106603488380168454820927220360012878679207958575 98\_$$
$$92915222706082371930628 08643.$$

So, it follows from the above factorization results that

Corollary 3.3.1. The composite number (i.e., the modulus) N used in the RSA cryptosystem should have more than 155 decimal digits.

Exercise 3.3.6. Below is an encrypted message (consisting of two blocks C_1 and C_2):

4660 4906 4350 6009 6392 3911 2238 7112
0237 3603 9163 4700 8276 8243 4103 8329
6685 0734 6202 7217 9820 0029 7925 0670
8833 7283 5678 0453 2383 8911 4071 9579

6506 4096 9385 1106 9741 5283 1334 2475
3966 4897 8551 7358 1383 6777 9635 0373
8147 2092 8779 3861 7878 7818 9741 5743
9185 7183 6081 9612 4160 0934 3883 0158

The public key used to encrypt the message is (e, N), where $e = 9137$ and N is the following RSA-129:

$$1143816257578888676692357799761466120102182967212\_$$
$$4236256256184293570693524573389783059712356395870\_$$
$$5058989075147599290026879543541.$$

Decrypt the message. (Note that in the encryption process if $\gcd(M_i, N) \neq 1$ for $i = 1, 2$, some dummy letter may be added to the end of M_i to make $\gcd(M_i, N) = 1$.)

Let us now consider a more general and more realistic case of secure communications in a computer network with n nodes. It is apparent that there are

$$\binom{n}{2} = n(n-1)/2$$

ways of communicating between two nodes in the network. Suppose one of the nodes (users), say, Alice (A), wants to send a secure message M to another node, say, Bob (B), or vice versa. Then A uses B's encryption key e_B to encrypt her message M_A

$$C_A = M_A^{e_B} \bmod N_B \tag{3.62}$$

and sends the encrypted message C to B; on receiving A's message M_A, B uses his own decryption key d_B to decrypt A's message C:

$$M_A = C_A^{d_B} \bmod N_B. \tag{3.63}$$

Since only B has the decryption key d_B, only B (at least from a theoretical point of view) can recover the original message. B can of course send a secure message M to A in a similar way. Figure 3.10 shows diagrammatically the idea of secure communication between any two parties, say, for example, Alice and Bob.

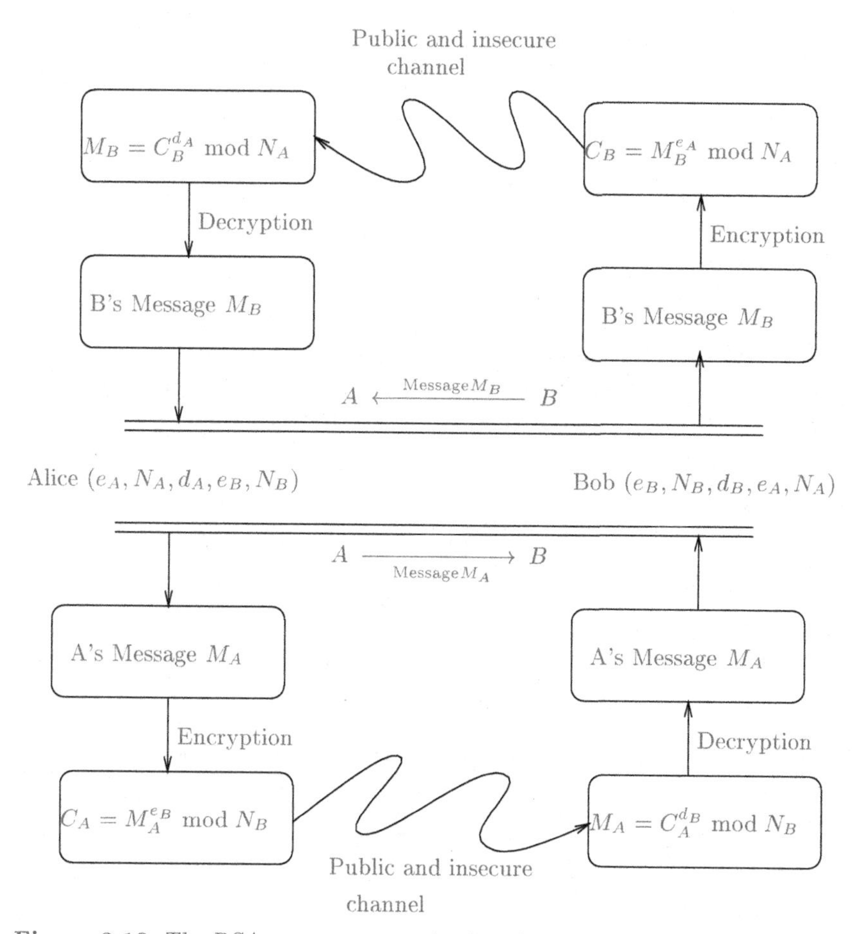

Figure 3.10. The RSA secure communications between two parties

A better example of a trap-door one-way function of the form used in the RSA cryptosystem would use Carmichael's λ-function rather than Euler's ϕ-function, and is as follows:

$$y = f(x) \equiv x^k \pmod{N} \tag{3.64}$$

where

$$\left.\begin{aligned} &N = pq \quad (p \text{ and } q \text{ are two large primes}), \\ &k > 1, \ \ \gcd(k, \lambda) = 1, \\ &\lambda(N) = \operatorname{lcm}(p-1, \ q-1) = \frac{(p-1)(q-1)}{\gcd(p-1, \ q-1)}. \end{aligned}\right\} \tag{3.65}$$

We assume that k and N are publicly known but p, q and $\lambda(N)$ are not. The inverse function of $f(x)$ is defined by

$$x = f^{-1}(y) \equiv y^{k'} \pmod N \ \text{ with } \ kk' \equiv 1 \pmod \lambda. \tag{3.66}$$

To show it works, we see

$$\begin{aligned} x &\equiv y^{k'} \equiv (x^k)^{k'} \equiv x^{kk'} \equiv x^{m\ \lambda(N)+1} \\ &\equiv (x^{\lambda(N)})^m \cdot x \equiv 1^m \cdot x \quad \text{(by Carmichael's theorem)} \\ &\equiv x \end{aligned}$$

It should be easy to compute $f^{-1}(y) \equiv y^{k'} \pmod N$ if k' is known, provided that $f^{-1}(y)$ exists (note that $f^{-1}(y)$ may not exist). The assumption underlying the RSA cryptosystem is that it is hard to compute $f^{-1}(y)$ without knowing k'. However, the knowledge of p, q or $\lambda(N)$ makes it easy to compute k'.

Example 3.3.9. Suppose we wish to encrypt the plaintext message

NATURAL NUMBERS ARE MADE BY GOD.

We first translate all the letters in the message into their numerical equivalents as in Table 3.4. Then we split the message into, for example, four message blocks, each with 15 digits as follows:

$\langle$140020211801120, 014211302051800, 011805001301040, 500022500071504$\rangle$.

and perform the following computation steps:

(1) Select two primes p and q, compute $N = pq$ and $\lambda(N)$:

$p = 440334654777631,$

$q = 145295143558111$

$N = pq = 63978486879527143858831415041$

$\lambda(N) = 710872076439183980322589770.$

(2) Determine the keys k and k': we try to factorize $m\lambda(N) + 1$ for $m = 1, 2, 3, \cdots$ until we find a "good" factorization that can be used to obtain suitable k and k':

$$\begin{aligned}
\lambda(N)+1 &= 1193 \cdot 2990957 \cdot 209791523 \cdot 17107 \cdot 55511\\
2\lambda(N)+1 &= 47 \cdot 131 \cdot 199 \cdot 3322357 \cdot 1716499 \cdot 203474209\\
3\lambda(N)+1 &= 674683 \cdot 1696366781 \cdot 297801601 \cdot 6257\\
4\lambda(N)+1 &= 17 \cdot 53 \cdot 5605331 \cdot 563022035211575351\\
5\lambda(N)+1 &= 17450633 \cdot 13017248387079301 \cdot 15647\\
6\lambda(N)+1 &= 1261058128567 \cdot 49864411 \cdot 2293 \cdot 29581\\
7\lambda(N)+1 &= 19 \cdot 261900238688120413803059389\\
8\lambda(N)+1 &= 15037114930441 \cdot 378195992902921\\
9\lambda(N)+1 &= 11 \cdot 13200581 \cdot 8097845885549501 \cdot 5441\\
10\lambda(N)+1 &= 7108720764391839803225897701\\
11\lambda(N)+1 &= 2131418173 \cdot 7417510211 \cdot 494603657\\
12\lambda(N)+1 &= 4425033337657 \cdot 1927774158146113\\
13\lambda(N)+1 &= 23 \cdot 6796296973884340591 \cdot 59120027\\
14\lambda(N)+1 &= 14785772846857861 \cdot 673093599721\\
15\lambda(N)+1 &= 500807 \cdot 647357777401277 \cdot 17579 \cdot 1871.
\end{aligned}$$

Suppose now we wish to use the 15th factorization $15\lambda(N)+1$ to obtain

$$\langle k, k' \rangle = \langle 17579,\ 606580644324919489438469\rangle$$

such that $kk' = 1 + 15\lambda(N)$.

(3) Encrypt the message $x \mapsto x^k \bmod N = y$ (using the fast modular exponentiation method, for example, Algorithm 2.1.1):

$$\begin{aligned}
140020211801120^{17579} \bmod N &= 60379537366647508826042726177\\
014211302051800^{17579} \bmod N &= 47215464067987497433568498485\\
011805001301040^{17579} \bmod N &= 20999327573397550148935085516\\
500022500071504^{17579} \bmod N &= 37746963038639759803119392704.
\end{aligned}$$

(4) Decrypt the message $y \mapsto y^{k'} \bmod N = x^{kk'} \bmod N = x$ (again using, for example, Algorithm 2.1.1):

$$\begin{aligned}
60379537366647508826042726177^{k'} \bmod N &= 140020211801120\\
47215464067987497433568498485^{k'} \bmod N &= 014211302051800\\
20999327573397550148935085516^{k'} \bmod N &= 011805001301040\\
37746963038639759803119392704^{k'} \bmod N &= 500022500071504
\end{aligned}$$

where $k' = 606580644324919489438469$.

Remark 3.3.6. Compared with the conventional cryptosystems such as the Data Encryption Standard (DES), the RSA system is *very* slow. For example, the DES, when implemented with special-purpose chips, can be run at speeds of tens of millions of bits per second, and even in software on modest size machines can encrypt on the order of 10^5 bits per second, whereas the RSA system, when implemented with the best possible special purpose chips, can only encrypt at the rate of 10^4 or $2 \cdot 10^4$ bits per second, and software implementations are limited to something on the order of 10^2 bits per second. Thus, the RSA system is about 100 to 1000 times slower than conventional cryptosystems.

Now we are in a position to give a brief discussion of the existence of the inverse function $f^{-1}(y)$ defined in (3.66) for all y. Let us first introduce a useful result (Riesel [207]):

Theorem 3.3.1. If N is a product of distinct primes, then for all a,

$$a^{\lambda(N)+1} \equiv a \pmod{N}. \tag{3.67}$$

Note that if N contains multiple prime factors, then (3.67) need no longer be true; say, for example, let $N = 12 = 2^2 \cdot 3$, then $9^{\lambda(12)+1} = 9^3 \equiv 9 \pmod{12}$, but $10^{\lambda(12)+1} = 10^3 \equiv 4 \not\equiv 10 \pmod{12}$. Now, let k and N have been chosen suitably as follows:

$$N = pq, \quad \text{with } p, q \text{ distinct primes} \tag{3.68}$$

$$a^{kk'} \equiv a \pmod{N}, \quad \text{for all } a. \tag{3.69}$$

Then, by Theorem 3.3.1, the inverse function $f^{-1}(y)$, defined in (3.66), exists for all y. It follows immediately from (3.67) that

$$a^{m\lambda(N)+1} \equiv a \pmod{N}, \tag{3.70}$$

which is exactly the form needed in a RSA cryptosystem. For an arbitrary integer N and $m \geq 1$, a necessary and sufficient condition for (3.70) to have a solution a is that (private communications with William Freeman)

$$\gcd(a^2, N) \mid a, \tag{3.71}$$

or equivalently,

$$\gcd(a,\ N/d) = 1, \quad \text{where } d = \gcd(a, N). \tag{3.72}$$

More generally (private communications with Peter Pleasants and Carl Pomerance), a necessary and sufficient condition for

$$a^{m\lambda(N)+k} \equiv a^k \pmod{N} \tag{3.73}$$

is

$$\gcd(a^{k+1}, N) \mid a^k, \tag{3.74}$$

or equivalently,

$$\gcd(a,\ N/d) = 1, \quad \text{where } d = \gcd(a^k, N). \tag{3.75}$$

The proof for the more general case is as follows: Let p be prime and $p^\alpha \parallel N$. Let β be such that $p^\beta \parallel a$. We assume that $p \mid N$, that is $\alpha > 0$. There are three cases:

(1) $\beta = 0$: we have $a^{m\lambda(N)+k} \equiv a^k \pmod{p^\alpha}$, by Euler's theorem,

(2) $0 < k\beta < \alpha$: we have $a^t \not\equiv a^k \pmod{p^\alpha}$ for all $t > k$, obviously,

(3) $k\beta \geq \alpha$: we have $a^t \equiv a^k \pmod{p^\alpha}$ for all $t > k$, obviously.

We conclude that $a^{m\lambda(N)+k} \equiv a^k \pmod{N}$ if and only if we are never in the second case for all primes $p \mid N$. Never being in the second case is equivalent to the condition $\gcd(a^{k+1}, N) \mid a^k$.

Now let us return to the construction of a good trapdoor function (Brent [37]) used in RSA:

Algorithm 3.3.1 (Construction of trapdoor functions). This algorithm constructs the trapdoor function and generates both the public and the secret keys suitable for RSA cryptography:

[1] Use Algorithm 3.3.3 or Algorithm 3.3.2 to find two large primes p and q, each with at least 100 digits such that:

[1-1] $|p - q|$ is large;

[1-2] $p \equiv -1 \pmod{12}, q \equiv -1 \pmod{12}$;

[1-3] The following values of p', p'', q' and q'' are all primes:

$$\begin{aligned} p' &= (p-1)/2, \\ p'' &= (p+1)/12, \\ q' &= (q-1)/2, \\ q'' &= (q+1)/12. \end{aligned}$$

[2] Compute $N = pq$ and $\lambda = 2p'q'$.

[3] Choose a random integer k relatively prime to λ such that $k - 1$ is not a multiple of p' or q'.

[4] Apply the extended Euclidean algorithm to k and λ to find k' and λ' such that $0 < k' < \lambda$ and

$$kk' + \lambda\lambda' = 1.$$

[5] Destroy all evidence of p, q, λ and λ'.

[6] Make (k, N) public but keep k' secret.

It is clear that the most important task in the construction of RSA cryptosystems is to find two large primes, say each with at least 100 digits. An algorithm for finding two 100 digit primes can be described as follows:

Algorithm 3.3.2 (Large prime generation). This algorithm generates prime numbers with 100 digits; it can be modified to generate any length of the required prime numbers:

[1] (Initialization) Randomly generate an odd integer n with say, for example, 100 digits;

[2] (Primality Testing – Probabilistic Method) Use a combination of the Miller–Rabin test and a Lucas test to determine if n is a probable prime. If it is, goto Step [3], else goto Step [1] to get another 100-digit odd integer.

[3] (Primality Proving – Elliptic Curve Method) Use the elliptic curve method to verify whether or not n is indeed a prime. If it is, then report that n is prime, and save it for later use; or otherwise, goto Step [1] to get another 100-digit odd integer.

[4] (done?) If you need more primes, goto Step [1], else terminate the algorithm.

How many primes with 100 digits do we have? By Chebyshev's inequality (1.167), if N is large, then

$$0.92129\frac{N}{\ln N} < \pi(N) < 1.1056\frac{N}{\ln N}. \tag{3.76}$$

Hence

$$0.92129\frac{10^{99}}{\ln 10^{99}} < \pi(10^{99}) < 1.1056\frac{10^{99}}{\ln 10^{99}},$$

$$0.92129\frac{10^{100}}{\ln 10^{100}} < \pi(10^{100}) < 1.1056\frac{10^{100}}{\ln 10^{100}}.$$

The difference $\pi(10^{100}) - \pi(10^{99})$ will give the number of primes with exactly 100 digits, we have

$$3.596958942 \cdot 10^{97} < \pi(10^{100}) - \pi(10^{99}) < 4.076949099 \cdot 10^{97}.$$

The above algorithm for large prime generation depends on primality testing and proving. However, there are methods which do not rely on primality testing and proving. One such method is based on Pocklington's theorem (Theorem 2.2.19), that can automatically lead to primes, say with 100 digits (Ribenboim [199]). We re-state the theorem in a slightly different way as follows:

Theorem 3.3.2. Let p be an odd prime, k a natural number such that p does not divide k and $1 < k < 2(p+1)$ and let $N = 2kp + 1$. Then the following conditions are equivalent:

(1) N is prime.

(2) There exists a natural number a, $2 \le a < N$, such that

$$a^{kp} \equiv -1 \pmod{N}, \text{ and} \tag{3.77}$$

$$\gcd(a^k + 1, N) = 1. \tag{3.78}$$

Algorithm 3.3.3 (Large prime number generation). This algorithm, based on Theorem 3.3.2, generates large prime numbers without the use of primality testing:

[1] Choose, for example, a prime p_1 with $d_1 = 5$ digits. Find $k_1 < 2(p_1+1)$ such that $p_2 = 2k_1p_1 + 1$ has $d_2 = 2d_1 = 10$ digits or $d_2 = 2d_1 - 1 = 9$ digits and there exists $a_1 < p_2$ satisfying the conditions $a_1^{k_1p_1} \equiv -1 \pmod{p_2}$ and $\gcd(a_1^{k_1} + 1, p_2) = 1$. By Pocklington's Theorem, p_2 is prime.

[2] Repeat the same procedure starting from p_2 to obtain the primes $p_3, p_4, \cdots$. In order to produce a prime with 100 digits, the process must be iterated five times. In the last step, k_5 should be chosen so that $2k_5p_5 + 1$ has 100 digits.

As pointed out in Ribenboim [199], for all practical purposes, the above algorithm for producing primes of a given size will run in polynomial time, even though this has not yet been supported by a proof.

According to the Prime Number Theorem, the probability that a randomly chosen integer in $[1, N]$ is prime is $\sim 1/\ln N$. Thus, the expected number of random trials required to find p (or p', or p''; assume that p, p', and p'' are independent) is conjectured to be $\mathcal{O}\left((\log N)^3\right)$. Based on this assumption, the expected time required to construct the above one-way trapdoor function is $\mathcal{O}\left((\log N)^6\right)$.

Finally, in this subsection, we shall give a brief account of some possible attacks on the RSA cryptosystem. We restrict ourselves to the simplified version of RSA system. Let N, the RSA modulus, be the product of two primes p and q. Let also e and d be two positive integers satisfying $ed \equiv 1 \pmod{\phi(N)}$, where $\phi(N) = (p-1)(q-1)$ is the order of the multiplicative group $(\mathbb{Z}/N\mathbb{Z})^*$. Recall that the RSA system works as follows:

$$\left.\begin{array}{l} C \equiv M^e \pmod{N} \\ M \equiv C^d \pmod{N} \end{array}\right\}$$

where $\langle N, e\rangle$ is the public key for encryption, and $\langle N, d\rangle$ the private key for decryption. From an cryptanalytic point of view we would like to know that given the triple $\langle N, e, C\rangle$, how hard (or how many ways) an enemy cryptanalyst can break the RSA system. In what follows, we shall present some possible ways of cracking the RSA scheme.

(1) Factoring N. The most obvious way of breaking the RSA system is to factor N, since if an enemy cryptanalyst could factor N, then he could determine $\phi(N) = (p-1)(q-1)$ and hence the private key d. But this is not easy, since integer factorization is a computationally intractable problem.

(2) Computing $\phi(N)$ without factoring N. It is also obvious that if an enemy cryptanalyst could compute $\phi(N)$ then he could break the system by computing d as the multiplicative inverse of e modulo $\phi(N)$. However, the knowledge of $\phi(N)$ can lead to an easy way of factoring N, since

$$
\begin{aligned}
p + q &= n - \phi(N) + 1, \\
(p-q)^2 &= (p+q)^2 - 4n, \\
p &= \frac{1}{2}\left[(p+q) + (p-q)\right], \\
q &= \frac{1}{2}\left[(p+q) - (p-q)\right].
\end{aligned}
$$

Thus, breaking the RSA system by computing $\phi(N)$ is no easier than breaking the system by factoring N.

(3) Determining d without factoring n or computing $\phi(N)$. If N is large and d is chosen from a large set, then a cryptanalyst should not be able to determine d any easier than he can factor N. Again, a knowledge of d enables N to be factored, since once d is known, $ed - 1$ (a multiple of $\phi(N)$) can be calculated; N can be factored using any multiple of $\phi(N)$.

(4) Computing the eth root of C modulo N. Clearly, the RSA decryption process is just the computation of the eth root of C modulo N. That is, the decryption problem is just the *root finding problem*. It is evident that in the following congruence

$$C \equiv M^e \pmod{N},$$

once $\langle N, e, C \rangle$ is given, we could try substituting $M = 0, 1, 2, \cdots$ until a correct M is found. In theory, it is possible to enumerate all elements of $(\mathbb{Z}/N\mathbb{Z})^*$, since $(\mathbb{Z}/N\mathbb{Z})^*$ is a finite set, but in practice, it is impossible when N is large. However, if $\phi(N)$ is known, then we can compute the eth root of C modulo N fairly easily (see Algorithm 2.4.8 in Chapter 2).

So, all the above obvious methods of breaking the RSA system are closely related to the integer factorization problem. In fact, Rivest, Shamir and Adleman [209] conjectured that

Conjecture 3.3.1 (RSA conjecture). Any method of breaking the RSA cryptosystem must be as difficult as factoring.

There are some other possible attacks on the RSA cryptosystem, which include:

(1) Wiener's attack [253] on the short RSA private-key. It is important that the private-key d should be large (nearly as many bits as the modulus N); otherwise, there is an attack due to Wiener and based on properties of continued fractions, that can find the private-key d in time polynomial in the length of the modulus N, and hence decrypt the message.

(2) Iterated encryption or fixed-point attack (Meijer [154] and Pinch [184]): Suppose e has order r in the multiplicative group modulo $\lambda(N)$. Then $e^r \equiv 1 \pmod{\lambda(N)}$, so $M^{e^r} \equiv M \pmod{N}$. This is just the rth iterate of the encryption of M. So we must ensure that r is large.

It is interesting to note that the attacks discovered so far mainly illustrate the pitfalls to be avoided when implementing RSA. RSA will be still secure if the parameters such as p, q, e, and d are properly chosen. Readers who wish to know more information about the attacks on the RSA cryptosystem are suggested to consult Boneh's recent paper "Twenty Years of Attacks on the RSA Cryptosystem" [30], as well as an earlier paper by Rivest [208].

3.3.7 Quadratic Residuosity Cryptosystems

The RSA cryptosystem discussed in the previous subsection is *deterministic* in the sense that under a fixed public-key, a particular plaintext M is always encrypted to the same ciphertext C. Some of the drawbacks of a deterministic scheme are:

(1) It is not secure for all probability distributions of the message space. For example, in RSA encryption, the messages 0 and 1 always get encrypted to themselves, and hence are easy to detect.

(2) It is easy to obtain some partial information of the secret key (p, q) from the public modulus n (assume that $n = pq$). For example, when the least-significant digit of n is 3, then it is easy to obtain the partial information that the least-significant digits of p and q are either 1 and 3 or 7 and 9, as indicated as follows:

$$\begin{array}{ll} 183 = 3 \cdot 61 & 253 = 11 \cdot 23 \\ 203 = 7 \cdot 29 & 303 = 3 \cdot 101 \\ 213 = 3 \cdot 71 & 323 = 17 \cdot 19. \end{array}$$

(3) It is sometimes easy to compute partial information about the plaintext M from the ciphertext C. For example, given (C, e, n), the Jacobi symbol of M over n can be easily deduced from C:

$$\left(\frac{C}{n}\right) = \left(\frac{M^e}{n}\right) \left(\frac{M}{n}\right)^e = \left(\frac{M}{n}\right).$$

(4) It is easy to detect when the same message is sent twice.

Probabilistic encryption, or randomized encryption, however, utilizes randomness to attain a strong level of security, namely, the *polynomial security* and *semantic security*, defined as follows:

Definition 3.3.2. A public-key encryption scheme is said to be *polynomially* secure if no passive adversary can, in expected polynomial time, select two plaintexts M_1 and M_2 and then correctly distinguish between encryptions of M_1 and M_2 with probability significantly greater that 1/2.

Definition 3.3.3. A public-key encryption scheme is said to be *semantically* secure if, for all probability distributions over the message space, whatever a passive adversary can compute in expected polynomial time about the plaintext given the ciphertext, it can also be computed in expected polynomial time without the ciphertext.

Intuitively, a public-key encryption scheme is semantically secure if the ciphertext does not leak any partial information whatsoever about the plaintext that can be computed in expected polynomial time. That is, given (C, e, n), it should be intractable to recover any information about M. Clearly, a public-key encryption scheme is semantically secure if and only if it is polynomially secure.

In this subsection, we shall introduce a semantically secure cryptosystem based on the *quadratic residuosity problem.* Recall that an integer a is a quadratic residue modulo n, denoted by $a \in Q_n$, if $\gcd(a, n) = 1$ and there exists a solution x to the congruence $x^2 \equiv a \pmod{n}$, otherwise a is a quadratic nonresidue modulo n, denoted by $a \in \overline{Q}_n$. The Quadratic Residuosity Problem may be stated as:

Given positive integers a and n, decide whether or not $a \in Q_n$.

It is believed that solving QRP is equivalent to computing the prime factorization of n, so it is computationally infeasible. We have seen in Subsection 1.6.6 of Chapter 1 that if n is prime then

$$a \in Q_n \iff \left(\frac{a}{n}\right) = 1, \tag{3.79}$$

and if n is composite, then

$$a \in Q_n \implies \left(\frac{a}{n}\right) = 1, \tag{3.80}$$

but

$$a \in Q_n \not\Leftarrow \left(\frac{a}{n}\right) = 1, \tag{3.81}$$

however

$$a \in \overline{Q}_n \Longleftarrow \left(\frac{a}{n}\right) = -1. \tag{3.82}$$

Let $J_n = \{a \in (\mathbb{Z}/n\mathbb{Z})^* : (\frac{a}{n}) = 1\}$, then $\tilde{Q}_n = J_n - Q_n$. Thus, $\tilde{Q}_n$ is the set of all pseudosquares modulo n; it contains those elements of J_n that do not belong to Q_n. Readers may wish to compare this result to Fermat's little theorem discussed in Subsection 1.6.3 of Chapter 1 namely (assuming $\gcd(a, n) = 1$),

$$n \text{ is prime} \implies a^{n-1} \equiv 1 \pmod{n}, \tag{3.83}$$

but

$$n \text{ is prime} \not\Leftarrow a^{n-1} \equiv 1 \pmod{n}, \tag{3.84}$$

however

$$n \text{ is composite } \Longleftarrow a^{n-1} \not\equiv 1 \pmod{n}. \tag{3.85}$$

The Quadratic Residuosity Problem can then be further restricted to:

> Given a composite n and an integer $a \in J_n$, decide whether or not $a \in Q_n$.

For example, when $n = 21$, we have $J_{21} = \{1, 4, 5, 16, 17, 20\}$ and $Q_{21} = \{1, 4, 16\}$, thus $\tilde{Q}_{21} = \{5, 17, 20\}$. So, the QRP problem for $n = 21$ is actually to distinguish squares $\{1, 4, 16\}$ from pseudosquares $\{5, 17, 20\}$. The only method we know for distinguishing squares from pseudosquares is to factor n; since integer factorization is computationally infeasible, the QRP problem is computationally infeasible. In what follows, we shall present a cryptosystem whose security is based on the infeasibility of the Quadratic Residuosity Problem; it was first proposed by Goldwasser and Micali [88] in 1984, under the term *probabilistic encryption.*

Algorithm 3.3.4 (Quadratic residuosity based cryptography). This algorithm uses the randomized method to encrypt messages and is based on the quadratic residuosity problem (QRP). The algorithm divides into three parts: key generation, message encryption and decryption.

[1] **Key generation**: Both Alice and Bob should do the following to generate their public and secret keys:

[1-1] Select two large distinct primes p and q, each with roughly the same size, say, each with β bits.

[1-2] Compute $n = pq$.

[1-3] Select a $y \in \mathbb{Z}/n\mathbb{Z}$, such that $y \in \overline{Q}_n$ and $\left(\frac{y}{n}\right) = 1$. ($y$ is thus a pseudosquare modulo n).

[1-4] Make (n, y) public, but keep (p, q) secret.

[2] **Encryption**: To send a message to Alice, Bob should do the following:

[2-1] Obtain Alice's public-key (n, y).

[2-2] Represent the message m as a binary string $m = m_1 m_2 \cdots m_k$ of length k.

[2-3] For i from 1 to k do

[i] Choose at random an $x \in (\mathbb{Z}/n\mathbb{Z})^*$ and call it x_i.

[ii] Compute c_i:

$$c_i = \begin{cases} x_i^2 \bmod n, & \text{if } m_i = 0, \quad \text{(r.s.)} \\ y x_i^2 \bmod n, & \text{if } m_i = 1, \quad \text{(r.p.s.)}, \end{cases} \tag{3.86}$$

where r.s. and r.p.s. represent random square and random pseudosquare, respectively.

[iii] Send the k-tuple $c = (c_1, c_2, \cdots, c_k)$ to Alice. (Note first that each c_i is in integer with $1 \leq c_i < n$. Note also that since n is a 2β-bit integer, it is clear that the ciphertext c is a much longer string than the original plaintext m.)

[3] **Decryption**: To decrypt Bob's message, Alice should do the following:

[3-1] For i from 1 to k do

[i] Evaluate the Legendre symbols:

$$\left.\begin{array}{l} e_i' = \left(\dfrac{c_i}{p}\right) \\ e_i'' = \left(\dfrac{c_i}{q}\right) \end{array}\right\} \tag{3.87}$$

[ii] Compute m_i:

$$m_i = \begin{cases} 0, & \text{if } e_i' = e_i'' = 1 \\ 1, & \text{if otherwise.} \end{cases} \tag{3.88}$$

That is, $m_i = 0$ if $c_i \in Q_n$, otherwise, $m_i = 1$. otherwise, set $m_i = 1$.

[3-2] Finally, get the decrypted message $m = m_1 m_2 \cdots m_k$.

Remark 3.3.7. The above encryption scheme has the following interesting features:

(1) The encryption is random in the sense that the same bit is transformed into different strings depending on the choice of the random number x. For this reason, it is called *probabilistic* (or *randomized*) encryption.

(2) Each bit is encrypted as an integer modulo n, and hence is transformed into a 2β-bit string.

(3) It is semantically secure against any threat from a polynomially bounded attacker, provided that the QRP is hard.

Exercise 3.3.7. Show that Algorithm 3.3.4 takes $\mathcal{O}(\beta^2)$ time to encrypt each bit and $\mathcal{O}(\beta^3)$ time to decrypt each bit.

Example 3.3.10. In what follows we shall give an example of how Bob can send the message "HELP ME" to Alice using the above cryptographic method. We use the binary equivalents of letters as defined in Table 3.5. Now both Alice and Bob proceed as follows:

[1] **Key Generation**:

[1-1] Alice chooses $(n, y) = (21, 17)$ as a public key, where $n = 21 = 3 \cdot 7$ is a composite, and $y = 17 \in \tilde{Q}_{21}$ (since $17 \in J_{21}$ but $17 \notin Q_{21}$), so that Bob can use the public key to encrypt his message and send it to Alice.

[1-2] Alice keeps the prime factorization $(3, 7)$ of 21 as a secret; since $(3, 7)$ will be used a private decryption key. (Of course, here we just show an example; in practice, the prime factors p and q should be at last 100 digits.)

[2] **Decryption**:

[2-1] Bob converts his plaintext HELP ME to the binary stream $M = m_1 m_2 \cdots m_{35}$:

$$00111\ 00100\ 01011\ 01111\ 11010\ 01100\ 00100$$

(To save space, we only consider how to encrypt and decrypt $m_2 = 0$ and $m_3 = 1$; readers are suggested to encrypt and decrypt the whole binary stream).

[2-2] Bob randomly chooses integers $x_i \in (\mathbb{Z}/21\mathbb{Z})^*$. Suppose he chooses $x_2 = 10$ and $x_3 = 19$ which are elements of $(\mathbb{Z}/21\mathbb{Z})^*$.

[2-3] Bob computes the encrypted message $C = c_1 c_2 \cdots c_k$ from the plaintext $M = m_1 m_2 \cdots m_k$ using Equation (3.86). To get, for example, c_2 and c_3, Bob performs:

$$c_2 = x_2^2 \bmod 21 = 10^2 \bmod 21 = 16, \qquad \text{since } m_2 = 0,$$

$$c_3 = y \cdot x_3^2 \bmod 21 = 17 \cdot 19^2 \bmod 21 = 5, \quad \text{since } m_3 = 1.$$

(Note that each c_i is an integer reduced to 21, i.e., m_i is a bit, but its corresponding c_i is not a bit but an integer, which is a string of bits, determined by Table 3.5.)

[2-4] Bob then sends c_2 and c_3 along with all other c_i's to Alice.

[3] **Decryption**: To decrypt Bob's message, Alice evaluates the Legendre symbols $\left(\frac{c_i}{p}\right)$ and $\left(\frac{c_i}{q}\right)$. Since Alice knows the prime factorization (p, q) of n, it should be easy for her to evaluate these Legendre symbols. For example, for c_2 and c_3, Alice performs:

Table 3.5. The binary equivalents of letters

Letter	Binary Code	Letter	Binary Code	Letter	Binary Code
A	00000	B	00001	C	00010
D	00011	E	00100	F	00101
G	00110	H	00111	I	01000
J	01001	K	01010	L	01011
M	01100	N	01101	O	01110
P	01111	Q	10000	R	10001
S	10010	T	10011	U	10100
V	10101	W	10110	X	10111
Y	11000	Z	11001	␣	11010

[3-1] Evaluates the Legendre symbols $\left(\frac{c_i}{p}\right)$:

$$e'_2 = \left(\frac{c_2}{p}\right) = \left(\frac{16}{3}\right) = \left(\frac{4^2}{3}\right) = 1,$$

$$e'_3 = \left(\frac{c_3}{p}\right) = \left(\frac{5}{3}\right) = \left(\frac{2}{3}\right) = -1.$$

[3-2] Evaluates the Legendre symbols $\left(\frac{c_i}{q}\right)$:

$$e''_2 = \left(\frac{c_2}{q}\right) = \left(\frac{16}{8}\right) = 1,$$

$$e''_3 = \left(\frac{c_3}{q}\right) = \left(\frac{5}{7}\right) = -1.$$

[3-3] Further by Equation (3.88), Alice gets

$$m_2 = 0, \quad \text{since } e'_2 = e''_2 = 1,$$

$$m_3 = 1, \quad \text{since } e'_3 = e''_3 = -1.$$

Remark 3.3.8. The scheme introduced above is a good extension of the public-key idea, but encrypts messages bit by bit. It is completely secure with respect to semantic security as well as bit security[15]. However, a major disadvantage of the scheme is the message expansion by a factor of $\log n$ bit. To improve the efficiency of the scheme, Blum and Goldwasser [28] proposed another randomized encryption scheme, in which the ciphertext is only longer than the plainext by a constant number of bits; this scheme is comparable to the RSA scheme, both in terms of speed and message expansion.

Exercise 3.3.8. RSA encryption scheme is deterministic and not semantically secure, but it can be made semantically secure by adding randomness to the encryption process (Bellare and Rogaway, [22]). Develop an RSA based probabilistic (randomized) encryption scheme that is semantically secure.

Several other cryptographic schemes, including digital signature schemes and authentication encryption schemes are based on the quadratic residuosity problem (QRP); interested readers are referred to, for example, Chen [47] and Nyang [175] for some recent developments and applications of the quadratic residuosity based cryptosystems.

---

[15] Bit security is a special case of semantic security. Informally, bit security is concerned with not only that the whole message is not recoverable but also that individual bits of the message are not recoverable. The main drawback of the scheme is that the encrypted message is much longer than its original plaintext.

3.3.8 Elliptic Curve Public-Key Cryptosystems

We have discussed some novel applications of elliptic curves in primality testing and integer factorization in Chapter 2. In this subsection, we shall introduce one more novel application of elliptic curves in public-Key cryptography. More specifically, we shall introduce elliptic curve analogues of several well-known public-key cryptosystems, including the Diffie–Hellman key exchange system and the RSA cryptosystem.

(I) Brief History of Elliptic Curve Cryptography. Elliptic curves have been extensively studied by number theorists for more than one hundred years, only for their mathematical beauty, not for their applications. However, in the late 1980s and early 1990s many important applications of elliptic curves in both mathematics and computer science were discovered, notably applications of elliptic curves in primality testing (see Kilian [120] and Atkin and Morain [12]) and integer factorization (see Lenstra [140]), both discussed in Chapter 2. Applications of elliptic curves in cryptography were not found until the following two seminal papers were published:

(1) Victor Miller, "Uses of Elliptic Curves in Cryptography", 1986. (See [163].)

(2) Neal Koblitz[16], "Elliptic Curve Cryptosystems", 1987. (See [126].)

Since then, elliptic curves have been studied extensively for the purpose of cryptography, and many practically more secure encryption and digital signature schemes have been developed based on elliptic curves. Now elliptic curve cryptography (ECC) is a standard term in the field and there is a textbook by Menezes [155] that is solely devoted to elliptic curve cryptography. There is even a computer company in Canada, called Certicom, which is a leading provider of cryptographic technology based on elliptic curves. In the subsections that follow, we shall discuss the basic ideas and computational methods of elliptic curve cryptography.

16

Neal Koblitz received his BSc degree in mathematics from Harvard University in 1969, and his PhD in arithmetic algebraic geometry from Princeton in 1974. From 1979 to the present, he has been at the University of Washington in Seattle, where he is now a professor in mathematics. In recent years his research interests have been centered around the applications of number theory in cryptography. He has published a couple of books in related to number theory and cryptography, two of them are as follows: *A Course in Number Theory and Cryptography* [128], and *Algebraic Aspects of Cryptography* [129]. His other interests include pre-university math education, mathematical development in the Third World, and snorkeling. (Photo by courtesy of Springer-Verlag.)

(II) Precomputations of Elliptic Curve Cryptography. To implement elliptic curve cryptography, we need to do the following precomputations:

[1] Embed Messages on Elliptic Curves: Our aim here is to do cryptography with elliptic curve groups in place of $\mathbb{F}_q$. More specifically, we wish to embed plaintext messages as points on an elliptic curve defined over a finite field $\mathbb{F}_q$, with $q = p^r$ and $p \in$ Primes. Let our message units m be integers $0 \leq m \leq M$, let also κ be a large enough integer for us to be satisfied with an error probability of $2^{-\kappa}$ when we attempt to embed a plaintext message m. In practice, $30 \leq \kappa \leq 50$. Now let us take $\kappa = 30$ and an elliptic curve $E : y^2 = x^3 + ax + b$ over $\mathbb{F}_q$. Given a message number m, we compute a set of values for x:

$$x = \{m\kappa + j,\ j = 0, 1, 2, \cdots\} = \{30m,\ 30m + 1,\ 30m + 2,\ \cdots\} \quad (3.89)$$

until we find $x^3 + ax + b$ is a square modulo p, giving us a point $(x, \sqrt{x^3 + ax + b})$ on E. To convert a point (x, y) on E back to a message number m, we just compute $m = \lfloor x/30 \rfloor$. Since $x^3 + ax + b$ is a square for approximately 50% of all x, there is only about a $2^{-\kappa}$ probability that this method will fail to produce a point on E over $\mathbb{F}_q$. In what follows, we shall give a simple example of how to embed a message number by a point on an elliptic curve. Let E be $y^2 = x^3 + 3x$, $m = 2174$ and $p = 4177$ (in practice, we select $p > 30m$). Then we calculate $x = \{30 \cdot 2174 + j,\ j = 0, 1, 2, \cdots\}$ until $x^3 + 3x$ is a square modulo 4177. We find that when $j = 15$:

$$\begin{aligned} x &= 30 \cdot 2174 + 15 \\ &= 65235 \\ x^3 + 3x &= (30 \cdot 2174 + 15)^3 + 3(30 \cdot 2174 + 15) \\ &= 277614407048580 \\ &\equiv 1444 \bmod 4177 \\ &\equiv 38^2 \end{aligned}$$

So we get the message point for $m = 2174$:

$$(x,\ \sqrt{x^3 + ax + b}) = (65235, 38).$$

To convert the message point $(65235, 38)$ on E back to its original message number m, we just compute

$$m = \lfloor 65235/30 \rfloor = \lfloor 2174.5 \rfloor = 2174.$$

[2] Multiply Points on Elliptic Curves over $\mathbb{F}_q$: We have discussed the calculation of $kP \in E$ over $\mathbb{Z}/N\mathbb{Z}$. In elliptic curve public-key cryptography, we are now interested in the calculation of $kP \in E$ over $\mathbb{F}_q$, which can be done in $\mathcal{O}(\log k (\log q)^3)$ bit operations by the *repeated doubling method.*

If we happen to know N, the number of points on our elliptic curve E and if $k > N$, then the coordinates of kP on E can be computed in $\mathcal{O}(\log q)^4$ bit operations [128]; recall that the number N of points on E satisfies $N \leq q+1+2\sqrt{q} = \mathcal{O}(q)$ and can be computed by René Schoof's algorithm in $\mathcal{O}(\log q)^8$ bit operations.

[3] Compute Discrete Logarithms on Elliptic Curves: Let E be an elliptic curve over $\mathbb{F}_q$, and B a point on E. Then the *discrete logarithm* on E is the problem: given a point $P \in E$, find an integer $x \in \mathbb{Z}$ such that $xB = P$ if such an integer x exists. It is likely that the discrete logarithm problem on elliptic curves over $\mathbb{F}_q$ is more intractable than the discrete logarithm problem in $\mathbb{F}_q$. It is this feature that makes cryptographic systems based on elliptic curves even more secure than that based on the discrete logarithm problem. In the rest of this subsection, we shall discuss elliptic curve analogues for some of the important public-key cryptosystems.

(III) Elliptic Curve Analogues of Some Public-Key Cryptosystems. In what follows, we shall introduce elliptic curve analogues of four widely used public-key cryptosystems, namely the Diffie–Hellman key exchange system, the Massey–Omura, the ElGamal and the RSA public-key cryptosystems.

(1) Analogue of the Diffie–Hellman Key Exchange System:

[1] Alice and Bob publicly choose a finite field $\mathbb{F}_q$ with $q = p^r$ and $p \in$ Primes, an elliptic curve E over $\mathbb{F}_q$, and a random *base* point $P \in E$ such that P generates a large subgroup of E, preferably of the same size as that of E itself.

[2] To agree on a secret key, Alice and Bob choose two secret random integers a and b. Alice computes $aP \in E$ and sends aP to Bob; Bob computes $bP \in E$ and sends bP to Alice. Both aP and bP are, of course, public but a and b are not.

[3] Now both Alice and Bob compute the secret key $abP \in E$, and use it for further secure communications.

There is no known fast way to compute abP if one only knows P, aP and bP – this is the discrete logarithm problem on E.

(2) Analogue of the Massey–Omura Cryptosystem:

[1] Alice and Bob publicly choose an elliptic curve E over $\mathbb{F}_q$ with q large, and we suppose also that the number of points (denoted by N) is publicly known.

[2] Alice chooses a secret pair of numbers $(e_A,\ d_A)$ such that $d_A e_A \equiv 1 \pmod N$. Similarly, Bob chooses $(e_B,\ d_B)$:

[3] If Alice wants to send a secret message-point $P \in E$ to Bob, the procedure is as follows:

[3-1] Alice sends $e_A P$ to Bob,

[3-2] Bob sends $e_B e_A P$ to Alice,

[3-3] Alice sends $d_A e_B e_A P = e_B P$ to Bob,

[3-4] Bob computes $d_B e_B P = P$.

Note that an eavesdropper would know $e_A P$, $e_B e_A P$, and $e_B P$. So if he could solve the discrete logarithm problem on E, he could determine e_B from the first two points and then compute $d_B = e_B^{-1} \bmod N$ and hence get $P = d_B(e_B P)$.

(3) Analogue of the ElGamal Cryptosystem:

[1] Alice and Bob publicly choose an elliptic curve E over $\mathbb{F}_q$ with $q = p^r$ and $p \in \text{Primes}$, and a random *base* point $P \in E$.

[2] Alice chooses a random integer r_a and computes $r_a P$; Bob also chooses a random integer r_b and computes $r_b P$.

[3] To send a message-point M to Bob, Alice chooses a random integer k and sends the pair of points $(kP,\ M + k(r_b P))$.

[4] To read M, Bob computes

$$M + k(r_b P) - r_b(kP) = M. \tag{3.90}$$

An eavesdropper who can solve the discrete logarithm problem on E can, of course, determine r_b from the publicly known information P and $r_b P$. But as everybody knows, there is no efficient way to compute discrete logarithms, so the system is secure.

(4) Analogue of the RSA Cryptosystem:

RSA, the most popular cryptosystem in use, also has the following elliptic curve analogue:

[1] $N = pq$ is a public key which is the product of the two large secret primes p and q.

[2] Choose two random integers a and b such that $E: \ y^2 = x^3 + ax + b$ defines an elliptic curve both mod p and mod q.

[3] To encrypt a message-point P, just perform $eP \bmod N$, where e is the public (encryption) key. To decrypt, one needs to know the number of points on E modulo both p and q.

The above are some elliptic curve analogues of certain public-key cryptosystems. It should be noted that almost every public-key cryptosystem has an elliptic curve analogue; it is of course possible to develop new elliptic curve cryptosystems which do not rely on the existing cryptosystems.

Exercise 3.3.9. Work back from the descriptions of the elliptic curve analogues of the ElGamal and the Massey–Omura cryptosystems discussed above, to give complete algorithmic descriptions of the original ElGamal and the original Massey–Omura public-key cryptosystems.

(IV) Menezes-Vanstone Elliptic Curve Cryptosystem. A serious problem with the above mentioned elliptic curve cryptosystems is that the plaintext message units m lie on the elliptic curve E, and there is no convenient method known of deterministically generating such points on E. Fortunately, Menezes[17] and Vanstone[18] had discovered a more efficient variation [156]; in this variation which we shall describe below, the elliptic curve is used for "masking", and the plaintext and ciphertext pairs are allowed to be in $\mathbb{F}_p^* \times \mathbb{F}_p^*$ rather than on the elliptic curve.

[1] Preparation: Alice and Bob publicly choose an elliptic curve E over $\mathbb{F}_p$ with $p > 3$ is prime and a random *base* point $P \in E(\mathbb{F}_p)$ such that P generates a large subgroup H of $E(\mathbb{F}_p)$, preferably of the same size as that of $E(\mathbb{F}_p)$ itself. Assume that randomly chosen $k \in \mathbb{Z}_{|H|}$ and $a \in \mathbb{N}$ are secret.

[2] Encryption: Suppose now Alice wants to sent message

$$m = (m_1, m_2) \in (\mathbb{Z}/p\mathbb{Z})^* \times (\mathbb{Z}/p\mathbb{Z})^* \tag{3.91}$$

to Bob, then she does the following:

[2-1] $\beta = aP$, where P and β are public.

17

Alfred J. Menezes is a professor of mathematics in the Department of Combinatorics and Optimization at the University of Waterloo, where he teaches courses in cryptography, coding theory, finite fields, and discrete mathematics. He is actively involved in cryptographic research, and consults on a regular basis for Certicom Corp., He completed the Bachelor of Mathematics and M.Math degrees in 1987 and 1989 respectively, and a Ph.D. in Mathematics from the University of Waterloo (Canada) in 1992.

18

Scott A. Vanstone is one of the founders of Certicom, the first company to develop elliptic curve cryptography commercially. He devotes much of his research to the efficient implementation of the elliptic curve cryptography for the provision of information security services in hand-held computers, smart cards, wireless devices, and integrated circuits. Vanstone has published more than 150 research papers and several books on topics such as cryptography, coding theory, finite fields, finite geometry, and combinatorial designs. Recently, he was elected a Fellow of the Royal Society of Canada. Vanstone received a Ph.D. in mathematics from the University of Waterloo in 1974.

[2-2] $(y_1, y_2) = k\beta$
[2-3] $c_0 = kP$.
[2-4] $c_j \equiv y_j m_j \pmod{p}$ for $j = 1, 2$.
[2-5] Alice sends the encrypted message c of m to Bob:

$$c = (c_0, c_1, c_2). \tag{3.92}$$

[3] Decryption: Upon receiving Alice's encrypted message c, Bob calculates the following to recover m:
[3-1] $ac_0 = (y_1, y_2)$.
[3-1] $m = \left(c_1 y_1^{-1} \pmod{p},\ c_2 y_2^{-1} \pmod{p}\right)$.

Example 3.3.11. The following is a nice example of Menezes-Vanstone cryptosystem, taken from [165].

[1] Key generation: Let E be the elliptic curve given by $y^2 = x^3 + 4x + 4$ over $\mathbb{F}_{13}$, and $P = (1, 3)$ be a point on E. Choose $E(\mathbb{F}_{13}) = H$ which is cyclic of order 15, generated by P. Let also the private keys $k = 5$ and $a = 2$, and the plaintext $m = (12, 7) = (m_1, m2)$.

[2] Encryption: Alice computes:

$$\begin{aligned}
&\beta = aP = 2(1,3) = (12,8)\\
&(y_1, y_2) = k\beta = 5(12,8) = (10,11)\\
&c_0 = kP = 5(1,3) = (10,2)\\
&c_1 \equiv y_1 m_1 \equiv 10 \cdot 2 \equiv 3 \pmod{13}\\
&c_2 \equiv y_2 m_2 \equiv 11 \cdot 7 \equiv 12 \pmod{13}.
\end{aligned}$$

Then Alice sends

$$c = (c_0, c_1, c_2) = ((10,2), 3, 12)$$

to Bob.

[3] Decryption: Upon receiving Alice's message, Bob computes:

$$\begin{aligned}
&ac_0 = 2(10,2) = (10,11) = (y_1, y_2)\\
&m_1 \equiv c_1 y_1^{-1} \equiv 12 \pmod{13}\\
&m_2 \equiv c_2 y_2^{-1} \equiv 7 \pmod{13}.
\end{aligned}$$

Thus, Bob recovers the message $m = (12, 7)$.

We have introduced so far the most popular public-key cryptosystems, such as Diffie-Hellman-Merkle, RSA, Elliptic curve and probabilistic cryptosystems. There are, of course, many other types of public-key cryptosystems in use, such as Rabin, McEliece and Knapsack cryptosystems. Readers who are interested in the cryptosystems which are not covered in this book are suggested to consult Menezes et al. [157].

3.3.9 Digital Signatures

The idea of public-key cryptography (suppose we are using the RSA public-key scheme) can also be used to obtain digital signatures. Recall that in public-key cryptography, we perform

$$C = E_{e_k}(M), \tag{3.93}$$

where M is the message to be encrypted, for message encryption, and

$$M = D_{d_k}(C), \tag{3.94}$$

where C is the encrypted message needed to be decrypted, for decryption. In digital signatures, we perform the operations in exactly the opposite direction. That is, we perform (see also Figure 3.11)

$$S = D_{d_k}(M), \tag{3.95}$$

where M is the message to be signed, for signature generation,

$$M = E_{e_k}(S), \tag{3.96}$$

where S is the signed message needed to be verified, for signature verification. Suppose now Alice wishes to send Bob a secure message as well as a digital

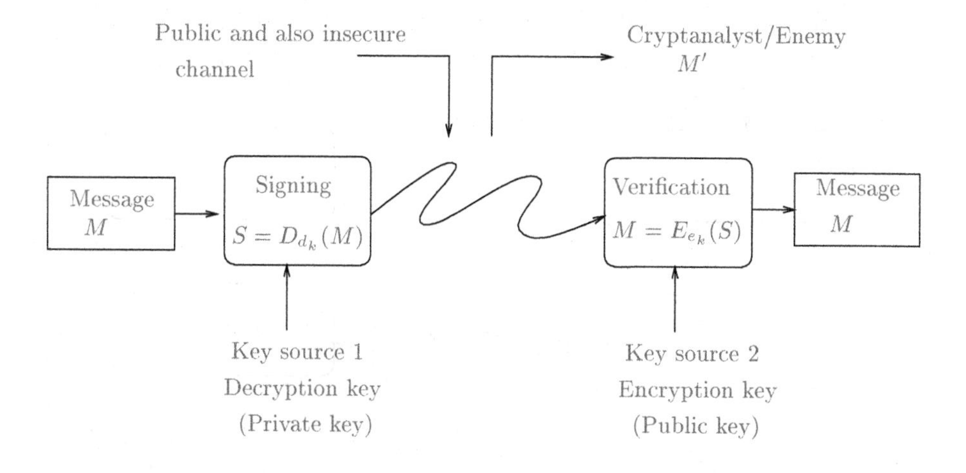

Figure 3.11. Digital signatures

signature. Alice first uses Bob's public key to encrypt her message, and then, she uses her private key to encrypt her signature, and finally sends out her

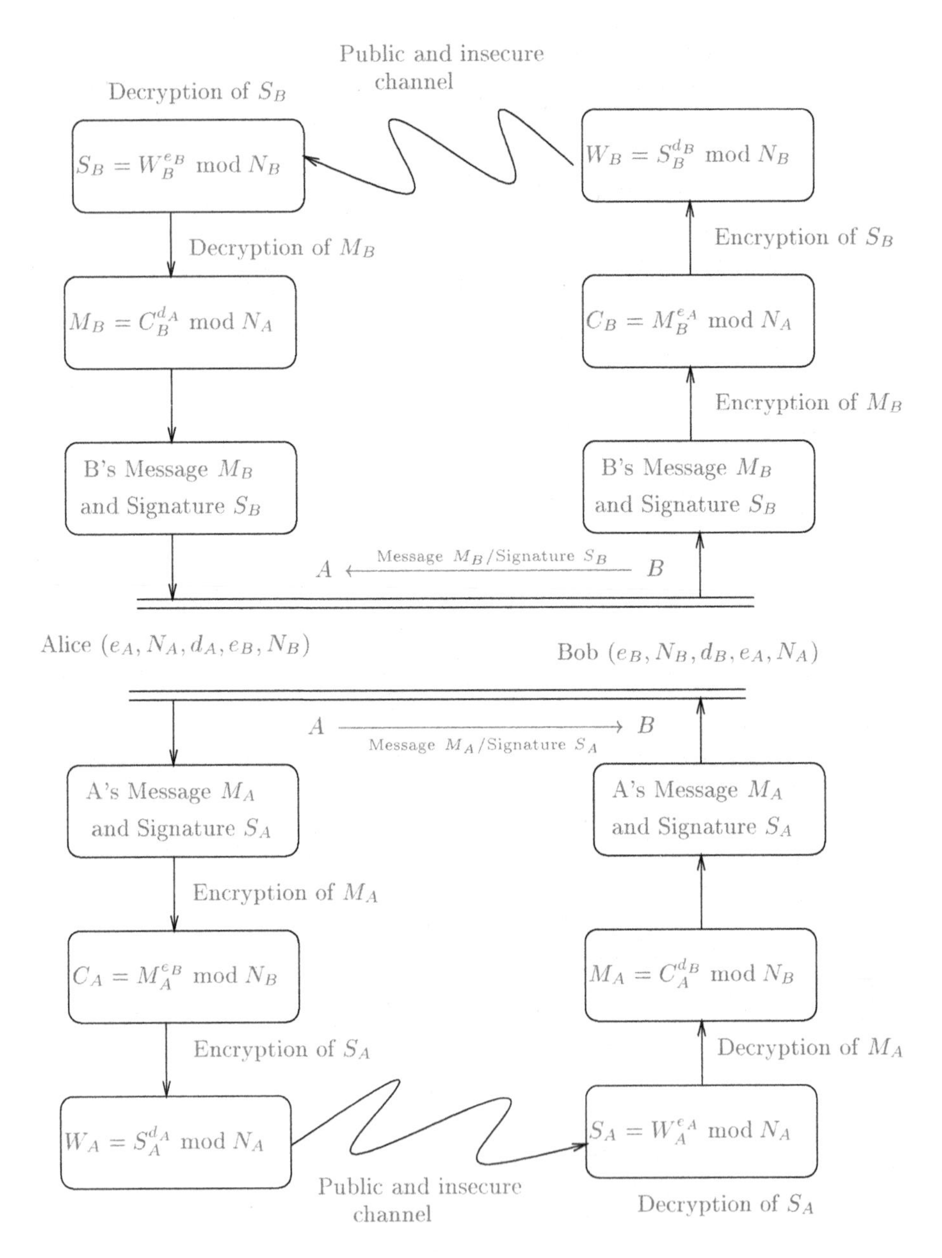

Figure 3.12. Sending encrypted messages and signatures using the RSA scheme (the encrypted message and the signature are two different texts)

message and signature to Bob. At the other end, Bob first uses Alice's public key to decrypt Alice's signature, and then uses his private key to decrypt Alice's message. Figure 3.12 shows how A (Alice) and B (Bob) can send secure message/signature to each other over the insecure channel.

Example 3.3.12 (Digital Signature). To verify that the $100 offer in Example 3.3.8 actually came from RSA, the following signature was added:

$$S = 16717861150380844246015271389168398245436901032358311217835038446929062655448792237114490509578608655662496577974840004057020373.$$

It was encrypted by $S = M^d \pmod{N}$, where d is the secret key, as in Example 3.3.8. To decrypt the signature, we use $M = S^e \pmod{N}$ by performing the following procedure (also the same as in Example 3.3.8):

```
C ← 1
e = (10001100101111)_2
for i from 1 to 14 do
    C ← C^2 mod N
    if e[i] = 0 then C ← C * M mod N
    print C
```

which gives the following decrypted text:

6091819200019151222051800230914190015140500082114_
041805040004151212011819.

It translates to

FIRST SOLVER WINS ONE HUNDRED DOLLARS

Since this signature was encrypted by RSA's secret key, it cannot be forged by an eavesdropper or even by RSA people themselves.

In Example 3.3.12, the signature is a different text from the message, and usually is appended to the encrypted message. We can, of course directly sign the signature on the message. This can be done in the following way. Suppose A (Alice) wants to send B (Bob) a signed message. Suppose also that

[1] Alice (A) has her own public and secret keys $(e_A, N_A; d_A)$ as well as B's public key e_B and N_B from a public domain;

[2] Bob (B) has his own public and secret keys $(e_B, N_B; d_B)$ as well as A's public key e_A and N_A from a public domain.

To send a signed message from A to B:

[1] Alice uses B's public key e_B and N_B to encrypt her message M_A:

$$C_A = M_A^{e_B} \bmod N_B. \tag{3.97}$$

[2] Alice signs the message using her own secret key d_A directly on the encrypted message:

$$S_A = C_A^{d_A} \bmod N_A, \tag{3.98}$$

and sends this signed message to B over the network.

Upon receiving A's signed message,

[1] B uses A's public key e_A to decrypt A's signature:

$$C_A = S_A^{e_A} \bmod N_A. \tag{3.99}$$

[2] B further uses his own secret key d_B to decrypt A's encrypted message:

$$M_A = C_A^{d_B} \bmod N_B. \tag{3.100}$$

In this way, Bob can make sure that the message he has just received indeed comes from A, since the signature of A's message is encrypted by A's own secret key, which is only known to A. Once the message is sent out, A cannot deny the message. Similarly, Bob can send a signed message to Alice. The above process is shown in Figure 3.13.

Example 3.3.13 (Digital Signature). Suppose now Alice wants to send Bob the signed message "Number Theory is the Queen of Mathematics". The process can be as follows:

[1] Suppose Alice has the following information at hand:

$$\begin{aligned}
M_A =\ & 14211302051800200805151825000919002008050017210505140 01_ \\
& 506001301200805130120090319 \\
N_A =\ & 180708208868740480595165616440590556627810251676940 1349_ \\
& 1701270214500566625402440483873411275908123033717818879_ \\
& 66563182013214880557 \text{ (130 digits)} \\
=\ & 3968599945959745429016112616288378606757644911281006483_ \\
& 2555157243 \cdot 4553449864673597218840368689727440886435630 1_ \\
& 2632050696009990 44599 \\
e_A =\ & 2617 \\
d_A =\ & 9646517683975179648125577614348681987353875490740747744_ \\
& 7102309852757971788848801635711139144032242624779107574_ \\
& 0923050236448593109
\end{aligned}$$

and suppose Bob has the following information at hand:

$$\begin{aligned}
N_B =\ & 1143816257578888676692357799761466120102182967212423625_ \\
& 6256184293570693524573389783059712356395870505898907514_ \\
& 7599290026879543541 \text{ (129 digits)} \\
=\ & 3490529510847650949147849619903898133417764638493387843_ \\
& 990820577 \cdot 3276913299326670954996198819083446141317764 29_ \\
& 6799294253979828853 3
\end{aligned}$$

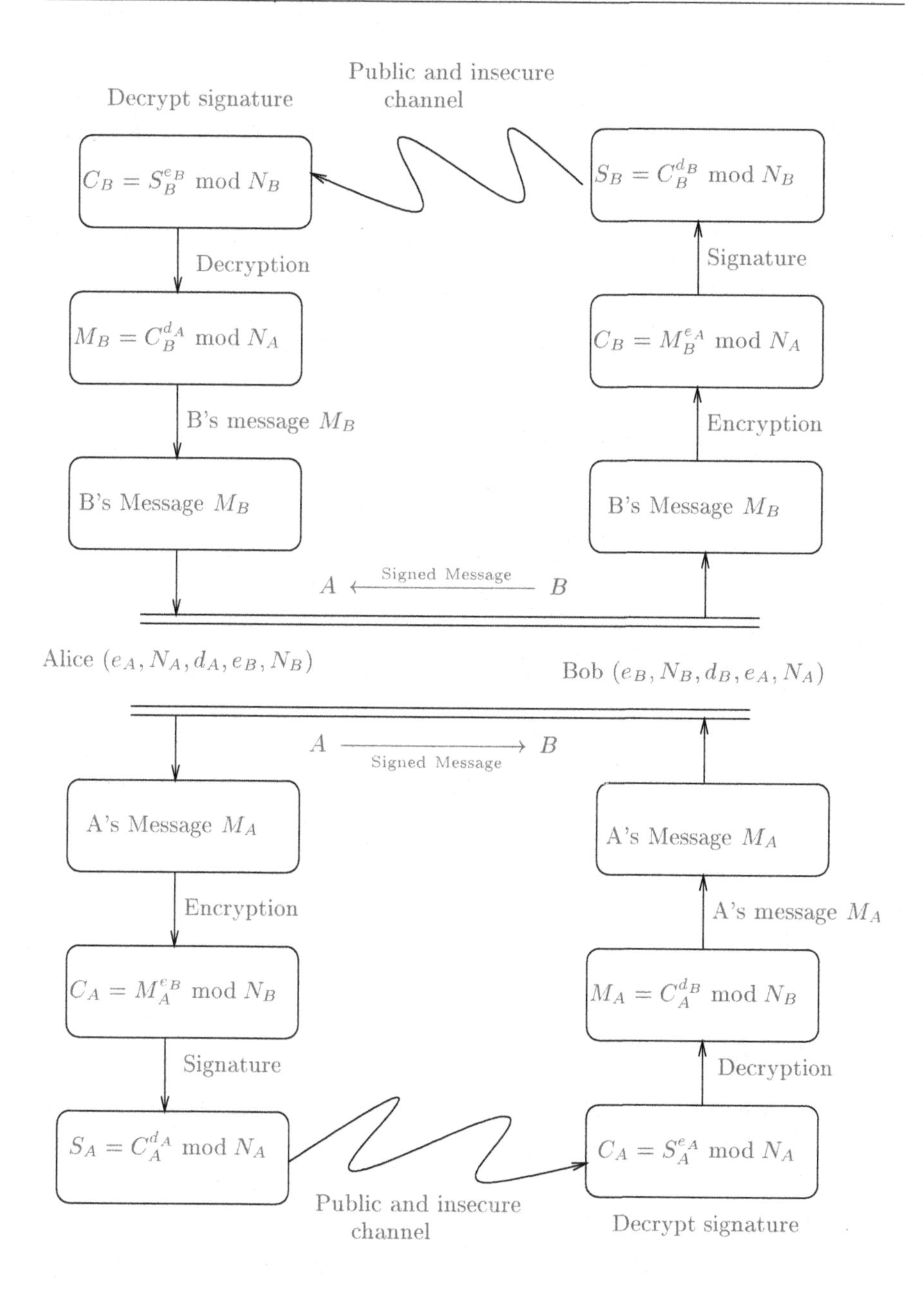

Figure 3.13. Sending encrypted and signed messages using the RSA scheme (the signatures are directly made on the encrypted message)

$e_B =$ 9007
$d_B =$ 1066986143685780244428687713289201547807099066339378628_
0122622449663106312591177447087334016859746230655396854_
4513277109053606095

[2] Alice first encrypts the message M_A using e_B and N_B to get

$$C_A = M_A^{e_B} \bmod N_B$$

by the following process:

$C_A \leftarrow 1$
$e_B \leftarrow (10001100101111)_2$
for i from 1 to 14 do
  $C_A \leftarrow C_A^2 \bmod N_B$
  if $e_B[i] = 1$ then $C_A \leftarrow C_A \cdot M_A \bmod N_B$
Save C_A

[3] Alice then signs the message C_A using d_A and N_A to get $S_A = C_A^{d_A} \bmod N_A$ via the following process:

$S_A \leftarrow 1$
$d_A \leftarrow (10110010 \cdots 11010101)_2$
for i from 1 to 429 do
  $S_A \leftarrow S_A^2 \bmod N_A$
  if $d_A[1, i] = 1$ then $S_A \leftarrow S_A \cdot C_A \bmod N_A$
Send S_A

[4] Upon receiving Alice's message, Bob first decrypts Alice's signature using e_A and N_A to get $C_A = S_A^{e_A} \bmod N_A$ via the following process:

$C_A \leftarrow 1$
$e_A \leftarrow (101000111001)_2$
for i from 1 to 12 do
  $C_A \leftarrow C_A^2 \bmod N_A$
  if $e_A[i] = 1$ then $C_A \leftarrow C_A \cdot S_A \bmod N_A$
Save C_A

[5] Bob then decrypts Alice's message using d_B and N_B to get $M_A = C_A^{d_B} \bmod N_B$ via the following process:

$M_A \leftarrow 1$
$d_B := (1001110110 \cdots 1001111)_2$
for i from 1 to 426 do
  $M_A \leftarrow M_A^2 \bmod N_B$
  if $d_B[i] = 0$ then $M_A \leftarrow M_A \cdot C_A \bmod N_B$
print M_A

Remark 3.3.9. Suppose Bob is sending an encrypted message to Alice. Normally, the encrypted message consists of a number of blocks; one of the blocks is Bob's signature. Alice can easily identify which block is the ***signature block***, since the ordinary decryption procedure for that block yields gibberish. In practice, there are two ways for constructing Bob's encrypted signature (Denson [60]), depending on the values of the moduli n_B and n_A:

(1) If $n_A < n_B$, then

$$S_B = \left(M^{d_B} \bmod n_B\right)^{e_A} \bmod n_A, \quad M_B = \left(S^{d_A} \bmod n_A\right)^{e_B} \bmod n_B.$$

The inequality $n_A < n_B$ ensures that the expression in the parentheses is not too large to be encrypted by Alice's encryption key.

(2) If $n_A < n_B$, then

$$S_B = (M^{e_A} \bmod n_A)^{d_B} \bmod n_B, \quad M_B = (S^{e_B} \bmod n_B)^{d_A} \bmod n_A.$$

The inequality $n_A > n_B$ ensures that the expression in the parentheses is not too large to be encrypted by Bob's decryption key.

The above mentioned signature scheme is based on RSA cryptosystem. Of course, a signature scheme can be based on other cryptosystem. In what follows, we shall introduce a very influential signature scheme based of ElGamal's cryptosystem [69]; the security of such a signature scheme depends on the intractability of discrete logarithms over a finite field.

Algorithm 3.3.5 (ElGamal Signature Scheme). This algorithm tries to generate digital signature $S = (a, b)$ for message m. Suppose that Alice wishes to send a signed message to Bob.

[1] [ElGamal key generation] Alice does the following:

[1-1] Choose a prime p and two random integers g and x, such that both g and x are less than p.

[1-2] Compute $y \equiv g^x \pmod p$.

[1-3] Make (y, g, p) public (both g and p can be shared among a group of users), but keep x as a secret.

[2] [ElGamal signature generation] Alice does the following:

[2-1] Choose at random an integers k such that $\gcd(k, p-1) = 1$.

[2-2] Compute

$$\left.\begin{aligned} a &\equiv g^x \pmod p, \\ b &\equiv k^{-1}(m - xa) \pmod{(p-1)}. \end{aligned}\right\} \qquad (3.101)$$

Now Alice has generated the signature (a, b). She must keep the random integer, k, as secret.

[3] [ElGamal signature verification] To verify Alice's signature, Bob confirms that

$$y^a a^b \equiv g^m \pmod{p}. \tag{3.102}$$

3.3.10 Digital Signature Standard (DSS)

In August 1991, the U.S. government's National Institute of Standards and Technology (NIST) proposed an algorithm for digital Signatures. The algorithm is known as DSA, for Digital Signature Algorithm. The DSA has become the U.S. Federal Information Processing Standard 186 (FIPS 186). It is called the Digital Signature Standard (DSS), and is the first digital signature scheme recognized by any government. The role of DSA/DSS is expected to be analogous to that of the Data Encryption Standard (DES). The DSA/DSS is similar to a signature scheme proposed by Schnorr [220]; it is also similar to a signature scheme of ElGamal [69]. The DSA is intended for use in electronic mail, electronic funds transfer, electronic data interchange, software distribution, data storage, and other applications which require data integrity assurance and data authentication. The DSA/DSS consists of two main processes:

(1) Signature generation (using the private key),

(2) Signature verification (using the public key).

A one-way hash function is used in the signature generation process to obtain a condensed version of data, called a message digest. The message digest is then signed. The digital signature is sent to the intended receiver along with the signed data (often called the message). The receiver of the message and the signature verifies the signature by using the sender's public key. The same hash function must also be used in the verification process. In what follows, we shall give the formal specifications of the DSA/DSS.

Algorithm 3.3.6 (Digital Signature Algorithm, DSA). This is a variation of ElGamal signature scheme. It generates a signature $S = (r, s)$ for the message m.

[1] [DSA key generation] To generate the DSA key, the sender performs the following:

[1-1] Find a 512-bit prime p (which will be public).

[1-2] Find a 160-bit prime q dividing evenly into $p-1$ (which will be public).

[1-3] Generate an element $g \in \mathbb{Z}/p\mathbb{Z}$ whose multiplicative order is q, i.e., $g^q \equiv 1 \pmod{p}$.

[1-4] Find a one-way function H mapping messages into 160-bit values.

[1-5] Choose a secret key x, with $0 < x < q$,

[1-6] Choose a public key y, where $y \equiv g^x \pmod{p}$.

Clearly, the secret x is the discrete logarithm of y, modulo p, to the base g.

[2] [DSA Signature Generation] To sign the message m, the sender produces his signature as (r, s), by selecting a random integer $k \in \mathbb{Z}/q\mathbb{Z}$ and computing

$$\left.\begin{aligned} r &\equiv \left(g^k \pmod{p}\right) \pmod{q}, \\ s &\equiv k^{-1}(H(m) + xr) \pmod{q}. \end{aligned}\right\} \tag{3.103}$$

[3] [DSA Signature Verification] To verify the signature (r, s) for the message m from the sender, the receiver first computes:

$$t \equiv s^{-1} \pmod{q}, \tag{3.104}$$

and then accepts the signature as valid if the following congruence holds:

$$r \equiv \left(g^{H(m)t} y^{rt} \pmod{p}\right) \pmod{q}. \tag{3.105}$$

If the congruence (3.105) does not hold, then the message either may have been incorrectly signed, or may have been signed by an impostor. In this case, the message is considered to be invalid.

There are, however, many responses solicited by the (US) Association of Computing Machinery (ACM) [45], positive and negative, to the NIST's DSA. Some positive aspects of the DSA include:

(1) The U.S. government has finally recognized the utility and the usefulness of public-key cryptography. In fact, the DSA is the only signature algorithm that has been publicly proposed by any government.

(2) The DSA is based on reasonable familiar number-theoretic concepts, and it is especially useful to the financial services industry.

(3) Signatures in DSA are relatively short (only 320 bits), and the key generation process can be performed very efficiently.

(4) When signing, the computation of r can be done even before the message m is available, in a "precomputation" step.

Whilst some negative aspects of the DSA include:

(1) The DSA does not include key exchanges, and cannot be used for key distribution and encryption.

(2) The key size in DSA is too short; it is restricted to a 512-bit modulus or key size, which is too short and should be increased to at least 1024 bits.

(3) The DSA is not compatible with existing international standards; for example, the international standards organizations such as ISO, CCITT and SWIFT all have accepted the RSA as a standard.

Nevertheless, the DSA is the only one publicly known government digital signature standard.

We have already noted that almost every public-key cryptosystem has an elliptic curve analogue. It should also be noted that digital signature schemes can also be represented by elliptic curves over $\mathbb{F}_q$ with q a prime power or over $\mathbb{Z}/n\mathbb{Z}$ with $n = pq$ and $p, q \in$ Primes. In exactly the same way as that for public-key cryptography, several elliptic curve analogues of digital signature schemes have already been proposed (see, for example, Meyer and Müller [160]). In what follows we shall describe an elliptic curve analogue of the DSA/DSS, called ECDSA.

Algorithm 3.3.7 (Elliptic Curve Digital Signature Algorithm). Let E be an elliptic curve over $\mathbb{F}_p$ with p prime, and let P be a point of prime order q (note that the q here is just a prime number, not a prime power) in $E(\mathbb{F}_p)$. Suppose Alice wishes to send a signed message to Bob.

[1] [ECDSA key generation] Alice does the following:

[1-1] select a random integer $x \in [1, \ q-1]$,

[1-2] compute $Q = xP$,

[1-3] make Q public, but keep x as a secret.

Now Alice has generated the public key Q and the private key x.

[2] [ECDSA signature generation] To sign a message m, Alice does the following:

[2-1] select a random integer $k \in [1, \ q-1]$,

[2-2] compute $kP = (x_1, y_1)$, and $r \equiv x_1 \pmod{q}$. If $r = 0$, go to step [2-1].

[2-3] compute $k^{-1} \bmod q$.

[2-4] compute $s \equiv k^{-1}(H(m) + xr) \pmod{q}$, where $H(m)$ is the hash value of the message. If $s = 0$, go to step [2-1].

The signature for the message m is the pair of integers (r, s).

[3] [ECDSA signature verification] To verify Alice's signature (r, s) of the message m, Bob should do the following:

[3-1] obtain an authenticated copy of Alice's public key Q;

[3-2] verify that (r, s) are integers in the interval $[1,\ q-1]$, computes $kP = (x_1, y_1)$, and $r \equiv x_1 \pmod q$.

[3-3] compute $w \equiv s^{-1} \pmod q$ and $H(m)$.

[3-4] compute $u_1 \equiv H(m)w \pmod q$ and $u_2 \equiv rw \pmod q$.

[3-5] compute $u_1P + u_2Q = (x_0, y_0)$ and $v \equiv x_0 \pmod q$.

[3-6] accept the signature if and only if $v = r$.

Exercise 3.3.10. Try to develop an elliptic curve analogue of an existing signature scheme that you are familiar with for obtaining and checking digital signatures.

3.3.11 Database Security

Databases pose a special challenge to the designer of secure information systems. Databases are meant to be shared. The sharing is often complex. In many organizations, there are many "rules" concerning the access to different *fields* (or parts) of a database. For example, the payroll department may have access to the name, address and salary fields, while the insurance office may have access to the health field of an individual. In this subsection, we shall introduce a method for database protection; it encrypts the entire database but the individual fields may be decrypted and read without affecting the security of other fields in the database.

Let

$$D = \langle F_1, F_2, \cdots, F_n \rangle \tag{3.106}$$

where D is the database and each F_i is an individual file (or record). As in RSA encryption, each file in D can be regarded as an integer. To encrypt D, we first select n distinct primes $m_1, m_2, \cdots, m_n$, where $m_i > F_i$, for $i = 1, 2, \cdots, n$. Then by solving the following system of congruences:

$$\left.\begin{array}{l} C \equiv F_1 \pmod{m_1}, \\ C \equiv F_2 \pmod{m_2}, \\ \quad \cdots\cdots \\ \quad \cdots\cdots \\ C \equiv F_n \pmod{m_n}, \end{array}\right\} \tag{3.107}$$

we get C, the encrypted text of D. According to the Chinese Remainder Theorem, such a C always exists and can be found. Let

$$\left.\begin{array}{l} M = m_1 m_2 \cdots m_n, \\ M_i = M/m_i, \\ e_i = M_i \left[M_i^{-1} \bmod m_i\right], \end{array}\right\} \tag{3.108}$$

for $i = 1, 2, \cdots, n$. Then C can be obtained as follows:

$$C = \sum_{i=1}^{n} e_j F_j \pmod{M}, \quad 0 \leq C < M. \tag{3.109}$$

The integers $e_1, e_2, \cdots, e_n$ are used as the *write-keys*. To retrieve the i-th file F_i from the encrypted text C of D, we simply perform the following operation:

$$F_i \equiv C \pmod{m_i}, \quad 0 \leq F_i < m_i. \tag{3.110}$$

The moduli $m_1, m_2, \cdots, m_n$ are called the *read-keys*. Only people knowing the read-key m_i can read file F_i, but not other files. To read other files, for example, F_{i+2}, it is necessary to know a read-key other than m_i. We present in the following an algorithm for database encryption and decryption.

Algorithm 3.3.8 (Database protection). Given $D = \langle F_1, F_2, \cdots, F_n \rangle$, this algorithm will first encrypt the database D into its encrypted text C. To retrieve information from the encrypted database C, the user uses the appropriate read-key m_i to read file F_i:

Part I: Database Encryption. The database administrators (DBA) perform the following operations to encrypt the database D:

[1] Select n distinct primes $m_1, m_2, \cdots, m_n$ with $m_i > F_i$, for $i = 1, 2, \cdots, n$.

[2] Use the Chinese Remainder Theorem to solve the following system of congruences:

$$\left.\begin{array}{l} C \equiv F_1 \pmod{m_1}, \\ C \equiv F_2 \pmod{m_2}, \\ \cdots\cdots \\ \cdots\cdots \\ C \equiv F_n \pmod{m_n}, \end{array}\right\} \tag{3.111}$$

and get

$$C = \sum_{i=1}^{n} e_j F_j \pmod{M}, \quad 0 \leq C < M \tag{3.112}$$

where

$$M = m_1 m_2 \cdots m_n,$$
$$M_i = M/m_i,$$
$$e_i = M_i \left[M_i^{-1} \bmod m_i \right],$$

for $i = 1, 2, \cdots, n$.

[3] Distribute the read-key m_i to the appropriate database user U_i.

Part II: Database Decryption. At this stage, the database user U_i is supposed to have access to the encrypted database C as well as to have the read-key m_i, so he performs the following operation:

$$F_i \equiv C \pmod{m_i}, \quad 0 \le F_i < m_i. \tag{3.113}$$

The required file F_i should be now readable by user U_i.

Example 3.3.14 (Database Encryption and Decryption). Let

$$\begin{aligned} D &= \langle F_1, F_2, F_3, F_4, F_5 \rangle \\ &= \langle 198753, 217926, 357918, 377761, 391028 \rangle. \end{aligned}$$

Choose five primes m_1, m_2, m_3, m_4 and m_5 as follows:

$$\begin{aligned} m_1 &= 350377 > F_1 = 198753, \\ m_2 &= 364423 > F_2 = 217926, \\ m_3 &= 376127 > F_2 = 357918, \\ m_4 &= 389219 > F_4 = 377761, \\ m_5 &= 391939 > F_5 = 391028. \end{aligned}$$

According to (3.111), we have:

$$\begin{aligned} C &\equiv F_1 \pmod{m_1} \Longrightarrow C \equiv 198753 \pmod{350377} \\ C &\equiv F_2 \pmod{m_2} \Longrightarrow C \equiv 217926 \pmod{364423} \\ C &\equiv F_3 \pmod{m_3} \Longrightarrow C \equiv 357918 \pmod{376127} \\ C &\equiv F_4 \pmod{m_4} \Longrightarrow C \equiv 377761 \pmod{389219} \\ C &\equiv F_5 \pmod{m_5} \Longrightarrow C \equiv 391028 \pmod{391939}. \end{aligned}$$

Using the Chinese Remainder Theorem to solve the above system of congruences, we get

$$C = 5826262707691801601352277219.$$

Since $0 \le C < M$ with

$$\begin{aligned} M &= 350377 \cdot 364423 \cdot 376127 \cdot 389219 \cdot 391939 \\ &= 7326362302832726883024522697, \end{aligned}$$

C is the required encrypted text of D. Now suppose user U_2 has the read-key $m_2 = 364423$. Then he can simply perform the following computation and get F_2:

$$F_2 \equiv C \pmod{m_i}.$$

Now

$$\begin{aligned} C \pmod{m_2} &= 5826262707691801601352277219 \bmod 364423 \\ &= 217926 \\ &= F_2, \end{aligned}$$

which is exactly what the user U_2 wanted. Similarly, a user can read F_5 if he knows m_5, since

$$\begin{aligned} C \pmod{m_5} &= 5826262707691801601352277219 \bmod 391939 \\ &= 391028 \\ &= F_5. \end{aligned}$$

Remark 3.3.10. In Example 3.3.14, we have not explicitly given the computing processes for the write keys e_i and the encrypted text C; we give now the detailed computing processes as follows:

$$\begin{aligned} e_1 &= M_1 \cdot (M_1^{-1} \bmod m_1) \\ &= 20909940729079611056161 \cdot (20909940729079611056161^{-1} \bmod 350377) \\ &= 3040577211237653482509539493 \\ e_2 &= M_2 \cdot (M_2^{-1} \bmod m_2) \\ &= 20104006341072673467439 \cdot (20104006341072673467439^{-1} \bmod 364423) \\ &= 2830382740740598479460334493 \; e_3 = M_3 \cdot (M_3^{-1} \bmod m_3) \\ &= 19478426975018349873911 \cdot (19478426975018349873911^{-1} \bmod 376127) \\ &= 1991883420892351476456012771 \\ e_4 &= M_4 \cdot (M_4^{-1} \bmod m_4) \\ &= 18823239109171769320163 \cdot (18823239109171769320163^{-1} \bmod 389219) \\ &= 6068028768384594103971626147 \\ e_5 &= M_5 \cdot (M_5^{-1} \bmod m_5) \\ &= 18692608550903908217923 \cdot (18692608550903908217923^{-1} \bmod 391939) \\ &= 7218524644102562236515324 91. \end{aligned}$$

So

$$\begin{aligned} C &= (e_1F_1 + e_2F_2 + e_3F_3 + e_4F_4 + e_5F_5) \bmod M \\ &= (3040577211237653482509539493 \cdot 198753 \\ &\quad + 2830382740740598479460334493 \cdot 217926 \\ &\quad + 1991883420892351476456012771 \cdot 357918 \\ &\quad + 6068028768384594103971626147 \cdot 377761 \\ &\quad + 7218524644102562236515324 91 \cdot 391028) \\ &\quad \bmod 7326362302832726883024522697 \\ &= 5826262707691801601352277219. \end{aligned}$$

Exercise 3.3.11. Let the database D be

$$\begin{aligned} D &= \langle F_1, F_2, F_3, F_4 \rangle \\ &= \langle 9853, 6792, 3761, 5102 \rangle. \end{aligned}$$

and the four read keys be

$$\begin{aligned} m_1 &= 9901 > F_1 = 9853, \\ m_2 &= 7937 > F_2 = 6792, \\ m_3 &= 5279 > F_3 = 3761, \\ m_4 &= 6997 > F_4 = 5102. \end{aligned}$$

(1) What are the four write keys e_1, e_2, e_3 and e_4 used in the encryption process?

(2) What is the encrypted text C corresponding to D?

(3) If F_1 is changed from $F_1 = 9853$ to $F_1 = 9123$, what is the new value of the encrypted text C?

To protect a database, we can encrypt it by using encryption keys. To protect encryption keys, however, we will need some different methods. In the next subsection, we shall introduce a method for protecting the cryptographic keys.

3.3.12 Secret Sharing

Liu [145] considers the following problem: eleven scientists are working on a secret project. They wish to lock up the documents in a cabinet such that the cabinet can be opened if and only if six or more of the scientists are present. What is the smallest number of locks needed? What is the smallest number of keys to the locks each scientist must carry? The minimal solution uses 462 locks and 252 keys. It is clear that these numbers are impractical, and they become exponentially worse when the number of scientists increases. In this section, we shall introduce an interesting method to solve similar problems. It is called secret sharing and was first proposed by Shamir in 1979 (see Mignotte [161] and Shamir [225]). The method can be very useful in the management of cryptographic keys and the keys for accessing the password file in a computer system.

Definition 3.3.4. A (k, n)-*threshold scheme* is a method for n people (or parties) $P_1, P_2, \cdots, P_n$ to share a secret S in such a way that the following properties hold:

(1) $k < n$,

(2) each P_i has some information I_i,

(3) knowledge of any k of the $\{I_1, I_2, \cdots, I_n\}$ enables one to find S easily,

(4) knowledge of less than k of the $\{I_1, I_2, \cdots, I_n\}$ does not enable one to find S easily.

Of course, there might be several ways to construct such a threshold scheme, but perhaps the simplest is the one based on congruence theory and the Chinese Remainder Theorem. It can be shown (Krana [134]) by the Chinese Remainder Theorem that:

Theorem 3.3.3. *For all $2 \le k \le n$, there exists a (k,n)-threshold scheme.*

In what follows, we shall introduce an algorithm for constructing a (k,n)-threshold scheme.

Algorithm 3.3.9 (Secret sharing). This algorithm is divided into two parts: the first part aims to construct a secret set $\{I_1, I_2, \cdots, I_n\}$, whereas the second part aims to find out the secret S by any k of the $\{I_1, I_2, \cdots, I_n\}$. Throughout the algorithm, S denotes the secret.

Part I: Construction of the secret set $\{I_1, I_2, \cdots, I_n\}$.

[1] Let the threshold sequence $m_1, m_2, \cdots, m_n$ be positive integers > 1 such that $\gcd(m_i, m_j) = 1$ for $i \neq j$ and

$$m_1 m_2 \cdots m_k > m_n m_{n-1} \cdots m_{n-k+2}. \tag{3.114}$$

[2] Determine the secret S in such a way that

$$\max(k-1) < S < \min(k) \tag{3.115}$$

where

$$\left.\begin{aligned} \min(k) &= m_1 m_2 \cdots m_k, \\ \max(k-1) &= m_n m_{n-1} \cdots m_{n-k+2}. \end{aligned}\right\} \tag{3.116}$$

[3] Compute $\{I_1, I_2, \cdots, I_n\}$ in the following way:

$$\left.\begin{array}{l} S \equiv I_1 \pmod{m_1}, \\ S \equiv I_2 \pmod{m_2}, \\ \cdots\cdots \\ \cdots\cdots \\ S \equiv I_n \pmod{m_n}. \end{array}\right\} \tag{3.117}$$

[4] Compute $M = m_1 m_2 \cdots m_n$.

[5] Send I_i and (m_i, M) to each P_i.

Part II: Recovering S from any k of these $I_1, I_2, \cdots, I_n$: Suppose now parties $\{P_{i_1}, P_{i_2}, \cdots, P_{i_k}\}$ want to combine their knowledge $\{I_{i_1}, I_{i_2}, \cdots, I_{i_k}\}$ to find out S. (Each $P_{i_j}, j = 1, 2, \cdots, n$ has the triple (I_{i_j}, m_{i_j}, M) at hand).

[1] Each P_{i_j}, $j = 1, 2, \cdots, k$ computes his own secret recovering key S_{i_j} as follows:

$$\left.\begin{aligned} M_{i_j} &= M / m_{i_j}, \\ N_{i_j} &= M_{i_j}^{-1} \pmod{m_{i_j}}, \\ S_{i_j} &= I_{i_j} M_{i_j} N_{i_j}. \end{aligned}\right\} \tag{3.118}$$

[2] Combine all the S_{i_j} to get the secret S:

$$S = \sum_{j=1}^{k} S_{i_j} \left(\bmod \prod_{j=1}^{k} m_{i_j} \right). \tag{3.119}$$

(By the Chinese Remainder Theorem, this computed S will be the required secret).

Example 3.3.15. Suppose we wish to construct a (k, n)-threshold scheme with $k = 3$ and $n = 5$. The scheme administrator of a security agency first defines the following threshold sequence m_i:

$$\begin{aligned} m_1 &= 97, \\ m_2 &= 98, \\ m_3 &= 99, \\ m_4 &= 101, \\ m_5 &= 103, \end{aligned}$$

and computes:

$$\begin{aligned} M &= m_1 m_2 m_3 m_4 m_5 = 9790200882 \\ \min(k) &= m_1 m_2 m_3 = 941094 \\ \max(k-1) &= m_4 m_5 = 10403. \end{aligned}$$

He then defines the secret S to be in the range

$$10403 < S = 671875 < 941094$$

and calculates each I_i for each P_i:

$$\begin{aligned} S &\equiv I_1 \pmod{m_1} \Longrightarrow I_1 = 53 \\ S &\equiv I_2 \pmod{m_2} \Longrightarrow I_2 = 85 \\ S &\equiv I_3 \pmod{m_3} \Longrightarrow I_3 = 61 \\ S &\equiv I_4 \pmod{m_4} \Longrightarrow I_4 = 23 \\ S &\equiv I_5 \pmod{m_5} \Longrightarrow I_5 = 6. \end{aligned}$$

Finally he distributes each I_i as well as m_i and M to each P_i, so that each P_i who shares the secret S has the triple (I_i, m_i, M).

Suppose now P_1, P_2 and P_3 want to combine their knowledge $\{I_1, I_2, I_3\}$ to find out S. They first individually compute:

$$\begin{aligned} M_1 &= M/m_1 = 100929906 \\ M_2 &= M/m_2 = 99900009 \\ M_3 &= M/m_3 = 98890918 \end{aligned}$$

and

$$\begin{aligned} N_1 &\equiv M_1^{-1} \pmod{m_1} \Longrightarrow N_1 = 95 \\ N_2 &\equiv M_2^{-1} \pmod{m_2} \Longrightarrow N_2 = 13 \\ N_3 &\equiv M_3^{-1} \pmod{m_3} \Longrightarrow N_3 = 31. \end{aligned}$$

Hence, they get

$$\begin{aligned} S &\equiv I_1 \cdot M_1 \cdot N_1 + I_2 \cdot M_2 \cdot N_2 + I_3 \cdot M_3 \cdot N_3 \pmod{m_1 \cdot m_2 \cdot m_3} \\ &\equiv 53 \cdot 100929906 \cdot 95 + 85 \cdot 99900009 \cdot 13 + 61 \cdot 98890918 \cdot 31 \\ &\qquad \pmod{97 \cdot 98 \cdot 99} \\ &\equiv 805574312593 \pmod{941094} \\ &= 671875. \end{aligned}$$

Suppose, alternatively, P_1, P_4 and P_5 wish to combine their knowledge $\{I_1, I_4, I_5\}$ to find out S. They do the similar computations as follows:

$$\begin{aligned} M_1 &= M/m_1 = 100929906 \\ M_4 &= M/m_4 = 96932682 \\ M_5 &= M/m_5 = 95050494 \end{aligned}$$

and

$$\begin{aligned} N_1 &\equiv M_1^{-1} \pmod{m_1} \Longrightarrow N_1 = 95 \\ N_4 &\equiv M_4^{-1} \pmod{m_4} \Longrightarrow N_4 = 61 \\ N_5 &\equiv M_5^{-1} \pmod{m_5} \Longrightarrow N_5 = 100. \end{aligned}$$

Therefore,

$$\begin{aligned} S &\equiv I_1 \cdot M_1 \cdot N_1 + I_4 \cdot M_4 \cdot N_4 + I_5 \cdot M_5 \cdot N_5 \pmod{m_1 \cdot m_4 \cdot m_5} \\ &\equiv 53 \cdot 100929906 \cdot 95 + 23 \cdot 96932682 \cdot 61 + 6 \cdot 95050494 \cdot 100 \\ &\qquad \pmod{97 \cdot 101 \cdot 103} \\ &\equiv 701208925956 \pmod{1009091} \\ &= 671875. \end{aligned}$$

However, knowledge of less than 3 of these I_1, I_2, I_3, I_4, I_5 is insufficient to find out S. For example, you cannot expect to find out S just by combining I_1 and I_4:

$$\begin{aligned} S' &\equiv I_1 \cdot M_1 \cdot N_1 + I_4 \cdot M_4 \cdot N_4 \pmod{m_1 \cdot m_4} \\ &\equiv 53 \cdot 100929906 \cdot 95 + 23 \cdot 96932682 \cdot 61 \pmod{97 \cdot 101} \\ &\equiv 644178629556 \pmod{9791} \\ &= 5679. \end{aligned}$$

Clearly, this is not the correct value of S. Of course, you can find out S by any 3 or more of the I_1, I_2, I_3, I_4, I_5.

Exercise 3.3.12. In the above context, find out S if P_1, P_3, P_4, P_5 wish to combine their knowledge $\{I_1, I_3, I_4, I_5\}$ to find out S.

Exercise 3.3.13. Suppose a security agency defines a $(5, 7)$-threshold scheme and sends each triple (I_i, m_i, M) defined as follows to each person P_i for $i = 1, 2, \cdots, 7$, who shares the secret S:

$$
\begin{aligned}
(I_1, m_1) &= (824, 1501) \\
(I_2, m_2) &= (1242, 1617) \\
(I_3, m_3) &= (1602, 1931) \\
(I_4, m_4) &= (1417, 5573) \\
(I_5, m_5) &= (3090, 6191) \\
(I_6, m_6) &= (281, 7537) \\
(I_7, m_7) &= (6261, 9513) \\
M &= 1501 \cdot 1617 \cdot 1917 \cdot 3533 \cdot 9657 \cdot 10361 \cdot 53113 \\
&= 1159414813752079260508694 1
\end{aligned}
$$

Now suppose parties P_1, P_3, P_5, P_6, P_7 wish to combine their knowledge $\{I_1, I_3, I_5, I_6, I_7\}$ to find out S. What is the S? Suppose also parties P_2, P_3, P_4, P_5, P_6 wish to combine their knowledge $\{I_2, I_3, I_4, I_5, I_6\}$ to find out S. What is the S then? (The two S's should be the same.)

3.3.13 Internet/Web Security and Electronic Commerce

> It is easy to run a secure computer system. You merely have to disconnect all dial-up connections and permit only direct-wired terminals, put the machine and its terminals in a shielded room, and post a guard at the door.
>
> GRAMPP AND MORRIS
> UNIX Operating System Security [91]

The security mentioned in the above quotation is unfortunately not what we need, though it is easy to achieve; an isolated and disconnected computer system is essentially a useless system in modern days. We would like such a (local network) system which is fully connected to the Internet but still be as secure as a disconnected system. How can we achieve such a goal? The first method to secure the local system is to introduce a firewall (security gateway) to protect a local system against intrusion from outside sources. An Internet firewall serves the same purpose as firewalls in buildings: to protect a certain area from the spread of fire and a potentially catastrophic explosion. It is used to examine the Internet addresses on packets or ports requested on incoming connections to decide what traffic is allowed into the local network. The simplest form of a firewall is the packer filter, as shown in Figure 3.14. It basically keeps a record of allowable sources and destination IP addresses and deletes all packets which do not have these addresses. Unfortunately, this firewalling technique suffers from the fact that IP addresses[19] can be easily forged. For example, a "hacker" might determine the list of good source addresses and then add one of these addresses to any packets which are addressed into the local network. Although some extra layers of security can

[19] An Internet Protocol address (IP address), or just Internet address, is a unique 32-bit binary number assigned to a host and used for all communication with the host.

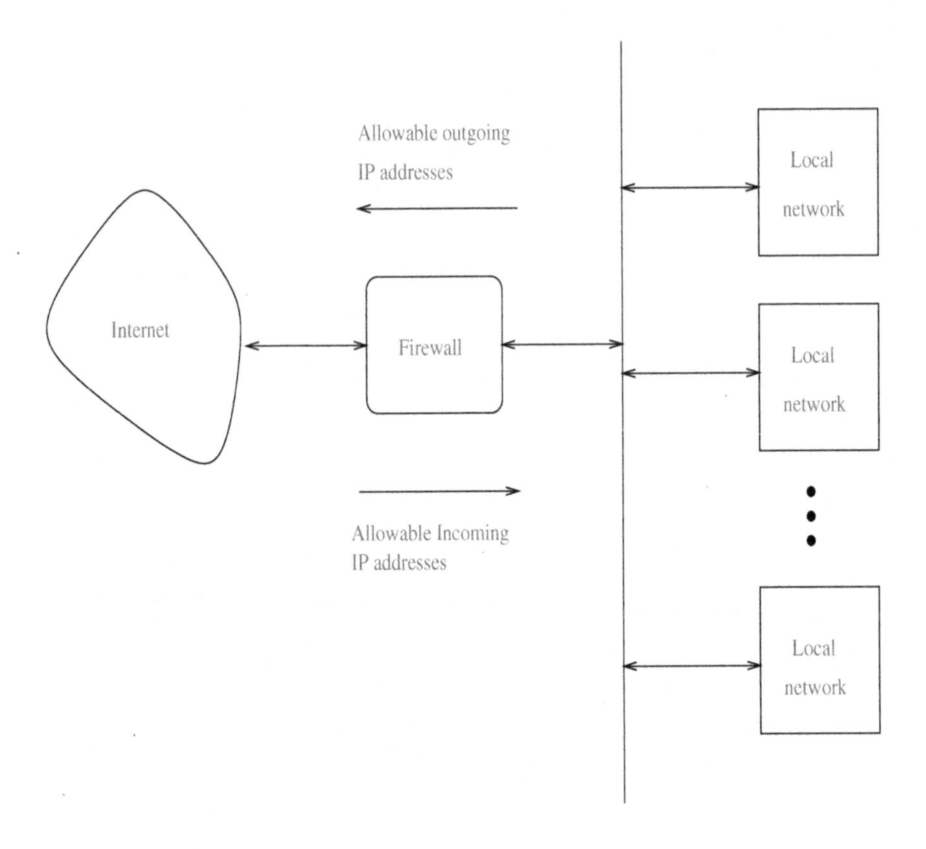

Figure 3.14. Packet filter firewalls

be added into a firewall, it is generally still not powerful enough to protect a local system against intrusion from outside unfriendly users in the Internet. It is worthwhile pointing out that all networked systems have holes in them by which someone else can slip into. For example, recently the U.S. Federal Bureau of Investigation (FBI) estimated that $7.5 billion are lost annually to electronic attack and the U.S. Department of Defence (DOD) says that in 96% of the cases where the crackers got in, they went undetected. The best method of protection for a local network system is to encrypt all the information stored in the local system and to decrypt it whenever an authorized user wants to use the information. This method has an an important application in secure communications – to encrypt the data leaving the local network and then to decrypt it on the remote site; only friendly sites will have the required encryption/decryption key to receive or to send data, and only the routers which connect to the Internet require to encrypt/decrypt. This technique is known as the cryptographic tunnels (see Figure 3.15), which has the extra advantage that data cannot be easily *tapped-into* (Buchanan [42]). A further

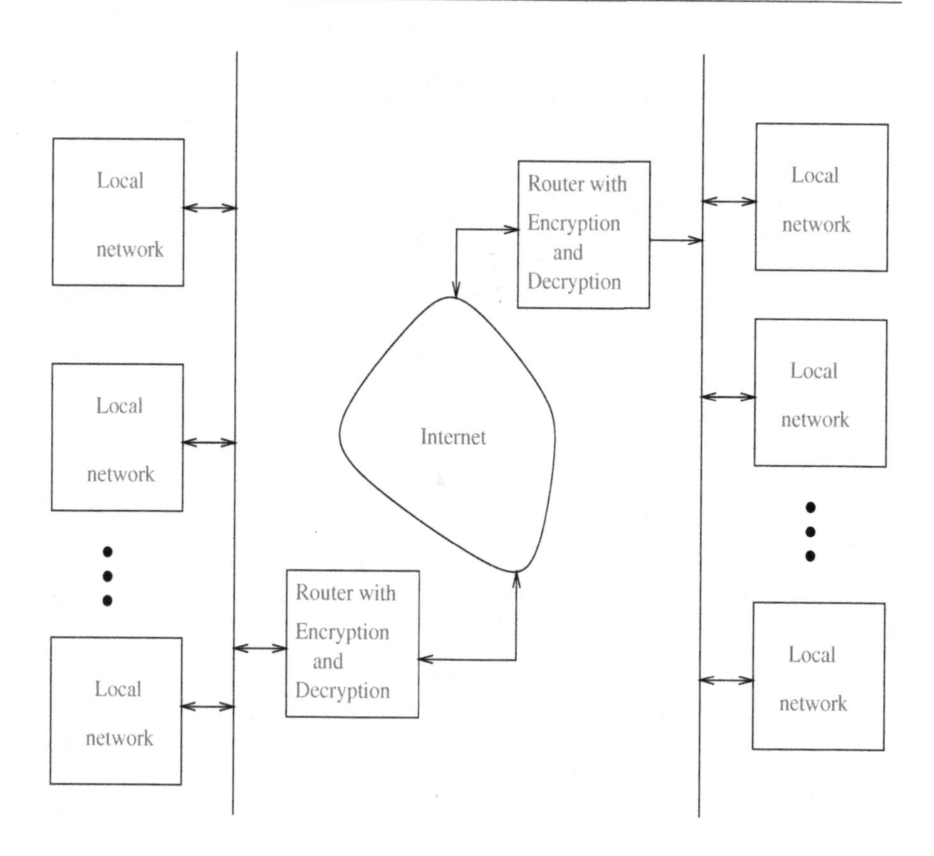

Figure 3.15. Cryptographic tunnels

development of the cryptographic tunnels is the Virtual Private Networks technologies [264], which use tunneling to create a private network so as to keep communication private.

Cryptographic tunnels have important applications in secure communications and digital payments, or more generally, the electronic commerce over the insecure Internet/World Wide Web. For example, if Bob wants to order a book from Alice's bookshop (see Figure 3.16), he uses the secure tunnel to send Alice his credit card number; on receiving Bob's credit card number, Alice sends Bob the required book. It is worthwhile pointing out that a great deal of effort has been put into commercial cryptographic-based Internet/Web security in recent years. Generally speaking, there are two categories of commercial cryptographic systems used for securing the Internet/Web communications. The first group are programs and protocols that are used for encryption of e-mail messages. These programs take a plaintext message, encrypt it and either store the encrypted message on a local machine or transmit it

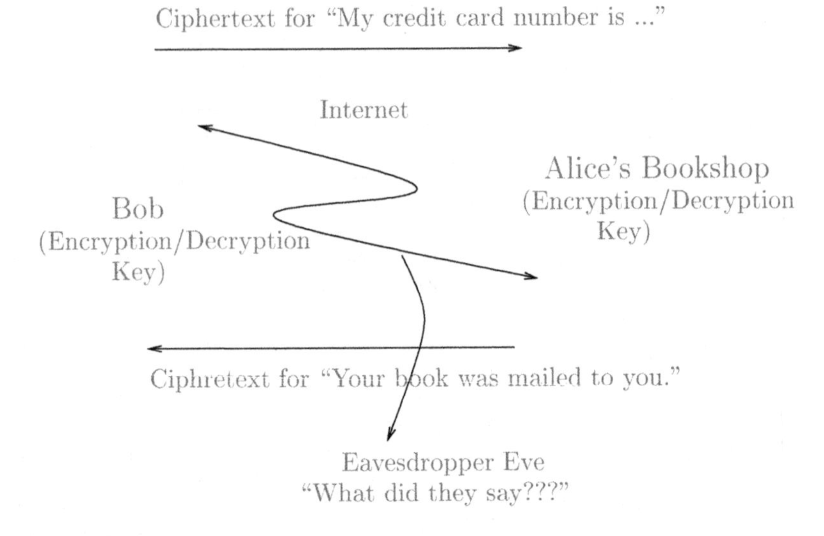

Figure 3.16. Electronic book ordering

to another user over the Internet. Some popular systems that fall into this category include the following:

(1) Pretty Good Privacy (PGP): PGP is a program created by Philip Zimmermann to encrypt e-mails using public-key cryptography. PGP was electronically published as free software in 1991. It has now become the worldwide de facto standard for e-mail encryption.

(2) Secure/Multipurpose Internet Mail Extensions (S/MIME): S/MIME is a security enhancement to the MIME Internet e-mail format standard, based on technology from RSA Data Security. Although both PGP and S/MIME are on an IETF (Internet Engineering Task Force) standards track, it appears likely that S/MIME will emerge as the industry standard for commercial and organizational use, while PGP will remain the choice for personal e-mail security for many users.

The second category of cryptographic systems are network protocols used for providing confidentiality, authentication, integrity, and nonrepudiation in a networked environment. These systems require real-time interplay between a client and a server to work properly. Listed below are some systems falling into this category:

(1) Secure Sockets Layer protocol (SSL): SSL is developed by Netscape Communications, and supported by Netscape and Microsoft browsers. It provides a secure channel between client and server which ensures privacy of data, authentication of the session partners and message integrity.

(2) Private Communication Technology protocol (PCT): PCT, proposed by Microsoft, is a slightly modified version of SSL. The Internet Engineering

Task Force (IETF) is in the process of creating a Transport Secure Layer (TSL) to merge the SSL and PCT.

(3) Secure HyperText Transport Protocol (S-HTTP): S-HTTP is developed by Enterprise Integration Technologies (EIT). It uses a modified version of HTTP clients and the server to allow negotiation of privacy, authentication and integrity characteristics.

(4) Secure Transaction Technology Protocol (STT): STT is a standard developed jointly by Microsoft and Visa International to enable secure credit card payment and authorisation over the web.

(5) Secure Electronic Payment Protocol (SEPP): SEPP is another electronic payments scheme, sponsored by MasterCard and developed in association with IBM, Netscape, CyberCash and GTE. Both STT and SEPP have been superseded by SET (Secure Electronic Transactions), proposed jointly by MasterCard and Visa.

Exercise 3.3.14. Try to order a copy of a book, e.g., the present book, from Springer-Verlag by using your SSL-aware web browser to create an encrypted connection to the Springer-Verlag web server:

`https://www.springer.de`

Now we are in a position to discuss a real-world commercial cryptographic protocol, the SET protocol for secure credit card payment over the insecure Internet. It is a simplified version of the SET, based on a description given in [87].

Algorithm 3.3.10 (SET protocol). This algorithm describes a cryptographic protocol for credit card payment over the Internet. Suppose that Alice wants to purchase a book from Bob (an Internet bookshop) using the credit card issued by Lisa (a bank), but Alice does not want Bob to see her credit card number, however she wants Bob to send her the book and Lisa to send Bob the payment. And of course, Alice also wants that the communications between Bob, Lisa and herself is kept confidential even if someone is eavesdropping over the Internet.

[1] Alice first prepares two documents: a purchase order O stating she wants to order a book from Bob, and a payment slip P, providing Lisa the card number to be used in the transaction, and the amount to be charged. Then she computes the digests:

$$\left.\begin{aligned} o &= H(O) \\ p &= H(P) \end{aligned}\right\} \tag{3.120}$$

and produces a digital signature S for the digest of the concatenation of o and p:

$$S = D_A(H(o \parallel p)) = D_A(H(H(O) \parallel H(P))) \tag{3.121}$$

where D_A is the function used by Alice to sign, based on her private key. Alice encrypts the concatenation of o, P and S with Lisa's public key, which yields the ciphertext:

$$C_L = E_L(o \parallel P \parallel S). \tag{3.122}$$

She also encrypts with Bob's public key the concatenation of O, p and S and gets the ciphertext:

$$C_B = E_B(O \parallel p \parallel S). \tag{3.123}$$

She then sends C_L and C_B to Bob.

[2] Bob retrieves O, p and S by decrypting C_B with his private key. He verifies the authenticity of the purchase order O with Alice's public key by checking that

$$E_A(S) = H(H(O \parallel p)) \tag{3.124}$$

and forwards C_L to Lisa.

[3] Lisa retrieves o, P and S by decrypting C_L with private key. She verifies the authenticity of the payment slip P with Alice's public key by checking that

$$E_A(S) = H(o \parallel H(P)) \tag{3.125}$$

and verifies that P indicates a payment to Bob. She then creates an authorization message M that consists of a transaction number, Alice's name, and the amount she agreed to pay. Lisa computes the signature T of M, encrypts the pair (M, T) with Bob's public key to get the ciphertext:

$$C_M = E_B(M \parallel T) \tag{3.126}$$

and sends it to Bob.

[4] Bob retrieves M and T by decrypting C_M and verifies the authenticity of the authorization message M with Lisa's public key, by checking that

$$E_L(T) = M. \tag{3.127}$$

He verifies that the name in M is Alice's, and that the amount is the correct price of the book. He fulfills the order by sending the book to Alice and requests the payment from Lisa by sending her the transaction number encrypted with Lisa's public key.

[5] Lisa pays Bob and charges Alice's credit card account.

3.3.14 Steganography

Cryptography means "secret writing". A closely related area to cryptography is *steganography*, which literally means *covered writing* as derived from Greek and deals with the hiding of messages so that the potential monitors do not even know that a message is being sent. It is different from cryptography where they know that a secret message is being sent. Figure 3.17 shows a schematic diagram of a typical steganography system. Generally, the sender

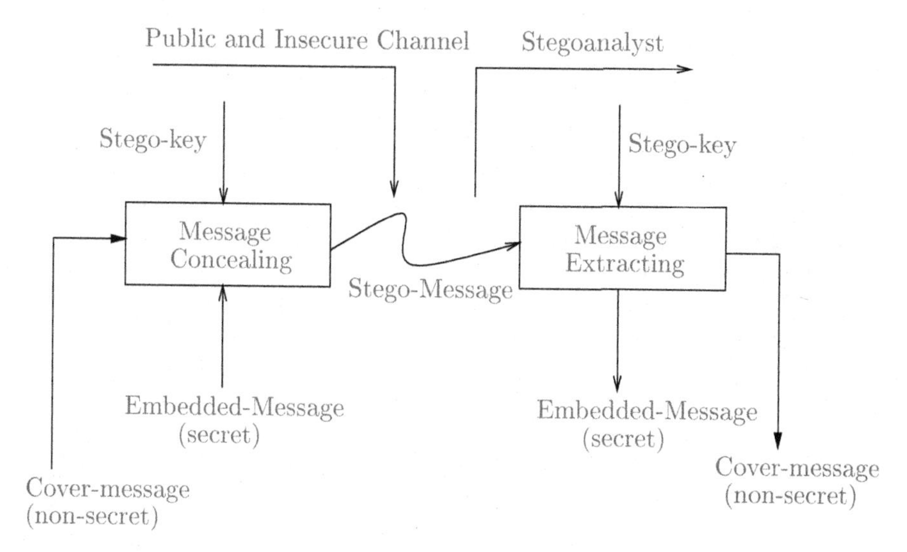

Figure 3.17. A steganographic system

performs the following operations:

(1) write a non-secret cover-message,

(2) produce a stego-message by concealing a secret embedded message on the cover-message by using a stego-key,

(3) send the stego-message over the insecure channel to the receiver.

At the other end, on receiving the stego-message, the intended receiver extracts the secret embedded message from the stego-message by using a pre-agreed stego-key (often the same key as used in the message concealing). Historical tricks include invisible inks, tiny pin punctures on selected characters, minute differences between handwritten characters, etc. For example, Kahn tells of a classical Chinese practice of embedding a code ideogram at a prearranged place in a dispatch (Kahn [117]). More recently, people have hidden secret messages in graphic images by replacing the least significant bits of the image with a secret message (Schneier [218]).

Note that the procedures of message concealing and message extracting in steganography are more or less the same as the message encryption and message decryption in cryptography. It is this reason that steganography is often used together with cryptography. For example, an encrypted message may be written using invisible ink. Note also that a steganographic system can either be secret or public. In a public-key steganographic system, different keys are used for message concealing and message extracting. Readers interested in steganography are suggested to consult the workshop proceedings on *Information Hiding* (Anderson [9] and Aucsmith [13]).

3.3.15 Quantum Cryptography

In Chapter 2, we introduced some quantum algorithms for factoring large integers and computing discrete logarithms. It is evident that if a quantum computer is available, then all the public-key cryptographic systems based on the difficulty of integer factorization and discrete logarithms will be insecure. However, the cryptographic systems based on quantum mechanics will still be secure even if a quantum computer is available. To make this book as complete as possible, we shall introduce in this subsection some basic ideas of quantum cryptography. More specifically, we shall introduce a quantum analog of the Diffie-Hellman key exchange/distribution system, proposed by Bennett and Brassard in 1984.

First let us define four *polarizations* as follows:

$$\{0^\circ,\ 45^\circ,\ 90^\circ,\ 135^\circ\} \stackrel{\text{def}}{=} \{\rightarrow,\ \nearrow,\ \uparrow,\ \nwarrow\}. \tag{3.128}$$

The quantum system consists of a transmitter, a receiver, and a quantum channel through which polarized photons can be sent [25]. By the law of quantum mechanics, the receiver can either distinguish between the *rectilinear polarizations* $\{\rightarrow,\ \uparrow\}$, or reconfigure to discriminate between the diagonal polarizations $\{\nearrow,\ \nwarrow\}$, but in any case, he cannot distinguish both types. The system works in the following way:

[1] Alice uses the transmitter to send Bob a sequence of photons, each of them should be in one of the four polarizations $\{\rightarrow,\ \nearrow,\ \uparrow,\ \nwarrow\}$. For instance, Alice could choose, at random, the following photons

$$\uparrow \quad \nearrow \quad \rightarrow \quad \nwarrow \quad \rightarrow \quad \rightarrow \quad \nearrow \quad \uparrow \quad \uparrow$$

to be sent to Bob.

[2] Bob then uses the receiver to measure the polarizations. For each photon received from Alice, Bob chooses, at random, the following type of measurements $\{+,\ \times\}$:

[3] Bob records the result of his measurements but keeps it secret:

[4] Bob publicly announces the type of measurements he made, and Alice tells him which measurements were of correct type:

[5] Alice and Bob keep all cases in which Bob measured the correct type. These cases are then translated into bits $\{0, 1\}$ and thereby become the key:

[6] Using this secret key formed by the quantum channel, Bob and Alice can now encrypt and send their ordinary messages via the classic public-key channel.

An eavesdropper is free to try to measure the photons in the quantum channel, but, according to the law of quantum mechanics, he cannot in general do this without disturbing them, and hence, the key formed by the quantum channel is secure.

3.4 Bibliographic Notes and Further Reading

We interpret *applied number theory* in this book as the application of number theory to computing and information technology, and thus this chapter is mainly concerned with these applications of number theory. Even with this restriction, we argue that it is impossible to discuss all the computing related applications of number theory in a single book. We have, in fact, only discussed the applications of number theory to the design of computer systems and cryptosystems.

Our first application of number theory in computing is the design of computer systems; these include residue number systems and residue computers, complementary arithmetic and fast adders, error detections and corrections, the construction of hash functions (particularly minimal perfect hash functions), and the generation of random numbers/bits. Our aim was to show the applicability of number theory in computer systems design rather than the actual design of the computer (hardware or software) systems. There are plenty of books available on computer arithmetic (including residue number systems and complementary arithmetic) and fast computer architectures, but those by Koren [132], McClellan and Radar [149], Soderstrand et al. [243], and Szabo and Tanaka [247] are highly recommended. A standard reference that contains many applications of number theory in computer arithmetic, random number generation and hash functions (and many more) is Knuth's three volumes of *The Art of Computer Programming* [122], [123], and [124]. For error detection and correction codes, see, for example, Gallian [77], Hill [104], and Welsh [252].

Cryptography, particularly public-key cryptography, is an area that heavily depends on ideas and methods from number theory; of course, number theory is also useful in information systems security, including communication network security. In this chapter, we have provided a mathematical foundation for cryptography and information security. Those who desire a more detailed exposition in the field are invited to consult Bauer [20], Koblitz [128] and [129], and Pinch [184]; for elliptic curve public-key cryptography, see Menezes [155]. Readers may also find the following books useful in cryptography and computer security: Jackson [112], Kaufman et al. [118], Pfleeger [182], Salomaa [215], Smith [242], Stinson [246] and Welsh [252]. The books edited by Pomerance [190] and [44] contain a number of excellent survey papers on cryptology and random number generation.

The series of conferences proceedings entitled *Advances in Cryptology* published in Lecture Notes in Computer Science by Springer-Verlag is an important source for new developments in cryptography and information security.

There is a special section on *computer and network security* in the *Scientific American*, **279**, 4(1998), 69–89; it contains the following articles:

[1] C. P. Meinel, "How Hackers Break in ... and How They Are Caught", pp 70–77.

[2] "How Computer Security Works",

  [i] W. Cheswick and S. M. Bellovin, "Firewalls", pp 78–79.

  [ii] W. Ford, "Digital Certificates", page 80.

  [iii] J. Gosling, "The Java Sandbox", page 81.

[3] P. R. Zimmermann, "Cryptography for the Internet", pp 82–87.

[4] R. L. Rivest, "The Case Against Regulating Encryption Technology", pp 88–89.

An issue of the IEEE journal *Computer*, **31**, 9(1998), also has a special report on *computer and network security*, which contains the following six papers:

[1] P. W. Dowd and J. T. McHenry, "Network Security: It's Time to Take It Seriously", pp 24–28.

[2] B. Schneier, "Cryptographic Design Vulnerabilities", pp 29–33.

[3] A. D. Rubin and D. E. Geer Jr, "A Survey on Web Security, pp 34-42.
[4] R. Oppliger, "Security at the Internet Layer", pp 43–47.
[5] W. A. Arbaugh, et al., "Security for Virtual Private Intranets", pp 48–56.
[6] T. D. Tarman, et al., "Algorithm-Agile Encryption in ATM Networks", pp 57–64.

Note that the paper by Rubin and Geer [213] also discussed some interesting issues in mobile code security. All the above mentioned papers are easy to read and hence suitable for beginners in the field.

As by-products to cryptography, we have also introduced some basic concepts of steganography and quantum cryptography. There has been an increasing number of references in these two fields in recent years; interested readers are referred to, for example, Anderson [9], Aucsmith [13], Hughes [106], Inamori [110] and Lo [146], and the references therein.

In addition to computing and cryptography, number theory has also been successfully applied to many other areas such as physics, chemistry, acoustics, biology, engineering, dynamical systems, digital communications, digital signal processing, graphics design, self-similarity, and even music. For more information about these applications, readers are invited to consult Burr [44], Schroeder [222] and Waldschmidt, Moussa, Luck and Itzykson [250].

Bibliography

1. L. M. Adleman, "A Subexponential Algorithmic for the Discrete Logarithm Problem with Applications to Cryptography", *Proceedings of the 20th Annual IEEE Symposium on Foundations of Computer Science*, IEEE Press, 1979, 55–60.
2. L. M. Adleman, "Algorithmic Number Theory – The Complexity Contribution", *Proceedings of the 35th Annual IEEE Symposium on Foundations of Computer Science*, IEEE Press, 1994, 88–113.
3. L. M. Adleman, C. Pomerance, and R. S. Rumely, "On Distinguishing Prime Numbers from Composite Numbers", *Annals of Mathematics*, **117** (1983), 173–206.
4. L. M. Adleman and M. D. A. Huang, *Primality Testing and Abelian Varieties over Finite Fields*, Lecture Notes in Mathematics **1512**, Springer-Verlag, 1992.
5. A. V. Aho, J. E. Hopcroft and J. D. Ullman, *The Design and Analysis of Computer Algorithms*, Addison-Wesley, 1974.
6. W. Alford, G. Granville and C. Pomerance, "There Are Infinitely Many Carmichael Numbers", *Annals of Mathematics*, **140** (1994), 703–722.
7. R. Alter, "Computations and Generalizations of a Remark of Ramanujan", *Analytic Number Theory*, Proceedings, Lecture Notes in Mathematics **899**, Springer-Verlag, 1981, 183–196.
8. J. A. Anderson and J. M. Bell, *Number Theory with Applications*, Prentice-Hall, 1997.
9. R. Anderson (editor), *Information Hiding*, First International Workshop, Proceedings, Lecture Notes in Computer Science **1174**, Springer-Verlag, 1996.
10. G. E. Andrews, *Number Theory*, W. B. Sayders Company, 1971. Also Dover Publications, 1994.
11. T. M. Apostol, *Introduction to Analytic Number Theory*, Corrected 5th Printing, Undergraduate Texts in Mathematics, Springer-Verlag, 1998.
12. A. O. L. Atkin and F. Morain, "Elliptic Curves and Primality Proving", *Mathematics of Computation*, **61** (1993), 29–68.
13. D. Aucsmith (editor), *Information Hiding*, Second International Workshop, Proceedings, Lecture Notes in Computer Science **1525**, Springer-Verlag, 1998.
14. E. Bach, M. Giesbrecht and J. McInnes, *The Complexity of Number Theoretical Algorithms*, Technical Report 247/91, Department of Computer Science, University of Toronto, 1991.
15. E. Bach, G. Miller and J. Shallit, "Sums of Divisors, Perfect Numbers and Factoring", *SIAM Journal on Computing*, **15** (1989), 1143–1154.

16. E. Bach and J. Shallit, *Algorithmic Number Theory I – Efficient Algorithms*, MIT Press, 1996.
17. A. Backer, *A Concise Introduction to the Theory of Numbers*, Cambridge University Press, 1984.
18. R. J. Baillie and S. S. Wagstaff, Jr., "Lucas Pseudoprimes", *Mathematics of Computation*, **35** (1980), 1391–1417.
19. S. Battiato and W. Borho, "Breeding Amicable Numbers in Abundance II", *Mathematics of Computation*, **70** (2001), 1329-1333.
20. F. L. Bauer, *Decrypted Secrets – Methods and Maxims of Cryptology*, 2nd Edition, Springer-Verlag, 2000.
21. B. Beckett, *Introduction to Cryptology and PC Security*, McGraw-Hill, 1997.
22. M. Bellare and P. Gogaway, "Optimal Asymmetric Encryption", *Advances in Cryptography*, CRYPTO '94, Proceedings, Lecture Notes in Computer Science **950**, Springer-Verlag, 1995, 92–111.
23. P. Benioff, "The Computer as a Physical System – A Microscopic Quantum Mechanical Hamiltonian Model of Computers as Represented by Turing Machines", *Journal of Statistical Physics*, **22** (1980), 563–591.
24. C. H. Bennett, "Quantum Information and Computation", *Physics Today*, October 1995, 24–30.
25. C. H. Bennett, G. Brassard and A. K. Ekert, "Quantum Cryptography", *Scientific American*, October 1992, 26–33.
26. C. H. Bennett, "Strengths and Weakness of Quantum Computing", *SIAM Journal on Computing*, **26** (5)1997, 1510–1523.
27. E. Bernstein and U. Vazirani, "Quantum Complexity Theory", *SIAM Journal on Computing*, **26** 5(1997), 1411–1473.
28. M. Blum and S. Goldwasser, "An Efficient Probabilistic Public-key Encryption Scheme that Hides all Partial Information", *Advances in Cryptography*, CRYPTO '84, Proceedings, Lecture Notes in Computer Science **196**, Springer-Verlag, 1985, 289–302.
Boll:1986 B. Bollobás (editor), *Littlewood's Miscellany*, Cambridge University Press, 1986.
29. E. Bombieri, *Problems of the Millennium: The Riemann Hypothesis*, Institute for Advanced Study, Princeton, 2000.
30. D. Boneh, "Twenty Years of Attacks on the RSA Cryptosystem", *Notices of the AMS*, **46** 2(1999), 203–213.
31. W. Borho, "Über die Fixpunkte der k-fach iterierten Teilersummenfunktion", *Mitt. Math. Gesellsch. Hamburg*, **9** 5(1969), 34–48.
32. W. Borho and H. Hoffmann, "Breeding Amicable Numbers in Abundance", *Mathematics of Computation*, **46** (1986), 281–293.
33. G. Brassard, "A Quantum Jump in Computer Science", *Computer Science Today – Recent Trends and Development*, Lecture Notes in Computer Science **1000**, Springer-Verlag, 1995, 1–14.
34. R. P. Brent, "Irregularities in the Distribution of Primes and Twin Primes", *Mathematics of Computation*, **29** (1975), 43–56.
35. R. P. Brent, "An Improved Monte Carlo Factorization Algorithm", *BIT*, **20** (1980), 176–184.

36. R. P. Brent, "Some Integer Factorization Algorithms using Elliptic Curves", *Australian Computer Science Communications*, **8** (1986), 149–163.

37. R. P. Brent, "Primality Testing and Integer Factorization", *Proceedings of Australian Academy of Science Annual General Meeting Symposium on the Role of Mathematics in Science*, Canberra, 1991, 14–26.

38. R. P. Brent, "Uses of Randomness in Computation", Report TR-CS-94-06, Computer Sciences Laboratory, Australian National University, 1994.

39. R. P. Brent, G. L. Cohen and H. J. J. te Riele, "Improved Techniques for Lower Bounds for Odd Perfect Numbers", *Mathematics of Computation*, **57** (1991), 857–868.

40. D. M. Bressoud, *Factorization and Primality Testing*, Undergraduate Texts in Mathematics, Springer-Verlag, 1989.

41. E. F. Brickell, D. M. Gordon and K. S. McCurley, "Fast Exponentiation with Precomputation" (Extended Abstract), *Advances in Cryptography*, EUROCRYPT '92, Proceedings, Lecture Notes in Computer Science **658**, Springer-Verlag, 1992, 200–207.

42. W. Buchanan, *Mastering the Internet*, Macmillan, 1997.

43. J. P. Buhler (editor), *Algorithmic Number Theory*, Third International Symposium, ANTS-III, Proceedings, Lecture Notes in Computer Science **1423**, Springer-Verlag, 1998.

44. S. A. Burr (editor), *The Unreasonable Effectiveness of Number Theory*, Proceedings of Symposia in Applied Mathematics **46**, American Mathematical Society, 1992.

45. CACM, "The Digital Signature Standard Proposed by NIST and Responses to NIST's Proposal", *Communications of the ACM*, **35**, 7(1992), 36–54.

46. J. R. Chen, "On the Representation of a Large Even Integer as the Sum of a Prime and the Product of at most Two Primes", *Scientia Sinica*, **XVI**, 2(1973), 157–176.

47. K. Chen, "Authenticated Encryption Scheme Based on Quadratic Residue", *Electronics Letters*, **34**, 22(1998), 2115–2116.

48. S. S. Chern, "Mathematics in the 21st Century", *Advances in Mathematics (China)*, **21**, 4(1992), 385–387.

49. L. Childs, *A Concrete Introduction to Higher Algebra*, Undergraduate Texts in Mathematics, Springer-Verlag, 1979.

50. H. Cohen, *A Course in Computational Algebraic Number Theory*, Graduate Texts in Mathematics **138**, Springer-Verlag, 1993.

51. J. H. Conway and R. K. Guy, *The Book of Numbers*, Springer-Verlag, 1996.

52. S. Cook, *The P versus NP Problem*, University of Toronto, April, 2000. (Manuscript prepared for the Clay Mathematics Institute for the Millennium Prize Problems; revised in November 2000.)

53. J. W. Cooley and J. W. Tukey, "An Algorithm for the Machine Calculation of Complex Fourier Series", *Mathematics of Computation*, **19** (1965), 297–301.

54. T. H. Cormen, C. E. Ceiserson and R. L. Rivest, *Introduction to Algorithms*, MIT Press, 1990.

55. R. Crandall, J. Doenias, C. Norrie and J. Young, "The Twenty-Second Fermat Number is Composite", *Mathematics of Computation*, **64** (1995), 863–869.

56. R. Crandall and C. Pomerance, *Prime Numbers – A Computational Perspective*, Springer-Verlag, 2001.

57. I. Damgård (editor), *Lectures in Data Security*, Lecture Notes in Computer Science **1561**, Springer-Verlag, 1999.
58. H. Davenport, *The Higher Arithmetic*, 7th Edition, Cambridge University Press, 1999.
59. M. Deleglise and J. Rivat, "Computing $\pi(x)$ – the Meissel, Lehmer, Lagarias, Miller, Odlyzko Method", *Mathematics of Computation*, **65** (1996), 235–245.
60. D. C. Denson, *The Moment of Proof – Mathematical Epiphanies*, Oxford University Press, 1997.
61. J. M. Deshouillers, G. Effinger, H. J. J. te Riele and D. Zinoviev, "A Complete Vinogradov 3-Prime Theorem under the Riemann Hypothesis", *Electronic Research Announcements of the AMS*, **3** (1997), 99–104.
62. J. M. Deshouillers, H. J. J. te Riele and Y. Saouter, *New Experimental Results Concerning the Goldbach Conjecture*, Technical Report MAS-R9804, Centre for Mathematics and Computer Science (CWI), Amsterdam, 1998.
63. D. Deutsch, "Quantum Theory, the Church–Turing Principle and the Universal Quantum Computer", Proceedings of the Royal Society of London, Series **A**, **400** (1985), 96–117.
64. K. Devlin, *Mathematics: The Science of Patterns*, Scientific American Library, 1997.
65. L. E. Dickson, *History of the Theory of Numbers I – Divisibility and Primality*, G. E. Stechert & Co., New York, 1934.
66. W. Diffie and E. Hellman, "New Directions in Cryptography", *IEEE Transactions on Information Theory*, **22**, 5(1976), 644–654.
67. W. Diffie and E. Hellman, "Privacy and Authentication: An Introduction to Cryptography", *Proceedings of the IEEE*, **67**, 3(1979), 393–427.
68. P. G. L. Dirichlet, *Lecturers on Number Theory, Supplements by* R. Dedekind, American Mathematics Society and London Mathematics Society, 1999.
69. T. ElGamal, "A Public Key Cryptosystem and a Signature Scheme based on Discrete Logarithms", *IEEE Transactions on Information Theory*, **31** (1985), 496–472.
70. G. Ellis, *Rings and Fields*, Oxford University Press, 1992.
71. S. S. Epp, *Discrete Mathematics with Applications*, 2nd Edition, PWS Publishing Company, Boston, 1995.
72. Euclid, *The Thirteen Books of Euclid's Elements*, Translated by T. L. Heath, *Great Books of the Western World* **11**, edited by R. M. Hutchins, William Benton Publishers, 1952.
73. Euclid, *The Thirteen Books of Euclid's Elements*, Second Edition, Translated by Thomas L. Heath, Dover Publications, 1956.
74. R. P. Feynman, "Simulating Physics with Computers", *International Journal of Theoretical Physics*, **21** (1982), 467–488.
75. R. P. Feynman, *Feynman Lectures on Computation*, Edited by A. J. G. Hey and R. W. Allen, Addison-Wesley, 1996.
76. J. B. Fraleigh, *A First Course in Abstract Algebra*, 5th Edition, Addison-Wesley, 1994.
77. J. A. Gallian, "Error Detection Methods", *ACM Computing Surveys*, **28**, 3(1996), 503–517.

78. M. Gardner, "Mathematical Games – A New Kind of Cipher that Would Take Millions of Years to Break", *Scientific American*, **237**, 2(1977), 120–124.

79. M. R. Garey and D. S. Johnson, *Computers and Intractability – A Guide to the Theory of NP-Completeness*, W. H. Freeman and Company, 1979.

80. S. Garfinkel, *Web Security and Commerce*, O'Reilly, 1997.

81. P. Garrett, *Making, Breaking Codes: An Introduction to Cryptology*, Prentice-Hall, 2001.

82. C. F. Gauss, *Disquisitiones Arithmeticae*, G. Fleischer, Leipzig, 1801. English translation by A. A. Clarke, Yale University Press, 1966. Revised English translation by W. C. Waterhouse, Springer-Verlag, 1975.

83. P. Giblin, *Primes and Programming – An Introduction to Number Theory with Computing*, Cambridge University Press, 1993.

84. S. Goldwasser, "The Search for Provably Secure Cryptosystems", *Cryptology and Computational Number Theory*, edited by C. Pomerance, Proceedings of Symposia in Applied Mathematics **42**, American Mathematical Society, 1990.

85. S. Goldwasser and J. Kilian, "Almost All Primes Can be Quickly Certified", *Proceedings of the 18th ACM Symposium on Theory of Computing*, Berkeley, 1986, 316–329.

86. S. Goldwasser and J. Kilian, "Primality Testing Using Elliptic Curves", *Journal of ACM*, **46**, 4(1999), 450–472.

87. M. T. Goodrich and R. Tamassia, *Algorithm Design: Foundations, Analysis, and Internet Examples*, John Wiley & Sons, 2001.

88. S. Goldwasser and S. Micali, "Probabilistic Encryption", *Journal of Computer and System Sciences*, **28** (1984), 270–299.

89. D. M. Gordon and K. S. McCurley, "Massively Parallel Computation of Discrete Logarithms", *Advances in Cryptography*, Crypto '92, Proceedings, Lecture Notes in Computer Science **740**, Springer-Verlag, 1992, 312–323.

90. D. M. Gordon, "Discrete Logarithms in GF(p) using the Number Field Sieve", *SIAM Journal on Discrete Mathematics*, **6**, 1(1993), 124–138.

91. F. T. Grampp and R. H. Morris, "UNIX Operating System Security", *AT&T Bell Laboratories Technical Journal*, **63** (1984), 1649–1672.

92. A. Granville, J. van de Lune and H. J. J. te Riele, "Checking the Goldbach Conjecture on a Vector Computer", *Number Theory and Applications*, edited by R. A. Mollin, Kluwer Academic Publishers, 1989, 423–433.

93. D. Gries and F. B. Schneider, *A Logical Approach to Discrete Math*, Texts and Monographs in Computer Science, Springer-Verlag, 1993.

94. R. K. Guy, *Unsolved Problems in Number Theory*, 2nd Edition, Springer-Verlag, 1994.

95. D. Guedj, *Numbers – The Universal Language*, Thames and Hudson, 1997.

96. F. Guterl, "Suddenly, Number Theory Makes Sense to Industry", *International Business Week*, 20 June 1994, pp. 62–64.

97. H. Halberstam and H. E. Richert, *Sieve Methods*, Academic Press, 1974.

98. G. H. Hardy, *A Mathematician's Apology*, Cambridge University Press, 1979.

99. G. H. Hardy and J. E. Littlewood, "Some Problems of 'Partitio Numerorum', III: On the Express of a Number as a Sum of Primes", *Acta Mathematica*, **44** (1923), 1–70.

100. G. H. Hardy and E. M. Wright, *An Introduction to Theory of Numbers*, 5th Edition, Oxford University Press, 1979.

101. D. R. Heath-Brown, "Odd Perfect Numbers", *Mathematical Proceedings of Cambridge Philosophy Society*, **115**, 1(1994), 191–196.

102. A. Heck, *Introduction to Maple*, 2nd Edition, Springer-Verlag, 1996.

103. I. N. Herstein, *Topics in Algebra*, 2nd Edition, Wiley, 1975.

104. R. Hill, *A First Course in Coding Theory*, Oxford University Press, 1991.

105. L. Hua, *Introduction to Number Theory*, English Translation from Chinese by P. Shiu, Springer-Verlag, 1980.

106. R. J. Hughes, "Cryptography, Quantum Computation and Trapped Ions", *Philosophic Transactions of the Royal Society London*, Series **A**, **356** (1998), 1853–1868.

107. R. M. Huizing, *An Implementation of the Number Field Sieve*, Note NM-R9511, Centre for Mathematics and Computer Science (CWI), Amsterdam, 1995.

108. T. W. Hungerford, *Abstract Algebra – An Introduction*, Saunders College Publishing, 1990.

109. D. Husemöller, *Elliptic Curves*, Graduate Texts in Mathematics **111**, Springer-Verlag, 1987.

110. H. Inamori, *A Minimal Introduction to Quantum Key Distribution*, Centre for Quantum Computation, Clarendon Laboratory, Oxford University, 1999.

111. K. Ireland and M. Rosen, *A Classical Introduction to Modern Number Theory*, 2nd Edition, Graduate Texts in Mathematics **84**, Springer-Verlag, 1990.

112. T. H. Jackson, *From Number Theory to Secret Codes*, A Computer Illustrated Text, Adam Hilger, Bristol, 1987.

113. G. Jaeschke, "Reciprocal Hashing: A Method for Generating Minimal Perfect Hashing Functions", *Communications of the ACM*, **24**, 12(1981), 829–833.

114. D. S. Johnson, "A Catalog of Complexity Classes", *Handbook of Theoretical Computer Science*, edited by J. van Leeuwen, MIT Press, 1990, 69–161.

115. R. Jozsa, "Quantum Factoring, Discrete Logarithms, and the Hidden Subgroup Problem", *Computing in Science and Engineering*, March/April 2001, 34–43.

116. B. S. Kaliski, "A Pseudo-Random Bit Generator Based on Elliptic Curve Logarithms", *Advances in Cryptography*, CRYPTO '86, Proceedings, Lecture Notes in Computer Science **263**, Springer-Verlag, 1986, 84–103.

117. D. Kahn, *The Codebreakers*, Macmillan, 1967.

118. C. Kaufman, R. Perlman and M. Speciner, *Network Security – Private Communication in a Public World*, Prentice-Hall, 1995.

119. A. Ya. Khinchin, *Continued Fractions*, English translation from Russian, Chicago University Press, 1964.

120. J. Kilian, *Uses of Randomness in Algorithms and Protocols*, MIT Press, 1990.

121. D. E. Knuth, "Computer Science and its Relation to Mathematics", *American Mathematical Monthly*, **81**, 4(1974), 323–343.

122. D. E. Knuth, *The Art of Computer Programming I – Fundamental Algorithms*, 3rd Edition, Addison-Wesley, 1997.

123. D. E. Knuth, *The Art of Computer Programming II – Seminumerical Algorithms*, 3rd Edition, Addison-Wesley, 1998.

124. D. E. Knuth, *The Art of Computer Programming III – Sorting and Searching*, 2nd Edition, Addison-Wesley, 1998.

125. C. Ko and Q. Sun, *Lecture Notes in Number Theory* (In Chinese), Higher Education Press, Beijing, 1984.

126. N. Koblitz, "Elliptic Curve Cryptography", *Mathematics of Computation*, **48** (1987), 203–209.

127. N. Koblitz, *Introduction to Elliptic Curves and Modular Forms*, 2nd Edition, Graduate Texts in Mathematics **97**, Springer-Verlag, 1993.

128. N. Koblitz, *A Course in Number Theory and Cryptography*, 2nd Edition, Graduate Texts in Mathematics **114**, Springer-Verlag, 1994.

129. N. Koblitz, *Algebraic Aspects of Cryptography*, Algorithms and Computation in Mathematics **3**, Springer-Verlag, 1998.

130. N. Koblitz, *Cryptography*, in: **Mathematics Unlimited – 2001 and Beyond**, Edited by B. Enguist and W. Schmid, Springer-Verlag, 2001, 749–769.

131. S. Konyagin and C. Pomerance, "On Primes Recognizable in Deterministic Polynomial Time", *The Mathematics of Paul Erdős*, edited by R. L. Graham and J. Nesetril, Algorithms and Combinatorics **13**, Springer-Verlag, 1997, 176–198.

132. I. Koren, *Computer Arithmetic Algorithms*, Prentice-Hall, 1993.

133. H. Krishna, B. Krishna, K. Y. Lin, and J. D. Sun, *Computational Number Theory and Digital Signal Processing*, CRC Press, 1994.

134. E. Kranakis, *Primality and Cryptography*, John Wiley & Sons, 1986.

135. R. Kumanduri and C. Ronero, *Number Theory with Computer Applications*, Prentice-Hall, 1998.

136. J. C. Lagarias, "Pseudorandom Number Generators", *Cryptology and Computational Number Theory*, edited by C. Pomerance, Proceedings of Symposia in Applied Mathematics **42**, American Mathematical Society, 1990, pp 115–143.

137. S. Lang, *Elliptic Functions*, 2nd Edition, Springer-Verlag, 1987.

138. J. van Leeuwen (editor), *Handbook of Theoretical Computer Science*, MIT Press, 1990.

139. R. S. Lehman, "Factoring Large Integers", *Mathematics of Computation*, **28** (1974), pp 637–646.

140. H. W. Lenstra, Jr., "Factoring Integers with Elliptic Curves", *Annals of Mathematics*, **126** (1987), 649–673.

141. A. K. Lenstra and H. W. Lenstra, Jr., *The Development of the Number Field Sieve*, Lecture Notes in Mathematics **1554**, Springer-Verlag, 1993.

142. H. R. Lewis and C. H. Papadimitriou, *Elements of the Theory of Computation*, 2nd Edition, Prentice-Hall, 1998.

143. P. Linz, *An Introduction to Formal Languages and Automata*, 2nd Edition, Jones and Bartlett Publishers, 1997.

144. J. E. Littlewood, *A Mathematician's Miscellany*, Methuen & Co. Ltd. London, 1953. (This book later became *Littlewood's Miscellany*, edited by B. Bollobás and published by Cambridge University Press in 1986.)

145. C L. Liu, *Introduction to Combinatorial Mathematics*, McGraw-Hill, 1968.

146. H. K. Lo, "Quantum Cryptography", *Introduction to Quantum Computation and Information*, edited by H. K. Lo, S. Popescu and T. Spiller, World Scientific, 1998, 76–119.

147. J. van de Lune, H. J. J. te Riele and D. T. Winter, "On the Zeros of the Riemann Zata Function in the Critical Strip IV", *Mathematics of Computation*, **46** (1986), 667–681.

148. R. S. Macgregor, A. Aresi and A. Siegert, *WWW.Security – How to Build a Secure World Wide Web Connection*, Prentice-Hall, 1996.

149. J. H. McClellan and C. M. Radar, *Number Theory in Digital Signal Processing*, Prentice-Hall, 1979.

150. K. S. McCurley, "The Discrete Logarithm Problem", *Cryptology and Computational Number Theory*, edited by C. Pomerance, Proceedings of Symposia in Applied Mathematics **42**, American Mathematics Society, 1990, pp 49–74.

151. K. S. McCurley, "Odds and Ends from Cryptology and Computational Number Theory", edited by C. Pomerance, Proceedings of Symposia in Applied Mathematics **42**, American Mathematics Society, 1990, pp 49–74.

152. R. J. McEliece, *Finite Fields for Computer Scientists and Engineers*, Kluwer Academic Publishers, 1987.

153. H. McKean and V. Moll, *Elliptic Curves – Function Theory, Geometry, Arithmetic*, Cambridge University Press, 1997.

154. A. R. Meijer, "Groups, Factoring, and Cryptography" *Mathematics Magazine*, **69**, 2(1996), 103–109.

155. A. J. Menezes, *Elliptic Curve Public Key Cryptosystems*, Kluwer Academic Publishers, 1993.

156. A. Menezes and S. A. Vanstone, "Elliptic curve cryptosystems and their implementation", *Journal of Cryptology*, **6** (1993), 209–224.

157. A. Menezes, P. C. van Oorschot and S. A. Vanstone, *Handbook of Applied Cryptosystems*, CRC Press, 1996.

158. R. C. Merkle, "Secure Communications over Insecure Channels" *Communications of the ACM*, **21** (1978), 294–299. (Submitted in 1975.)

159. J. F. Mestre, "Formules Explicites et Minoration de Conducteurs de Variétés algébriques" *Compositio Mathematica*, **58** (1986), 209–232.

160. B. Meyer and and V. Müller, "A Public Key Cryptosystem Based on Elliptic Curves over $\mathbb{Z}/n\mathbb{Z}$ Equivalent to Factoring", *Advances in Cryptology*, EUROCRYPT '96, Proceedings, Lecture Notes in Computer Science **1070**, Springer-Verlag, 1996, 49–59.

161. M. Mignotte, "How to Share a Secret", *Cryptography*, Workshop Proceedings, Lecture Notes in Computer Science **149**, Springer-Verlag, 1983, 371–375.

162. G. Miller, "Riemann's Hypothesis and Tests for Primality", *Journal of Systems and Computer Science*, **13** (1976), 300–317.

163. V. Miller, "Uses of Elliptic Curves in Cryptography", *Advances in Cryptology*, CRYPTO '85, Proceedings, Lecture Notes in Computer Science **218**, Springer-Verlag, 1986, 417–426.

164. R. A. Mollin, *Fundamental Number Theory with Applications*, CRC Press, 1998.

165. R. A. Mollin, *An Introduction to Cryptography*, Chapman & Hall/CRC, 2001.

166. P. L. Montgomery, "Speeding Pollard's and Elliptic Curve Methods of Factorization", *Mathematics of Computation*, **48** (1987), 243–264.

167. P. L. Montgomery, "A Survey of Modern Integer Factorization Algorithms", *CWI Quarterly*, **7**, 4(1994), 337–394.

168. F. Morain, *Courbes Elliptiques et Tests de Primalité*, Université Claude Bernard, Lyon I, 1990.

169. M. A. Morrison and J. Brillhart, "A Method of Factoring and the Factorization of F_7", *Mathematics of Computation*, **29** (1975), 183–205.

170. R. Motwani and P. Raghavan, *Randomized Algorithms*, Cambridge University Press, 1995.

171. C. J. Mozzochi, "A Simple Proof of the Chinese Remainder Theorem", *American Mathematical Monthly*, **74** (1967), 998.

172. M. B. Nathanson, *Elementary Methods in Number Theory*, Springer-Verlag, 2000.

173. NIST, "Data Encryption Standard", Federal Information Processing Standards Publication 46-3, National Institute of Standards and Technology, U.S. Department of Commerce, 1999.

174. I. Niven, H. S. Zuckerman and H. L. Montgomery, *An Introduction to the Theory of Numbers*, 5th Edition, John Wiley & Sons, 1991.

175. D. H. Nyang and J. S. Song, "Fast Digital Signature Scheme Based on the Quadratic Residue Problem", *Electronics Letters*, **33**, 3(1997), 205–206.

176. S. Pohlig and M. Hellman, "An Improved Algorithm for Computing Logarithms over GF(p) and its Cryptographic Significance", *IEEE Transactions on Information Theory*, **24** (1978), pp 106–110.

177. J. O'Connor and E. Robertson, *The MacTutor History of Mathematics Archive*, http://www.groups.dcs.st-and.ac.uk/~history/Mathematicians.

178. A. M. Odlyzko, "Discrete Logarithms in Finite Fields and their Cryptographic Significance", *Advances in Cryptography*, EUROCRYPT '84, Proceedings, Lecture Notes in Computer Science **209**, Springer-Verlag, 1984, 225–314.

179. T. Okamoto and K. Ohta, "Universal Electronic Cash", *Advances in Cryptography*, CRYPTO '91, Proceedings, Lecture Notes in Computer Science **576**, Springer-Verlag, 1991, 324–337.

180. Open University Course Team, *Number Theory*, Complex Analysis Unit **15**, Open University Press, 1974.

181. O. Ore, *Number Theory and its History*, Dover Publications, 1988.

182. C. P. Pfleeger, *Security in Computing*, Prentice-Hall, 1997.

183. R. G. E. Pinch, "Some Primality Testing Algorithms", *Notices of the American Mathematical Society*, **40**, 9(1993), 1203–1210.

184. R. G. E. Pinch, *Mathematics for Cryptography*, Queen's College, University of Cambridge, 1997.

185. R. G. E. Pinch, *The Carmichael Numbers up to* 10^{16}, Queen's College, University of Cambridge, 1997.

186. S. C. Pohlig and M. Hellman, "An Improved Algorithm for Computing Logarithms over GF(p) and its Cryptographic Significance", *IEEE Transactions on Information Theory*, **24** (1978), 106–110.

187. J. M. Pollard, "A Monte Carlo Method for Factorization", *BIT*, **15** (1975), 331–332.

188. J. M. Pollard, "Monte Carlo Methods for Index Computation (mod p)", *Mathematics of Computation*, **32** (1980), 918–924.

189. C. Pomerance, "Very Short Primality Proofs", *Mathematics of Computation*, **48** (1987), 315–322.

190. C. Pomerance (editor), *Cryptology and Computational Number Theory*, Proceedings of Symposia in Applied Mathematics **42**, American Mathematical Society, 1990.

191. C. Pomerance, "Cryptology and Computational Number Theory – An Introduction", *Cryptology and Computational Number Theory*, edited by C. Pomerance, Proceedings of Symposia in Applied Mathematics **42**, American Mathematical Society, 1990, 1–12.

192. C. Pomerance, J. L. Selfridge and S. S. Wagstaff, Jr., "The Pseudoprimes to $25 \cdot 10^9$", *Mathematics of Computation*, **35** (1980), 1003–1026.

193. V. R. Pratt, "Every Prime Has a Succinct Certificate", *SIAM Journal on Computing*, **4** (1975), 214–220.

194. W. H. Press and Teukolsky et al., *Numerical Recipes in C – The Art of Scientific Computing*, 2nd Edition, Cambridge University Press, 1992.

195. M. O. Rabin, "Probabilistic Algorithms for Testing Primality", *Journal of Number Theory*, **12** (1980), 128–138.

196. E. D. Reilly and F. D. Federighi, *PASCALGORITHMS – A Pascal-Based Introduction to Computer Science*, Houghton Mifflin, Boston, 1989.

197. D. Redmond, *Number Theory: An Introduction*, Marcel Dekker, New York, 1996.

198. P. Ribenboim, *The Little Book on Big Primes*, Springer-Verlag, 1991.

199. P. Ribenboim, "Selling Primes", *Mathematics Magazine*, **68**, 3(1995), 175–182.

200. P. Ribenboim, *The New Book of Prime Number Records*, Springer-Verlag, 1996.

201. J. Richstein, "Goldbach's Conjecture up to $4 \cdot 10^{14}$", *Mathematics of Computation*, **70**, (2001), 1745-1749.

202. E. Rieffel and W. Polak, "An Introduction to Quantum Computing for Non-Physicists", *ACM Computing Surveys*, **32**, 3(2000), 300–335.

203. H. J. J. te Riele, "New Very Large Amicable Pairs", Number Theory, Noordwijkerhout 1983, Proceedings, Lecture Notes in Mathematics **1068**, Springer-Verlag, 1984, 210–215.

204. H. J. J. te Riele, "A New Method for Finding Amicable Numbers", Reprint from *Mathematics of Computation 1943-1993, A Half-century of Computational Mathematics*, Vancouver, 9-13 August 1993.

205. H. J. J. te Riele, "Factorization of RSA-140 using the Number Field Sieve", `http://www.crypto-world.com/announcements/RSA140.txt`, 4 February 1999.

206. H. J. J. te Riele, "Factorization of a 512-bits RSA Key using the Number Field Sieve", `http://www.crypto-world.com/announcements/RSA155.txt`, 26 August 1999.

207. H. Riesel, *Prime Numbers and Computer Methods for Factorization*, Birkhäuser, Boston, 1990.

208. R. L. Rivest, "Remarks on a Proposed Cryptanalytic Attack on the M.I.T. Public-key Cryptosystem", *Cryptologia*, **2**, 1(1978), 62–65.

209. R. L. Rivest, A. Shamir and L. Adleman, A Method for Obtaining Digital Signatures and Public Key Cryptosystems, *Communications of the ACM*, **21**, 2(1978), 120–126.

210. H. E. Rose, *A Course in Number Theory*, 2nd Edition, Oxford University Press, 1994.

211. K. Rosen, *Elementary Number Theory and its Applications*, 4th Edition, Addison-Wesley, 2000.

212. J. J. Rotman *An Introduction to the Theory of Groups*, Springer-Verlag, 1994.

213. A. D. Rubin and D. E. Geer, Jr., Mobile Code Security, *IEEE Internet Computing*, **2**, 6(1998), 30–34.

214. G. Rozenberg and A. Salomaa, *Cornerstones of Undecidability*, Prentice-Hall, 1994.

215. A. Salomaa, *Public-Key Cryptography*, 2nd Edition, Springer-Verlag, 1996.

216. Y. Saouter, *Vinogradov's Theorem is True up to* 10^{20}, Publication Interne No. 977, IRISA, 1995.

217. V. Scarani, "Quantum Computing", *American Journal of Physics*, **66**, 11(1998), 956–960.

218. B. Schneier, *Applied Cryptography – Protocols, Algorithms, and Source Code in C*, 2nd Edition, John Wiley & Sons, 1996.

219. B. Schneier, John Kelsey, Doug Whiting, David Wagner, Chris Hall and Niels Ferguson, *The Twofish Encryption Algorithm*, John Wiley & Sons, 1999.

220. C. P. Schnorr, "Efficient Identification and Signatures for Smart Cards", *Advances in Cryptography*, CRYPTO '89, Proceedings, Lecture Notes in Computer Science **435**, Springer-Verlag, 1990, 239–252.

221. R. Schoof, "Elliptic Curves over Finite Fields and the Computation of Square Roots mod p", *Mathematics of Computation*, **44** (1985), 483–494.

222. M. R. Schroeder, *Number Theory in Science and Communication*, 3rd Edition, Springer Series in Information Sciences **7**, Springer-Verlag, 1997.

223. W. Schwarz and J. Wolfgang, "Some Remarks on the History of the Prime Number Theorem from 1896 to 1960" *Development of Mathematics 1900–1950*, edited by J.-P. Pier, Birkhäuser, 1994.

224. A. Shamir, "Factoring Numbers in $\mathcal{O}(\log n)$ Arithmetic Steps", *Information Processing Letters*, **8**, 1(1979), 28–31.

225. A. Shamir, "How to Share a Secret", *Communications of the ACM*, **22**, 11(1979), 612–613.

226. P. Shor, "Algorithms for Quantum Computation: Discrete Logarithms and Factoring", *Proceedings of 35th Annual Symposium on Foundations of Computer Science*, IEEE Computer Society Press, 1994, 124–134.

227. P. Shor, "Polynomial-Time Algorithms for Prime Factorization and Discrete Logarithms on a Quantum Computer", *SIAM Journal on Computing*, **26**, 5(1997), 1484–1509.

228. J. H. Silverman and J. Tate, *Rational Points on Elliptic Curves*, Undergraduate Texts in Mathematics, Springer-Verlag, 1992.

229. J. H. Silverman, *The Arithmetic of Elliptic Curves*, Graduate Texts in Mathematics **106**, Springer-Verlag, 1994.

230. J. H. Silverman, *A Friendly Introduction to Number Theory*, Second Edition, Prentice-Hall, 2001.

231. J. H. Silverman, "The Xedni Calculus and the Elliptic Curve Discrete Logarithm Problem", Dept of Mathematics, Brown University, 10 February 1999.

232. J. H. Silverman and J. Suzuki, "Elliptic Curve Discrete Logarithms and the Index Calculus", *Advances in Cryptology – ASIACRYPT '98*, Springer Lecture Notes in Computer Science **1514**, 1998, 110–125.

233. R. D. Silverman, 'The Multiple Polynomial Quadratic Sieve", *Mathematics of Computation*, **48** (1987), 329–339.

234. R. D. Silverman, "A Perspective on Computational Number Theory", *Notices of the American Mathematical Society*, **38**, 6(1991), 562–568.

235. R. D. Silverman, "Massively Distributed Computing and Factoring Large Integers", *Communications of the ACM*, **34**, 11(1991), 95–103.

236. D. R. Simon, "On the Power of Quantum Computation", *Proceedings of the 35th Annual IEEE Symposium on Foundations of Computer Science*, IEEE Press, 1994, 116–123.

237. S. Singh, *The Code Book – The Science of Secrecy from Ancient Egypt to Quantum Cryptography*, Fourth Estate, London, 1999.

238. S. Singh, *The Science of Secrecy – The Histroy of Codes and Codebreaking*, Fourth Estate, London, 2000.

239. M. K. Sinisalo, "Checking the Goldbach Conjecture up to $4 \cdot 10^{11}$", *Mathematics of Computation*, **61** (1993), 931–934.

240. M. Sipser, *Introduction to the Theory of Computation*, PWS Publishing Company, Boston, 1997.

241. D. Slowinski, "Searching for the 27th Mersenne Prime", *Journal of Recreational Mathematics*, **11**, 4(1978-79), 258–261.

242. R. E. Smith, *Internet Cryptography*, Kluwer Academic Publishers, 1997.

243. M. A. Soderstrand, W. K. Jenkins, G. A. Jullien and F. J. Taylor, *Residue Number System Arithmetic, Modern Applications in Digital Signal Processing*, IEEE Press, 1986.

244. R. Solovay and V. Strassen, "A Fast Monte-Carlo Test for Primality", *SIAM Journal on Computing*, **6**, 1(1977), 84–85. "Erratum: A Fast Monte-Carlo Test for Primality", *SIAM Journal on Computing*, **7**, 1(1978), 118.

245. I. Stewart, "Geometry Finds Factor Faster", *Nature*, **325**, 15 January 1987, 199.

246. D. R. Stinson, *Cryptography: Theory and Practice*, CRC Press, 1995.

247. N. S. Szabo and R. I. Tanaka, *Residue Arithmetic and its Applications to Computer Technology*, McGraw-Hill, 1967.

248. H. C. A. van Tilborg, *An Introduction to Cryptography*, Kluwer Academic Publishers, 1988.

249. I. Vardi, *Computational Recreations in Mathematica*, Addison-Wesley, 1991.

250. M. Waldschmidt, P. Moussa, J. M. Luck and C. Itzykson, *From Number Theory to Physics*, Springer-Verlag, 1992.

251. S. Wagon, "Primality Testing", *The Mathematical Intelligencer*, **8**, 3(1986), 58–61.

252. D. Welsh, *Codes and Cryptography*, Oxford University Press, 1989.

253. H. Wiener, "Cryptanalysis of Short RSA Secret Exponents", *IEEE Transactions on Information Theory*, **36**, 3(1990), 553–558.

254. A. Wiles, "Modular Elliptic Curves and Fermat's Last Theorem", *Annals of Mathematics*, **141** (1995), 443–551.

255. H. C. Williams, "The Influence of Computers in the Development of Number Theory", *Computers & Mathematics with Applications*, **8**, 2(1982), 75–93.

256. H. C. Williams, "Factoring on a Computer", *Mathematical Intelligencer*, **6**, 3(1984), 29–36.

257. H. C. Williams, *Édouard Lucas and Primality Testing*, John Wiley & Sons, 1998.

258. C. P. Williams and S. H. Clearwater, *Explorations in Quantum Computation*, The Electronic Library of Science (TELOS), Springer-Verlag, 1998.

259. H. Woll, "Reductions Among Number Theoretic Problems", *Information and Computation*, **72** (1987), 167–179.

260. S. Y. Yan, "Primality Testing of Large Numbers in Maple", *Computers & Mathematics with Applications*, **29**, 12(1995), 1–8.

261. S. Y. Yan, *Perfect, Amicable and Sociable Numbers – A Computational Approach*, World Scientific, 1996.

262. S. Y. Yan, *An Introduction to Formal Languages and Machine Computation*, World Scientific, 1998.

263. J. Young, "Large Primes and Fermat Factors", *Mathematics of Computation*, **67** (1998), 1735–1738.

264. R. Yuan and W. T. Strayer, *Virtual Private Networks – Technologies and Solutions*, Addison-Wesley, 2001.

265. K. C. Zeng, C. H. Yang, D. Y. Wei and T. R. N. Rao, "Pseudorandom Bit Generators in Stream-Cipher Cryptography", *Computer*, **24**, 2(1991), 8–17.

Index